FUNDAMENTALS OF OPTICAL FIBER COMMUNICATION

By
Sapna Katiyar
Assistant Professor
Deptt. of Electronics & Comm. Engg.
Greater Noida Institute of Technology,
Greater Noida (UP)

S.K. KATARIA & SONS®
Educational Publishers
4885/109, Prakash Mahal, Dr. Subhash Bhargav Lane,
Opposite Delhi Medical Association, Daryaganj, New Delhi-110002 (INDIA)
Phone: +91-11-23243489, 43551243; Mobile: +91-9871775858
e-mail: katariabook@gmail.com; katariabooks@yahoo.com
Website: www.skkatariaandsons.com
(e-books, subscriptions and bulk buying also available on our website)

Published by:

S.K. Kataria & Sons®
4885/109, Prakash Mahal, Dr. Subhash Bhargav Lane,
Opposite Delhi Medical Association, Daryaganj, New Delhi-110002 (INDIA)
Phone: +91-11-23243489, 43551243; Mobile: +91-9871775858
e-mail: katariabook@gmail.com; katariabooks@yahoo.com
Website: www.skkatariaandsons.com

First Edition: 2010, Reprint: 2023

Price: ₹ 495/-

Laser Typesetting by :
AISA ENTERPRISES, DELHI

PREFACE

The optical fiber communication is emerging technology and growing at a faster rate over last two decades because of the dramatic reduction in transmission loss of optical fibers coupled with equally important developments in the area of light sources and detectors. The birth of optical fiber communications coincided with the fabrication of low pass optical fibers and operation of room temperature semiconductor lasers in 1970. The development of optical fiber technology for communication networks, medical applications, and other areas represents unique confluences of physics, electronics and mechanical engineering disciplines.

This book deals with important topics like electroluminescence sources, photo detectors, optical fibers, transmission and receiver optical systems, description of various types of optical transmitter and receiver.

This text book is intended for students perusing an undergraduate course in electronics and communication engineering. The aim is to cover the field of fiber optics and its applications. It also contains solved and unsolved problems.

Chapter 1 deals with a perspective on optical fiber communication and its performance with respect to microwave and lower frequency waves. This chapter also describes the block diagram of optical communication system and its advantages.

Chapter 2 introduces the fundamentals of ray and wave theory. It also explains the propagation of light in step index and graded index optical fibers.

Chapter 3 describes the various optical phenomenons like total internal reflection, refraction, dispersion, polarization etc. and various types of fibers used in optical fiber communication with illustrations. Multipath dispersion has been calculated for step index and graded index fibers.

Chapter 4 addresses the different types of losses like attenuation behaviour, waveguide imperfections etc. of optical fibers that exists.

Chapter 5 is devoted to the origins and effects of nonlinear processes in optical fibers. Some of these nonlinear effects degrade system performance and need to be controlled, whereas others can have beneficial uses.

Chapter 6 describes the manufacturing process of fiber cables. The fabrication methods and, the different types of optical cables and connectors are also discussed in this chapter.

Chapter 7 explains the principle of optical transmitter. It also explains the semiconductor materials used and the concept of hetrojunctions.

Chapter 8 describes the various types of electroluminescent optical sources which are used in optical communication. High power light emitting diodes, laser diodes and solid state lasers, and fiber lasers are frequently employed in optical communication. Working principles of surface and edge LEDs are explained and their performance been compared.

Chapter 9 explains the p-i-n and avalanche photodiode receiver. The signal to noise ratio for direct detection receivers is derived and the concept of a shot noise limited receiver is also introduced.

Chapter 10 explains the principle of optical detectors. The photoconductor and photo transistor are also discussed in this chapter. The noise associated with detection and their mathematical values are given for better understanding of the subject.

Chapter 11 is devoted to WDM lightwave systems by considering the architectural aspects, optical components requires for their implementation and the performance issues.

Chapter 12 show how the elements are put together to form networks. Optical networking concepts for long hour, metro, local area and access networks are explained here. The optical TDM and wavelength division multiplexing (WDM) have been described in details. Finally the WDM components and subscriber multiplexing has been presented.

AUTHOR

ACKNOWLEDGEMENT

I take this opportunity to express my gratitude to the Prof. S. N. Sharan (Director, GNIT Gr. Noida), Dr. L. N. Paliwal (DG, VIET Ghaziabad), Dr. S. Prasad (Academic Advisor GLBITM Gr. Noida), Dr. R. P. S.Gangwar (Professor & Head, GBPUAT Pantnagar), Prof. S. K. Mahajan (GNIT Gr. Noida) and Mr. Dhiraj Gupta (HOD, E&C GNIT Gr. Noida) for providing encouraging environment without which this book would not have come into existence.

I would like to express my sincere thanks to all my colleagues of E&C deptt. GNIT Gr. Noida for their continuous help and valuable suggestions.

I am grateful to my family members for their constant motivation and guidance that has helped me to complete this project. I would also like to express my gratefulness for their patience and understanding during the time I devoted to writing the book.

I take this opportunity to express my thanks to Mr. Sanjeev Kumar Kataria of S. K. Kataria and Sons for publishing this book in such a short time. I am also thankful to Mr. Sharda Singh and Mr. Sachin for their assistance.

SAPNA KATIYAR

Dedicated to My Son

"Arnav"

CONTENTS

Introduction

Chapter 1

INSIDE THIS CHAPTER

1.1. Introduction

Sending and receiving of messages from one place to another has always been a matter of great importance. The elements involved in such a system namely, **information source**, **transmitter**, **transmission channel** and **receiver** and the means to continually improve their performance are therefore of great interest.

The use of light to send messages is not new. Fires were used for signaling in Biblical times, smoke signals have been used for thousands of years and flashing lights have been used to communicate between warships at sea since the days of Lord Nelson.

In electronic communication, optical fibers are preferred over copper wires because they are :

- extremely light
- small
- can be accommodated in a small space.

By using optical fibers the number of signals that can be transmitted simultaneously is enhanced.

The idea of using glass fiber to carry an optical communication signal originated with **Alexander Graham Bell.** However this idea had to wait some 80 years for better glasses and low-cost electronics for it to become useful in practical situations. Development of fibers and devices for optical communications began in the early 1960s and continues strongly today. But the real change came in the 1980s. During this decade optical communication in public communication networks developed from the status of a curiosity into being the dominant technology.

Among the tens of thousands of developments and inventions that have contributed to this progress **four stands out as milestones :**

1. The invention of the LASER (in the late 1950's)
2. The development of low loss optical fiber (1970's)
3. The invention of the optical fiber amplifier (1980's)
4. The invention of the in-fiber Bragg grating (1990's).

The continuous development of semiconductor technology is quite fundamental but of course not specifically optical. The predominant use of optical technology is as very fast "electric wire". Optical fibers replace electric wire in communications systems and nothing much else changes. Perhaps this is not quite fair. The speed and quality of optical communications systems has itself predicated the development of a new type of electronic communications itself designed to be run on optical connections. ATM and SDH technologies are good examples of the new type of systems.

It is important to realize that optical communications is not like electronic communications. While it seems that light travels in a fiber much like electricity does in a wire, this is very misleading. Light is an electromagnetic wave and optical fiber is a waveguide. Everything to do with transport of the signal even to simple things like coupling (joining) two fibers into one is very different from what happens in the electronic world. The two fields (electronics and optics) while closely related employ different principles in different ways.

In 1998 the "happening" area in optical communications is **Wavelength Division Multiplexing (WDM).** This is the ability to send many (perhaps up to 1000) independent optical channels on a single fiber. The first fully commercial WDM products appeared on the market in 1996. WDM is a major step toward fully optical networking.

In optical region of electromagnetic spectrum, the transmission of information is carried out not by frequency modulation of the carrier, but by varying the intensity of the optical power, this is known as optical communication. Optical communication systems use high carrier frequencies in the visible or near-infrared region of the electromagnetic spectrum. They are sometimes called **light wave systems** to distinguish them from microwave systems, whose carrier frequency is typically smaller.

1.2. A Brief History of Fiber Optics

The use of light for communication purposes dates back to antiquity if we interpret optical communications in a broad sense. Most civilizations have used mirrors, fire beacons, or smoke signals to convey a single piece of information (such as victory in a war). Essentially the same idea was used up to the end of the eighteenth century through signaling lamps, flags, and other semaphore devices. The idea was extended further, following a suggestion of Claude Chappe in 1792, to transmit mechanically coded messages over long distances *(around* 100 km) by the use of intermediate relay stations, acting as **regenerators or repeaters** in the modern-day language. The first such "optical telegraph" was put in service between Paris and Lillie (two French cities about 200 km apart) in July 1794. By 1830, the network had expanded throughout Europe. The role of light in such systems was simply to make the coded signals visible so that they could be intercepted by the relay stations. The opto-mechanical communication systems of the nineteenth century were inherently slow. In modern-day terminology, the effective bit rate of such systems was less than 1 bit per second (B < 1 b/s).

The Nineteenth Century

In 1870, John Tyndall, using a jet of water that flowed from one container to another and a beam of light, demonstrated that light used internal reflection to follow a specific path. As water poured out through the spout of the first container, Tyndall directed a beam of sunlight at the path of the water. The light, as seen by the audience, followed a zigzag path inside the curved path of the water. This is shown in Fig. 1.1 and it marked the first research into the guided transmission of light.

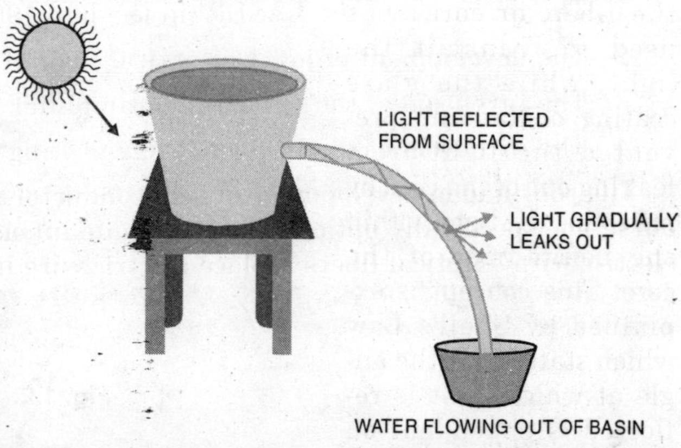

Fig. 1.1. John Tyndall's Experiment.

William Wheeling, in 1880, patented a method of light transfer called "piping light." Wheeling believed that by using mirrored pipes branching off from a single source of illumination, *i.e.* a bright electric arc; he could send the light to many different rooms in the same way that water, through plumbing, is carried throughout buildings today. Due to the ineffectiveness of Wheeling's idea and to the concurrent introduction of Edison's highly successful incandescent light bulb, the concept of piping light never took off.

In the same year, Alexander Graham Bell developed an optical voice transmission system known as **photo phone**. The photo phone used free-space light to carry the human voice 200 meters. Specially placed mirrors reflected sunlight onto a diaphragm attached within the mouthpiece of the photo phone. At the other end, mounted within a parabolic reflector, was a light-sensitive selenium resistor. This resistor was connected to a battery that was, in turn, wired to a telephone receiver. As one spoke into the photo phone, the illuminated diaphragm vibrated, casting various intensities of light onto the selenium resistor. The changing intensity of light altered the current that passed through the telephone receiver which then converted the light back into speech. Bell believed this invention was superior to the telephone because it did not need wires to connect the transmitter and receiver. Today, free-space optical links find extensive use in metropolitan applications.

The Twentieth Century

Fiber optic technology experienced a phenomenal rate of progress in the second half of the twentieth century. Early success came during the 1950's with the development of the fiberscope. This image-transmitting device, which used the first practical all-glass fiber, was concurrently devised by Brian O'Brien at the American Optical Company and Narinder Kapany (who first coined the term "fiber optics" in 1956) and colleagues at the Imperial College of Science and Technology in London. Early all-glass fibers experienced excessive optical loss, the loss of the light signal as it travelled the fiber, limiting transmission distances.

This motivated scientists to develop glass fibers that included a separate glass coating. The innermost region of the fiber, or core, was used to transmit the light, while the glass coating, or cladding, prevented the light from leaking out of the core by reflecting the light within the boundaries of the core. This concept is explained by Snell's Law which states that the angle at which light is reflected is dependent on the refractive indices of the two materials - in this case, the core and the cladding. **The lower refractive index of the cladding (with respect to the core) causes the light to be angled back into the core as illustrated in Fig. 1.2.**

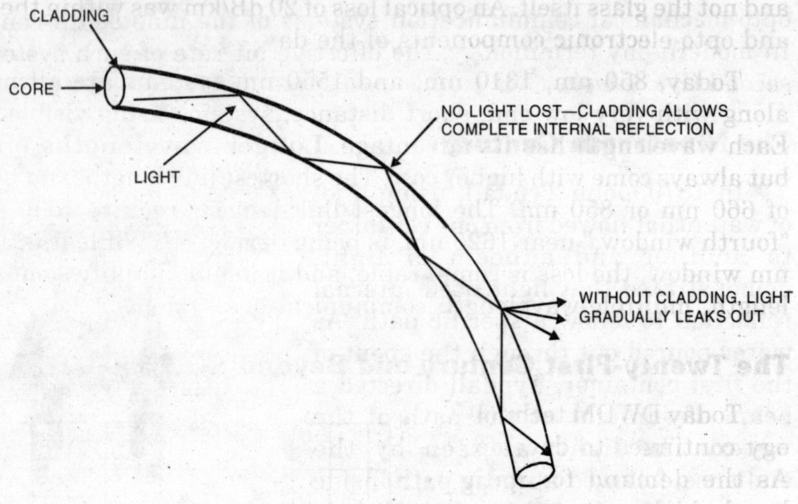

Fig. 1.2. Optical Fiber With Cladding.

The fiberscope quickly found application inspecting welds inside reactor vessels and combustion chambers of jet aircraft engines as well as in the medical field. Fiberscope technology has evolved over the years to make laparoscopic surgery one of the great medical advances of the twentieth century.

The development of laser technology was the next important step in the establishment of the industry of fiber optics. Only the laser diode (LD) or its lower-power cousin, the light-emitting diode (LED), had the potential to generate large amounts of light in a spot tiny enough to be useful for fiber optics. In 1957, Gordon Gould popularized the idea of using lasers when, as a graduate student at Columbia University, he described the laser as an intense light source. Shortly after, Charles Townes and Arthur Schawlow at Bell Laboratories supported the laser in scientific circles. Lasers went through several generations including the development of the ruby laser and the helium-neon laser in 1960. Semiconductor lasers were first realized in 1962; these lasers are the type most widely used in fiber optics today.

Because of their higher modulation frequency capability, the importance of lasers as a means of carrying information did not go unnoticed by communications engineers. Light has an information-carrying capacity 10,000 times that of the highest radio frequencies being used. However, the laser is unsuited for open-air transmission because it is adversely affected by environmental conditions such as rain, snow, hail, and smog. Faced with the challenge of finding a transmission medium other than air, Charles Kao and Charles Hockham, working at the Standard Telecommunication Laboratory in England in 1966, published a landmark paper proposing that optical fiber might be a suitable transmission medium if its attenuation could be kept under 20 decibels per kilometer (dB/km). At the time of this proposal, optical fibers exhibited losses of 1,000 dB/ km or more. At a loss of only 20 dB/km, 99% of the light would be lost over only 3,300 feet. In other words, only 1/100th of the optical power that was transmitted reached the receiver. Intuitively, researchers postulated that the current, higher optical losses were the result of impurities in the glass and not the glass itself. An optical loss of 20 dB/km was within the capability of the electronics and opto-electronic components of the day.

Today, 850 nm, 1310 nm, and 1550 nm systems are all manufactured and deployed along with very low-end, short distance, systems using visible wavelengths near 660 nm. Each wavelength has its advantage. **Longer wavelengths offer higher performance,** but always come with higher cost. The shortest link lengths can be handled with wavelengths of 660 nm or 850 nm. The longest link lengths require 1550 nm wavelength systems. A "fourth window," near 1625 nm, is being developed. While it is not lower loss than the 1550 nm window, the loss is comparable, and it might simplify some of the complexities of long-length, multiple-wavelength communications systems.

The Twenty-First Century and Beyond

Today DWDM technology continues to develop. As the demand for data bandwidth increases, driven by the phenomenal growth of the Internet, the move to optical networking is the focus of new technologies. At this writing, nearly half a billion people have Internet access and use it regularly. Some 40 million or more households are "wired." The world wide web already hosts over 2 billion web pages, and according to estimates people upload more than 3.5 million new web pages everyday.

The important factor in these developments is the increase in fiber transmission capacity, which has grown by a factor of 200 in the last decade. Fig. 1.4 illustrates this trend.

Because of fiber optic technology's immense potential bandwidth, 50 THz or greater, there are extraordinary possibilities for future fiber optic applications. Already, the

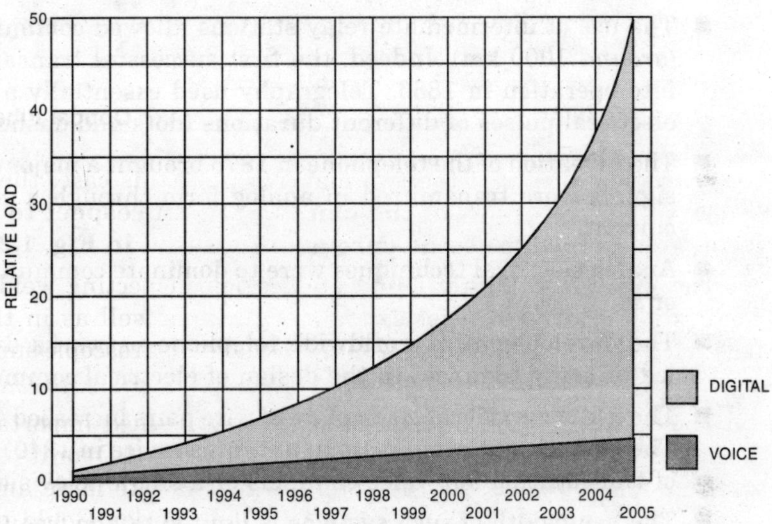

Fig. 1.3. Projected internet traffic increases.

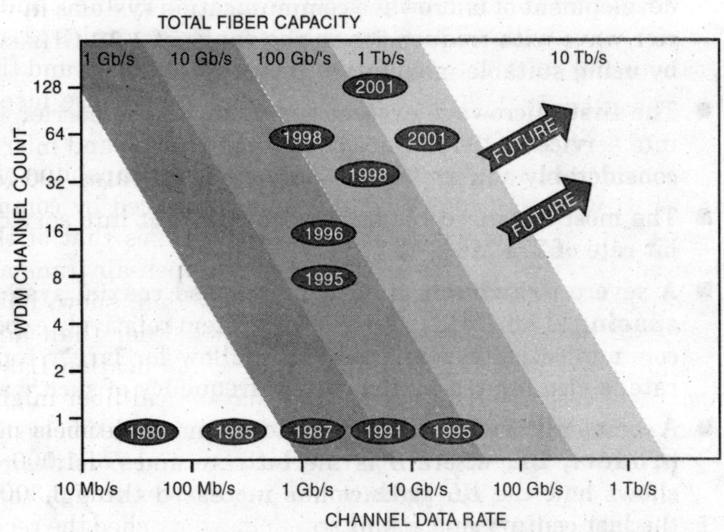

Fig. 1.4. The Growth of Optical Fiber Transmission Capacity.

push to bring broadband services, including data, audio, and especially video, into the home is well underway.

Broadband service available to a mass market opens up a wide variety of interactive communications for both consumers and businesses, bringing to reality interactive video networks, interactive banking and shopping from the home, and interactive distance learning. The "last mile" for optical fiber goes from the curb to the television set top, known as fiber-to-the-home (FTTH) and fiber-to-the-curb (FTTC), allowing video on demand to become a reality.

1.3. Need for Fiber Optic Communications

- The advent of telegraphy in the 1830s replaced the use of light by electricity and began the era of electrical communications. The bit rate *B* could be increased to 10 b/s by the use of new coding techniques, such as the **Morse code.**
- The use of intermediate relay stations allowed communication over long distances *(around* 1000 km). Indeed, the first successful transatlantic telegraph cable went into operation in 1866. Telegraphy used essentially a digital scheme through two electrical pulses of different durations (dots and dashes of the Morse code).
- The invention of the telephone in 1876 brought a major change in as much as electric signals were transmitted in analog form through a continuously varying electric current.
- Analog electrical techniques were to dominate communication systems for a century or so.
- The development of worldwide telephone networks during the twentieth century led to many advances in the design of electrical communication systems.
- The use of coaxial cables in place of wire pairs increased system capacity considerably. The first coaxial-cable system, put into service in 1940, was a 3 MHz system capable of transmitting 300 voice channels or a single television channel.
- The bandwidth of such systems is limited by the frequency-dependent cable losses, which increase rapidly for frequencies beyond 10 MHz. This limitation led to the development of microwave communication systems in which an electromagnetic carrier wave with frequencies in the range of 1-10 GHz is used to transmit the signal by using suitable modulation techniques.
- The first microwave system operating at the carrier frequency of 4 GHz was put into service in 1948. Since then, both coaxial and microwave systems have evolved considerably and are able to operate at bit rates 100 Mb/s.
- The most advanced coaxial system was put into service in 1975 and operated at a bit rate of 274 Mb/s.
- A severe **drawback** of such high-speed coaxial systems is their small **repeater spacing** (1 km), which makes the system relatively expensive to operate. Microwave communication systems generally allow for larger repeater spacing, but their bit rate is also limited by the carrier frequency of such waves.
- A commonly used figure of merit for communication systems is the **bit rate-distance product, BL**, where *B* is the bit rate and *L* is the repeater spacing. Figure 1.5 shows how the *BL* product has increased through technological advances during the last century and a half.
- Communication systems with *BZ*, 100 (Mb/s)-km were available by 1970 and were

limited to such values because of fundamental limitations. It was realized during the second half of the twentieth century that an increase of several orders of magnitude in the *BL* product would be possible if optical waves were used as the carrier. However, neither a coherent optical source nor a suitable transmission medium was available during the 1950s.

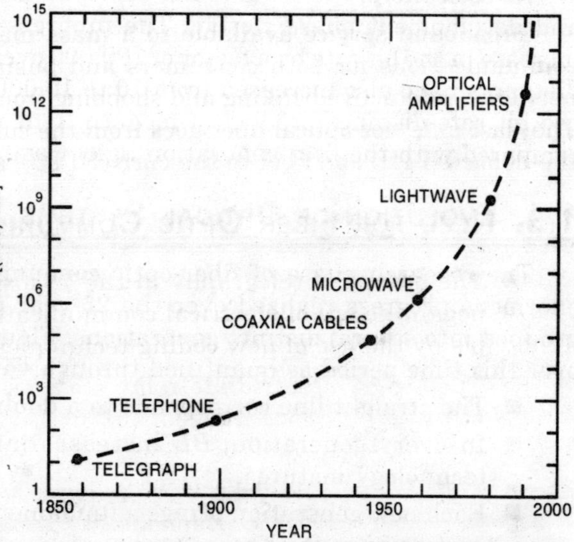

Fig. 1.5. Increase in Bit Rate-Distance Product *EL* During the Period 1850-2000. The Emergence of a New Technology is Marked by a Solid Circle.

- The invention of the laser and its demonstration in 1960 solved the first problem. Attention was then focused on finding ways for using laser light for optical communications.
- Many ideas were advanced during the 1960s, the most noteworthy being the idea of light confinement using a sequence of gas lenses.
- It was suggested in 1966 that optical fibers might be the best choice, as they are capable of guiding the light in a manner similar to the guiding of electrons in copper wires.
- The **main problem** was the high losses of optical fibers—fibers available during the 1960s had losses in excess of 1000 dB/km. A breakthrough occurred in 1970 when fiber losses could be reduced to below 20 dB/km in the wavelength region near 1 m. At about the same time, GaAs semiconductor lasers, operating continuously at room temperature, were demonstrated.
- The simultaneous availability *of compact* optical sources and a **low-loss optical fibers** led to a worldwide effort for developing fiber-optic communication systems.
- Figure 1.6 shows the increase in the capacity of light wave systems realized after 1980 through several generations of development.

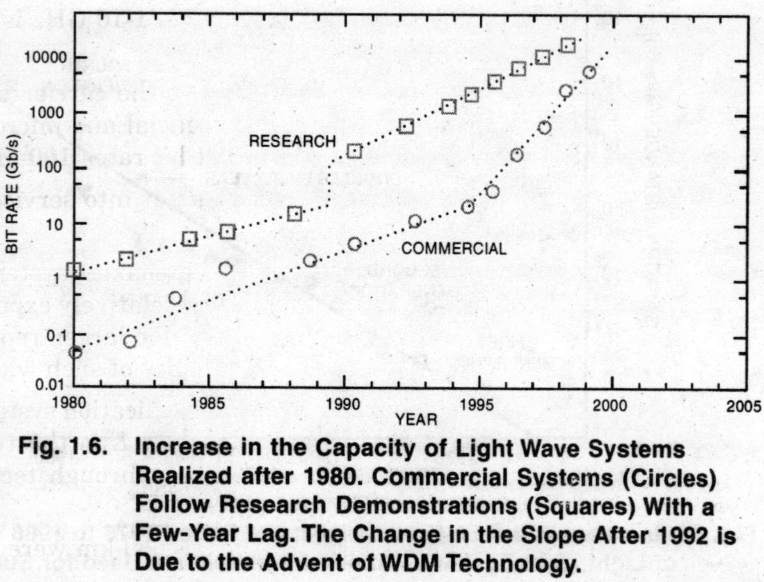

Fig. 1.6. Increase in the Capacity of Light Wave Systems Realized after 1980. Commercial Systems (Circles) Follow Research Demonstrations (Squares) With a Few-Year Lag. The Change in the Slope After 1992 is Due to the Advent of WDM Technology.

As seen there, the commercial deployment of light wave systems followed the research and development phase closely. The progress has indeed been rapid as evident from an increase in the bit rate by a factor of 100,000 over a period of less than 25 years. Transmission distances have also increased from 10 to 10,000 km over the same time period. As a result, the bit rate-distance product of modern light wave systems can exceed by a factor of 10^7 compared with the first-generation light wave systems.

1.4. Evolution of Optical Systems

The research phase of fiber-optic communication systems started around 1975. The enormous progress realized over the 25-year period extending from 1975 to 2000 can be grouped into several distinct generations. Figure 1.7 shows the increase in the *BL* product over this time period as quantified through various laboratory experiments.

- The straight line corresponds to a doubling of the *BL* product every year.
- In every generation, *BL* increases initially but then begins to saturate as the technology matures.
- Each new generation brings a fundamental change that helps to improve the system performance further.

First Generation

The first generation of light wave systems operated near 0.8 μm and used GaAs semiconductor lasers. After several field trials during the period 1977-79, such systems became available commercially in 1980. They operated at a bit rate of 45 Mb/s and allowed repeater spacing of up to 10 km. The larger repeater spacing compared with 1 km spacing of coaxial systems was an important motivation for system designers because it decreased the installation and maintenance costs associated with each repeater. It was clear during the 1970s that the repeater spacing could be increased considerably by operating the light wave system in the wavelength region near 1.3 μm, where fiber loss is below 1 dB/km.

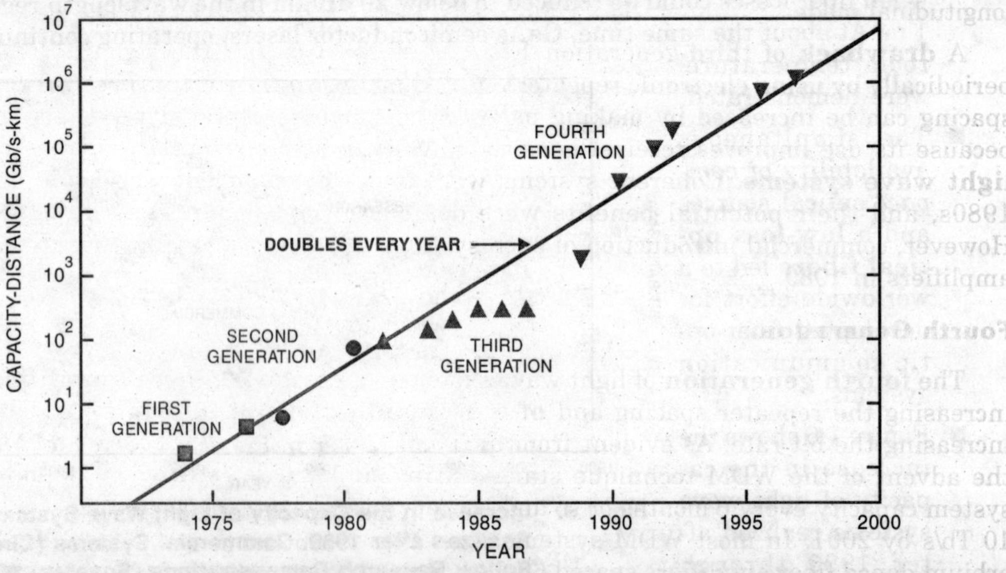

Fig. 1.7. Increase in the *BL* Product Over the Period 1975 to 1980 Through Several Generations of Light Wave Systems. Different Symbols are Used for Successive Generations.

Optical fibers also exhibit minimum dispersion in this wavelength region. This realization led to a worldwide effort for the development of In GaAsP semiconductor lasers and detectors operating near 1.3 μm.

Second Generation

The second generation of fiber-optic communication systems became available in the early 1980s, but the bit rate of early systems was limited to below 100 Mb/s because of dispersion in multimode fibers. This limitation was overcome by the use *of single-mode* fibers. A laboratory experiment in 1981 demonstrated transmission at 2 Gb/s over 44 km of single-mode fiber. The introduction of commercial systems soon followed. By 1987, second-generation light wave systems, operating at bit rates of up to 1.7 Gb/s with a repeater spacing of about 50 km, were commercially available. The repeater spacing of the second-generation light wave systems was limited by the fiber losses at the operating wavelength of 1.3 μm (typically 0.5 dB/km). Losses of silica fibers become minimum near 1.55 μm. Indeed, a 0.2 dB/km loss was realized in 1979 in this spectral region.

However, the introduction of third-generation light wave systems operating at 1.55 μm was considerably delayed by large fiber dispersion near 1.55 μm. conventional In GaAsP semiconductor lasers could not be used because of pulse spreading occurring as a result of simultaneous oscillation of several longitudinal modes. The dispersion problem can be overcome either by using dispersion-shifted fibers designed to have minimum dispersion near 1.55 μm or by limiting the laser spectrum to a single longitudinal mode. Both approaches were followed during the 1980s. By 1985, laboratory experiments indicated the possibility of transmitting information at bit rates of up to 4 Gb/s over distances in excess of 100 km.

Third Generation

Third-generation light wave systems operating at 2.5 Gb/s became available commercially in 1990. Such systems are capable of operating at a bit rate of up to 10 Gb/s. The best performance is achieved using dispersion-shifted fibers in combination with lasers oscillating in a single longitudinal mode.

A **drawback** of third-generation 1.55 μm systems is that the signal is regenerated periodically by using electronic repeaters spaced apart typically by 60-70 km. The repeater spacing can be increased by making use of a homodyne or heterodyne detection scheme because its use improves receiver sensitivity. Such systems are referred to as **coherent light wave systems.** Coherent systems were under development worldwide during the 1980s, and their potential benefits were demonstrated in many system experiments. However, commercial introduction of such systems was postponed with the advent of fiber amplifiers in 1989.

Fourth Generation

The **fourth generation** of light wave systems makes use of **optical amplification** for increasing the repeater spacing and **of wavelength-division multiplexing (WDM)** for increasing the bit rate. As evident from different slopes in Fig. 1.7 before and after 1992, the advent of the WDM technique started a revolution that resulted in doubling of the system capacity every 6 months or so and led to light wave systems operating at a bit rate of 10 Tb/s by 2001. In most WDM systems, fiber losses are compensated periodically using erbium-doped fiber amplifiers spaced 60-80 km apart. Such amplifiers were developed after 1985 and became available commercially by 1990. A 1991 experiment showed the possibility of data transmission over 21,000 km at 2.5 Gb/s, and over 14,300 km at 5 Gb/s, using a recirculation-loop configuration. This performance indicated that an amplifier-based, all-optical, submarine transmission system was feasible for intercontinental communication. By 1996, not only transmission over 1 1,300 km at a bit rate of 5 Gb/s had been demonstrated by using actual submarine cables, but commercial transatlantic and transpecific cable systems also became available. Since then, a large number of submarine light wave systems have been deployed worldwide.

1.5. Why Are Optical Fibers So Good For Communications ?

The information-carrying capacity of an optical fiber is far greater than it is for its competitors: wires, coaxial cables, and microwave links. In addition, optical fibers are :

- inexpensive to produce
- do not conduct electricity (which makes them immune to disturbance by lightning storms, and other electromagnetic signals—except nuclear radiation)
- do not corrode
- small size.

The primary reason that optical fibers have very much larger information-carrying capacity than other media is that they carry light: this might seem a trivially obvious observation but it has fundamental significance. The frequency of the light beams that travel along optical fibers is in the vicinity of two hundred trillion cycles per second (Hz). Compare this with the frequency of the latest generation of personal communication service (PCS) cellular wireless systems - approximately two billion cycles per second (2 GHz). Consider the frequencies that must be transmitted for voice communications, which cover the range (bandwidth) from about 50 Hz to 20,000 Hz (20 kHz). Indeed, since there is very little need to include the high frequencies for understandable voice communications, the actual bandwidth needed is really only about 4 kHz. It is possible, in principle, to carry about 50 billion voice conversations on a single laser beam in an optical fiber. This capacity results from the very simple calculation :

$$2 \times 10^{14} \text{ Hz}/4 \times 10^{3} \text{ Hz} = 50 \times 10^{9}$$

The entire population of the Earth could be on the phone on a single fiber at the same time. The corresponding capacity of a PCS link is about 500,000 simultaneous voice channels. In practice, it has not proven possible to achieve these maximum capacities, although in current links the optical fiber wins by a huge margin. The domination of optical fiber as a means for carrying information is apparent when we note that such fibers are being manufactured worldwide at a rate of 2000 miles per hour!

Whether an optical or microwave link has the ability to use its full capacity depends on the way in which the information is encoded, and how different messages are mixed together *(multiplexed)* without them all getting mixed up. To understand in more detail how this is done, we must digress and discuss the nature of digital representations of information and how this influences the way in which information, whether this is voice, video, or computer data, is encoded and transmitted.

1.6. Basic Concepts of Communication System

This section introduces a few basic concepts common to all communication systems. We begin with a description of analog and digital signals and describe how an analog signal can be converted into digital form. We then consider time- and frequency division multiplexing of input signals, and conclude with a discussion of various modulation formats.

1.6.1. Analog and Digital Signals

In any communication system, information to be transmitted is generally available as an electrical signal that may take *analog* or *digital* form.

In the **analog case**, the signal (e.g., electric current) varies continuously with time, as shown schematically in Fig. 1.8(*a*). Familiar examples include audio and video signals resulting when a microphone converts voice or a video camera converts an image into an electrical signal.

By contrast, the **digital signal** takes only a few discrete values. In the *binary representation* of a digital signal only two values are possible. The simplest case of a binary digital signal is one in which the electric current is either on or off, as shown in Fig. 1.8(*b*). These two possibilities are called "bit 1" and "bit 0" (*bit* is a contracted form of binary *digit*). Each bit lasts for a certain period of time T_B, known as the **bit period** or **bit slot.**

Since one bit of information is conveyed in a time interval T_B, the bit rate *B*, defined as the number of bits per second, is simply

$$B = T_b^{-1}$$

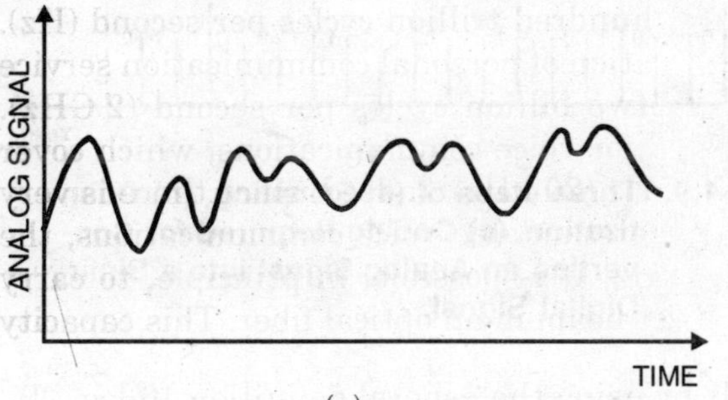

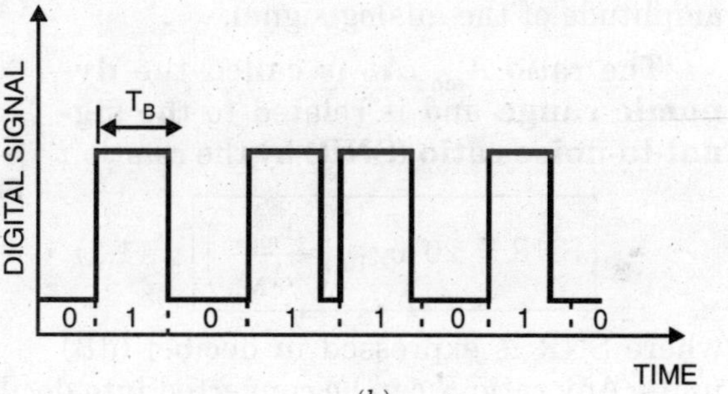

Fig. 1.8. Representation of (a) an Analog Signal and (b) a Digital Signal.

A well-known example of digital signals is provided by computer data. Each letter of the alphabet together with other common symbols (decimal numerals, punctuation marks, etc.) is assigned a code number (ASCII code) in the range 0-127 whose binary representation corresponds to a 7-bit digital signal. The original ASCII code has been extended to represent 256 characters transmitted through 8-bit bytes. Both analog and digital signals are characterized by their bandwidth, which is a measure of the spectral contents of the signal.

The **signal bandwidth** represents the range of frequencies contained within the signal and is determined mathematically by its Fourier transform. An analog signal can be converted into digital form by sampling it at regular intervals of time.

Conversion Method

Figure 1.9 shows the conversion method schematically. The sampling rate is determined by the bandwidth Δf of the analog signal. According to the **sampling theorem,** a bandwidth-limited signal can be fully represented by discrete samples, without any loss of information, provided that the sampling frequency f_s satisfies the **Nyquist criterion,**

$$f_s \geq 2\Delta f$$

The first step consists of sampling the analog signal at the right frequency. The sampled values can take any value in the range $0 \leq A \leq A_{max}$,

where, A_{max} is the maximum amplitude of the given analog signal.

Let us assume that A_{max} is divided into M discrete (not necessarily equally spaced) intervals. Each sampled value is quantized to correspond to one of these discrete values. Clearly, this procedure leads to additional noise, known as **quantization noise**, which adds to the noise already present in the analog signal. The effect of quantization noise can be minimized by choosing the number of discrete levels such that $M > A_{max}/A_N$, where A_N is the root-mean-square noise amplitude of the analog signal.

The ratio A_{max}/A_N is called the **dynamic range** and is related to the **signal-to-noise ratio (SNR)** by the relation

$$\text{SNR} = 20 \log_{10}\left(\frac{A_{max}}{A_N}\right) \quad ...(1.1)$$

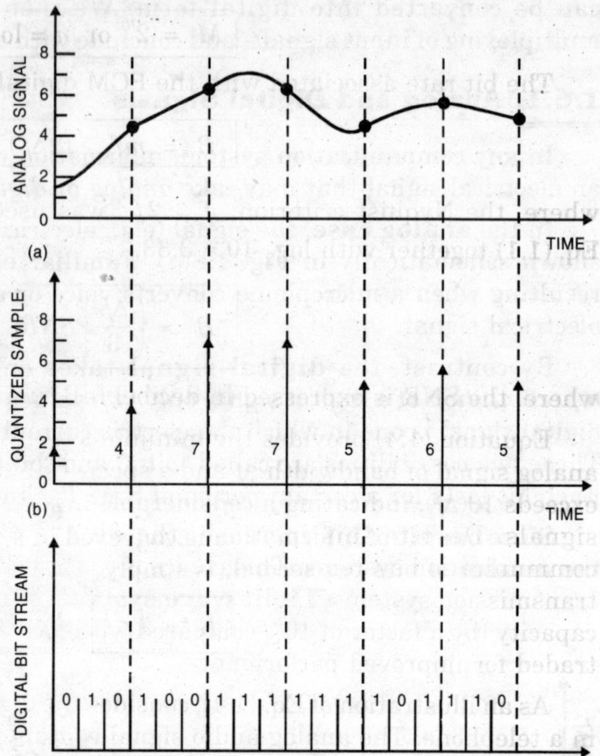

Fig. 1.9. Three steps of (a) Sampling, (b) Quantization, (c) Coding Required for Converting an Analog Signal into a Binary Digital Signal.

where SNR is expressed in decibel (dB) units. Any ratio R can be converted into decibels by using the general definition 10 $\log_{10}$ R. Equation (1.1) contains a factor of 20 in place of 10 simply because the SNR for electrical signals is defined with respect to the electrical power, whereas A is related to the electric current (or voltage).

The quantized sampled values can be converted into digital format by using a suitable conversion technique.

- In one scheme, known as **pulse-position modulation,** pulse position within the bit slot is a measure of the sampled value.
- In another, known as **pulse-duration modulation,** the pulse width is varied from bit to bit in accordance with the sampled value.
- These techniques are rarely used in practical optical communication systems, since it is difficult to maintain the pulse position or pulse width to high accuracy during propagation inside the fiber. The technique used almost universally, known as **pulse-code modulation (PCM)**, is based on a binary scheme in which information is conveyed by the absence or the presence of pulses that are otherwise identical. A binary code is used to convert each sampled value into a string of 1 and 0 bits.

The number of bits m needed to code each sample is related to the number of quantized signal levels M by the relation

$$M = 2^m \text{ or } m = \log_2 M \qquad \text{...(1.2)}$$

The bit rate associated with the PCM digital signal is thus given by

$$B = mf_s \geq (2\Delta f) \log_2 M \qquad \text{...(1.3)}$$

where, the Nyquist criterion, $f_s \geq 2f$, was used. By noting that $M > A_{max}/A_N$ and using Eq. (1.1) together with $\log_2 10 \approx 3.33$,

$$B > \left(\frac{\Delta f}{3}\right) SNR \qquad \text{...(1.4)}$$

where, the SNR is expressed in decibel (dB) units.

Equation (1.4) provides the minimum bit rate required for digital representation of an analog signal of bandwidth Δf and a specific SNR. When SNR > 30 dB, the required bit rate exceeds 10(Δf), indicating a considerable increase in the bandwidth requirements of digital signals. Despite this increase, the digital format is almost always used for optical communication systems. This choice is made because of the superior performance of digital transmission systems. **Light wave systems** offer such an enormous increase in the system capacity (by a factor of 10^5) compared with microwave systems that some bandwidth can be traded for improved performance.

As an illustration of Eq. (1.4), consider the digital conversion of an audio signal generated in a telephone. The analog audio signal contains frequencies in the range 0.3-3.4 kHz with a bandwidth Δf = 3.1 kHz and has a SNR of about 30 dB. Equation (1.4) indicates that $B >$ 31 kb/s. In practice, a digital audio channel operates at 64 kb/s. The analog signal is sampled at intervals of 125 μs (sampling rate f_s = 8 kHz), and each sample is represented by 8 bits. The required bit rate for a digital video signal is higher by more than a factor of 1000. The analog television signal has a bandwidth 4 MHz with a SNR of about 50 dB. The minimum bit rate from Eq. (1.4) is 66 Mb/s. In practice, a digital video signal requires a bit rate of 100 Mb/s or more unless it is compressed by using a standard format (such as MPEG-2).

1.6.2. Channel Multiplexing

As seen in the preceding discussion, a digital voice channel operates at 64 kb/s. Most fiber-optic communication systems are capable of transmitting at a rate of more than 1 Gb/s. To utilize the system capacity fully, it is necessary to transmit many channels simultaneously through multiplexing. This can be accomplished through **time-division multiplexing (TDM)** or **frequency-division multiplexing (FDM)**.

In the case of **TDM**, bits associated with different channels are interleaved in the time domain to form a composite bit stream. For example, the bit slot is about 15 ms for a single voice channel operating at 64 kb/s. Five such channels can be multiplexed through TDM if the bit streams of successive channels are delayed by 3 ms. Figure 1.10(a) shows the resulting bit stream schematically at a composite bit rate of 320 kB/s.

In the case of **FDM**, the channels are spaced apart in the frequency domain.

Each channel is carried by its own carrier wave. The carrier frequencies are spaced more than the channel bandwidth so that the channel spectra do not overlap, as seen Fig. 1.10(b).

- FDM is suitable for both analog and digital signals and is used in broadcasting of radio and television channels.
- TDM is readily implemented for digital signals and is commonly used for telecommunication networks.
- It is important to realize that TDM and FDM can be implemented in both the electrical and optical domains; optical FDM is often referred to as WDM.
- The concept of TDM has been used to form **digital hierarchies.**
- In North America and Japan, the **first level** corresponds to multiplexing of 24 voice channels with a composite bit rate of 1.544 Mb/s (hierarchy DS-1), whereas in Europe 30 voice channels are multiplexed, resulting in a composite bit rate of 2.048 Mb/s. The bit rate of the multiplexed signal is slightly larger than the simple product of 64 kB/s with the number of channels because of extra control bits that are added for separating (demultiplexing) the channels at the receiver end.

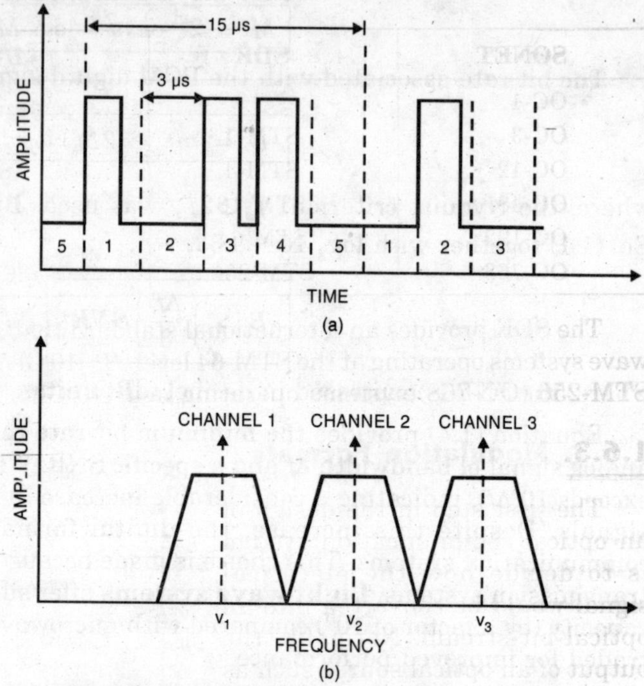

Fig. 1.10. (a) Time-Division Multiplexing of Five Digital Voice Channels Operating at 64 kB/s; (b) Frequency-Division Multiplexing of Three Analog Signals.

- The **second-level** hierarchy is obtained by multiplexing 4 DS-1 TDM channels. This results in a bit rate of 6.312 Mb/s (hierarchy DS-2) for North America or Japan and 8.448 Mb/s for Europe.
- This procedure is continued to obtain higher-level hierarchies. For example, at the fifth level of hierarchy, the bit rate becomes 565 Mb/s for Europe and 396 Mb/s for Japan.

The lack of an international standard in the telecommunication industry during the 1980s led to the advent of a new standard, first called the **synchronous optical network (SONET)** and later termed the **synchronous digital hierarchy (SDH).** It defines a synchronous frame structure for transmitting TDM digital signals.

- The basic building block of the SONET has a bit rate of 51.84 Mb/s. The corresponding optical signal is referred to as OC-1, where OC stands for **optical carrier.**
- The basic building block of the SDH has a bit rate of 155.52 Mb/s and is referred to as STM-1, where STM stands for a **synchronous transport module.**
- A useful feature of the SONET and SDK is that higher levels have a bit rate that is an exact multiple of the basic bit rate.

Table 1.1 lists the correspondence between SONET and SDK bit rates for several levels.

Table 1.1. SOXET/SDH Bit Rates.

SONET	SDK	*B* (Mb/s)	Channels
OC-1		51.84	672
OC-3	STM-1	155.52	2,016
OC-12	STM-4	622.08	8,064
OC-48	STM-16	2,488.32	32,256
OC-192	STM-64	9,953.28	129,024
OC-768	STM-256	39,813.12	516,096

The SDK provides an international standard that appears to be well adopted. Indeed, light wave systems operating at the STM-64 level ($B \approx 10$ Gb/s) are available since 1996. Commercial STM-256 (OC-768) systems operating near 40 Gb/s became available by 2002.

1.6.3. Modulation Formats

The first step in the design of an optical communication system is to decide how the electrical signal would be converted into an optical bit stream. Normally, the output of an optical source such as a semiconductor laser is modulated by applying the electrical signal either directly to the optical source or to an external modulator.

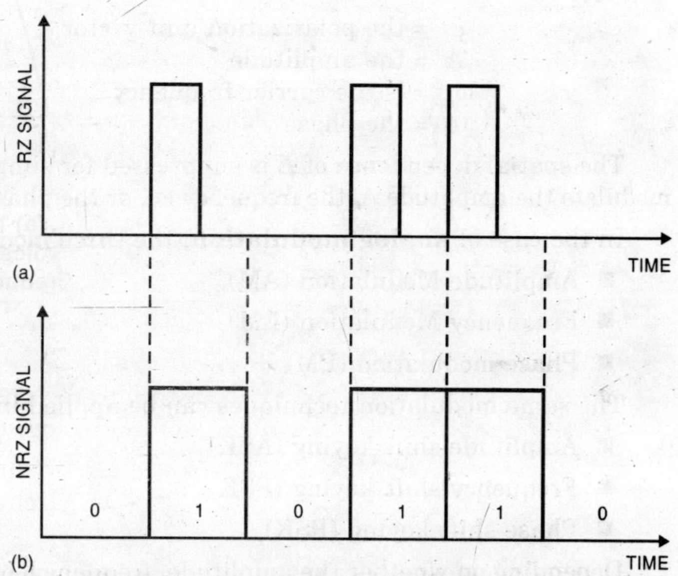

Fig. 1.11. Digital Bit Stream 010110... Coded by using : (a) Return-to-zero (RZ), (b) Non-Return-to-Zero (NRZ) Formats.

There are two choices for the modulation format of the resulting optical bit stream. These are shown in Fig. 1.11 and are known as the

- **return-to-zero (RZ)** format
- **non return-to-zero (NRZ)** format

In the **RZ format**, each optical pulse representing bit 1 is shorter than the bit slot, and its amplitude returns to zero before the bit duration is over. In the **NRZ format**, the optical pulse remains on throughout the bit slot and its amplitude does not drop to zero between two or more successive 1 bits. As a result, pulse width varies depending on the bit pattern, whereas it remains the same in the case of RZ format.

An **advantage of the NRZ format** is that the bandwidth associated with the bit stream is smaller than that of the RZ format by about a factor of 2 simply because on-off transitions occur fewer times. However, its use requires tighter control of the pulse width and may lead to bit-pattern-dependent effects if the optical pulse spreads during transmission. The NRZ format is often used in practice because of a smaller signal bandwidth associated with it. The use of the RZ format in the optical domain began to attract attention around 1999 after it was found that its use may help the design of high-capacity light wave systems.

An example of the RZ format is provided by the dispersion-managed soliton systems where a chirped pulse propagates inside the fiber link in a periodic fashion, and the average dispersion is used to counteract the buildup of the nonlinear effects. In an interesting variant of the RZ format, known as the **chirped RZ** (or **CRZ) format**, optical pulses in each bit slot are chirped before they are launched into the fiber link but the system is operated in a quasi-linear regime.

In other schemes, modulation formats well known in the field of microwave communications are applied to the optical domain. Such formats are known as **carrier-suppressed RZ (CSRZ),** single-sideband, or vestigial-sideband formats. Such RZ formats benefit from a reduced bandwidth compared to the standard RZ format. An *important* issue is related to the choice of the physical variable that is modulated to encode the data on the optical carrier. The optical carrier wave before modulation is of the form

$$\boxed{E(t) = \hat{e}A\cos(\omega_0 t + \phi)} \quad ...(1.5)$$

where
E = the electric field vector
$\hat{e}$ = the polarization unit vector
A = the amplitude
ω_0 = is the carrier frequency
ϕ = the phase

The spatial dependence of $\boldsymbol{E}$ is suppressed for simplicity of notation. One may choose to modulate the amplitude A, the frequency ω_0, or the phase ϕ.

In the case of **analog modulation**, the three modulation choices are known as :

- Amplitude Modulation (AM)
- Frequency Modulation (FM)
- Phase modulation (PM)

The same modulation techniques can be applied in the digital case and are called as :

- Amplitude-shift keying (ASK)
- Frequency-shift keying (FSK)
- Phase-shift keying (PSK),

Depending on whether the amplitude, frequency, or phase of the carrier wave is shifted between the two levels of a binary digital signal.

The simplest technique consists of simply changing the signal power between two levels, one of which is set to zero, and is often called **on-off keying (OOK)** to reflect the on-off nature of the resulting optical signal. Most digital light wave systems employ OOK in combination with PCM.

Components of an Optical Communication System

The basic components of an optical communication system are shown in Fig. 1.12.

- A serial bit stream in electrical form is presented to a modulator, which encodes the data appropriately for fiber transmission.
- A light source (laser or Light Emitting Diode - LED) is driven by the modulator and the light focused into the fiber.

- The light travels down the fiber (during which time it may experience dispersion and loss of strength).
- At the receiver end the light is fed to a detector and converted to electrical form.
- The signal is then amplified and fed to another detector, which isolates the individual state changes and their timing. It then decodes the sequence of state changes and reconstructs the original bit stream.

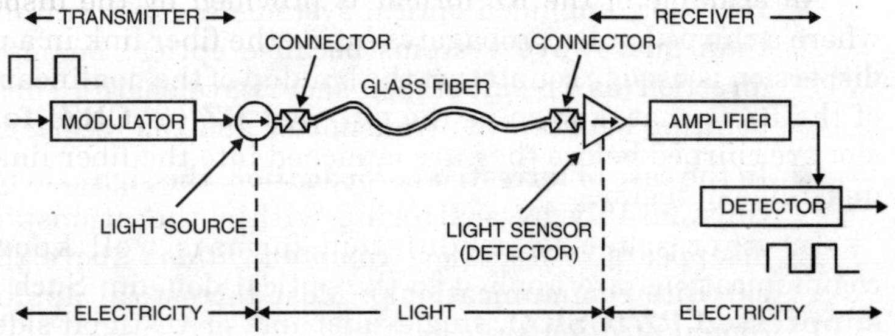

Fig. 1.12. Optical Transmission.

- The timed bit stream so received may then be fed to a using device. A detailed diagram is shown in Fig. 1.12.

1.7. Block Diagram of Optical Communication System

Optical communication systems differ in principle from microwave systems only in the frequency range of the carrier wave used to carry the information. The optical carrier frequencies are typically 200 THz, in contrast with the microwave carrier frequencies (1 GHz). An increase in the information capacity of optical communication systems by a factor of up to 10,000 is expected simply because of such high carrier frequencies used for light wave systems. This increase can be understood by noting that the bandwidth of the modulated carrier can be up to a few percent of the carrier frequency. Taking, for illustration, 1% as the limiting value, optical communication systems have the potential of carrying information at bit rates 1 Tb/s. It is this enormous potential bandwidth of optical communication systems that is the driving force behind the worldwide development and deployment of light wave systems. Current state-of-the-art systems operate at bit rates 10 Gb/s, indicating that there is considerable room for improvement.

Figure 1.13 shows a generic block diagram of an optical communication system. It consists of a transmitter, a communication channel, and a receiver, the three elements common to all communication systems.

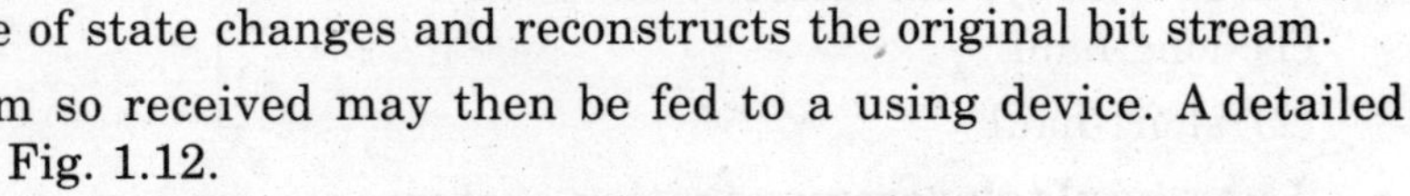

Fig. 1.13. Generic Optical Communication System.

Optical communication systems can be classified into two broad categories :

(1) guided

(2) unguided

- As the name implies, in the case of **guided light wave systems,** the optical beam emitted by the transmitter remains spatially confined. Since all guided optical communication systems currently use optical fibers, the commonly used term for them is fiber-optic communication systems.
- The term **light wave system** is also sometimes used for fiber-optic communication systems, although it should generally include both guided and unguided systems.
- In the case of **unguided optical communication systems,** the optical beam emitted by the transmitter spreads in space, similar to the spreading of microwaves.

- However, unguided optical systems are less suitable for broadcasting applications than microwave systems because optical beams spread mainly in the forward direction (as a result of their short wavelength). Their use generally requires accurate pointing between the transmitter and the receiver.
- In the case of terrestrial propagation, the signal in unguided systems can deteriorate considerably by scattering within the atmosphere. This problem, of course, disappears *in free-space communications* above the earth atmosphere (e.g., inter satellite communications). Most terrestrial applications make use *of fiber-optic communication systems.*

The **telecommunication applications** can be broadly classified into two categories depending on whether the optical signal is transmitted over relatively long or short distances compared with typical intercity distances :

(1) long-haul

(2) short-haul

Long-haul telecommunication systems require high-capacity trunk lines and benefit most by the use of fiber-optic light wave systems. Indeed, the technology behind optical fiber communication is often driven by long-haul applications. Each successive generation of light wave systems is capable of operating at higher bit rates and over longer distances. Periodic regeneration of the optical signal by using repeaters is still required for most long-haul systems. However, more than an order-of-magnitude increase in both the repeater spacing and the bit rate compared with those of coaxial systems has made the use of light wave systems very attractive for long-haul applications. Furthermore, transmission distances of thousands of kilometers can be realized by using optical amplifiers.

Short-haul telecommunication applications cover intracity and local-loop traffic. Such systems typically operate at low bit rates over distances of less than 10 km. The use of single-channel light wave systems for such applications is not very cost-effective, and multi channel networks with multiple services should be considered. The concept of a broadband integrated-services digital network requires a high-capacity communication system capable of carrying multiple services. The asynchronous transfer mode (ATM) technology also demands high bandwidths. Only fiber-optic communication systems are likely to meet such wideband distribution requirements.

1.8. Optical System Components

The basic block diagram of Fig. 1.12 applies to a fiber-optic communication system, the only difference being that the communication channel is an optical fiber cable. The other two components, the optical transmitter and the optical receiver, are designed to meet the needs of such a specific communication channel. In this section we discuss the general issues related to the role of optical fiber as a communication channel and to the design of transmitters and receivers. The objective is to provide an introductory overview.

1.8.1. Optical Fiber a Communication Channel

The role of a communication channel is to transport the optical signal from transmitter to receiver without distorting it. Most light wave systems use optical fibers as the communication channel because silica fibers can transmit light with losses as small as 0.2 dB/km. Even then, optical power reduces to only 1% after 100 km. For this reason, fiber losses remain an important design issue and determine the repeater or amplifier spacing of a long-haul light wave system.

Another important design issue *is* **fiber dispersion,** which leads to broadening of individual optical pulses with propagation. If optical pulses spread significantly outside their allocated bit slot, the transmitted signal is severely degraded. Eventually, it becomes impossible to recover the original signal with high accuracy. The problem is most severe in the case of multimode fibers, since pulses spread rapidly (typically at a rate of 10 ns/km) because of different speeds associated with different fiber modes. It is for this reason that most optical communication systems use single-mode fibers.

Material dispersion (related to the frequency dependence of the refractive index) still leads to pulse broadening (typically < 0.1 ns/km), but it is small enough to be acceptable for most applications and can be reduced further by controlling the spectral width of the optical source. But material dispersion sets the ultimate limit on the bit rate and the transmission distance of fiber-optic communication systems.

1.8.2. Optical Transmitters

The role of an **optical transmitter** is to convert the electrical signal into optical form and to launch the resulting optical signal into the optical fiber. Figure 1.14 shows the block diagram of an optical transmitter. It consists of an optical source, a modulator and a channel coupler.

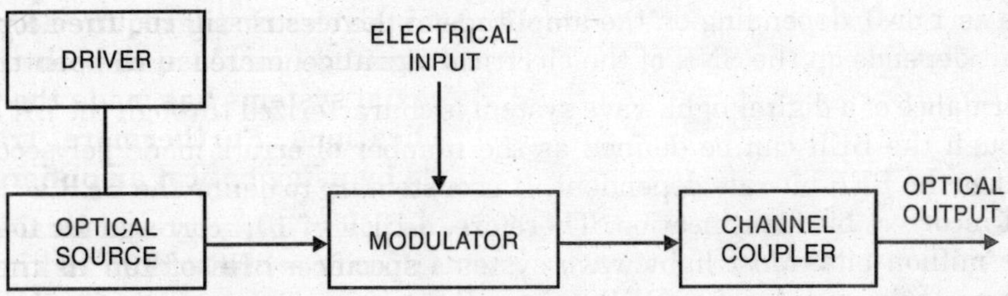

Fig. 1.14. Components of an optical transmitter.

Semiconductor lasers or light-emitting diodes are used as optical sources because of their compatibility with the optical-fiber communication channel. The optical signal is generated by modulating the optical carrier wave. Although an **external modulator** is sometimes used, it can be dispensed with in some cases, since the output of a semiconductor optical source can be modulated directly by varying the injection current. Such a scheme simplifies the transmitter design and is generally cost-effective. The coupler is typically a micro lens that focuses the optical signal onto the entrance plane of an optical fiber with the maximum possible efficiency.

The **launched power** is an important design parameter. One can increase the amplifier (or repeater) spacing by increasing it, but the onset of various *nonlinear effects* limits how much the input power can be increased. The launched power is often expressed in "dBm" units with 1 mW as the reference level.

$$\text{Power (dBm)} = 10 \log_{10}\left(\frac{\text{Power}}{1\text{ mW}}\right)$$

Thus, 1 mW is 0 dBm, but 1 μW corresponds to –30 dBm. The launched power is rather low (< –10 dBm) for light-emitting diodes but semiconductor lasers can launch powers 10 dBm. As light-emitting diodes are also limited in their modulation capabilities, most light wave systems use semiconductor lasers as optical sources. The bit rate of optical transmitters is often limited by electronics rather than by the semiconductor laser itself. With proper design, optical transmitters can be made to operate at a bit rate of up to 40 Gb/s.

1.8.3. Optical Receivers

An **optical receiver** converts the optical signal received at the output end of the optical fiber back into the original electrical signal. Figure 1.15 shows the block diagram of an optical receiver. It consists of a coupler, a photodetector, and a demodulator.

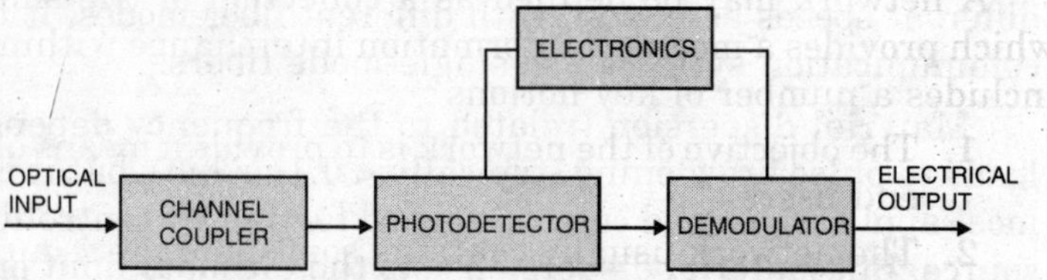

Fig. 1.15. Components of an Optical Receiver.

The **coupler** focuses the received optical signal onto the photodetector. Semiconductor photodiodes are used as photodetectors because of their compatibility with the whole system.

The design of the demodulator depends on the modulation format used by the light wave system. Most light wave systems employ a scheme referred to as **"intensity modulation with direct detection" (IM/DD).** Demodulation in this case is done by a decision circuit that identifies bits as 1 or 0, depending on the amplitude of the electric signal. The accuracy of the decision circuit depends on the SNR of the electrical signal generated at the photodetector.

The performance of a digital light wave system is characterized through the **bit error rate (BER).** Although the BER can be defined as the number of errors made per second, such a definition makes the BER bit-rate dependent. It is customary to define the BER as the average probability of incorrect bit identification. Therefore, a BER of 10^{-6} corresponds to on average one error per million bits. Most light wave systems specify a BER of 10^{-9} as the operating requirement; some even require a BER as small as 10^{-14}. The error-correction codes are sometimes used to improve the raw BER of a light wave systems.

An important parameter for any receiver is the **receiver sensitivity.** It is usually defined as **the minimum average optical power required to realize a BER of 10^{-9}.** Receiver sensitivity depends on the SNR, which in turn depends on various noise sources that corrupt the signal received.

Even for a perfect receiver, some noise is introduced by the process of photo detection itself. This is referred to as the **quantum noise** or the **shot noise,** as it has its origin in the particle nature of electrons. Optical receivers operating at the shot-noise limit are called **quantum-noise-limited receivers.** No practical receiver operates at the quantum-noise limit because of the presence of several other noise sources. Some of the noise sources such as **thermal noise** are internal to the receiver. Others originate at the transmitter or during propagation along the fiber link.

For instance, any amplification of the optical signal along the transmission line with the help of optical amplifiers introduces the so-called **amplifier noise** that has its origin in the fundamental process of spontaneous emission. Chromatic dispersion in optical fibers can add additional noise through phenomenon such as inter symbol interference and mode-partition noise.

The receiver sensitivity is determined by a cumulative effect of all possible noise mechanisms that degrade the SNR at the decision circuit. In general, it also depends on the bit rate as the contribution of some noise sources (e.g., shot noise) increases in proportion to the signal bandwidth.

1.9. Optical Networking

A network may be defined as a collection of transmission links and other equipment which provides a means of information interchange within a group of end users. The concept includes a number of key notions :

1. The objective of the network is to provide a means of information interchange between end users.
2. The network usually contains some shared resources *i.e.* links and switching nodes are shared between many users.
3. Most networks have a common (often centralized) network management system.
4. Information may be exchanged between any user and any other user. This concept while common in communication networks is far from being universal. Many networks exist for the purpose of connecting a few thousand devices to only one (or a few) central device(s). For example in a network of cash dispensers there is no communication (and no conceivable need for communication) between the cash dispensers - all communication is to/from a central computer.
5. A single user may communicate with multiple other users either simultaneously or serially (one after another) using only a single connection to the network. Again, this is a very common characteristic of networks but it is by no means universal. Some networks provide fixed connections between pre-determined end users with no possibility for quick changes. The characteristic here that qualifies such a system to be called a network would be the sharing of intermediate switching nodes and/or transmission links within the network.
6. The term "network" also usually implies a geographic separation between end users. This is not always true in the sense that communicating end users may be across the room or across the world.

There are many types of networks :

- Perhaps the simplest kind provides a fixed communication path (or rather, a collection of fixed paths) between end users.
- A slightly more complex type is where there is only one connection allowed per user but users may make arbitrary connections at will with other end users. The telephone network is a good example here.
- More complex are so-called "packet switching" networks where information is carried between end users in the form of packets (a.k.a. frames or cells). In these networks a single end user is usually capable of communicating with a large number of other end users at the same time.
- Packet switched networks themselves come in many types. In a "connection-oriented" network paths through the network are defined before information is transferred. Information is always transferred along the predefined path. In a "connectionless" network frames of data are routed through the network based on a destination address carried within the data itself. In this case there is no necessary correlation between the path through the network taken by one particular frame of information and the next frame of data sent by the same user to the same destination.

- Another major distinction between networks comes from the method by which switching is performed.
- Most networks (especially wide area ones) contain nodal points where information is switched from link to link. The nodes typically consist of a computer-like device.
- Other networks switch information by placing frames onto a shared medium (such as a bus) to which many end users are connected. Each end user filters the information and pays attention to only information addressed to it. This type is typical of Local Area Networks (LANs).

Networks may be further characterized by their geographic extent such as :

1. Local Area Network (LAN)
2. Metropolitan Area Network (MAN)
3. Wide Area Network (WAN).

These names don't only delimit or suggest geographic extent but also denote quite different types of networks which have evolved to meet the requirements of the different environments. Optical networking is just beginning! In some areas (such as the LAN) quite large experimental (but nevertheless operational) networks have been built. In other areas (such as packet networks) researchers are still wrestling with the problems of applying the concepts in an optical world.

The **simplest type of network** is where a single link is shared by many different communications between different end users. This is achieved in commercial operation today in **Wavelength Division Multiplexing (WDM)** systems.

A simple extension of WDM allows for long links where channels are tapped off to service individual end users along the way. Networks using this principle **(add/drop multiplexing)** are possible today and a number of very large networks are planned. Wide area switched networks (similar to the telephone network) could be built today as all of the required components are available. It is expected that networks of this kind will develop and become common within the next few years.

Optical packet switching is a long way from reality and indeed may not ever be achievable. However, to realize the benefits of optical technology it will be necessary to build fully optical networks. It is self-defeating to convert the signal from optical to electronic form every time it needs to be routed or switched. It seems likely that fully optical networks will not be just optical copies of concepts well explored in the electronic networking world. Over time we can expect new types of networks to evolve which may have some similarities with electronic networks but which will make use of the properties and characteristics of unique optical components.

1.10. Fiber Optic System Topology

Most of the discussion on fiber optic data links are based on point-to-point links. A point-to-point fiber optic data link consists of an optical transmitter, optical fiber, and an optical receiver. In addition, any splices or connectors used to join individual

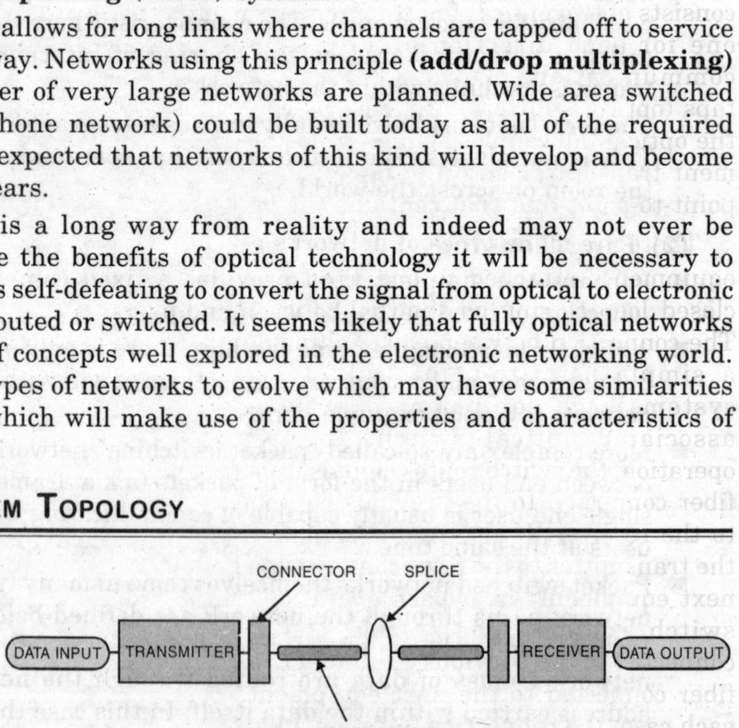

Fig. 1.16. A Schematic Diagram of a Point-to-point Fiber Optic Data Link.

optical fiber sections to each other and to the transmitter and the receiver are included. Figure 1.16 provides a schematic diagram of a point-to-point fiber optic data link.

A common fiber optic application is the full duplex link. This link consists of two simple point-to-point links. The links transmit in opposite directions between the equipments. This application may be configured using only one fiber. If configured with one fiber, fiber optic splitters are used at each end to couple the transmit signal onto the fiber and receive signal to the detector. All fiber optic systems are simply sets of point-to-point fiber optic links. Different system topologies arise from the different ways that point-to-point fiber optic links can be connected between equipments.

The term topology, as used here, refers to the configuration of various equipments and the fiber optic components interconnecting them. This equipment may be computers, workstations, consoles, or other equipments. Point-to-point links are connected to produce systems with linear bus, ring, star, or tree topologies. Point-to-point fiber optic links are the basic building block of all fiber optic systems.

(1) Linear Bus Topology : It consists of a single transmission line that is shared by a number of equipments (Fig. 1.17). Generally the transmission line in a fiber optic linear bus consists of two optical lines, one for each direction of communication. Optical taps (optical splitters) are used by each equipment to connect to each line. For each line, the optical tap couples signals from the line to the equipment receiver and from the equipment transmitter onto the line. The connection between any two equipments is a simple point-to-point link that contains the optical tap for each equipment.

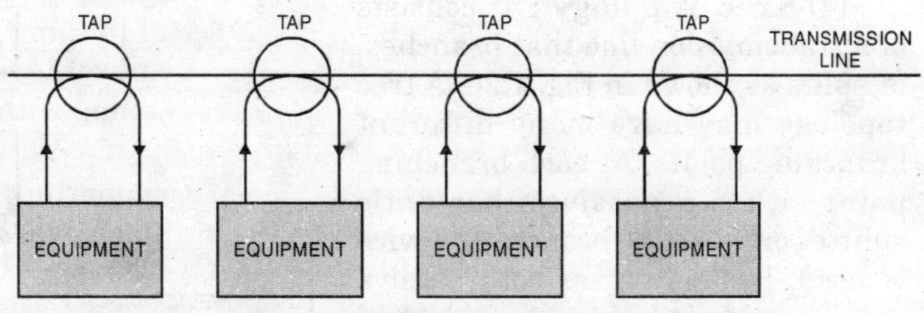

Fig. 1.17. Linear Bus Topology.

(2) Ring Topology : It consists of equipments attached to one another in a closed loop or ring as shown in Fig. 1.18. The connection between each equipment is a simple point-to-point link. In some systems each equipment may have an associated optical switch. In normal operation, the switch routes signals from the fiber connected to the previous equipment to the receiver. It also routes signals from the transmitter to the fiber connected to the next equipment. In bypass operation, the switch routes signals from the fiber connected to the previous equipment to the fiber connected to the next equipment. In each case, the connection between adjacent equipments on the ring is a simple point-to-point link through fiber, connectors, and switches.

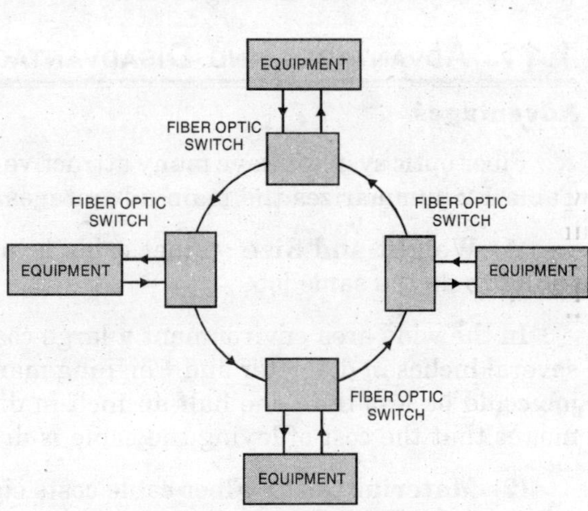

Fig. 1.18. Ring Topology.

(3) Star Topology : In this each equipment is connected to a common center hub as shown in Fig. 1.19. The center hub can be a passive fiber optic star coupler or active equipment. If the center hub is a passive star coupler, each equipment transmitter is connected to an input port of the coupler and an output port of the coupler is connected to each equipment receiver. The connection between any two equipments is a simple point-to-point link through the star coupler. If the center hub is active equipment, the connection between any two equipments consists of two point-to-point links. Each connection consists of one link from the first equipment to the center hub and a second link from the center hub to the second equipment.

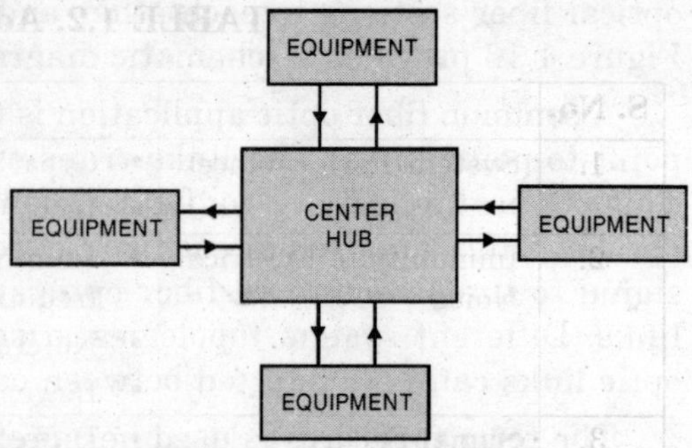

Fig. 1.19. Star Topology.

(4) Tree Topology : It consists of a transmission line that branches, or splits as shown in Fig. 1.20. A tree topology may have many different branching points. At each branching point either a passive fiber optic splitter or an active branching device is used. In many cases both passive couplers and active branching devices are used within a particular system. Regardless of the branching method, each connection within the tree is a simple point-to-point link through splitters or multiple point-to-point links through active branching devices.

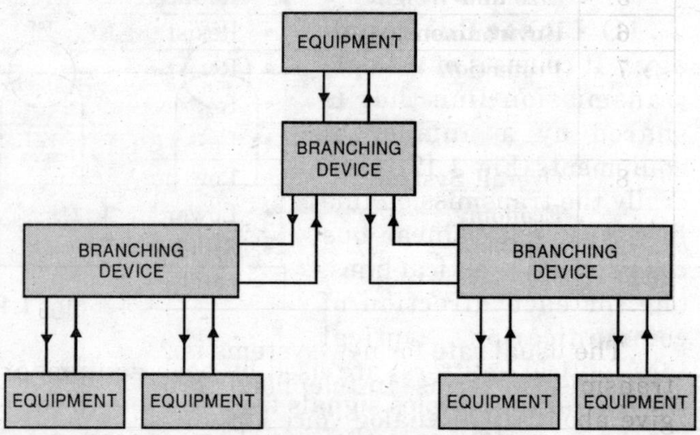

Fig. 1.20. Tree Topology.

1.11. Advantages and Disadvantages of Optical Fiber Communication

Advantages

Fiber optic systems have many attractive features that are superior to electrical systems. Table 1.2 summarizes the main advantages of fiber optic systems.

(1) Weight and Size : Fiber cable is significantly smaller and lighter than electrical cables to do the same job.

In the wide area environment a large coaxial cable system can easily involve a cable of several inches in diameter and weighing many pounds per foot. A fiber cable to do the same job could be less than one half an inch in diameter and weigh a few ounces per foot. This means that the cost of laying the cable is dramatically reduced.

(2) Material Cost : Fiber cable costs significantly less than copper cable for the same transmission capacity.

(3) Information Capacity : The data rate of systems in use in 1998 is generally 150 or 620 Mbps on a single (unidirectional) fiber. This is because these systems were installed in past years.

TABLE 1.2. Advantages of Fiber Optics.

S. No.		
1.	System Performance	• Greatly increased bandwidth and capacity • Lower signal attenuation (loss)
2.	Immunity to Electrical Noise	• Immune to noise (electromagnetic interference [EMI] and frequency interference [RFI] • No crosstalk • Lower bit error rates
3.	Signal Security	• Difficult to tap • Nonconductive (does not radiate signals)
4.	Electrical Isolation	• No common ground required • Freedom from short circuit and sparks
5.	Size and Weight	• Reduced size and weight cables
6.	Environmental	• Resistant to radiation and corrosion
7.	Protection	• Resistant to temperature variations • Improved ruggedness and flexibility • Less restrictive in harsh environments
8.	Overall System Economy	• Low per-channel cost • Lower installation cost • Silica is the principle, abundant, and inexpensive material is sand)

The usual rate for new systems is 2.4 Gbps or even 10 Gbps. This is very high in digital transmission terms. In telephone transmission terms the very best coaxial cable systems give about 2,000 analog voice circuits. A 150 Mbps fiber connection gives just over 2,000 digital telephone (64 Kbps) connections. But the 150 Mbps fiber is at a very early stage in the development of fiber optical systems.

(4) No Electrical Connection : This is an obvious point but nevertheless a very important one. Electrical connections have problems.

- In electrical systems there is always the possibility of "ground loops" causing a serious problem, especially in the LAN or computer channel environment.
  When we communicate electrically, we often have to connect the grounds to one another or at least go to a lot of trouble to avoid making this connection. One little known problem is that there is often a voltage potential difference between "ground" at different locations. It has been observed that as much as 3 volts difference in ground potential between adjacent buildings (this was a freak situation). It is normal to observe 1 or 2 volt differences over distances of a kilometer or so. With shielded cable there can be a problem if we earth the shields at both ends of the connection.
- Optical connection is very safe. Electrical connections always have to be protected from high voltages because of the danger to people touching the wire.
- In some tropical regions of the world, lightning poses a severe hazard even to buried telephone cables. Of course, optical fiber isn't subject to lightning problems but it must be remembered that sometimes optical cables carry wires within them for strengthening or to power repeaters. These wires can be a target for lightning.

(5) No Electromagnetic Interference : Because the connection is not electrical, we can neither pick up nor create electrical interference (the major source of noise). This is one reason that optical communication has so few errors.

There are very few sources of things that can distort or interfere with the signal :

- In a building this means that fiber cables can be placed almost anywhere electrical cables would have problems, (for example near a lift motor or in a cable duct with heavy power cables).
- In an industrial plant such as a steel mill, this gives much greater flexibility in cabling than previously available.
- In the wide area networking environment there is much greater flexibility in route selection. Cables may be located near water or power lines without risk to people or equipment.

(6) Distances between Regenerators : As a signal travels along a communication line it loses strength (is attenuated) and picks up noise. The traditional way to regenerate the signal, restoring its power and removing the noise, is to use a either a repeater or an amplifier.

In long-line optical transmission cables current in use by the telephone companies, the repeater spacing is typically 40 kilometers. The number of required repeaters and their spacing is a major factor in system cost. Some recently installed systems (1997) have spacing of up to 120 kilometers.

(7) Open Ended Capacity : The maximum theoretical capacity of installed fiber is very great (almost infinite). This means that additional capacity can be had on existing fibers as new technology becomes available. All that must be done is change the equipment at either end and change or upgrade the regenerators.

(8) Better Security : It is possible to tap fiber optical cable. But it is very difficult to do and the additional loss caused by the tap is relatively easy to detect. There is an interruption to service while the tap is inserted and this can alert operational staff to the situation.

Disadvantages

Despite the many advantages of fiber optic systems, there are some disadvantages. Because of the relative newness of the technology, fiber optic components are expensive. Fiber optic transmitters and receivers are still relatively expensive compared to electrical interfaces. Some limitations are :

(1) Joining Cables : The best way of joining cables is to use **fusion splicing**. This is where fiber ends are fused to one another by melting the glass. Making such splices in a way that will ensure minimal loss of signal is a skilled task that requires precision equipment. In the early days of fiber optical systems (the early 1980s) connectors which allowed cables to be plugged and unplugged were unreliable and caused a large amount of signal loss (as much as 3 dB per connector). Of course, the larger the core diameter of the fiber, the easier it is to make a low-loss connector. In the last few years connector systems for fibers with thicker cores (multimode fibers) have improved to the point where the use of a large number of connectors in a LAN system can be very reliable.

(2) Bending Cables : As light travels along the fiber, it is reflected from the interface between the core and cladding whenever it strays from the path straight down the center.

When the fiber is bent, the light only stays in the fiber because of this reflection. But the reflection only works if the angle of incidence is relatively low. If we bend the fiber too much the light escapes.

The amount of allowable bending is specific to particular cables because it depends on the difference in refractive index, between core and cladding. The bigger the difference in refractive index the tighter the allowable bend radius. There is a tradeoff here because there are many other reasons that we would like to keep this difference as small as possible.

(3) Slow Standards Development : This is nobody's fault. Development is happening so quickly, and getting worldwide agreement to a standard is necessarily so slow that standards setting just can't keep up. Things are improving considerably and very quickly. Now there are firm standards for optical link connections in LAN protocols (token-ring, Ethernet and FDDI), SDH (for the wide area) and in ATM for both the LAN and the wide area.

(4) Optics for Transmission Only : Until very recently there was no available optical amplifier. The signal had to be converted to electrical form and put through a complex repeater in order to boost its strength. Recently, optical amplifiers have emerged and look set to solve this problem. However, optical logic processing and/or switching systems seem to be a few years off yet.

(5) Gamma Radiation : Gamma radiation comes from space and is always present. It can be thought of as a high-energy X-ray. Gamma radiation can cause some types of glass to emit light (causing interference) and also gamma radiation can cause glass to discolor and hence attenuate the signal.

(6) Electrical Fields : Very high-voltage electrical fields also affect some glasses in the same way as gamma rays. One proposed route for fiber communication cables is wrapped around high-voltage electrical cables on transmission towers. This actually works quite well where the electrical cables are only of 30,000 volts or below.

1.12. Applications of Optical Fiber Communication

- The U.S. military moved quickly to use fiber optics for improved communications and tactical systems. In the early 1970's, the U.S. Navy installed a fiber optic telephone link aboard the U.S.S. Little Rock. The Air Force followed suit by developing its Airborne Light Optical Fiber Technology (ALOFT) program in 1976. Encouraged by the success of these applications, military R&D programs were funded to develop stronger fibers, tactical cables, ruggedized, high-performance components, and numerous demonstration systems ranging from aircraft to undersea applications.
- Commercial applications followed soon after. In 1977, both AT&T and GTE installed fiber optic telephone systems in Chicago and Boston respectively. These successful applications led to the increase of fiber optic telephone networks.

  By the early 1980's, single-mode fiber operating in the 1310 nm and later the 1550 nm wavelength windows became the standard fiber installed for these networks.
- The broadcast industry also embraced fiber optic transmission. In 1980, broadcasters of the Winter Olympics, in Lake Placid, New York, requested a fiber optic video transmission system for backup video feeds. The fiber optic feed, because of its quality and reliability, soon became the primary video feed. Later, fiber optics transmitted the first ever digital video signal, an application that continues to evolve today.

- In the mid-1980's the United States government deregulated telephone service, allowing small telephone companies to compete with the giant, AT&T. Companies like MCI and Sprint quickly went to work installing regional fiber optic telecommunications networks throughout the world.

  Taking advantage of railroad lines, gas pipes, and other natural rights of way, these companies laid miles fiber optic cable, allowing the deployment of these networks to continue throughout the 1980's. However, this created the need to expand fiber's transmission capabilities.
- In 1990, Bell Labs transmitted a 2.5 Gb/s signal over 7,500 km without regeneration. The system used a soliton laser and an erbium-doped fiber amplifier (EDFA) that allowed the light wave to maintain its shape and density.
- In 1998, they went one better as researchers transmitted 100 simultaneous optical signals, each at a data rate of 10 gigabits (giga means billion) per second for a distance of nearly 250 miles (400 km).

SUMMARY

1. **Fiber Optics** is the branch of optical technology concerned with the transmission of radiant power (light energy) through fibers.
2. A **Fiber Optic Data Link** has three basic functions :
   - to convert an electrical input signal to an optical signal.
   - to send the optical signal over an optical fiber.
   - to convert the optical signal back to an electrical signal.
3. Fiber optic data link consists of three parts :
   - transmitter
   - optical fiber
   - receiver.
4. The TRANSMITTER consists of two parts :
   - an interface circuit
   - a source drive circuit.
5. The transmitter converts the electrical input signal to an optical signal by varying the current flow through the light source.
6. The RECEIVER consists of two parts:
   - the optical detector
   - signal conditioning circuits.
7. The receiver converts the optical signal exiting the fiber back into the original form of the electrical input signal.
8. Fiber optics uses light to send information (data). More formally, fiber optics is the branch of optical technology concerned with the transmission of radiant power (light energy) through fibers.
9. There are two basic types of optical fibers :
   - multimode fibers
   - single mode fibers.
10. Fiber performance depends on the amount of loss and signal distortion introduced by the fiber when it is operating at a specific wavelength.

11. Long-haul systems require single mode optical fibers with very high performance. Single mode fibers tend to have lower loss and produce less signal distortion.
12. Short-distance and military systems tend to use only multimode technology. Examples of short-distance systems include process control and local area networks (LANs).
13. Fiber optic systems have many attractive features that are superior to electrical systems. These include improved system performance, immunity to electrical noise, signal security, and improved safety and electrical isolation.
14. **Scattering, Absorption** and **Dispersion Mechanisms** in the fiber waveguides cause the optical signal launched into the fiber to become weakened and distorted.
15. **Noise** is any disturbance that obscures or reduces the quality of the signal.
16. **Signal Loss** is the decrease in the amount of light reaching the end of the fiber. Impurities in the fiber material cause the signal toss in optical fibers. By removing these impurities, construction of low toss optical fibers was possible.
17. A **Low-loss Multimode Optical Fiber** was developed in 1970.
18. A **Small Area Light-emitting Diode** (LED) was developed in 1971. This light source was suitable for low-loss coupling to optical fibers.
19. **Fiber Optic System Design** has centered on long-haul communications and the subscriber loop plant. Limited work has also been done on short-distance applications and some military systems.
20. **Fiber Performance** depends on the amount of loss and signal distortion introduced by the fiber when it is operating at a specific wavelength. Single mode fibers tend to have lower loss and produce less distortion than multimode fibers.
21. The **Larger Fiber Core** and the **Higher Numerical Aperture** (NA) of multimode fibers reduce the amount of loss at fiber connections.
22. In **Military** and **Subscriber-loop Applications**, system designers consider tradeoffs in the following areas: fiber properties, types of connections, optical sources, and detector types.
23. The **advantages** of fiber optic systems include improved system performance, immunity to electrical noise, signal security, and electrical isolation. Advantages also include reduced size and weight, environmental protection, and overall system economy.
24. The **disadvantages** of fiber optic systems include problems with the relative newness of the technology, the relatively expensive cost, and the lack of component and system standardization. However, these disadvantages are already being eliminated because of increased use and acceptance of fiber optic technology.
25. In electronic communication, optical fibers are preferred over copper wires because they are extremely light, small and can be accommodated in a small space.
26. In any communication system, information to be transmitted is generally available as an electrical signal that may take *analog* or *digital form.*
27. The ratio A_{max}/A_N is called the dynamic range and is related to the *signal-to-noise ratio* (SNR) by the relation SNR = 20 $\log_{10} (A_{max}/A_N)$
28. The number of bits m needed to code each sample is related to the number of quantized signal levels M by the relation $M = 2m$ or $m = \log_2 M$.
29. The concept of TDM has been used to form *digital hierarchies.* In North America and Japan, the first level corresponds to multiplexing of 24 voice channels with a composite bit rate of 1.544 Mb/s.

30. There are two choices for the modulation format of the resulting optical bit stream. These are :

- *return-to-zero* (RZ)
- *non-return-to-zero* (NRZ) formats

31. Optical communication systems differ in principle from microwave systems only in the frequency range of the carrier wave used to carry the information. The optical carrier frequencies are typically 200 THz, in contrast with the microwave carrier frequencies (1 GHz).

32. Optical communication systems can be classified into two broad categories :

- *guided*
- *unguided*

33. *Receiver sensitivity* is usually defined as the minimum average optical power required to realize a BER of 10^{-9}. Receiver sensitivity depends on the SNR, which in turn depends on various noise sources that corrupt the signal received.

34. Networks may be further characterized by their geographic extent such as :

- Local Area Network (LAN)
- Metropolitan Area Network (MAN)
- Wide Area Network (WAN)

35. The term topology, as used here, refers to the configuration of various equipments and the fiber optic components interconnecting them. This equipment may be computers, workstations, consoles, or other equipments.

36. Point-to-point links are connected to produce systems with linear bus, ring, star, or tree topologies. Point-to-point fiber optic links are the basic building block of all fiber optic systems.

REVIEW QUESTIONS

1. Discuss the major trends in the development of optical fiber communication networks.
2. Briefly explain the evolution and applications of optical communication.
3. List the various advantages and disadvantages of fiber optic communications over other communication technologies.
4. What do you understand by fiber optic system topology ? Explain its various types also.
5. Why are optical fibers so good for communications ?
6. Discuss the various generation developments of optical fiber communication system.
7. Discuss the basic components of an optical communication system.
8. What are the various modulation formats used in fiber optic communication system ?
9. Draw and discuss the basic block diagram of an optical transmitter and optical receiver.
10. What do you meant by optical networking ? How many types of networks are there ?
11. What is the need for fiber optic communications ?
12. Draw the block diagram of a fiber optic communication system and explain the function of each component separately.
13. How the power levels are measured in communication system ?
14. Why optical fiber is used in place of copper wire or coaxial cable in communication systems ?

❑❑❑

Optical Fibers

INSIDE THIS CHAPTER

2.1. INTRODUCTION

An optical fiber is a dielectric waveguide and ideally has a cylindrical shape as shown in Fig. 2.1.

It consists of a core made up of a dielectric material which is surrounded by a **cladding** made up of a dielectric material of lower refractive index. Even though light will propagate along the fiber core without the layer of cladding material, the cladding does perform some necessary functions. The index of refraction of the cladding material is less than that of the core material. The cladding is generally made of glass or plastic.

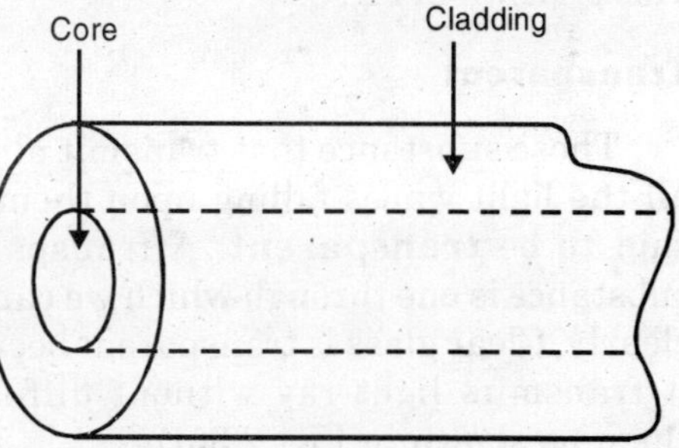

Fig. 2.1. Basic Structure of an Optical Fiber.

The cladding performs the following functions

- Reduce loss of light from the core into the surrounding air.
- Reduce scattering loss at the surface of the core.
- Protects the fiber from absorbing surface contaminants.
- Adds mechanical strength.

For extra protection, the cladding is enclosed in an additional layer called the **coating** or **buffer**. The coating or buffer is a layer of material used to protect an optical fiber from physical damage. The material used for a buffer is a type of plastic. The buffer is elastic is nature and prevents abrasion. The buffer also prevents the optical fiber from scattering losses caused by microbends. **Microbends** occur when an optical fiber is placed on a rough and distorted surface.

The invention of the clad waveguide structure led to the first serious proposal by **Kao and Hockman,** and **Werts** in 1996 to utilize optical fibers as a communication medium.

Traditionally three waveguides of operation centered around 850 nm, 1300 nm and 1550 nm have been used for communications through optical fibers. Optical fibers could broadly be classified into :

- Step index fibers.
- Graded index fibers.

2.2. Principles of Light Transmission in a Fiber

Fiber optics deals with the transmission of light energy through transparent fibers. How an optical fiber guides light depends on the nature of light and structure of the optical fiber.

A light wave is a form of energy that is moved by wave motion. Wave motion can be defined as the recuring disturbance advancing through space with or without the use of a physical medium. In fiber optics, wave motion is the movement of light energy through an optical fiber.

2.2.1. Properties of Light

When light waves, which travel in straight lines, encounter any substance, they are either reflected, absorbed, transmitted or refracted. This is shown in Fig. 2.2.

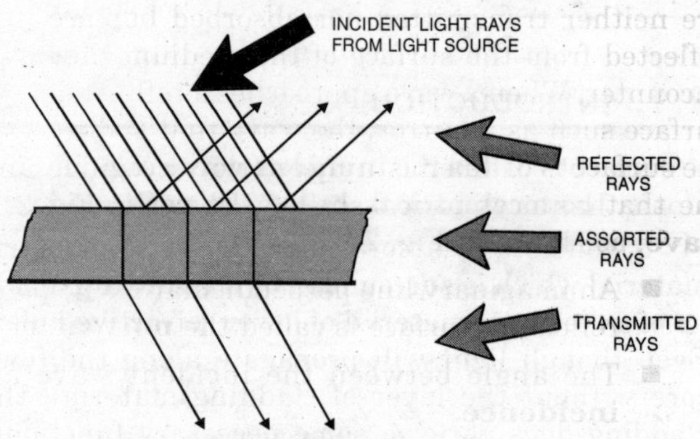

Fig. 2.2. Light Waves Reflected, Absorbed and Transmitted.

Transparent

Those substance that transmit almost all the light waves falling upon them are said to be **transparent**. A transparent substance is one through which we can see clearly. Clear glass is transparent because it transmits light ray without diffusing them us shown in Fig. 2.3(*a*).

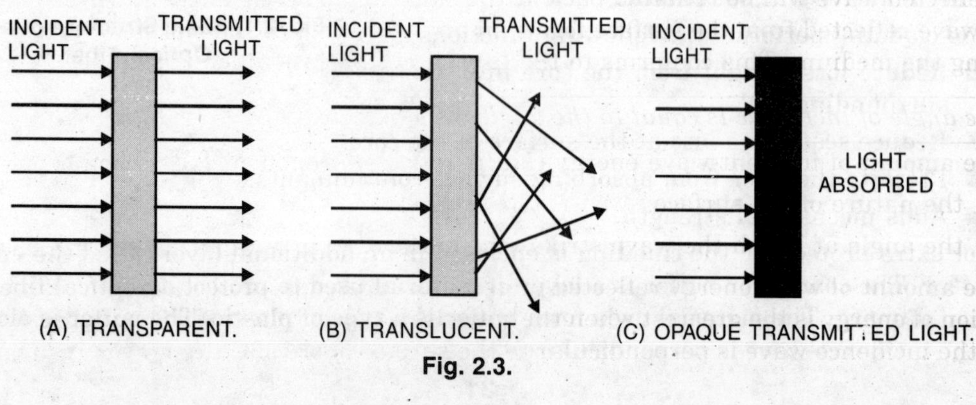

Fig. 2.3.

Translucent

There is no substance known that is perfectly transparent, but many substances are nearly, so substance through which some light ray can pass but through which objects cannot be seen clearly because the rays are diffused, are known as **translucent**, as shown in Fig. 2.3(*b*).

The frosted glass of a light bulb and a piece of oiled paper are examples of translucent materials.

Opaque

Those substances that are unable to transmit ray are called **opaque**, as shown in Fig. 2.3(*c*). Opaque substances either reflect or absorb all the light rays that full upon than.

All the substances that are not light source are visible only because they reflect all or some part of the light reaching than from some luminous source. Examples of luminous sources include the sun, a gas flame, and an electrical light filament because they are source of light energy. If light is neither transmitted or reflected, it is absorbed or taken up by the medium. When light strikes a substance, some absorption and some reflections always take place. No substance completely transmits, reflects or absorbs all the light rays that reach its surface.

2.2.2. Reflection

Reflected waves are simply those waves that are neither transmitted nor absorbed but are reflected from the surface of the medium they encounter. When a wave approaches a reflecting surface such as a mirror, the wave that strikes the surface is called the **incident wave** and the one that bounces back is called the **reflected wave**, as shown in Fig. 2.4.

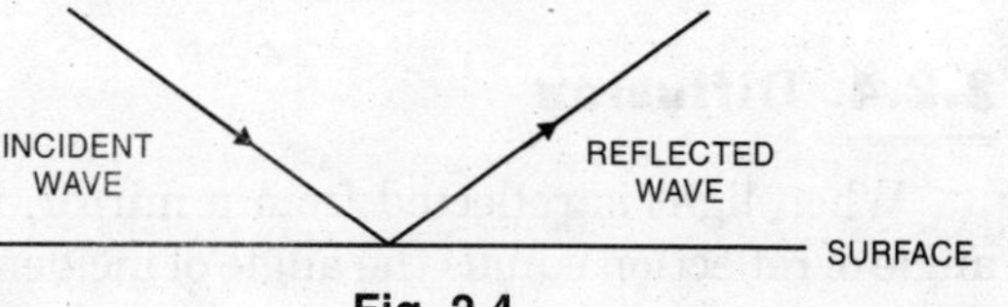

Fig. 2.4.

- An imaginary line perpendicular to the point at which the incident wave strikes the reflecting surface is called the **normal**, or the **perpendicular.**
- The angle between the incident wave and the normal is called the **angle of incidence**.
- The angle between the reflected wave and normal is called **angle of reflection.**

If the surface of the medium contacted by the incident wave is smooth and polished, each reflected wave will be reflected back at the same angle as the incident wave. The path of the wave reflected from the surface forms an angle equal to the one formed by its path in reaching the medium. This conforms to the **law of reflection** which states :

The angle of incidence is equal to the angle of reflection.

The amount of incident-wave energy that is reflected from a surface depends on :

- the nature of the surface
- the angle at which the wave strikes the surface.

The amount of wave energy reflected increases as the angle of incidence increases. The reflection of energy is the greatest when the wave is nearly parallel to the reflecting surface. When the incidence wave is perpendicular to the surface, more of the energy is transmitted into the substance and reflection of energy is at its least. At any incident angle, a mirror reflects almost all of the wave energy, while a dull, black surface reflects very little.

Light waves obey the law of reflection.

2.2.3. Refraction

When a light wave passes from one medium into a medium having a different velocity of propagation, a change in the direction of the wave will occur. This change of direction as the wave enters the second medium is called **refraction.** As the wave passes through the boundary, it is bent either toward or away from the normal. The angle between the normal and the path of the wave through the second medium is the **angle of refraction.** Refraction of wave is shown in Fig. 2.5.

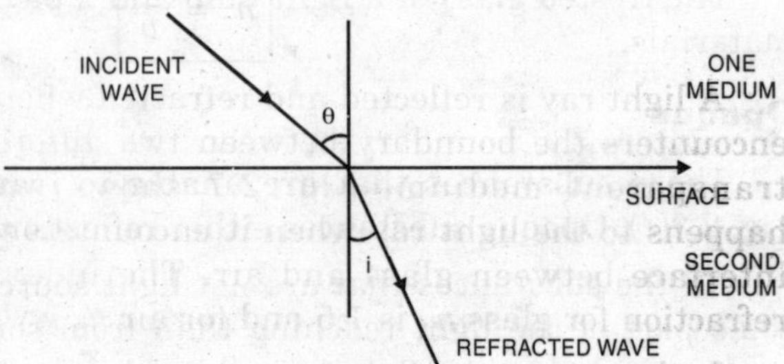

Fig. 2.5.

As shown in Fig. 2.5, if the wave passes from a less dense to a more dense medium, it is bent toward the normal and the angle of refraction (*r*) is less than the angle of incidence (*i*). Likewise, if the wave passes from a more dense to a less dense medium, it is bent away from the normal and the angle of refraction (*r*) is greater than the angle of incidence (*i*).

2.2.4. Diffusion

When light is reflected from a mirror, the angle of reflection equals the angle of incidence. When light is reflected from a piece of plain white paper, however the reflected beam is scattered or diffused as shown in Fig. 2.6.

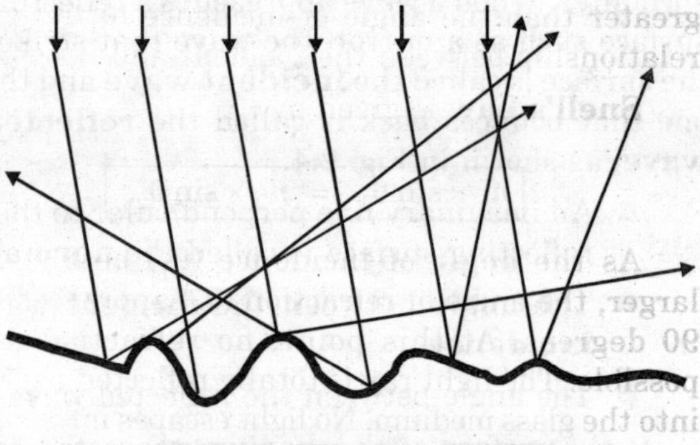

Fig. 2.6. Diffusion of Light.

Since the surface of paper is not smooth, the reflected light is broken up into many, light beams that are reflected in all directions.

Absorption : A light beam is reflected and diffused when it falls onto a piece of white paper. If the light beam falls upon a piece of black paper, the black paper absorbs most of the light rays and very little light is reflected from the paper. If the surface upon which the light beam falls is perfectly black, there is no reflection *i.e.* light is totally absorbed. It does not depend upon the type of surface light falls upon, some of light is absorbed.

2.3. Ray Theory Transmission

2.3.1. Total Internal Reflection

The basic optical property of a optical fibers material is the index of refraction. The **index of refraction (*n*)** measures the speed of light in an optical medium. The index of refraction of a material is the ratio of the speed of light in a vacuum to the speed of light in the material itself. The speed of light (c) in free space is 3×10^8 m/sec. The speed of light is the frequency (f) of light multiplied by the wavelength of light (λ). When light enters the fiber material, the light travels slower at a speed (v). Light will always travel slower in the fiber material than in air. The index of refraction is given by

$$n = \frac{c}{v} \qquad ...(2.1)$$

A light ray is reflected and refracted when it encounters the boundary between two different transparent medium. Fig. 2.7 shows what happens to the light ray when it encounters the interface between glass and air. The index of refraction for glass n_1 is 1.5 and for air n_2 is 1.0.

Let's assume the light ray or incident ray is travelling through the glass. When the light ray encounters the glass-air boundary, there are two results :

- The first result is that, part of the ray is reflected back into the glass.
- The second result is that, part of the ray is refracted (bent) as it enters the air.

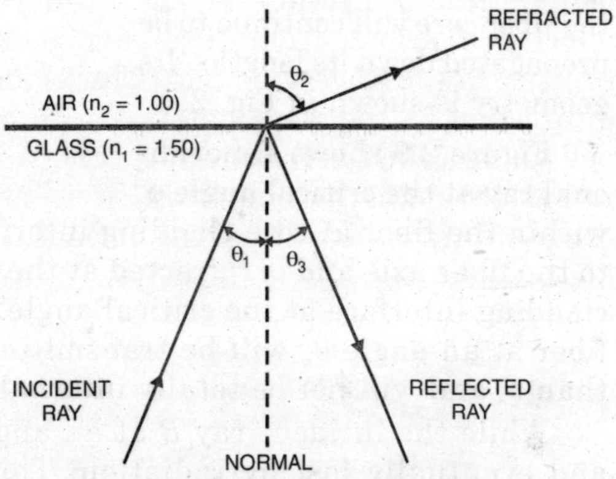

Fig. 2.7. Light Reflection and Refraction at a Glass-Air Boundary.

The bending of the light at the glass-air interface is the result of the difference between the index of refractions. Since n_1 is greater than n_2, the angle of refraction (θ_2) will be greater than the angle of incidence (θ_1). Snell's law of refraction is used to describe the relationship between the incident and the refracted rays at the boundary.

Snell's Law is given by :

$$n_1 \times \sin \theta_1 = n_2 \times \sin \theta_2$$

As the angle of incidence (θ_1) becomes larger, the angle of refraction (θ_2) approaches 90 degrees. At this point, no refraction is possible. The light ray is totally reflected back into the glass medium. No light escapes into the air. This condition is called **total internal reflection**. The angle at which total internal reflection occurs is called the critical angle of incidence. The critical angle of incidence (ϕ_c) is shown in Fig. 2.8. At any angle of incidence (θ_1) greater than the critical angle, light is totally reflected back into the glass medium. The critical angle of incidence is determined by using Snell's Law. The critical angle is given by

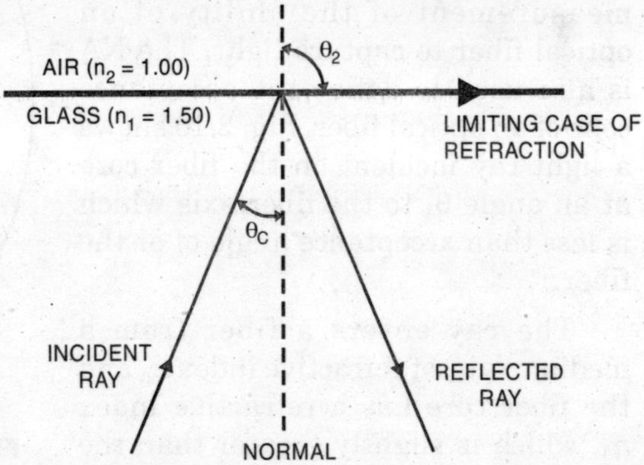

Fig. 2.8. Critical Angle of Incidence.

$$\sin \phi_c = \frac{n_2}{n_1}$$

The condition of total internal reflection is an ideal situation. However, in reality, there is always some light energy that penetrates the boundary. This situation is explained by the **mode theory**, or the **electromagnetic wave theory of light.**

2.3.2. Acceptance Angle

At the core cladding interface, rays with a sufficiently shallow grazing angle are transmitted by total internal reflection. Thus all the rays entering the fiber core will continue to be propagated down its lengths. Its geometry is shown in Fig. 2.7.

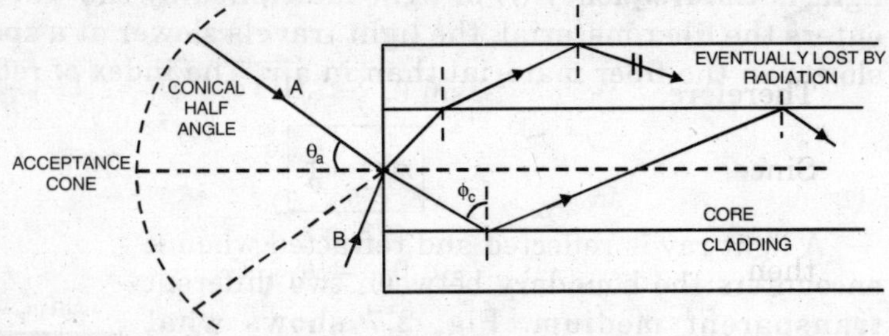

Fig. 2.9. Acceptance Angle, θ_a.

Figure 2.9 shows a meridional ray at the critical angle ϕ_c within the fiber at core-cladding interface. This ray enters the fiber core at an angle θ_a to the fiber axis and is refracted at the air-core interface before transmission to the core cladding interface at the critical angle. Therefore any rays which are incidence into the fiber at an angle θ_a will be transmitted to the core-cladding interface at an angle less than ϕ_c and will not be totally internally reflected.

While the incident ray *B* at an angle greater than θ_a is refracted into the cladding and eventually lost by radiation. Thus the rays to be transmitted by total internal reflection within the fiber core they must be incident an the fiber core with an acceptance cone defined by the **conical half angle θ_a.**

Here θ_a is the maximum angle to the axis at which light may enter the fiber in order to be propagated and is known as **acceptance angle** for the fiber.

2.3.3. Numerical Aperture

The numerical aperture (NA) is a measurement of the ability of an optical fiber to capture light. The NA is also used to define the acceptance cone of an optical fiber. Fig. 2.10 shows a light ray incident on the fiber core at an angle θ_1 to the fiber axis which is less than acceptance angle of or the fiber.

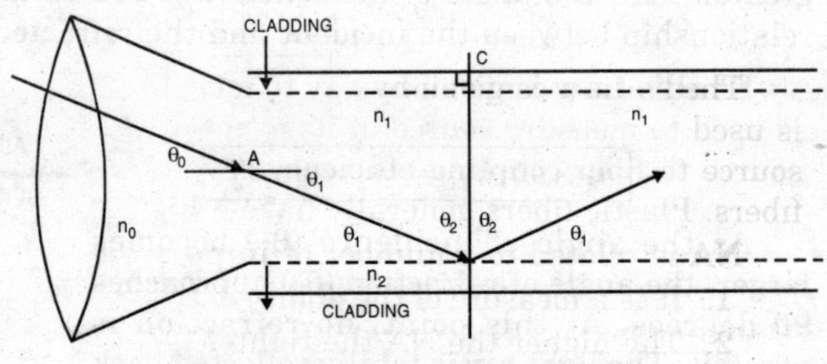

n_0 **= refractive index of air (n_0 = 1)**

n_1 **= refractive index of core**

n_2 **= refractive index of cladding**

Fig. 2.10. How a Light Ray Enters an Optical Fiber.

The ray enters a fiber from a medium (air) of refractive index n_0 and the fiber core has a refractive index n_1, which is slightly greater than the cladding refractive index n_2. It is clear from the Fig. 2.10 that there is a cone of acceptance. If a ray enters the fiber at an angle within the cone then it will be captured and propagate as a bound mode. We know that

$$\sin \theta_2 = \frac{n_2}{n_1}$$

where, θ_2 = Critical angle

From the Snell's law

$$n_0 \sin \theta_0 = n_1 \sin \theta_1 \qquad ...(2.2)$$

Now, $\cos\theta_1 = \sin\theta_2 = \frac{n_2}{n_1}$

Therefore, $\sin\theta_1 = \sqrt{1-\cos^2\theta_1} = \sqrt{1-\left(\frac{n_2}{n_1}\right)^2}$ ...(2.3)

Since, $n_0 = 1$

then, $\sin\theta_0 = n_1\sqrt{1-\left(\frac{n_2}{n_1}\right)^2}$

Therefore, $\boxed{NA = \sqrt{n_1^2 - n_2^2}}$ ...(2.4)

where, n_1 = Refractive index of core

n_2 = Refractive index of cladding.

The **numerical aperture** may also be given in terms of **relative refractive index difference, Δ** between the core and cladding, which is defined as

$$\Delta = \frac{n_1^2 - n_2^2}{2n_1^2} \quad ...(2.5)$$

$$\boxed{\Delta = \frac{n_1 - n_2}{n_1} \quad \text{for } \Delta \leq 1}$$

Hence combining Eqs. (2.4) and (2.5) we get

$$\boxed{NA = n_1 (2\Delta)^{1/2}} \quad ...(2.6)$$

The NA is a convenient way to measure the light gathering ability of an optical fiber. It is used to measure source to fiber power coupling efficiencies. A high NA indicates a high source to fiber coupling efficiency. Typical values of NA range from 0.20 to 0.29 for glass fibers. Plastic fibers generally have a higher NA.

NA is related to a number of **important fiber characteristics.**

1. It is a measure of the ability of the fiber to gather light at the input end.
2. The higher the NA the tighter (smaller radius) in the fiber.
3. The higher the NA the more modes we have.

2.3.4. Skew Rays

A possible path of propagation of skew rays is shown in Fig. 2.11. Skew rays propagate without passing through the center axis of the fiber. The acceptance angle for skew rays is larger than the acceptance angle of meridional rays. Skew rays are often used in the calculation of light acceptance in an optical fiber. The addition of skew rays increases the amount of light capacity of a fiber. In large *NA* fibers, the increase may be significant.

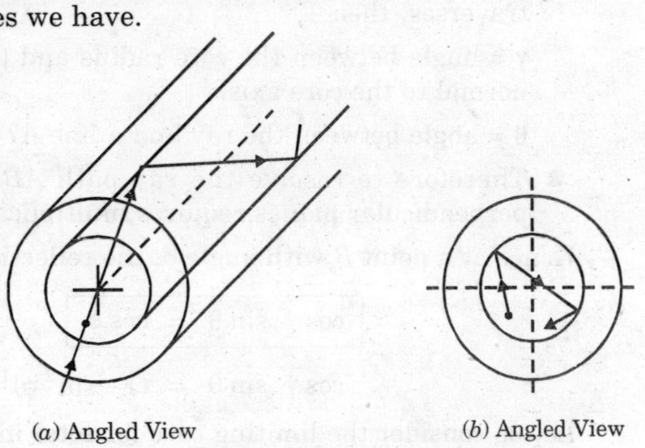

(*a*) Angled View (*b*) Angled View

Fig. 2.11. Skew Ray Propagation : (a) Angled View, (b) Front View.

The addition of skew rays also increases the amount of loss in a fiber. Skew rays tend to propagate near the edge of the fiber core. A large portion of the number of Skew rays that are trapped in the fiber core are considered to be leaky rays. Leaky rays are predicted to be totally reflected at the core-cladding boundary. However, these rays are partially refracted because of the curved nature of the fiber boundary.

Calculation of Acceptance Angle

To calculate the acceptance angle for a skew ray it is necessary to define the direction of the ray in two perpendicular planes. The geometry of the situation is shown in Fig. 2.12.

- The skew ray shown in Fig. 2.12 is incident on the fiber core at point *A*, at an angle θ_s to the normal at the fiber end face.
- The ray is refracted at the air-core interface before travelling to the point *B* in the same plane.
- The angle of incidence and reflection at the point *B* are ϕ, which is greater than the critical angle for the core cladding interface.

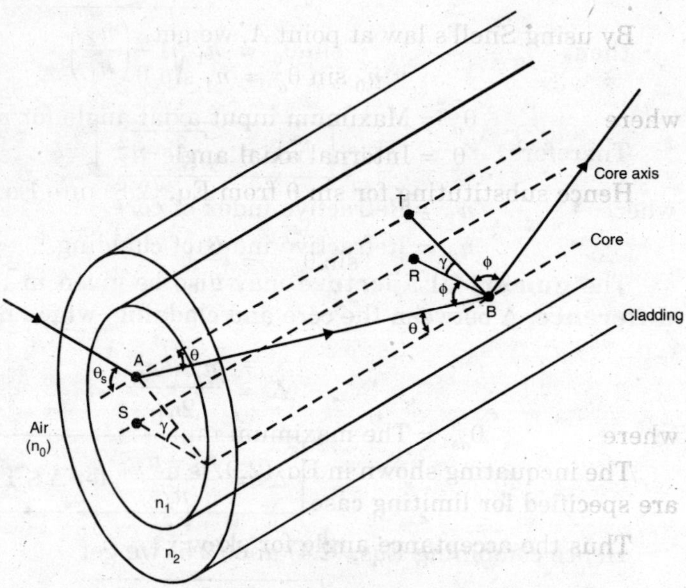

Fig. 2.12. The Ray Path Within the Fiber Core for a Skew Ray Incident at an Angle θ_s to the Normal at the Air Core Interface.

Let us consider the ray between *A* and *B* : It is necessary to resolve the direction of the ray path *AB* to the core radius at point *B*.

- The incident and reflected rays at the point *B* are in same phase, it is simply cos ϕ.
- But if we consider the two perpendicular planes through which the ray path *AB* traverses, then

  γ = angle between the core radius and the projection of the ray onto a plane BRS normal to the core axis.

  θ = angle between the ray and a line *AT* drawn parallel to the core axis.
- Therefore to resolve the ray path *AB* relative to the radius *BR* in these two perpendicular planes requires multiplication by cos γ and sin θ.

Hence at a point *B* with angle ϕ, the reflection may be given by

$$\boxed{\cos\gamma \,.\, \sin\theta \;=\; \cos\phi} \qquad \text{...(2.7)}$$

$$\cos\gamma \,.\, \sin\theta \;=\; (1 - \sin^2\phi)^{1/2} \qquad \text{...(2.8)}$$

Let us consider the limiting case for total internal reflection.

The θ becomes equal to the critical angle ϕ_c for core-cladding interface.

and $$\sin \phi_c = \frac{n_2}{n_1}$$

The above equation may be written as

$$\cos\gamma . \sin\theta \leq \cos\phi_c = \left(1 - \frac{n_2}{n_1}\right)^{1/2} \quad ...(2.9)$$

By using Snell's law at point A, we get,

$$n_0 \sin\theta_a = n_1 \sin\theta$$

where θ_a = Maximum input axial angle for meridional rays

θ = Internal axial angle

Hence substituting for $\sin\theta$ from Eq. (2.8) into Eq. (2.9) gives

$$\sin\theta_{as} = \frac{n_1}{n_0}\frac{\cos\phi_c}{\cos\gamma}$$

$$= \frac{n_1}{n_0 . \cos\gamma}\left(1 - \frac{n_2^2}{n_1^2}\right)^{1/2} \quad ...(2.10)$$

where θ_{as} = The maximum input angle or acceptance angle for skew rays.

The inequating shown in Eq. (2.9) is no longer necessary because the terms in Eq. (2.10) are specified for limiting case.

Thus the acceptance angle for skew rays are :

$$n_0 \sin\theta_{as} . \cos\gamma = (n_1^2 - n_2^2)^{1/2}$$

$$n_0 \sin\theta_{as} . \cos\gamma = \text{NA} \quad ...(2.11)$$

In the case of fiber in air, ($n_0 = 1$);

$$\sin\theta_{as} . \cos\gamma = \text{NA} \quad ...(2.12)$$

Thus, $$\theta_{as} = \sin^{-1}\left(\frac{NA}{\cos\gamma}\right) \quad ...(2.13)$$

2.4. Geometrical Optics Description

According to the number of modes that they can propagate :

- Single mode fibers can propagate only the fundamental mode.
- Multimode fibers can propagate hundreds of modes.

However, the classification of an optical fiber depends on more than the number of modes that a fiber can propagate.

An optical fiber's refractive index profile and core size further distinguish single mode and multimode fibers. The refractive index profile describes the value of refractive index as a function of radial distance at any fiber diameter. Fiber refractive index profiles classify single mode and multimode fibers as follows:

- Multimode step-index fibers
- Multimode graded-index fibers
- Single mode step-index fibers
- Single mode graded-index fibers

2.4.1. Step Index Fibers

In a step-index fiber, the refractive index of the core is uniform and undergoes an abrupt change at the core-cladding boundary. Step-index fibers obtain their name from this abrupt change called the step change in refractive index.

Single mode and multimode fibers can have a step-index or graded-index refractive index profile. **The performance of multimode graded-index fibers is usually superior to multimode step-index fibers.** However, each type of multimode fiber can improve system design and operation depending on the intended application. Performance advantages for single mode graded-index fibers compared to single mode step-index fibers are relatively small. Therefore, single mode fiber production is almost exclusively step-index.

Since light propagates differently in each fiber type. Figure 2.13 shows the propagation of light along each fiber.

- Standard core sizes for multimode step-index fibers are 50 μm and 100 μm.
- Standard core sizes for multimode graded-index fibers are 50 μm, 62.5 μm, 85 μm and 100 μm.
- Standard core sizes for single mode fibers are between 8 μm and 10 μm.

2.4.1.1. Multimode Step-index Fibers

Figure 2.13 shows the refractive index profile for a multimode step-index fiber.

A multimode step-index fiber has a core of radius (a) and a constant refractive index n_1. A cladding of slightly lower refractive index n_2 surrounds the core. The refractive index profile $n(r)$ for this type of fiber.

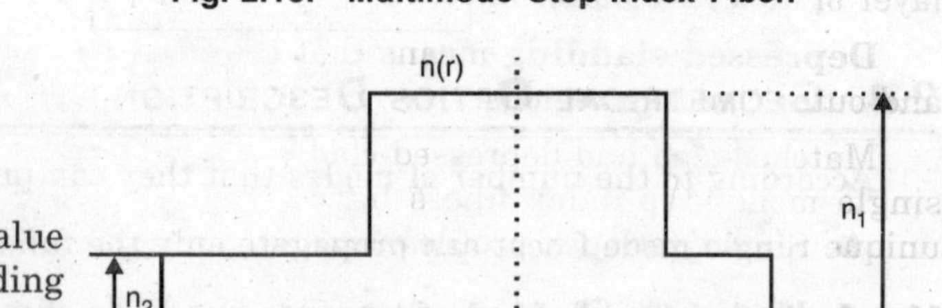

Fig. 2.13. Multimode Step-Index Fiber.

$$n(r) = \begin{cases} n_1 & r < a \text{ (core)} \\ n_2 & r \geq a \text{ (cladding)} \end{cases}$$

in both cases.

Notice the step decrease in the value of refractive index at the core-cladding interface. This step decrease occurs at a radius equal to distance (a). The difference in the core and cladding refractive index is the parameter Δ :

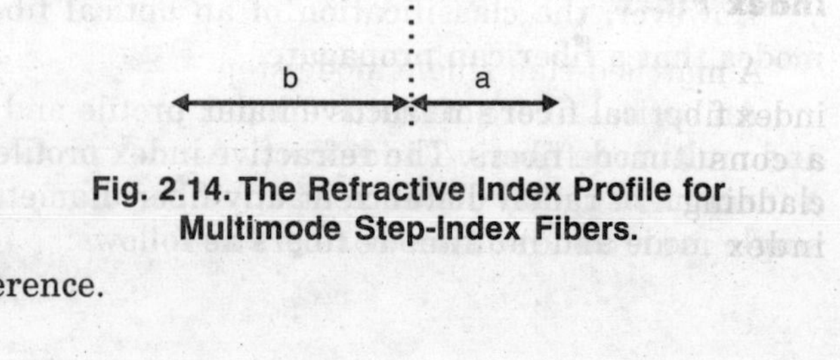

Fig. 2.14. The Refractive Index Profile for Multimode Step-Index Fibers.

$$\Delta = \frac{n_1^2 - n_2^2}{2n_1^2}$$

Δ is the relative refractive index difference.

The ability of the fiber to accept optical energy from a light source is related to Δ. Δ also relates to the numerical aperture by

$$NA \simeq n_1 \sqrt{2\Delta}. \quad ...(2.14)$$

- The number of modes that multimode step-index fibers propagate depends on Δ and core radius (a) of the fiber.
- The number of propagating modes also depends on the wavelength (λ) of the transmitted light.
- In a typical multimode step-index fiber, there are hundreds of propagating modes.

Multimode step-index fibers have relatively large core diameters and large numerical apertures A large core size and a large numerical aperture make it easier to couple light from a light-emitting diode (LED) into the fiber. Multimode step-index fiber core size is typically 50 µm or 100 µm. Unfortunately, multimode step-index fibers have limited bandwidth capabilities. Dispersion, mainly modal dispersion, limits the bandwidth or information-carrying capacity of the fiber. System designers consider each factor when selecting an appropriate fiber for each particular application.

Multimode step-index selection depends on system application and design. Short-haul, limited bandwidth, low-cost applications typically use multimode step-index fibers.

2.4.1.2. Single Mode Step-Index Fibers

Figure 2.15 also shows the refractive index profile for a single mode step-index fiber.

There are two **basic types of single mode step-index fibers :**

- Matched clad
- Depressed clad

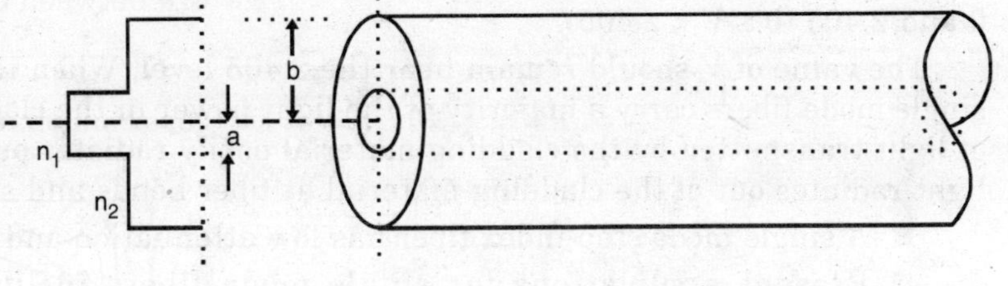

Fig. 2.15. Single Mode Step-Index Fiber.

Matched cladding means that the fiber cladding consists of a single homogeneous layer of dielectric material.

Depressed cladding means that the fiber cladding consists of two regions: the inner and outer cladding regions.

Matched-clad and depressed-clad single mode step-index fibers have unique refractive index profiles.

Matched Clad Single Mode Step Index Fiber

A matched-clad single mode step-index fiber has a core of radius (a) and a constant refractive index n_1. A cladding of slightly lower refractive index surrounds the core. The

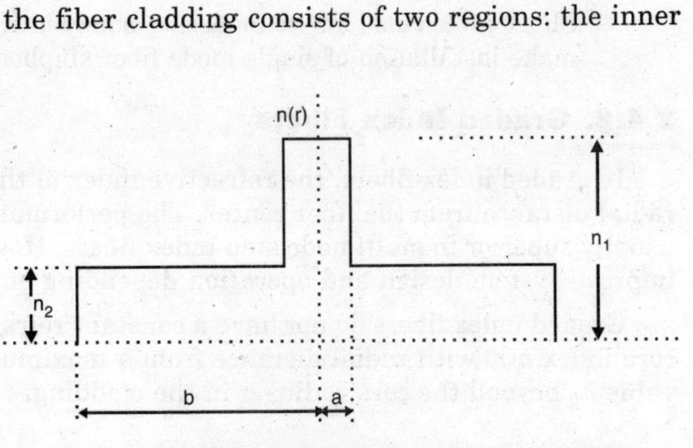

Fig. 2.16. Matched-Clad Refractive Index Profile.

cladding has a refractive index n_2. Figure 2.15 shows the refractive index profile $n(r)$ for the matched-clad single mode fiber.

Depressed Clad Single Mode Step Index Fiber

Figure 2.17 shows the refractive index profile $n(r)$ for the depressed-clad single mode fiber. A depressed-clad single mode step-index fiber has a core of radius (a) with a constant refractive index n_1. A cladding, made of two regions, surrounds the core. An inner cladding region surrounds the core of the fiber and has a refractive index of n_2. The inner cladding refractive index n_2 is lower than the core's refractive index n_1. An outer cladding region surrounds the inner cladding region and has a higher refractive index n_3 than the inner cladding region. However, the outer cladding refractive index n_3 is lower than the core's refractive index n_1.

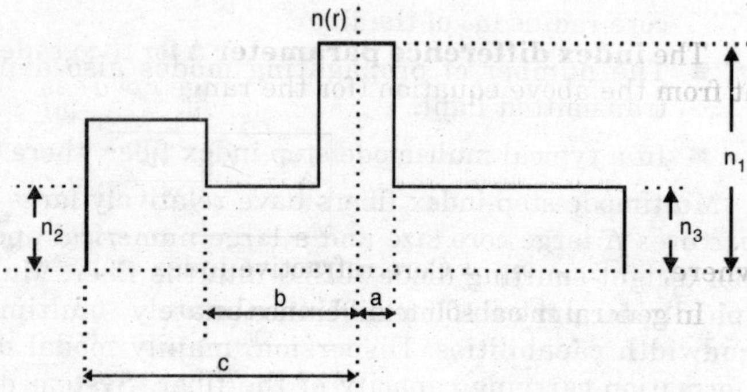

Fig. 2.17. Depressed-Clad Refractive Index Profile.

Single mode step-index fibers propagate only one mode, called the **fundamental mode.** Single mode operation occurs when the value of the fiber's normalized frequency is between 0 and 2.405 ($0 \le V < 2.405$).

The value of V should remain near the 2.405 level. When the value of V is less than 1, single mode fibers carry a majority of the light power in the cladding material. The portion of light transmitted by the cladding material easily radiates out of the fiber. For example, light radiates out of the cladding material at fiber bends and splices.

- A single mode step-index fiber has low attenuation and high bandwidth properties.
- Present applications for single mode fibers include long-haul, high-speed telecommunication systems.
- Future applications include single mode fibers for sensor systems. However, the current state of single mode technology makes installation of single mode systems expensive and difficult.
- Short cable runs, low to moderate bandwidth requirements, and high component cost make installation of single mode fiber shipboard systems impractical at this time.

2.4.2. Graded Index Fibers

In graded index fibers, the refractive index of the core varies gradually as a function of radial distance from the fiber center. The performance of multimode graded index fibers is usually superior to multimode step index fibers. However, each type of multimode fiber can improve system design and operation depending on the intended application.

Graded index fibers do not have a constant refractive index in the core but a decreasing core index $n(r)$ with radial distance from a maximum value of n_1 at the axis to a constant value n_2 beyond the core radius a in the cladding.

The **index variation** may be represented as

$$n(r) = \begin{cases} n_1\left[1-2\Delta\left(\dfrac{r}{a}\right)^{\alpha}\right]^{1/2} & 0 \le r \le a \\ n_2 = n_1\,(1-2\Delta)^{1/2} & r > a \end{cases} \quad \text{...(2.15)}$$

The **index difference parameter** Δ for the graded index fibers could be readily arrived at from the above equation (for the range $r > a$) as

$$\boxed{\Delta = \frac{n_1^2 - n_2^2}{n_1^2}} \quad \text{...(2.16)}$$

where, n_1 = Core refractive index at $r = 0$

In general the absolute difference between n_1 and n_2 is very small, Δ could be approximated

$$\boxed{\Delta \cong \frac{n_1 - n_2}{n_1}}$$

Since the refractive index of the core is a function of the radial distance r from its center, the NA as a function of r is given by

$$NA(r) = \sqrt{n^2(r) - n_2^2} \quad \text{...(2.17)}$$

where $n(r)$ is for r within the core.

Substituting the value of $n(r)$ and n_2 from Eq. (2.14) and simplifying we get

$$NA(r) = n_1\sqrt{2\Delta}\left[1-\left(\frac{r}{a}\right)^{\alpha}\right]^{1/2}$$

In graded index fiber, the number of bound modes is given by :

$$Mg = \frac{\alpha}{\alpha+2}(n_1 ka)^2\,\Delta$$

But the normalized frequency V for the fiber when $\Delta \le 1$ is approximately given by

$$V = n_1 ka(2\Delta)^{1/2}$$

Therefore, $$Mg \cong \frac{\alpha}{\alpha+2}\left(\frac{V^2}{2}\right)$$

Hence for a parabolic refractive index profile core fiber ($\alpha = 2$),

$$Mg = \frac{V^2}{4}$$

which is half the number supported by a step index fiber ($\alpha = \infty$) with the same V value.

Figure 2.18 shows the refractive index profile for a multimode graded index fiber.

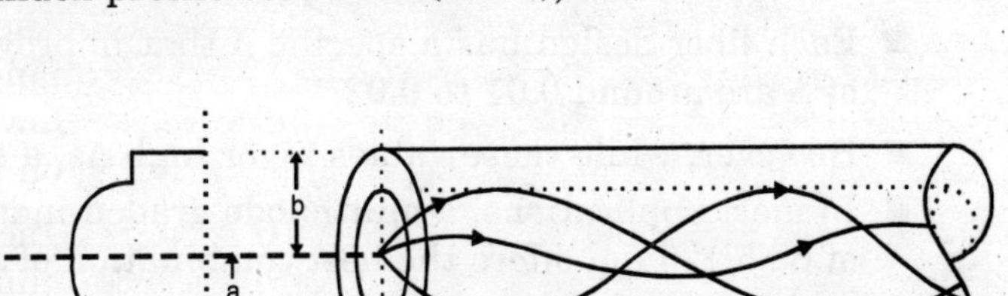

Fig. 2.18. Multimode Graded-Index Fiber.

2.4.2.1. Multimode Graded-Index Fibers

A multimode graded-index fiber has a core of radius (a). Unlike step-index fibers, the value of the refractive index of the core (n_1) varies according to the radial distance (r). The value of n_1 decreases as the distance (r) from the center of the fiber increases. The value of n_1 decreases until it approaches the value of the refractive index of the cladding (n_2). The value of n_1 must be higher than the value of n_2 to allow for proper mode propagation.

Figure 2.19 shows a possible refractive index profile $n(r)$ for a multimode graded-index fiber. Notice the parabolic refractive index profile of the core.

- The profile parameter (a) determines the shape of the core's profile. As the value of a increases, the shape of the core's profile changes from a triangular shape to step as shown in Fig. 2.20.
- Most multimode graded-index fibers have a parabolic refractive index profile. Multimode fibers with near parabolic graded-index profiles provide the best performance. Unless otherwise specified, when discussing multimode graded-index fibers, assume that the core's refractive index profile is parabolic ($\alpha = 2$).

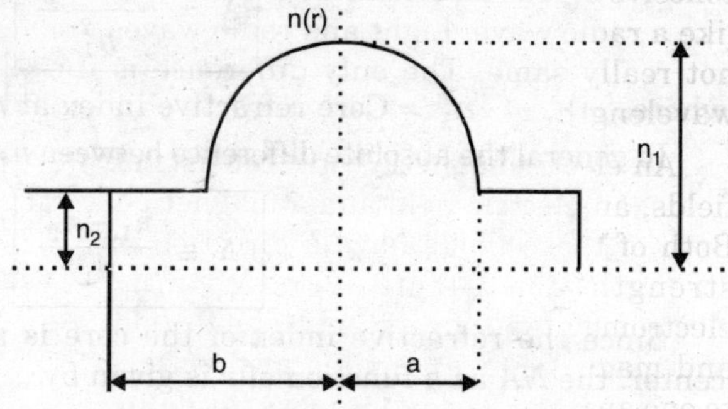

Fig. 2.19. The Refractive Index Profile for Multimode Graded-Index Fibers.

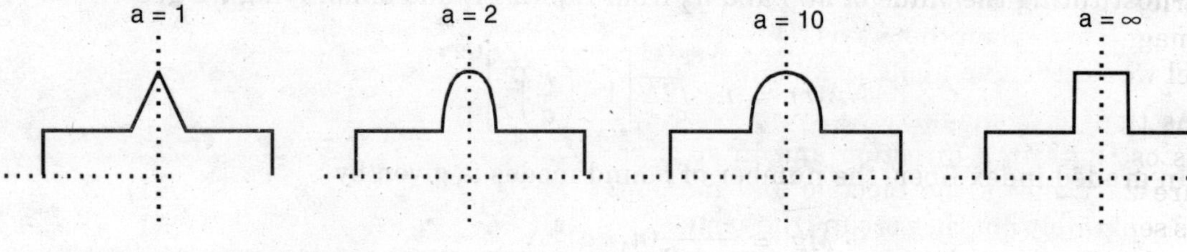

Fig. 2.20. The Refractive Index Profiles for Different Values of α.

- Light propagates in multimode graded-index fibers according to refraction and total internal reflection.
- Most present day applications that use multimode fiber use graded-index fibers.
- The basic design parameters are the fiber's core and cladding size and Δ. Standard multimode graded-index fiber core and cladding sizes are 50/125 μm, 62.5/125 μm, 85/125 μm, and 100/140 μm.
- Each fiber design has a specific Δ that improves fiber performance. Typical values of Δ are around 0.01 to 0.02.
- However, while these values favor high Δs, a smaller A improves fiber bandwidth.
- In most applications, a multimode graded-index fiber with a core and cladding size of 62.5/125 μm offers the best combination of the following properties :

1. Relatively high source-to-fiber coupling efficiency.
2. Low loss.

3. Low sensitivity to microbending and macrobending.
4. High bandwidth.
5. Expansion capability.

2.5. Wave Propagation

2.5.1. Electromagnetic Waves

One way to understand about light is to conceive of it as an electromagnetic wave just like a radio wave. Light and radio waves are not really same. The only difference is the wavelength.

An electromagnetic wave consists of two fields, an electric field and a magnetic field. Both of these fields have a direction and a strength (or amplitude). Within the electromagnetic wave the two fields (electric and magnetic) are oriented at precisely 90° to one another. The fields move (by definition at the speed of light) in a direction at 90° to both of them.

In three dimensions we can consider the electric field to be oriented on the y-axis, and the magnetic field on the x-axis. Direction of travel would then be along the z-direction.

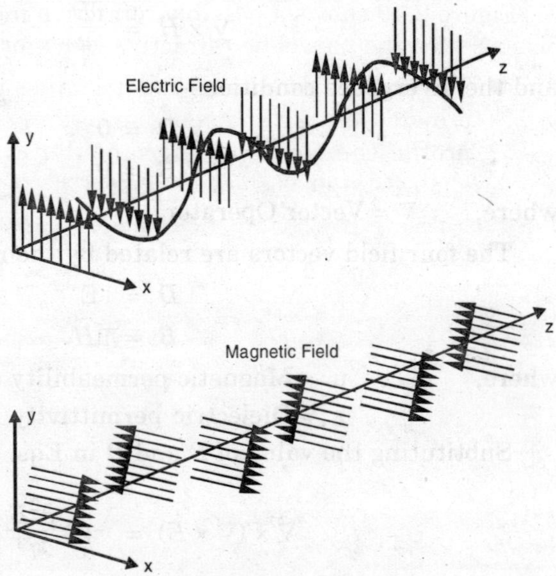

Fig. 2.21. The Structure of an Electromagnetic Wave.

As the electromagnetic wave moves the fields oscillate in z direction and in strength. Figure 2.21 shows the electric and magnetic fields separately but they occupy the same space. They should be overlayed on one another and are only drawn this way for clarity. We could consider the z-direction in the figure to represent passing time or it could represent a wave travelling in space at a single instant in time.

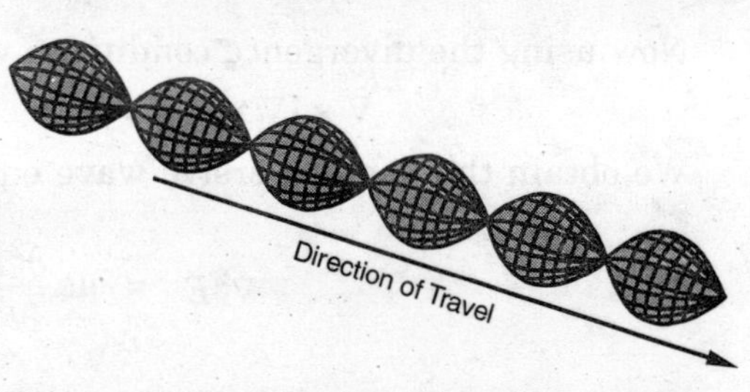

Fig. 2.22. Amplitude Fluctuation in an Electromagnetic Wave.

The curved line is intended to represent field strength. The field might start at a maximum in one direction, decay to a zero and then build up in the other direction until it reaches a maximum in the other direction. The field strength changes sinusoidally.

The important thing is that we have two fields and they oscillate in phase. That is the electric and magnetic fields reach their peaks and their nulls at exactly the same time (and place). The rate of oscillation is the frequency of the wave. The distance travelled during one period of oscillation is the wavelength.

So we can visualise a "particle" or ray of light travelling through space as two interlocking force fields (electric and magnetic). These fields centre on a point (over time this is of course a line in the direction of propagation) and decay exponentially as we move away from that point (or line).

We begin with the **Maxwell's equation** as applied to dielectric materials. For a medium with zero conductivity there vector relationships may be written in terms of the electric field *E*, magnetic field *H*, electric flux density *D* and magnetic flux density *B* as the core equations :

$$\nabla \times E = -\frac{\partial B}{\partial t} \quad ...(2.18)$$

$$\nabla \times H = \frac{\partial D}{\partial t} \quad ...(2.19)$$

and the divergence conditions :

$$\nabla . D = 0 \quad \text{(No free charges)}$$

$$\nabla . B = 0 \quad \text{(No free poles)}$$

where, ∇ = Vector Operator.

The four field vectors are related by the relations :

$$D = \varepsilon E \quad ...(2.20a)$$

$$B = \mu H \quad ...(2.20b)$$

where, μ = Magnetic permeability of medium

ε = Dielectric permittivity

Subtituting the value of *B* and *D* in Eqs. (2.18) and (2.19), we get

$$\nabla \times (\nabla \times E) = -\mu\varepsilon \frac{\partial^2 E}{\partial t^2} \quad ...(2.21a)$$

$$\nabla \times (\nabla \times H) = -\mu\varepsilon \frac{\partial^2 H}{\partial t^2} \quad ...(2.21b)$$

Now using the divergence conditions with vector identity

$$\nabla \times (\nabla \times Y) = \nabla (\nabla . Y) - \nabla^2(Y)$$

We obtain the non-dispersive wave equations

$$\nabla^2 E = \mu\varepsilon \frac{\partial^2 E}{\partial t^2} \quad ...(2.22a)$$

and

$$\nabla^2 H = \mu\varepsilon \frac{\partial^2 H}{\partial t^2} \quad ...(2.22b)$$

where ∇^2 = Laplacian operator.

For **rectangular cartesian and cylinderical polar coordinates** the above wave equations hold for each component of the field vector, every component satisfying the scalar wave equation.

$$\nabla^2 \psi = \frac{1}{v_p^2} \frac{\partial^2 \psi}{\partial t^2} \quad ...(2.23)$$

where, ψ = Component of the *E* or *H* field

v_p = Phase velocity in dielectric medium.

and
$$v_p = \frac{1}{(\mu\varepsilon)^{1/2}}$$

$$v_g = \frac{1}{(\mu_r \, \mu_0 \, \varepsilon_r \, \varepsilon_0)^{1/2}} \quad \text{...(2.24)}$$

where μ_r, μ_0 = Relative permeability of dielectric and free space

$\varepsilon_r, \varepsilon_0$ = Relative permittivity.

The velocity of light in free space, C is given by

$$C = \frac{1}{(\mu_0 \, \varepsilon_0)^{1/2}} \quad \text{...(2.25)}$$

If rectangular cartesian coordinates (x, y, z) or circular fibers described by or cylindrical polar coordinates (r, ϕ, z) are considered then the laplacian operator takes the form :

$$\nabla^2\psi = \frac{\partial^2\psi}{\partial x^2} + \frac{\partial^2\psi}{\partial y^2} + \frac{\partial^2\psi}{\partial z^2} \quad \text{...(2.26)}$$

or
$$\nabla^2\psi = \frac{\partial^2\psi}{\partial r^2} + \frac{1}{r} \cdot \frac{\partial\psi}{\partial r} + \frac{1}{r^2}\frac{\partial^2\psi}{\partial\phi^2} + \frac{\partial^2\psi}{\partial z^2} \quad \text{...(2.27)}$$

The basic solution of the wave equation is a sinusoidal wave, the most important form of which is a uniform plane wave given by

$$\psi = \psi_0 \exp j(wt - K \, . \, r) \quad \text{...(2.28)}$$

where, w = Angular frequency of field

t = Time

K = Propagation vector.

2.5.2. Modes in a Planar Guide

Mode Theory

The mode theory along with the ray theory is used to describe the propagation of light along an optical fiber. The mode theory is used to describe the properties of light that ray theory is unable to explain. The mode theory uses electromagnetic wave behavior to describe the propagation of light along a fiber. A set of guided electromagnetic waves is called the **modes of the fiber.**

Plane Waves

The mode theory suggests that a light wave can be represented as a plane wave. A plane wave is described by its direction, amplitude, and wavelength of propagation. A plane wave is a wave whose surfaces of constant phase are infinite parallel planes normal to the direction of propagation. The planes having the same phase are called the **wavefronts.** The wavelength (λ) of the plane wave is given by :

$$\text{Wavelength } (\lambda) = \frac{c}{fn} \quad \text{...(2.29)}$$

where, c = speed of light in a vacuum

f = frequency of the light

n = index of refraction of the plane-wave medium.

Figure 2.23 shows the direction and wavefronts of plane-wave propagation. Plane waves, or wavefronts, propagate along the fiber similar to light rays. While all the wavefronts are not incident on the fiber at an angle less than or equal to the critical angle of light acceptance propagate along the fiber. Wavefronts may undergo a change in phase that prevents the successful transfer of light along the fiber.

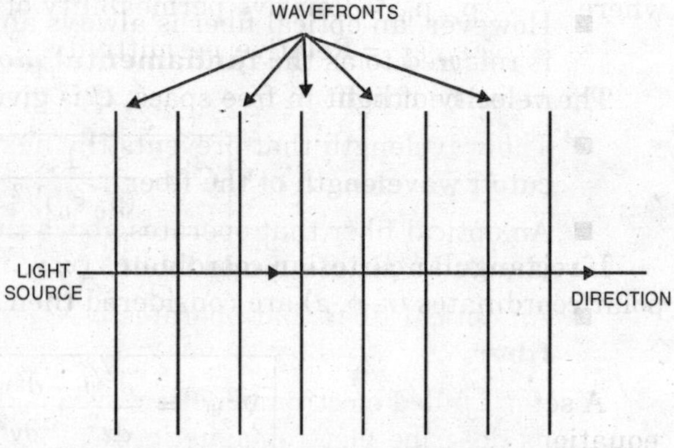

Fig. 2.23. Plane-Wave Propagation.

Wavefronts are required to remain in phase for light to be transmitted along the fiber. Consider the wavefront incident on the core of an optical fiber as shown in Fig. 2.24.

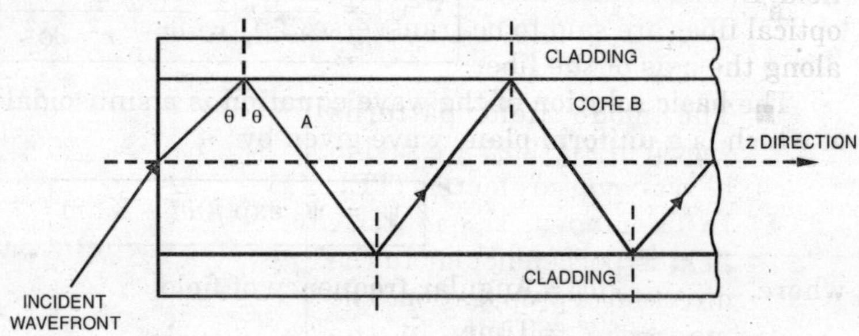

Fig. 2.24. Wavefront Propagation Along an Optical Fiber.

- Only those wavefronts incident on the fiber at angles less than or equal to the critical angle may propagate along the fiber.
- The wavefront undergoes a gradual phase change as it travels down the fiber. Phase changes also occur when the wavefront is reflected.
- The wavefront must remain in phase after the wavefront transverses the fiber twice and is reflected twice. The distance transversed is shown between point A and point B on Fig. 2.24.
- The reflected waves at point A and point B are in phase if the total amount of phase collected is an integer multiple of 2π radian.
- If propagating wavefronts are not in phase, they eventually disappear. Wavefronts disappear because of destructive interference.
- The wavefronts that are in phase interfere with the wavefronts that are out of phase. This interference is the reason **why only a finite number of modes can propagate along the fiber.**
- The plane waves repeat as they travel along the fiber axis. The direction the plane waves travel is assumed to be the z direction as shown in Fig. 2.24.
- The quantity β is defined as the **propagation constant** along the fiber axis. As the wavelength (λ) changes, the value of the propagation constant must also change.

For a given mode, a change in wavelength can prevent the mode from propagating along the fiber.

- The mode is no longer bound to the fiber. The mode is said to be **cutoff**. Modes that are bound at one wavelength may not exist at longer wavelengths.
- The wavelength at which a mode ceases to be bound is called the **cutoff wavelength** for that mode.
- However, an optical fiber is always able to propagate at least one mode. This mode is referred to as the **fundamental mode of the fiber.** The fundamental mode can never be cutoff.
- The wavelength that prevents the next higher mode from propagating is called the cutoff wavelength of the fiber.
- An optical fiber that operates above the cutoff wavelength (at a longer wavelength) is called a **single mode fiber.**
- An optical fiber that operates below the cutoff wavelength is called a **multimode fiber.**

A set of guided electromagnetic waves is called the modes of an optical fiber. Maxwell's equations describe electromagnetic waves or modes as having two components. The two components are the electric field, $E(x, y, z)$, and the magnetic field, $H(x, y, z)$. The electric field, E, and the magnetic field, H, are at right angles to each other. Modes traveling in an optical fiber are said to be transverse. The transverse modes, shown in Fig. 2.25, propagate along the axis of the fiber.

- The mode field patterns shown in Fig. 2.25 are said to be **transverse electric (TE)**. In TE modes, the electric field is perpendicular to the direction of propagation. The magnetic field is in the direction of propagation.
- Another type of transverse mode is the **transverse magnetic (TM) mode**. TM modes are opposite to TE modes. In TM modes, the magnetic field is perpendicular to the direction of propagation. The electric field is in the direction of propagation.

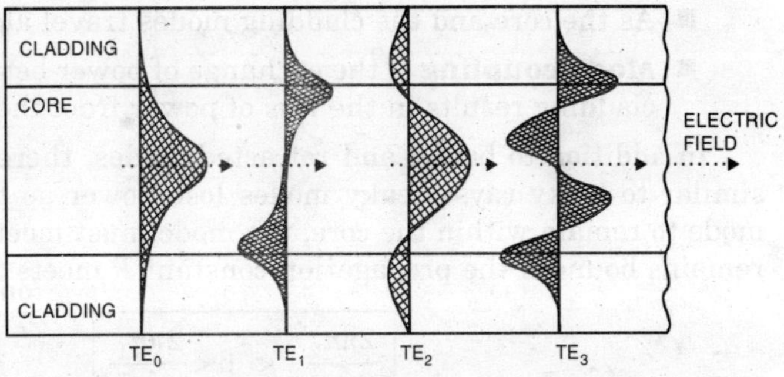

Fig. 2.25. Transverse Electric (TE) Mode Field Patterns.

The TE mode field patterns shown in Fig. 2.25 indicate the order of each mode. The order of each mode is indicated by the number of field maxima within the core of the fiber. For example, TE_0 has one field maxima. The electric field is a maximum at the center of the waveguide and decays toward the core-cladding boundary. TE_0 is considered the fundamental mode or the lowest order standing wave. As the number of field maxima increases, the order of the mode is higher. Generally, modes with more than a few (5-10) field maxima are referred to as **high-order modes.**

The order of the mode is also determined by the angle the wavefront makes with the axis of the fiber. Figure 2.26 illustrates light rays as they travel down the fiber. These light rays indicate the direction of the wavefronts. High-order modes cross the axis of the fiber at steeper angles. Low-order and high-order modes are shown in Fig. 2.26.

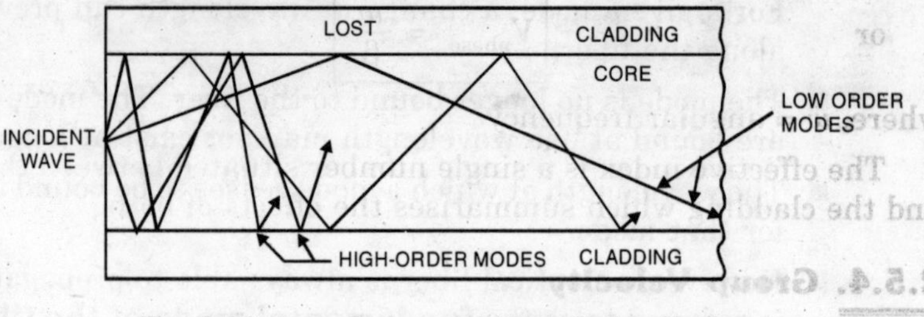

Fig. 2.26. Low-Order and High-Order Modes.

Before we progress, let us refer back to Fig. 2.24. Notice that the modes are not confined to the core of the fiber.

- The modes extend partially into the cladding material. Low-order modes penetrate the cladding only slightly. In low-order modes, the electric and magnetic fields are concentrated near the center of the fiber.
- However, high-order modes penetrate further into the cladding material. In high-order modes, the electrical and magnetic fields are distributed more toward the outer edges of the fiber.
- This penetration of low-order and high-order modes into the cladding region indicates that some portion is refracted out of the core.
- The refracted modes may become trapped in the cladding due to the dimension of the cladding region. The modes trapped in the cladding region are called **cladding modes.**
- As the core and the cladding modes travel along the fiber, mode coupling occurs.
- **Mode coupling** is the exchange of power between two modes. Mode coupling to the cladding results in the loss of power from the core modes.

In addition to bound and refracted modes, there are **leaky modes.** Leaky modes are similar to leaky rays. Leaky modes lose power as they propagate along the fiber. For a mode to remain within the core, the mode must meet certain boundary conditions. A mode remains bound if the propagation constant *P* meets the following boundary condition :

$$\frac{2\pi n_2}{\lambda} < \beta < \frac{2\pi n_1}{\lambda} \quad \text{...(2.30)}$$

where n_1 and n_2 are the index of refraction for the core and the cladding, respectively. When the propagation constant becomes smaller than $2\pi n_2 / \lambda$, power leaks out of the core and into the cladding. Generally, modes leaked into the cladding are lost in a few centimeters. However, leaky modes can carry a large amount of power in short fibers.

2.5.3. Phase Velocity

The phase velocity is the speed of propagation of an electromagnetic wave in some medium. In fiber optics it is the speed of the planar phase front of the mode as it propagates along the fiber. It is just the speed of light divided by the effective index of the fiber

$$V_{\text{phase}} = \frac{c}{n_{\text{effective}}} \quad \text{...(2.31)}$$

or $$V_{\text{phase}} = \frac{w}{\beta} \qquad ...(2.32)$$

where, w = angular frequency

The effective index is a single number situated between the refractive index of the core and the cladding which summarises the effects of both.

2.5.4. Group Velocity

Group velocity is the usual way of discussing the speed of propagation on a fiber. It is the speed of propagation of modulations along the fiber. It is generally a little less than the phase velocity.

The reason that group velocity is different from phase velocity is related to the amount of dispersion of the medium. If there is no dispersion in the medium then group velocity and phase velocity are the same.

When a group of waves with closely similar frequencies propagate then their resultant forms a packet of waves. This wave formation from a wave packet resulting from the combination of two waves of slightly different frequency propagate together. This wave packet does not travel at the phase velocity of the individual wave but move at a **group velocity, v_g.**

$$v_g = \frac{\delta w}{\delta \beta} \qquad ...(2.33)$$

In an infinite medium of refractive index n_1 **propagation constant, β** is defined as

$$\beta = n_1 \frac{2\pi}{\lambda} = \frac{n_1 w}{c} \qquad ...(2.34)$$

where, c= Velocity of light in free space

From the Eq. (2.33), we get

$$v_g = \frac{d\lambda}{d\beta} \cdot \frac{dw}{d\lambda}$$

$$= \frac{d}{d\lambda}\left(n_1 \cdot \frac{2\pi}{\lambda}\right)^{-1}\left(\frac{-w}{\lambda}\right)$$

$$= \frac{-w}{2\pi\lambda}\left[\frac{1}{\lambda}\cdot\frac{dn_1}{d\lambda} - \frac{n_1}{\lambda^2}\right]^{-1}$$

$$v_g = \frac{c}{\left(n_1 - \lambda \frac{dn_1}{d\lambda}\right)} = \frac{c}{N_g} \qquad ...(2.35)$$

The parameter N_g is known as the **group index** of the guide.

2.5.5. Maxwell's Equations

We begin with the Maxwell's equation as applied to dielectric materials. For a medium with zero conductivity there vector relationships may be written in terms of the electric field ***E***, magnetic field ***H***, electric flux density ***D*** and magnetic flux density ***B*** as the curl equations :

$$\nabla \times E = -\frac{\partial B}{\partial t} \quad \text{...(2.36)}$$

$$\nabla \times H = \frac{\partial D}{\partial t} \quad \text{...(2.37)}$$

and the divergence conditions :

$$\nabla . D = 0 \quad \text{(No free charges)}$$

$$\nabla . B = 0 \quad \text{(No free poles)}$$

where, ∇ = Vector Operator.

The four field vectors are related by the relations :

$$D = \varepsilon E \quad \text{...(2.38)}$$

$$B = \mu H \quad \text{...(2.39)}$$

where, μ = Magnetic permeability of medium

ε = Dielectric permittivity

Subtituting the value of B and D in Eqns. (2.36) and (2.37) we get

$$\nabla \times (\nabla \times E) = -\mu\varepsilon \frac{\partial^2 E}{\partial t^2} \quad \text{...(2.40)}$$

$$\nabla \times (\nabla \times H) = -\mu\varepsilon \frac{\partial^2 H}{\partial t^2} \quad \text{...(2.41)}$$

Now using the divergence conditions with vector identity

$$\nabla \times (\nabla \times Y) = \nabla (\nabla . Y) - \nabla^2(Y)$$

We obtain the non-dispersive wave equations

$$\nabla^2 E = \mu\varepsilon \frac{\partial^2 E}{\partial t^2} \quad \text{...(2.42)}$$

and

$$\nabla^2 H = \mu\varepsilon \frac{\partial^2 H}{\partial t^2} \quad \text{...(2.43)}$$

where ∇^2 = Laplacian operator.

For **rectangular cartesian and cylinderical polar coordinates** the above wave equations hold for each component of the field vector, every component satisfying the scalar wave equation.

$$\nabla^2 \psi = \frac{1}{v_p^2} \frac{\partial^2 \psi}{\partial t^2} \quad \text{...(2.44)}$$

where, ψ = Component of the E or H field

v_p = Phase velocity in dielectric medium.

and $$v_p = \frac{1}{(\mu\varepsilon)^{1/2}}$$

$$v_p = \frac{1}{(\mu_r\,\mu_0\,\varepsilon_r\,\varepsilon_0)^{1/2}} \quad ...(2.45)$$

where μ_r, μ_0 = Relative permeability of dielectric and free space

$\varepsilon_r, \varepsilon_0$ = Relative permittivity.

The velocity of light in free space, C is given by

$$C = \frac{1}{(\mu_0\,\varepsilon_0)^{1/2}} \quad ...(2.46)$$

If rectangular cartesian coordinates (*x, y, z*) or cylindrical coordinates (*x, y, z*) or cylindrical polar coordinates (r, ϕ, z) are considered then the laplacian operator takes the form :

$$\nabla^2\psi = \frac{\partial^2\psi}{\partial x^2} + \frac{\partial^2\psi}{\partial y^2} + \frac{\partial^2\psi}{\partial z^2} \quad ...(2.47)$$

or $$\nabla^2\psi = \frac{\partial^2\psi}{\partial r^2} + \frac{1}{r}\cdot\frac{\partial\psi}{\partial r} + \frac{1}{r^2}\frac{\partial^2\psi}{\partial y^2} + \frac{\partial^2\psi}{\partial z^2} \quad ...(2.48)$$

The basic solution of the wave equation is a sinusoidal wave, the most important form of which is a uniform plane wave given by

$$\psi = \psi_0 \exp j(wt - K \,.\, r) \quad ...(2.49)$$

where, w = Angular frequency of field

t = Time

K = Propagation vector = $\frac{2\pi}{\lambda}$

2.5.6. Goos-Haenchen Shift

The phase change because of the total internal reflection of a light beam on a planar dielectric interface may be understood from physical observation. The reflected beam is shifted laterally from the trajectory predicted by the simple ray theory, as shown in Fig. 2.27.

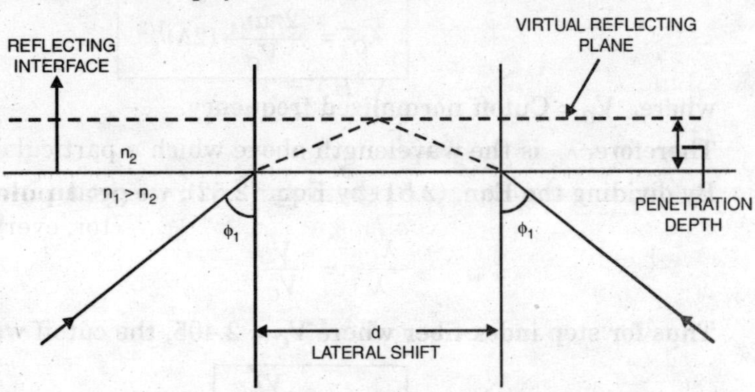

Fig. 2.27. Goos-Haenchen Shift (Lateral Displacement of Light Beam).

The geometric reflection occurs at a virtual reflecting plane which is parallel to the dielectric interface with a lower index medium. With the help of wave theory, lateral shift can be determined although it is very small and difficult to observe. This lateral shift is 0.06 to 0.10 μm for a silver glass interface at a wavelength of 0.55 μm.

2.5.7. Single Mode Fibers

Single mode fibers are designed by taking the dimension of the core radius to a few wavelength (usually 4-6 μm) and by having small index difference between core and cladding. The size of core in single mode fiber is very small, therefore to obtain a fiber of several kilometre is quite difficult.

In practical designing of single mode fiber, the core cladding index difference varies between 0.2% and 1.0%, and core diameter should be chosen to first below the cutoff of the first higher order mode. Thus *V* should be approximately 2.405. Example : A typical single mode fiber has a core radius of 3 μm and numerical aperature of 0.1 at a wavelength of 0.8 μm radius of 3 μm and numerical aperture of 0.1 at a wavelength of 0.8 μm.

Graded index fibers may also be designed for single mode operation. The **cutoff value** of **normalized frequency** V_c to support a single mode in a graded index fiber is given by

$$V_c = 2.405\left[1+\frac{2}{\alpha}\right]^{1/2} \qquad ...(2.50)$$

Therefore as in the case of the step index, it is possible to determine the fiber parameters which give single mode operation.

2.5.7.1. Cutoff Wavelength

The normalized frequency (V) in terms of NA and the relative refractive index difference Δ is given by

$$V = \frac{2\pi}{\lambda}\cdot a(NA) \qquad ...(2.51)$$

$$V = \frac{2\pi}{\lambda} an_1 (2\Delta)^{1/2} \qquad ...(2.52)$$

The single mode operation only occur above a theoretical cutoff wavelength λ_C and is given by

$$\lambda_C = \frac{2\pi a n_1}{V_C}(2\Delta)^{1/2} \qquad ...(2.53)$$

where, V_C = Cutoff normalized frequency.

Therefore, λ_C is the wavelength above which a particular fiber becomes single moded.

By dividing the Eqn. (2.51) by Eqn. (2.52), we get an inverse relationship

$$\frac{\lambda_C}{\lambda} = \frac{V}{V_C} \qquad ...(2.54)$$

Thus for step index fiber where V_C = 2.405, the cutoff wavelength is given by

$$\lambda_C = \frac{V\lambda}{2.405} \qquad ...(2.55)$$

CCITT defines an effective cutoff wavelength, which is obtained from a 2*m* length of fiber containing a single 14 cm radius loop.

2.5.7.2. Mode Field Diameter (MFD) and Spot Size

Optical power in single-mode fiber travels in both the core and the cladding. In many situations when we join fiber, we need a number that will give us a measure of the extent of the region that carries the optical signal. In single-mode fiber the core diameter is not sufficient.

Therefore the concept of the mode field has been introduced. The mode field can be considered as the effective core of the fiber although the real core size is quite smaller. One important point is that there is no abrupt boundary that defines the extent of the mode field. Even though the signal decays exponentially there is no exact cutoff point. Therefore we have to choose an arbitrary point as the boundary.

Figure 2.28 shows the distribution of optical power across the diameter of a typical single-mode fiber. The mode field is defined as the distance between the points where the strength of the electric field is decayed to 0.37 (1/*e*) of the peak.

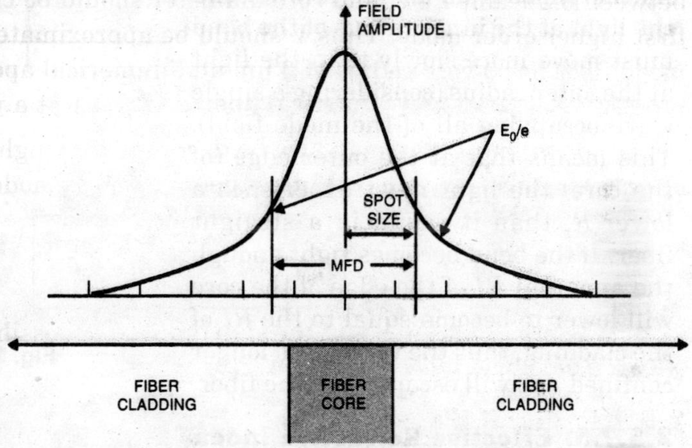

Fig. 2.28. Mode Field Definition.

The MFD of standard SMF at 1550 nm is between 10.5 microns and 11 microns depending on the fiber. In the 1310 nm band the MFD of standard fiber is 9.3 microns. Sometimes the spot size is also used to characterize single-mode fiber. The diameter of the spot is just the radius of the mode field.

2.5.7.3. Field Strength at the Fiber End

When the intensity of light is measured at the end of a fiber there are two places at which it can be done—right at the end face itself and some distance away from it.

Near Field

This is the electromagnetic field at the end face of the fiber itself. In a single-mode fiber this will usually be a profile of the bound mode but if the measurement is made close to the transmitter there may be cladding modes present as well.

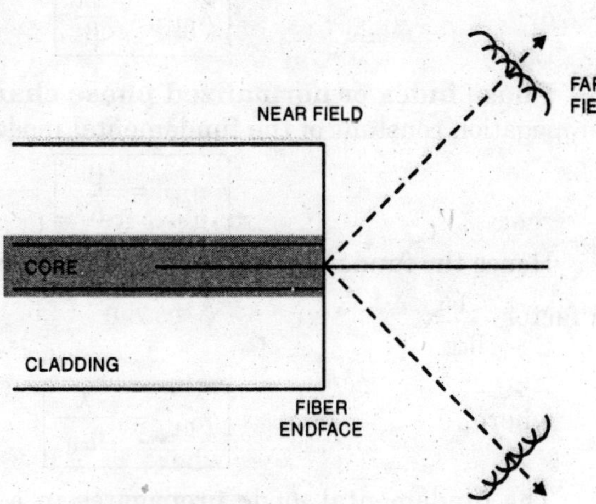

Fig. 2.29. Fields at the End of the Fiber.

Far Field

It is extremely difficult to measure the near field directly. Thus the far field is measured and used to calculate the characteristics of the near field such as mode field diameter etc. The far field consists of a series of lobes spread out away from the axis of the fiber. This is due to the fact that light leaving the end face diffracts in many directions.

2.5.7.4. Bend Loss in Single-Mode Fiber

Let us consider light in SM fiber propagating along the fiber as a wave. There is a phase front which moves along the fiber perpendicular to the direction of travel. The wavefront must be in-phase with itself across the diameter of the field. As the phase front moves into a bend the light at the inner radius of the bend must move more slowly than the light at the outer radius (considering a single wave occupying all of the mode field). This means that at the outer edge (of the core) the light must experience a lower R_τ than it would in a straight fiber. If the bend becomes tight enough the apparent *RI* of the edge of the core will lower to become equal to the *RI* of the cladding, thus the wave is no longer confined and will escape from the fiber.

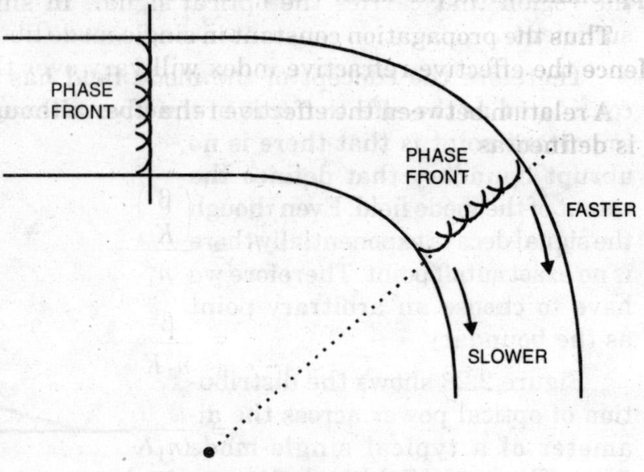

Fig. 2.30. Reason for Bend Loss in Single-Mode Fiber.

2.5.7.5. Effective Refractive Index

The **phase propagation constant** β is determined by the rate of change of phase of the fundamental LP_{01} mode propagating along a fiber. It is directly related to the wavelength of the LP_{01} mode λ_{01} by the factor 2π because β gives the increases in phase angle per unit length.

Hence, $$\beta\lambda_{01} = 2\pi$$

or $$\boxed{\lambda_{01} = \frac{2\pi}{\beta}} \quad ...(2.56)$$

Phase index or **normalized phase change** coefficient η_{eff} is defined by the ratio of the propagation constant of the fundamental mode to that of the vacuum propagation constant.

$$\boxed{\eta_{eff} = \frac{\beta}{K}} \quad ...(2.57)$$

Hence the fundamental mode (λ_{01}) wavelength is smaller than vacuum wavelength λ by a factor $\frac{1}{\eta_{eff}}$,

where, $$\boxed{\lambda_{01} = \frac{\lambda}{\eta_{eff}}} \quad ...(2.58)$$

The fundamental mode propagates in a medium with a refactive index $n(r)$ which is dependent on the distance r from the fiber axis. The refractive index can be considered as an average over the refractive index of this medium.

At long wavelength within a normally clad fiber, the mode field diameter is large compared to the core diameter and hence the electric field extends for into the cladding region. In this case the propagation constant β will be approximately equal to n_2K and the effective index will be similar to the refractive index of the cladding n_2.

Physically most of the power is transmitted in the cladding material. However at short wavelengths, the field is concentrated in the core region and the propagation constant β approximates to the max. wave number n_1K.

Thus the propagation constant in single mode fiber varies over the interval $n_2K < \beta < n_1K$. Hence the effective refractive index will vary over the range $n_2 < \eta_{eff} < n_1$.

A relation between the effective refractive index and the normalized propagation constant b is defined as

$$b = \frac{\left(\frac{\beta}{K}\right)^2 - n_2^2}{n_1^2 - n_2^2}$$

$$= \frac{\beta^2 - n_2^2 K^2}{n_1 K^2 - n_2^2 K^2} \quad ...(2.59)$$

$$= \frac{(\beta + n_2K)(\beta - n_2K)}{(n_1K + n_2K)(n_1K - n_2K)}$$

But $\beta \cong n_1K$, then

$$b = \frac{\beta - n_2K}{n_1K - n_2K}$$

$$\boxed{b = \frac{\beta/K - n_2}{n_1 - n_2} = \frac{\eta_{eff} - n_2}{n_1 - n_2}} \quad ...(2.60)$$

In single mode fibers, the parameter b varies between 0 and 1.

2.5.7.6. Group Delay and Mode Factor Delay

For a light pulse propagating along a unit length of fiber, the transit time or group delay τ_g is the inverse of the group velocity v_g. Therefore.

$$\tau_g = \frac{1}{v_g} = \frac{d\beta}{d\omega}$$

$$= \frac{1}{C} \cdot \frac{d\beta}{dK} \quad ...(2.61)$$

The group index of a uniform plane wave propagating in a homogeneous medium can be determined by the following equation

$$\boxed{N_g = \frac{C}{v_g}} \quad ...(2.62)$$

But for a single mode fiber effective group index N_{ge} is defined as

$$N_{ge} = \frac{C}{v_g} \quad ...(2.63)$$

where, v_g = Group velocity of the fundamental fiber mode.

The effective group index in terms of effective refractive index η_{eff} may be written as

$$N_{ge} = \eta_{eff} - \lambda . \frac{d\eta_{eff}}{d\lambda} \quad ...(2.64)$$

β may be expressed in terms of the relative index difference Δ and the normalized propagation constant b by the following approximate equation.

$$\beta = K\left[\left(n_1^2 - n_2^2\right)b + n_2^2\right]^{1/2}$$

$$\beta \simeq Kn_2\,(1 + b\Delta) \quad ...(2.65)$$

For a **weakly guiding fiber**, the relative refractive index as,

$$\frac{n_1 - n_2}{n_2} \simeq \frac{N_{g1} - N_{g2}}{N_{g2}} \quad ...(2.66)$$

where, N_{g1} and N_{g2} = Group indices for the fiber core and cladding regions respectively.

Subtituting the value of β in Eqn. (2.64) and by using Eqn. (2.65), the group delay, per unit distance is given by

$$\tau_g = \frac{1}{C}\left[N_{g2} + \left(N_{g1} + N_{g2}\right)\frac{d(V_b)}{dV}\right] \quad ...(2.67)$$

The wavelength dependence of Δ can be ignored because the dispersive properties of fiber core and the cladding are almost same.

Therefore the **group delay** can be written as

$$\tau_g = \frac{1}{C}\left[N_{g2} + n_2\Delta\frac{d(Vb)}{dV}\right] \quad ...(2.68)$$

- The first term is dependent on group delay an wavelength when a uniform plane wave is propagating in an infinite extended medium.
- The second term results from the waveguiding properties of the fiber only and is determined by the mode delay factor $\frac{d(Vb)}{dV}$ which describes the change in group delay.

SOLVED EXAMPLES

Example 2.1. A cylindrical waveguide is made with film refractive index of 3.5 and cladding refractive index of 3.0. Determine the maximum thickness of film that will allow no more than six guided modes to propagate for λ = 1.5 μm.

Solution. Given that,

Waveguide refractive index, $n_1 = 3.5$

Cladding refractive index, $n_2 = 3.0$

Number of modes, $v = 6$

$\lambda = 1.5$ μm

Since, the waveguide is symmetrical,

$$\theta_s = \theta_c = \sin^{-1}\left(\frac{n_2}{n_1}\right)$$

$$= \sin^{-1}\left(\frac{3.0}{3.5}\right) = 59^\circ.$$

Therefore for cutoff the number of modes, the thickness of film can be found from the relation

$$\frac{2\pi n_1 h \cos\theta_c}{\lambda} < 2\pi v$$

$$= 2 \times \pi \times 6 = 12\pi$$

$$h < \frac{12\pi\lambda}{2\pi n_1 \cos\theta_c}$$

$$h < \frac{12 \times 3.14 \times 1.5}{2 \times 3.14 \times 3.5 \times \cos 59^\circ}$$

$$\boxed{h < 5.0\ \mu m} \quad \textbf{Ans.}$$

Therefore the thickness of the film should be less than 5 μm so that only six modes will be guided along the fiber.

Example 2.2. For a multimode step index fiber the refractive indices of core and cladding are 1.53 and 1.48 respectively. Find out the angle of acceptance and critical angle when light is incident from air.

Solution. Given that, $n_1 = 1.53$

$n_2 = 1.48$

The angle of acceptance for the fiber is given by

$$\theta_A = \sin^{-1}\left[\frac{(n_1^2 - n_2^2)^{1/2}}{n_0}\right]$$

where, $n_0 = 1$ (for air)

$$\theta_A = \sin^{-1}\left[(1.53)^2 - (1.48)^2\right]^{1/2}$$

$$= \sin^{-1}\left[3.01 \times 0.05\right]^{1/2}$$

$$= \sin^{-1}\left[0.1505\right]^{1/2}$$

$$\boxed{\theta_A = 22.8^\circ} \quad \textbf{Ans.}$$

The critical angle is given by

$$\theta_c = \sin^{-1}\left(\frac{1.48}{1.53}\right)$$

$$\boxed{\phi_C = 75.3^\circ} \quad \textbf{Ans.}$$

Example 2.3. For a step index fiber the normalized frequency frequency (v) is 26.6 at a wavelength of 1300 nm. Determine the numerical if the core radius is 25 μm.

Solution. Given that,

$$\nu = 26.6$$
$$\lambda = 130 \text{ mm} = 1.3 \text{ μm}$$
$$a = 25 \text{ μm.}$$

The normalized frequency for a step index fiber is given by

$$\nu = \frac{2\pi a}{\lambda}(NA)$$

$$NA = \frac{v.\lambda}{2\pi a}$$

$$= \frac{26.6 \times 1.3}{2 \times 3.14 \times 25} = \frac{34.58}{157}$$

$$\boxed{NA = 0.22} \quad \textbf{Ans.}$$

Example 2.4. Determine the numerical aperture of a single mode fiber where, n_1 = 1.4675 and n_2 = 1.4622.

Solution. The numerical aperture of a single mode fiber is given by

$$NA = \sqrt{n_1^2 - n_2^2} = \sqrt{(1.4675)^2 - (1.4622)^2}$$

$$= \sqrt{2.9297 \times 0.0053} = \sqrt{0.01553}$$

$$\boxed{NA = 0.125} \quad \textbf{Ans.}$$

Example 2.5. For a step index fiber the following properties are given

a = 4 μm, n_1 = 1.5, n_2 = 1.47.

Determine the cutoff wavelength for the fiber.

Solution. The cutoff wavelength for the fiber is given by

$$\lambda_c = \frac{2\pi a\,(NA)}{2.405}$$

$$= \frac{2\pi a\,(n_1^2 - n_2^2)^{1/2}}{2.405}$$

$$= \frac{2 \times 3.14 \times 4 \text{ μm} \times \left[(1.5)^2 - (1.47)^2\right]^{1/2}}{2.405}$$

$$= \frac{25.12 \times \left[2.25 - 2.16\right]^{1/2}}{2.405}$$

$$= \frac{25.12 \times (0.09)^{1/2}}{2.405} = \frac{7.536}{2.405}$$

$$\boxed{\lambda_c = 3.12 \text{ μm}} \quad \textbf{Ans.}$$

Example 2.6. An optical fiber for single mode is to be designed for 1.55 µm communication system. Determine the maximum diameter of the core which can be allowed if n_1 = 1.55 and n_2 = 1.48.

Solution. Given that,

$$\lambda = 1.55 \text{ µm}$$

The core thickness of the fiber is given by

$$a = \frac{2.405\,\lambda}{2\pi\,(NA)}$$

$$= \frac{2.405\,\lambda}{2\pi\left[n_1^2 - n_2^2\right]^{1/2}}$$

$$= \frac{2.405 \times 1.55}{2 \times 3.14\left[(1.55)^2 - (1.48)^2\right]^{1/2}}$$

$$= \frac{3.73}{6.28 \times [3.03 \times 0.07]^{1/2}}$$

$$= \frac{0.5939}{(0.2121)^{1/2}} = 1.29 \times 10^{-6} \text{ m} = 1.29 \text{ µm}$$

Hence the diameter $\Delta = 2\alpha$

$$= 2 \times 1.29$$

$$\boxed{\Delta = 2.58 \text{ µm}} \quad \textbf{Ans.}$$

Example 2.7. An optical fiber is operating at 0.84 µm wavelength. Determine the number of modes propagating through fiber if a = 25 µm, n_1 = 1.48, n_2 = 0.01.

Solution. The number of modes in the optical fiber is given by

$$m = 2.\left(\frac{2\pi}{\lambda}\right)^2 . \frac{a^2}{2}(NA)^2 = 2.\left(\frac{2\pi}{\lambda}\right)^2 . \frac{a^2}{2} . \left(n_1^2 - n_2^2\right)$$

Substituting values we get

$$= \left(\frac{2 \times 3.14}{0.84}\right)^2 . (25)^2 . \left[(1.48)^2 - (0.01)^2\right]$$

$$= 55.95 \times 625 \times [1.49 \times 1.47]$$

$$\simeq 759.85$$

$$m \cong 760$$

and, $$m = \frac{v^2}{2}$$

where, v = Number of modes

Therefore, $$760 = \frac{v^2}{2} \Rightarrow v^2 = 1520$$

$$\boxed{v \simeq 39} \quad \textbf{Ans.}$$

Example 2.8. Determine the number of modes for a multimode fiber, if a = 20 μm, λ = 0.85 μm, n_1 = 1.475 and n_2 = 1.472. How many modes will propagate if λ is changed to 1.3 μm ?

Solution. For a single mode fiber,

$$v = 2.405$$

$$v = \frac{2\pi a}{\lambda}\sqrt{n_1^2 - n_2^2}$$

Substituting the values we get

$$v = \frac{2\pi \times 20}{0.85}\sqrt{(1.475)^2 - (1.472)^2}$$

$$= \frac{40 \times 3.14}{0.85} \times \sqrt{2.947 \times 0.003}$$

$$= 147.76 \times \sqrt{0.008841}$$

$$\boxed{v = 12.901} \quad \textbf{Ans.}$$

and number of modes, $$M = \frac{v^2}{2}$$

if λ is changed to 1.3 μm, then

$$v = \frac{2\pi \times 20}{1.3}\sqrt{(1.475)^2 - (1.472)^2}$$

$$= \frac{125.6}{1.3}\sqrt{2.947 \times 0.003}$$

$$= 96.62 \times \sqrt{0.008841}$$

$$= 9.089$$

and $$M = \frac{v^2}{2} = \frac{(9.089)^2}{2}$$

$$\boxed{M = 40} \quad \textbf{Ans.}$$

Example 2.9. Determine the numerical aperture of a step index fiber when the core refractive index n_1 = 1.5 and the cladding refractive index n_2 = 1.48. Find out the the maximum angle of entrance of light if the fiber is placed in air.

Solution. The numerical aperture (*NA*) is given by

$$NA = \sqrt{n_1^2 - n_2^2}$$

$$= \sqrt{(1.5)^2 - (1.48)^2}$$

$$= \sqrt{2.98 \times 0.02}$$

$$\boxed{NA = 0.24413} \quad \textbf{Ans.}$$

Example 2.10. A step index fiber is operating at a wavelength of 1.3 μm with the following parameters :

$$n_1 = 1.450, \; n_2 = 1.447$$

Determine : (*a*) Core radius; (*b*) *NA*; (*c*) Maximum acceptance angle.

Solution. (*a*) For a single mode fiber,

$$v \leq 2.405$$

$$v = \frac{2\pi a}{\lambda}\sqrt{n_1^2 - n_2^2}$$

$$2.405 = \frac{2 \times 3.14 \times a}{1.3}\sqrt{(1.450)^2 - (1.447)^2}$$

$$2.405 = \frac{0.5858}{1.3} a$$

$$a = \frac{2.405}{0.4506}$$

$$\boxed{a = 5.337 \text{ μm}} \quad \textbf{Ans.}$$

Therefore the radius of fiber core should be less than or equal to 5.337 μm.

(*b*) The numerical aperture is given by

$$NA = \sqrt{n_1^2 - n_2^2}$$

$$= \sqrt{(1.45)^2 - (1.447)^2}$$

$$= \sqrt{2.897 \times 0.003} = \sqrt{0.008691}$$

$$\boxed{NA = 0.0932} \quad \textbf{Ans.}$$

(*c*) Since,

$$NA = 1 \,.\, \sin \theta_{max}$$

$$\theta_{max} = \sin^{-1}(NA)$$

$$= \sin^{-1}(0.0932)$$

$$\boxed{\theta_{max} = 5.349^{\circ}} \quad \textbf{Ans.}$$

Example 2.11. A silica optical fiber with a core diameter large enough to be considered by ray diameter large enough to be considered by ray theory analysis has a core refractive index of 1.50 and a cladding refractive index 1.47. Determine

(*a*) Critical angle at core cladding interface.

(*b*) NA for the fiber.

(*c*) Acceptance angle.

Solution. Given that, $n_1 = 1.50$

$n_2 = 1.47$

(*a*) The critical angle at the core-cladding interface is given by

$$\theta_c = \sin^{-1}\left(\frac{n_2}{n_1}\right)$$

$$= \sin^{-1}\left(\frac{1.47}{1.50}\right) = \sin^{-1}(0.98)$$

$$\boxed{\phi_c = 78.5^\circ} \quad \textbf{Ans.}$$

(*b*) The numerical aperture is

$$NA = \sqrt{n_1^2 - n_2^2}$$

$$= \sqrt{(1.50)^2 - (1.47)^2}$$

$$= \sqrt{2.25 - 2.16} = \sqrt{0.09}$$

$$\boxed{NA = 0.30} \quad \textbf{Ans.}$$

(*c*) The acceptance angle θ_a the air is given by

$$\theta_a = \sin^{-1}(NA)$$

$$= \sin^{-1}(0.30)$$

$$\boxed{\theta_a = 17.4^\circ} \quad \textbf{Ans.}$$

Example 2.12. An optical fiber in air has an NA of 0.4. If the skew rays changes direction by 100° at each reflection, determine the acceptance angle for skew rays.

Solution. Given that,

$$NA = 0.4$$

Since skew rays changes direction by 100° at each reflection, therefore

$$r = 50^\circ$$

The acceptance angle for skew rays is given by

$$\theta_{as} = \sin^{-1}\left(\frac{NA}{\cos\gamma}\right)$$

$$= \sin^{-1}\left(\frac{0.4}{\cos 50^\circ}\right)$$

$$\boxed{\theta_{as} = 38.5^\circ} \quad \textbf{Ans.}$$

Example 2.13. A multimode step index fiber is operating at a wavelength of 0.85 μm with a core diameter of 80 μm and a relative index difference of 1.5%. If the refractive index of core is 1.48. Determine

(*a*) The normalized frequency for the fiber.

(*b*) The number of guided modes.

Solution. Given that,

$$\lambda = 0.85 \ \mu m$$

$$n_1 = 1.48$$

$$a = \frac{80}{2} \mu m = 40 \ \mu m$$

$$\Delta = 1.5\% = 0.015.$$

(*a*) The normalized frequency is given by

$$v = \frac{2\pi}{\lambda} a n_1 (2\Delta)^{1/2}$$

$$= \frac{2 \times 3.14}{0.85} \times 40 \times (0.015 \times 2)^{1/2} \times 1.48$$

$$= 147.76 \times 2 \times 1.48 \times (0.030)^{1/2}$$

$$= 437.37 \times (0.03)^{1/2}$$

$$\boxed{v = 75.8} \quad \textbf{Ans.}$$

(*b*) Total number of guided modes is given by

$$M = \frac{v^2}{2} = \frac{(75.8)^2}{2} = \frac{5745.6}{2}$$

$$\boxed{M \cong 2873} \quad \textbf{Ans.}$$

Therefore this fiber has $v \simeq 76$ and it gives nearly 3000 guided modes.

Example 2.14. A graded index fiber with a parabolic refractive index profile core has a refractive index at the core axis of 1.5 and a relative index differences of 1%. If the single mode fiber is operating at 1.3 μm, determine

(*a*) The cutoff value of normalized frequency.

(*b*) Maximum core radius.

Solution. Given that,

$$\lambda = 1.3 \ \mu m = 1.3 \times 10^{-6} \ m$$

$$n_1 = 1.5$$

$$\Delta = 1\% = 0.01$$

(*a*) The maximum value of normalized frequency

$$v = 2.4\left(1 + \frac{2}{\alpha}\right)^{1/2} = 2.4\left[1 + \frac{2}{2}\right]^{1/2}$$

$$v = 2.4\sqrt{2}$$ **Ans.**

(*b*) The maximum core radius may be obtained from

$$a = \frac{V\lambda}{2\pi n_1 (2\Delta)^{1/2}}$$

$$= \frac{2.4\sqrt{2} \times 1.3}{2 \times 3.14 \times 1.5 \times (2 \times 0.01)^{1/2}}$$

$$= \frac{4.412}{9.42 \times (0.02)^{1/2}} = \frac{0.468}{(0.02)^{1/2}}$$

$$a = 3.3\ \mu\text{m}$$ **Ans.**

Example 2.15. Determine the cutoff wavelength for a step index fiber to exhibit single mode operation when the core refractive index and radius are 1.46 and 4.5 μm respectively with the relative index differences of 0.25%. **[PTU 2007]**

Solution. Given that,

$$a = 4.5\ \mu\text{m}$$

$$n_1 = 1.46$$

$$\Delta = 0.25\% = 0.0025$$

For a single mode fiber the value of $v_c \leq 2.405$ cutoff wavelength for step index fiber is given by,

$$\lambda_c = \frac{2\pi a n_1 (2\Delta)^{1/2}}{2.405}$$

$$= \frac{2 \times 3.14 \times 1.46 \times (2 \times 0.0025)^{1/2}}{2.405}$$

$$= \frac{9.169 \times (0.005)^{1/2}}{2.405} = 3.812 \times (0.05)^{1/2}$$

$$\lambda_c = 1.214\ \mu\text{m}$$ **Ans.**

Example 2.16. Calculate the numerical aperture and acceptance angle for an optical fiber. Given that refractive indices of the core and the cladding are 1.45 and 1.40 respectively.

Solution. Given that $n_1 = 1.45$

$$n_2 = 1.40$$

Numerical aperture is given by,

$$NA = \sqrt{n_1^2 - n_2^2}$$

$$= \sqrt{(1.45)^2 - (1.40)^2}$$

$$NA = 0.3775$$ **Ans.**

Acceptance angle θ_m is given by

$$\theta_m = \sin^{-1}\left(\sqrt{n_1^2 - n_2^2}\right) = \sin^{-1}(\text{NA}) = \sin^{-1}(0.3775)$$

$$\boxed{\theta_m = 22.18^\circ} \quad \textbf{Ans.}$$

Example 2.17. A glass clad fiber is made with core glass of refractive index 1.5 and the cladding is doped to give a fractional index difference of 0.0005. Calculate

(*a*) The cladding index

(*b*) The critical internal reflection angle

(*c*) The external critical acceptance angle

(*d*) The numerical aperture.

Solution. Given that,

$$n_1 = 1.5$$
$$\Delta = 0.0005$$

(*a*) If n_2 is the refraction index of cladding

$$\Delta = \frac{n_1 - n_2}{n_1}$$

$$0.0005 = \frac{1.5 - n_2}{1.5}$$

$$n_2 = 1.5 - 1.5 \times 0.0005$$

$$\boxed{n_2 = 1.49925} \quad \textbf{Ans.}$$

(*b*) Let ϕ_c be the critical internal reflection angle

$$\sin \phi_c = \frac{n_2}{n_1}$$

$$\phi_c = \sin^{-1}\left(\frac{n_2}{n_1}\right) = \sin^{-1}\left(\frac{1.49925}{1.5}\right) = \sin^{-1}(0.9995)$$

$$\boxed{\phi_c = 88.2^\circ} \quad \textbf{Ans.}$$

(*c*) Let θ_m is the external critical acceptance angle

$$\sin \theta_m = \frac{\sqrt{n_1^2 - n_2^2}}{n_0}$$

where, $$n_0 = 1$$

$$\theta_m = \sin^{-1}\sqrt{n_1^2 - n_2^2} = \sin^{-1}\sqrt{(1.5)^2 - (1.49925)^2}$$

$$= \sin^{-1}(0.0474)$$

$$\boxed{\theta_m = 2.72^\circ} \quad \textbf{Ans.}$$

(*d*) Numerical aperture,

$$NA = n_1\sqrt{2\Delta} = 1.5\sqrt{2\times 0.0005} = 1.5\sqrt{0.001}$$

$$\boxed{NA = 0.0474} \quad \textbf{Ans.}$$

Example 2.18. An optical fiber has a numerical aperture of 0.20 and a cladding refractive index of 1.59. Determine the acceptance angle for the fiber in water, which has a refractive index of 1.33.

Solution. Given that,

$$NA = 0.20$$
$$n_2 = 1.59$$

For air, $n_0 = 1$

Numerical aperture is given by,

$$NA = \sqrt{n_1^2 - n_2^2}$$

$$(0.20) = \sqrt{n_1^2 - (1.59)^2}$$

$$n_1^2 = (0.20)^2 + (1.59)^2$$

$$\boxed{n_1 = 1.6025} \quad \textbf{Ans.}$$

In water, $n_0 = 1.33$

Now,

$$NA = \frac{\sqrt{n_1^2 - n_2^2}}{n_0} = \frac{\sqrt{(1.6)^2 - (1.59)^2}}{1.33}$$

$$\boxed{NA = 0.15} \quad \textbf{Ans.}$$

Acceptance angle is given by

$$\theta_m = \sin^{-1}(NA) = \sin^{-1}(0.15)$$

$$\boxed{\theta_m = 8.6^o} \quad \textbf{Ans.}$$

Example 2.19. An optical fiber is made of glass with a refractive index of 1.55 and is clad with another glass with a refractive index of 1.51. Launching takes place from air. Determine

(*a*) the numerical aperture does the fiber have

(*b*) acceptance angle.

Solution. Given that,

$$n_1 = 1.55$$
$$n_2 = 1.51$$

(*a*) Normalized difference between the indices is

$$\Delta = \frac{n_1 - n_2}{n_1} = \frac{1.55 - 1.51}{1.55}$$

$$\Delta = 0.0258.$$

The numerical aperture is,

$$NA = \sqrt{2\Delta} = 2\sqrt{0.0258}$$

$$\boxed{NA = 0.352} \quad \textbf{Ans.}$$

(*b*) The acceptance angle is given by

$$\theta_m = \sin^{-1}(NA) = \sin^{-1}(0.352)$$

$$\boxed{\theta_m = 20.6^o} \quad \textbf{Ans.}$$

Example 2.20. A step index fiber in air has a numerical aperture of 0.16, core refractive index 1.45 and core diameter 60 μm. Determine normalize frequency for fiber.

Solution. Given that,

Core diameter = 60 μm

Core refractive index = 1.45

NA = 0.16

Let us assume that $\lambda = 0.1$ μm

The normalized frequency for a step index fiber is given by

$$v = \frac{2\pi a}{\lambda}(NA)$$

$$= \frac{2 \times 3.14 \times 30 \times 10^{-6} \times 0.16}{0.1 \times 10^{-6}}$$

$$= 1884 \times 0.16$$

$$\boxed{v = 60.28} \quad \textbf{Ans.}$$

Example 2.21. The speed of light in vacuum and that in the core of a SI fiber is 3 × 10^8 ms^{-1} and 2 × 10^8 ms^{-1} respectively. When the fiber is placed in air, the critical angle at the core-cladding interface is 75°. Calculate :

(*a*) the NA of the fiber.

(*b*) multi path dispersion per unit length.

Solution. Given that, $c = 3 \times 10^8$ m/s

$v = 2 \times 10^8$ m/s

$\phi_c = 75°$

Here, $$n_1 = \frac{c}{v} = \frac{3 \times 10^8}{2 \times 10^8} = 1.5$$

and $$\sin \phi_c = \frac{n_2}{n_1}$$

$$\sin 75° = \frac{n_2}{1.5}$$

$$\Rightarrow \qquad n_2 = 1.5\left(\sin 75^\circ\right)$$

$$n_2 = 1.44$$

(*a*) Numerical aperture is given by,

$$\text{NA} = \sqrt{n_1^2 - n_2^2}$$

$$= \sqrt{(1.5)^2 - (1.44)^2}$$

$$= \sqrt{2.94 \times 0.06}$$

$$\boxed{\text{NA} = 0.42} \qquad \textbf{Ans.}$$

(*b*) Multipath dispersion per unit length is given by

$$\left(\frac{n_1}{n_2}\right)\left(\frac{\Delta n}{c}\right)$$

where,
$$\Delta n = n_1 - n_2$$
$$= 1.5 - 1.44$$
$$\Delta n = 0.06$$

Thus, $\quad$ Multipth dispersion $= \dfrac{1.5}{1.44} \times \dfrac{0.06}{3 \times 10^8}$

$$= 0.176 \times 10^{-9} \text{ s/m}$$

$$\boxed{\text{Dispersion multipath} = 0.176 \ \mu\text{s/km}} \qquad \textbf{Ans.}$$

Example 2.22. What should be the maximum thickness of the guide slab of a symmetrical SI planar wave guide, so that it supports only the fundamental TE mode ?

Take, $\quad n_1 = 3.6, \quad n_2 = 3.56, \quad \lambda = 0.85 \ \mu\text{m}.$ **[PTU 2001]**

Solution. Fundamental TE mode is TE_{10}

where, $\quad m = 1, \quad n = 0$

For step index fiber cutoff wavelength is given by

$$\lambda_c = \frac{v.\lambda}{2.405}$$

but,
$$\lambda_c = \frac{2}{\sqrt{\left(\frac{m}{a}\right)^2 + \left(\frac{n}{a}\right)^2}} = \frac{2}{\sqrt{\left(\frac{1}{a}\right)^2 + \left(\frac{0}{b}\right)^2}}$$

$$= 2a$$

Hence,
$$v_c = \frac{2\pi a}{\lambda} \sqrt{n_1^2 - n_2^2}$$

For planar guide, $v_c = 2.405$

$$\Rightarrow \quad a = \frac{v_c . \lambda}{2\pi \sqrt{n_1^2 - n_2^2}}$$

$$= \frac{2.405 \times 0.05 \times 10^{-6}}{2 \times 3.14 \times \sqrt{(3.6)^2 - (3.56)^2}}$$

$$= \frac{2.044 \times 10^{-6}}{6.28 \times \sqrt{7.16 \times 0.04}} = \frac{0.325 \times 10^{-6}}{\sqrt{0.2864}}$$

$$= 0.604 \times 10^{-6} \text{ m}$$

$$\boxed{a = 0.604 \ \mu\text{m}} \quad \textbf{Ans.}$$

Example 2.23. A multimode step index fiber has a relative refractive index difference of 1% and a core refractive index of 1.5. The number of modes propagating at a wavelength of 1.3 μm is 1100. Estimate the diameter of the fiber core. **[PTU 2003]**

Solution. Given that,

$$M = 1100$$

$$n_1 = 1.5$$

$$\Delta = 1\% = 0.01$$

$$\lambda = 1.3 \ \mu\text{m}$$

Total number of guided modes are given by

$$M = \frac{V^2}{2} \quad \Rightarrow \quad V = \sqrt{2M}$$

$$= \sqrt{2 \times 1100} = \sqrt{2200}$$

$$V \cong 47$$

and

$$V = \frac{2\pi}{\lambda} a n_1 (2\Delta)^{1/2}$$

$$47 = \frac{2 \times 3.14}{1.3 \times 10^{-6}} \times a \times 1.5 \, (2 \times 0.01)^{1/2}$$

$$a = \frac{47 \times 1.3 \times 10^{-6}}{6.28 \times 1.5 \times (0.02)^{1/2}}$$

$$= \frac{61.1 \times 10^{-6}}{3.899} = 15.67 \times 10^{-6} \text{ m}$$

$$a = 15.67 \ \mu\text{m}$$

Thus the diameter of fiber core = $2a$ = 2 × 15.67 μm

$$\boxed{\text{Diameter} = 31.34 \ \mu\text{m}} \quad \textbf{Ans.}$$

Example 2.24. A step index fiber has core cladding refractive index of 1.50 and 1.46 respectively. What is the value of critical angle and numerical aperture of fiber ? **[PTU 2002]**

Solution. Given that,

$$n_1 = 1.50$$
$$n_2 = 1.46$$

Critical angle is given by

$$\phi_c = \sin^{-1}\left(\frac{n_2}{n_1}\right)$$
$$= \sin^{-1}\left(\frac{1.46}{1.50}\right) = \sin^{-1}(0.973)$$
$$\phi_c = 76.74° \quad \textbf{Ans.}$$

Numerical operature of fiber,

$$NA = \sqrt{n_1^2 - n_2^2}$$
$$= \sqrt{(1.5)^2 - (1.46)^2}$$
$$= \sqrt{2.96 \times 0.04} = \sqrt{11.84 \times 10^{-2}}$$
$$\boxed{NA = 0.34} \quad \textbf{Ans.}$$

Example 2.25. An optical fiber has numerical aperture of 0.344. What is the acceptance angle for meridional rays ? Calculate the acceptance angle for skew rays which change direction by 100° at each reflection. **[PTU 2006]**

Solution. Given that,

$$NA = 0.344$$
$$2\gamma = 100°$$
$$\Rightarrow \quad \gamma = 50°$$

The acceptance angle for meridional ray is given by

$$\theta_a = \sin^{-1}(NA) = \sin^{-1}(0.344)$$
$$\theta_a = 20.12° \quad \textbf{Ans.}$$

Now acceptance angle for speed rays,

$$\theta_{as} = \sin^{-1}\left(\frac{NA}{\cos\gamma}\right) = \sin^{-1}\left(\frac{0.344}{\cos 50°}\right)$$
$$\boxed{\theta_{as} = 32.34} \quad \textbf{Ans.}$$

Example 2.26. A graded index fiber with index profile 1.90 has a core refractive index 1.5 and relative index difference 1.3%. The core diameter is 40 μm. Calculate the number of guided modes in the fiber when the transmitting light has

wavelength 1.55 μm. Also find the cutoff value of normalized frequency for single mode transmission in fiber. **[PTU 2001]**

Solution. Given that,

$$\alpha = 1.90$$

$$n_1 = 1.5$$

$$\Delta = 1.3\% = 0.013$$

$$2a = 40 \text{ μm} \Rightarrow a = 20 \text{ μm}$$

$$\lambda = 1.55 \text{ μm}$$

Number of guided modes,

$$M_g = \left(\frac{\alpha}{\alpha+2}\right)\left(n_1 \frac{2\pi}{\lambda} a\right)^2 \Delta$$

$$= \left(\frac{1.9}{1.9+2}\right)\left(1.5 \times \frac{2\times 3.14}{1.55} \times 20\right)^2 \times (0.013)$$

$$= \left(\frac{1.9}{3.9}\right)\left(\frac{188.4}{1.55}\right)^2 (0.013)$$

$$\boxed{M_g = 94} \quad \textbf{Ans.}$$

Now cutoff value of normalized frequency for single mode transmission,

$$V_c = 2.405\sqrt{1+\frac{2}{\alpha}}$$

$$= 2.405\sqrt{1+\frac{2}{1.9}} = 2.405\sqrt{\frac{3.9}{1.9}}$$

$$\boxed{V_c = 3.45} \quad \textbf{Ans.}$$

SUMMARY ....

1. A light wave is a form of energy that is moved by wave motion.
2. Wave motion is defined as a recurring disturbance advancing through space with or without the use of a physical medium.
3. Scientific experiments seem to show that light is composed of tiny particles, while other experiments indicate that light is made up of waves. Today, physicists have come to accept a theory concerning light that is a combination of particle (ray) theory and wave (mode) theory.
4. Transverse wave motion describes the up and down wave motion that is at right angle (transverse) to the outward motion of the waves.
5. Light rays, when they encounter any substance are either transmitted, refracted, reflected, or absorbed.
6. Reflection occurs when a wave strikes an object and bounces back (toward the source). The wave that moves from the source to the object is called the incident wave, and the wave that moves away from the object is called the reflected wave.

7. The law of reflection states that the angle of incidence is equal to the angle of reflection. Refraction occurs when a wave travelling through two different mediums passes through the boundary of the mediums and bends toward or away from the normal.
8. The Ray theory and the Mode theory describe how light energy is transmitted along an optical fiber.
9. The index of refraction is the basic optical material property that measures the speed of light in an optical medium.
10. Snell's law of refraction describes the relationship between the incident and the refracted rays when light rays encounter the boundary between two different transparent materials.
11. Total internal reflection occurs when light rays are totally reflected at the boundary between two different transparent materials. The angle at which total internal reflection occurs is called the critical angle of incidence.
12. The Core, Cladding, and Coating or Buffer are the three basic parts of an optical fiber.
13. The Ray theory describes how light rays propagate along an optical fiber. Meridional rays pass through the axis of the optical fiber. SKEW RAYS propagate through an optical fiber without passing through its axis.
14. Bound rays propagate through an optical fiber core by total internal reflection. Unbound rays refract out of the fiber core into the cladding and are eventually lost.
15. The Acceptance angle is the maximum angle to the axis of the fiber that light entering the fiber is bound or propagated. The light ray incident on the fiber core must be within the acceptance cone defined by the acceptance angle to be propagated along an optical fiber.
16. Numerical aperture (NA) is a measurement of the ability of an optical fiber to capture light.
17. The Mode theory uses electromagnetic wave behavior to describe the propagation of light along an optical fiber. A set of guided electromagnetic waves are called the modes of the fiber.
18. Modes traveling in an optical fiber are said to be transverse. Modes are described by their electric, $E(x, y, z)$, and magnetic, $H(x, y, z)$, fields. The electric field and magnetic field are at right angles to each other.
19. Normalized frequency determines how many modes a fiber can support. The number of modes is represented by the normalized frequency constant.
20. Single mode and Multimode fibers are classified by the number of modes that propagate along the optical fiber. Single mode fibers propagate only one mode because the core size approaches the operational wavelength. Multimode fibers can propagate over 100 modes depending on the core size and numerical aperture.
21. Attenuation is the loss of optical power as light travels along an optical fiber. Attenuation in an optical fiber is caused by absorption, scattering, and bending losses.
22. Absorption is the conversion of optical power into another energy form, such as heat.
23. Intrinsic absorption is caused by basic fiber-material properties. Extrinsic absorption is caused by impurities introduced into the fiber material.

24. Silica fibers are predominately used in fiber optic communications. They have low intrinsic material absorption at the wavelengths of operation.

25. The wavelength of operation in fiber optics is between 700 nm and 1600 nm. The wavelength of operation is between the ultraviolet (below 400 nm) and infrared (above 2000 nm) intrinsic absorption regions.

26. Extrinsic absorption occurs when impurities, such as hydroxyl ions (OH^-), are introduced into the fiber. OH^- absorption peaks define three regions or windows of preferred operation. The first window is centered at 850 nm. The second window is centered at 1300 nm. The third window is centered at 1550 nm.

27. Scattering losses are caused by the interaction of light with density fluctuations within a fiber. Rayleigh scattering is the main source of loss in commercial fibers operating between 700 nm and 1600 nm.

28. Microbends are small microscopic bends of the fiber axis that occur mainly when a fiber is cabled. Macrobends are bends having a large radius of curvature relative to the fiber diameter.

29. Optical fiber classification depends on more than the number of modes that a fiber can propagate. The optical fiber's refractive index profile and core size further distinguish different types of single mode and multimode fibers.

30. Multimode step index fibers have relatively large core diameters and large numerical apertures.

31. Unfortunately, multimode step-index fibers have limited bandwidth capabilities and poor bend performance. Short-haul, limited bandwidth, low-cost applications use multimode step-index fibers.

32. Multimode graded index fibers have a core of radius (α). Unlike step-index fibers, the value of the refractive index of the core (n_1) varies according to the radial distance (r). The value of n_1 decreases until it approaches the value of the refractive index of the cladding (n_2). Like the step-index fiber, the value of n_2 is constant and has a slightly lower refractive index than n_1.

33. The profile parameter (α) determines the shape of the core's refractive index profile. As the value of a increases, the shape of the core's profile changes from a triangular shape to a step.

34. Single mode fibers have low attenuation and high-bandwidth properties. Present applications for single mode fibers include long-haul, high-speed telecommunication systems.

35. Vapour phase oxidation and Direct melt process are two methods of fabricating multimode and single mode optical fibers.

36. Cable structures include buffers, strength members, and the jacket, or sheath.

37. Tight-buffered, Loose-tube, and Gel-filled Loose-tube are types of fiber optic buffering techniques.

REVIEW QUESTIONS

1. Explain the different types of fibers and how they are classified ?
2. Briefly discuss about the structures of fiber.
3. Explain Snell's law. What is total internal reflection ? Explain with suitable diagram.

4. Draw and discuss the structure of an optical fiber. What are the various types of fibers ? Explain their advantages and disadvantages also.

5. Differentiate between meridional and skew rays.

6. What do you understand by numerical aperture ? **[PTU 2007]**

7. Briefly explain the basic laws of ray optics. What do you meant by critical angle and total internal reflection ?

8. What is acceptance angle ? Derive an expression for it and show its relation with numerical aperture.

9. What do you understand by modes in optical fibers ? Compare a single mode and multimode fiber. **[PTU 2006, 2007]**

10. Using Ray theory describe the mechanism for the propagation of light in an optical fiber.

11. With the help of simple ray theory explain :

(*a*) Single mode step index fiber.

(*b*) Multimode step index fiber.

12. Briefly discuss the fiber classification based on modes of propagation and index profile. Draw index profile of various types of fibers.

13. What is normalized frequency of fiber ? Prove that, $M_g = \frac{v^2}{4}$ for graded index fiber.

14. Define the relative refractive index difference for an optical fiber and show how it may be related to the numerical aperture ?

15. Derive an expression for the acceptance angle for a skew ray which changes direction by an angle $2r$ at each reflection in a step index fiber in terms of fiber *NA* and *r*.

16. Determine the core radius necessary for single mode operation at 1320 nm of a step index fiber with n_1 = 1.480 and n_2 = 1.478. What are the numerical aperture and maximum acceptance angle of this fiber ?

17. A step index multimode fiber with a numerical aperture of 0.20 supports approximately 1000 modes at an 850 nm wavelength.

(*a*) What is the diameter of its core ?

(*b*) How many modes does the fiber support at 1320 nm ?

(*c*) How many modes does the fiber support at 1550 nm ?

18. Determine the normalized frequency at 820 nm for a step index fiber having a 25 μ core radius, n_1 = 1.48 and n_2 = 1.46.

(*a*) How many modes propagate in this fiber at 820 nm ?

(*b*) How many modes propagate in this fiber at 1320 nm ?

(*c*) What % of the optical power flows in the cladding in each case ?

19. Determine the maximum radius allowed for a fiber having core refractive index 1.47 and cladding refractive index 1.46. The fiber is to support only one mode at a wavelength of 1300 nm.

20. Consider a fiber with a 25 μm core radius and refractive index of core n_1 = 1.50 and Δ = 0.02. If λ = 1000 nm, then

(*a*) Calculate the value of *V*

(*b*) Maximum number of modes the fiber can support.

21. A step index multimode fiber supports approximately 1200 modes at an 900 nm wavelength. If the light gathering ability of the fiber = 0.20 then calculate the diameter of the core.

22. A graded index fiber with a parabolic refractive index profile core has a refractive index at the core axis of 1.5 and relative index difference of 1%. Estimate the maximum possible core diameter which allows single node operation at a wavelength of 1.3 μm

❑❑❑

Dispersion Effect

INSIDE THIS CHAPTER

3.1. Introduction of Dispersion

Dispersion spreads the optical pulse as it travels along the fiber. This spreading of the signal pulse reduces the system bandwidth or the information carrying capacity of the fiber. Dispersion limits how fast information is transferred as shown in Fig. 3.1.

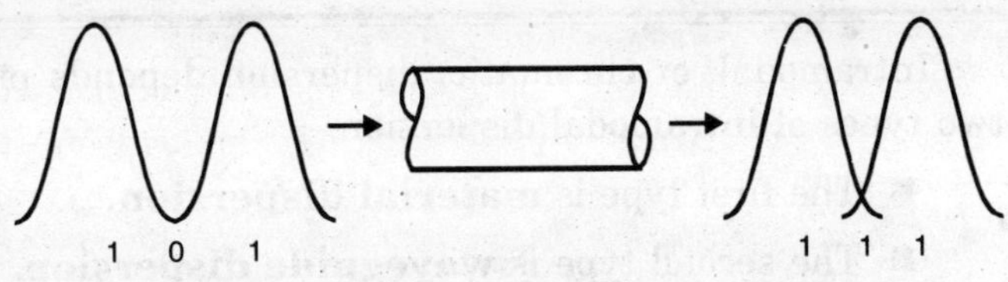

Fig. 3.1. Dispersion (Spreads the Pulse).

- An error occurs when the receiver is unable to distinguish between input pulses caused by the spreading of each pulse. The dispersion increases as the pulse travels the length of the fiber.
- Therefore dispersion in optical fiber results in a broadening in time domain of the information signal which modulates the intensity of light propagating through the fiber.
- Dispersion of the transmitted optical signal causes distortion for both digital and anolog transmission along optical fibers.
- When optical pulses injected into the fiber spread in time domain, they can interfere with the adjacent pulses resulting in what is commonly known as **Inter Symbol Interference or ISI** in digital communication.

Figure 3.2 shows the diagramatic representation of the consequences of dispersion.

An increasing number of errors may occur on the digital optical channel as the ISI becomes more pronounced. **The error rate is a function of**

- Signal attenuation on the link.
- Signal to noise ratio (SNR) at the receiver.

Signal dispersion limits the maximum possible bandwidth. For no overlapping of light pulses, digital bit rate must be less than the reciprocal of pulse duration (2τ). Hence

$$BT \leq \frac{1}{2\tau} \quad \text{...(3.1)}$$

where, τ = Input pulse duration.

If the light pulse at the output is of Gaussian shape with an rms width of σ, it allows certain amount of signal overlap on the channel. **The maximum bit rate is given by**

$$B_{\tau\,(max)} = \frac{0.2}{\sigma} \text{ bit/sec} \quad \text{...(3.2)}$$

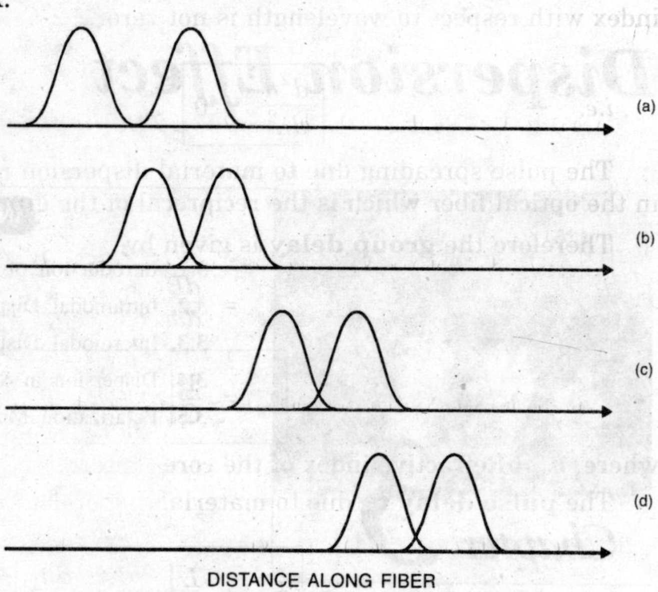

Fig. 3.2. Consequences of Dispersion and Overlapping Result In severe ISI is shown in (d).

3.2. Intramodal Dispersions

Intramodal, or chromatic, dispersion depends primarily on fiber materials. There are two types of intramodal dispersion.

- The first type is **material dispersion.**
- The second type is **waveguide dispersion.**

Intramodal dispersion occurs because different colors of light travel through different materials and different waveguide structures at different speeds.

In **multimode fibers, waveguide dispersion** and **material dispersion** are basically separate properties. Multimode waveguide dispersion is generally small compared to material dispersion. Waveguide dispersion is usually neglected. However, in single mode fibers, material and waveguide dispersion are interrelated.

The **total dispersion present in single mode fibers** may be minimized by trading material and waveguide properties depending on the wavelength of operation.

3.2.1. Material Dispersion

Material dispersion occurs because the spreading of a light pulse is dependent on the wavelengths interaction with the refractive index of the fiber core. Different wavelengths travel at different speeds in the fiber material. Different wavelengths of a light pulse that enter a fiber at one time exit the fiber at different times.

Material dispersion is a function of the source spectral width. The **spectral width** specifies the range of wavelengths that can propagate in the fiber. Material dispersion is less at longer wavelengths.

A material will exhibit material dispersion when the second differential of the refractive index with respect to wavelength is not zero.

i.e. $$\frac{d^2 n}{d\lambda^2} \neq 0$$

The pulse spreading due to material dispersion may be obtained by the group delay τ_g in the optical fiber which is the reciprocal of the group velocity v_g.

Therefore the **group delay** is given by

$$\tau_g = \frac{dB}{dw}$$

$$\tau_g = \frac{1}{C}\left[n_1 - \lambda . \frac{dn_1}{d\lambda}\right] \quad ...(3.3)$$

where, n_1 = Refractive index of the core material.

The **pulse delay τ_m** due to material dispersion in a fiber of length L is

$$\tau_m = \frac{L}{C}\left[n_1 - \lambda . \frac{dn_1}{d\lambda}\right] \quad ...(3.4)$$

For a source with **rms spectral width σ_λ** and a mean wavelength λ, the rms pulse broadening due to material dispersion σ_m may be obtained by the Taylor series expansion about λ. Where,

$$\sigma_m = \sigma_\lambda . \frac{d\tau_m}{d\lambda} + \sigma_\lambda . \frac{2d^2\tau_m}{d\lambda^2} + ... \quad ...(3.5)$$

For the source operating over the 0.8 to 0.9 μm wavelength, usually the first term dominates. Then,

$$\sigma_m = \sigma_\lambda . \frac{d\tau_m}{d\lambda} \quad ...(3.6)$$

Therefore pulse spreading may be calculated by considering the dependence of τ_m on λ. From Eqn. (3.3) we get

$$\frac{d\tau_m}{d\lambda} = \frac{L\lambda}{C}\left[\frac{dn_1}{d\lambda} - \frac{d^2 n_1}{d\lambda^2} - \frac{dn_1}{d\lambda}\right]$$

$$= -\frac{L\lambda}{C}\left[\frac{d^2 n_1}{d\lambda^2}\right] \quad ...(3.7)$$

Substituting the value of $\frac{d\tau_m}{d\lambda}$ from eqns. (3.16) into (3.15) we get

$$\sigma_m \simeq \frac{\sigma_\lambda . L}{C}\left|\lambda . \frac{d^2 n_1}{d\lambda^2}\right| \quad ...(3.8)$$

Therefore the **material dispersion parameter M** is defined as

$$M = \frac{1}{L} \cdot \frac{d\tau_m}{d\lambda}$$

or

$$M = \frac{\lambda}{C} \left| \frac{d^2 n_1}{d\lambda^2} \right| \quad ...(3.9)$$

Sometimes which is expressed in **units of *ps* nm^{-1} km^{-1}.**

3.2.2. Waveguide Dispersion

Waveguide dispersion occurs because the mode propagation constant (β) is a function of the size of the fiber's core relative to the wavelength of operation. Waveguide dispersion also occurs because light propagates differently in the core than in the cladding.

If a fiber can be operated so that intermodal dispersion and material dispersion both disappear (as for a single mode fiber operated near 1.3 μm), then a third dispersion medium will predominate. This is known as waveguide dispersion and results only from the guiding characteristics of the fiber.

All practical light sources contain light components of different wavelength distributed over a spectral bandwidth. A slight spectral shift to higher wavelength will slightly changes the path length between successive reflection points and increases the corresponding incidence angle for each supported mode. This will increase the corresponding group velocity. Therefore each supported mode will suffer a dispersion effect dependent on spectral width of the source so that, even if the other effects cancel, this could still remain. It will also disappear if a truely monochromatic light source could be developed, but that is not possible.

For an ideal single mode step index fiber the dispersion coefficient due to the waveguide dispersion mechanism has a peak value of about Δ_w = 6.6 ps/nm-km.

For a single mode whose propagation constant is β, the fiber exhibit waveguide dispersion

when, $\left(\frac{d^2\beta}{d\lambda^2}\right) \neq 0.$

For multimode fibers, the majority of modes propagate far from cutoff, are almost free of waveguide disperssion and generally it is negligible compared with material dispersion (≃ 0.1 to 0.2 ns km^{-1}).

3.3. Intermodel Dispersion

Intermodal or modal dispersion causes the input light pulse to spread. The input light pulse is made up of a group of modes. As the modes propagate along the fiber, light energy distributed among the modes is delayed by different amounts. The pulse spreads because each mode propagates along the fiber at different speeds. Since modes travel in different directions, some modes travel longer distances. Modal dispersion occurs because each mode travels a different distance over the same time span, as shown in Fig. 3.3. The modes of a light pulse that enter the fiber at one time exit the fiber a different times. This condition causes the light pulse to spread. As the length of the fiber increases, modal dispersion increases.

Modal dispersion is the dominant source of dispersion in multimode fibers. Modal dispersion does not exist in single mode fibers. Single mode fibers propagate only the fundamental mode. Therefore,

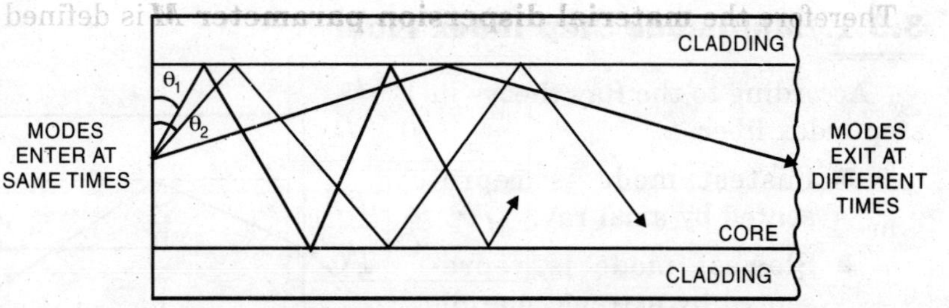

Fig. 3.3. Distance Travelled by Each Mode Over the Same Time Span.

- Single mode amplifiers exhibit the lowest amount of total dispersion.
- Single mode fibers also exhibit the highest possible bandwidth.

Let the total fiber length be Z. Since it is convenient to state losses and delays in terms of a standard unit length. Figure 3.3 show different meridional rays of different modes following their zigzag paths down as fiber is incident at an angle ϕ. The total zigzag path length for each ray is found as

$$Z_t = \frac{Z}{\sin\phi} \quad \text{...(3.10)}$$

In the **lowest order mode**, the maximum angle of incidence is 90° and in the **highest order mode** it is almost critical

$$\phi(\max) = \phi_c = \sin^{-1}\left(\frac{n_2}{n_1}\right) \quad \text{...(3.11)}$$

Now the **shortest path** is found to be

$$Z_t(\min) = \frac{Z}{\sin\phi(\max)} = \frac{Z}{\sin 90^\circ} = Z \quad \text{...(3.12)}$$

and the **longest path** is

$$Z_t(\max) = \frac{Z}{\sin\phi(\min)} = Z \cdot \frac{n_1}{n_2} \quad \text{...(3.13)}$$

The **maximum time dispersion** now becomes

$$\Delta Z = Z_t(\max) - Z_t(\min)$$

$$\Delta Z = Z\left(\frac{n_1}{n_2} - 1\right)$$

In terms of normalized index of refraction difference as

$$\Delta Z = Z\left(\frac{\Delta}{1-\Delta}\right) \quad \text{...(3.14)}$$

3.3.1. Multimode Step Index Fiber

According to the Ray theory in step index fiber

- Fastest mode is represented by axial ray.
- Slowest mode is represented by extreme meridional ray.

Figure 3.4 shows the path taken by these two rays in a structured step index fiber.

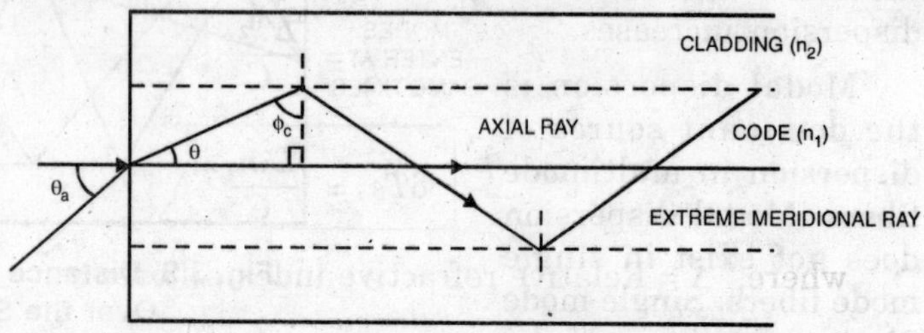

Fig. 3.4. Path Taken by Axial and Meridional Ray.

The delay difference these two rays gives estimation of the pulse broadening resulting from intermodal dispersion within the fiber. When both rays are travelling with same velocity within the fiber. When both lays are travelling with same velocity with in the constant refractive index fiber core, then the delay difference is directly related to their path length.

Therefore **minimum delay time** is given by

$$T_{min} = \frac{\text{Distance}}{\text{Velocity}} = \frac{L}{\left(\frac{C}{n_1}\right)} = \frac{L.n_1}{C} \quad ...(3.15)$$

where, n_1 = Refractive index of core.

C = Velocity of right in vacuum.

L = Length along the fiber.

The **extreme meridional ray** shows the maximum delay time, $\boldsymbol{T_{max}}$ where

$$T_{max} = \frac{L/\cos\theta}{C/n_1} = \frac{L.n_1}{C.\cos\theta} \quad ...(3.16)$$

According to the **Snell's law** at core-cladding interface we get

$$\sin\phi_C = \frac{n_2}{n_1} = \cos\theta \quad ...(3.17)$$

where, n_2 = Refractive index of cladding.

Putting the value of cos θ in Eqn. (3.17) we get,

$$\boxed{T_{max} = \frac{L.n_1^2}{C.n_2}} \quad ...(3.18)$$

The **delay difference** $\boldsymbol{\delta Ts}$ between the extreme meridional ray and the axial ray may be obtained by subtracting Eqns. (3.17) from (3.18).

Hence,

$$\delta Ts = T_{max} - T_{min}$$

$$= \frac{L.n_1^2}{C.n_2} - \frac{L.n_1}{C}$$

$$= \frac{L.n_1}{C}\left[\frac{n_1}{n_2} - 1\right]$$

$$= \frac{L.n_1}{C}\left[\frac{n_1 - n_2}{n_2}\right]$$

$$\boxed{\delta Ts \simeq \frac{L.n_1}{C}.\Delta} \quad ...(3.19) \text{ [when } \Delta \leq 1]$$

where, Δ = Relative refractive index difference.

where $\Delta \simeq \left(\frac{n_1 - n_2}{n_1}\right)$

In terms of **numerical aperture,**

$$NA = n_1 (2\Delta)^{1/2} \quad ...(3.20)$$

Substituting the value of Δ from Eqns. (3.19) to (3.20).

$$\boxed{\delta T_s \cong \frac{L(NA)^2}{2n_1 C}} \quad ...(3.21)$$

This analysis is based on the pulse broadening due to meridional rays and totally ignores skew rays with acceptance angles $\theta_{as} > \theta_a$.

Another consideration with perfect step index fiber, useful quantity with intermodal disperation is the rms pulse broadening results from dispersion mechanism.

If the optical input to the fiber is a pulse $P_i(t)$ of unit area, as shown in Fig. 3.5.

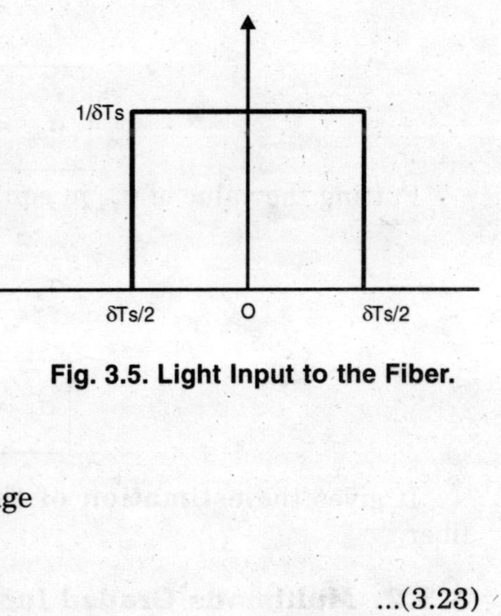

Fig. 3.5. Light Input to the Fiber.

$$\int_{-\infty}^{\infty} P_i(t).dt = 1 \quad ...(3.22)$$

$P_i(t)$ has a constant amplitude of $\frac{1}{\delta T_s}$, over the range

$$\boxed{-\frac{\delta T_s}{2} \leq P(t) \leq \frac{\delta T_s}{2}} \quad ...(3.23)$$

At the fiber output, the rms pulse for the multimode step index fiber may be given in terms of the variance, σ_s^2 as

$$\boxed{\sigma_s^2 = M_2 - M_1^2} \quad ...(3.24)$$

where, M_1 = First temporal moment it is equivalent to mean value of pulse.

M_2 = Second temporal moment it is equivalent to mean square value of pulse.

Hence, $$M_1 = \int_{-\infty}^{\infty} t\,P_i(t)\,.\,dt \qquad ...(3.25)$$

and $$M_2 = \int_{-\infty}^{\infty} t^2\,P_i(t)\,.\,dt \qquad ...(3.26)$$

The mean value M_1 for the unit input pulse is zero and assuming that it is maintained for the output pulse.

Then $$\sigma_s^2 = M_2 \int_{-\infty}^{\infty} t^2\,.\,P_i(t)\,.dt \qquad ...(3.27)$$

Integrating over the limits of input pulse and putting the value of $P_i(t)$ we get

$$\sigma_s^2 = \int_{-\delta T_s/2}^{\delta T_s/2} \frac{1}{\delta T_s}\,.\,t^2\,.\,dt$$

$$= \frac{1}{\delta T_s}\left[\frac{t^3}{3}\right]_{-\delta T_s/2}^{\delta T_s/2}$$

$$= \frac{1}{3}\left[\frac{\delta T_s}{2}\right]^2 \qquad ...(3.28)$$

$$\sigma_s = \frac{1}{\sqrt{3}}\cdot\left[\frac{\delta T_s}{2}\right] \qquad ...(3.29)$$

Putting the value of σ_{Ts} in eqn. (2.77) we get

$$\sigma_s = \frac{1}{\sqrt{3}}\left[\frac{L.n_1\,\Delta}{2C}\right] \qquad ...(3.30)$$

$$\sigma_s = \frac{L.(NA)^2}{4\sqrt{3}\,.\,n_1 C} \qquad ...(3.31)$$

It gives the **estimation of the rms impulse response** of a multimode step index fiber.

3.3.2. Multimode Graded Index Fiber

Intermodal dispersion in multimode fiber is reduced with the use of graded index fibers. Therefore it gives substantial bandwidth over multimode step index fibers.

Analytically, multimode graded index profile is given by

$$n(r) = \begin{cases} n_1\left[1 - 24\left(\dfrac{r}{a}\right)^2\right]^{1/2} & r < a \text{ (core)} \\ n_2\,(1-2\Delta)^{1/2} = n_2 & r \geq a \text{ (cladding)} \end{cases} \qquad ...(3.32)$$

Graded profile reduces the disparity in the mode transit time. According to the ray theory the **delay difference** is given by

$$\delta T_g \simeq \frac{L n_1 \Delta^2}{2C} \quad ...(3.33)$$

$$\cong \frac{(NA)^4}{\delta n_1^3 C} \quad ...(3.34)$$

Electromagnetic mode theory analysis gives an absolute temporal width at fiber output

$$\delta T_g = \frac{L.n_1 \Delta^2}{\delta C} \quad ...(3.35)$$

which indicates an increase in transmission time for the slowest mode of $\frac{\Delta^2}{8}$ over the fastest mode.

The **rms pulse broadening** is given by

$$\sigma_g = \frac{\Delta}{D} \sigma_s \quad ...(3.36)$$

where, D = Constant between 4 and 10 depending upon the evaluation and exact optimum profile.

The best minimum theoretical intermodal rms pulse broadening for a graded index fiber with an optimum characteristic refractive index profile for the core α_{op} is

$$\alpha_{op} = 2 - \frac{12\,\Delta}{5} \quad ...(3.37)$$

Therefore by combining Eqn. (3.33) and Eqn. (3.37) we get

$$\sigma_g = \frac{L.n_1 \Delta^2}{20\sqrt{3}\, C}$$

3.3.3. Mode Coupling

Under ideal conditions light travelling in one mode on a fiber cannot excite or transfer power to any other mode on the same fiber. However when there are irregularities in the fiber light can couple or power can transfer from one mode to another.

➢ **Connector Misalignment**

When a connector is not perfectly aligned some modes will leave the fiber altogether and be lost While other modes may transfer some or all of their power into different modes. Power will be lost from some modes and transferred to other bound modes.

➢ **Material Inconsistencies at a Join or Connector**

When two fibers are joined in a perfect fusion splice or with a perfect connector, there may be differences in the refractive indices of the two fibers at corresponding points on the interface. This is the case where the two fibers are of different specification even it can happen with fibers of identical specification but was drawn from different preforms. Whenever such a difference occurs some mode transformation occurs across the connection.

➢ **Material Irregularities**

This is because of the presence of impurities or just variations in density of the glass in the fiber.

➢ **Fiber Geometry Irregularities**

It covers the things such as :

(1) Changes in concentricity of the fiber

(2) Variations in thickness (diameter) of the fiber

(3) Local variations in the refractive index of the glass etc.

➢ **Bends and Microbends**

Bends in the fiber can cause some modes to be lost and others to couple power into different modes. Microbends are also an issue.

These are bends caused by putting fiber into direct contact with a rough surface of some kind and perhaps putting pressure onto it. The microbends so created will cause the fiber to lose power and also to couple power between modes.

3.3.4. Total Dispersion and Maximum Transmission Rates

Each of three dispersion mechanism contributes to pulse broadening during fiber transmission to a greater or lesser extent.

- In **multimode step index fiber** the **intermodal dispersion** will predominate,
- In a **single mode fiber** operated at the wavelength for minimum, **material dispersion** the **waveguide dispersion** will predominate.

The dispersion results in an approximately Gaussian pulse shape at the receiver end of the fiber 1 for which the indepedent dispersion effects combine in a root mean square manner to give a **total dispersion** of

$$\Delta_t(\text{total}) = \sqrt{\Delta_t^2\,(ind) + \Delta_t^2(md) + \Delta_t^2(wg)} \quad \text{...(3.38)}$$

This is the only dispersion effect and does not include the transmitted pulse width. If the transmitted pulse has an approximately Gaussian shape, then the **received pulse width t_r** can be approximately by the root mean square combination of the transmitted pulse width t_w and the total dispersion to give

$$t_r = \sqrt{t_w^2 + \Delta_t(total)^2} \quad \text{...(3.39)}$$

The maximum allowed bit transmission rate B can be found. If a power penalty of less than 1 dB is allowed to obtain the same error rate as at lower bit rates, then the bit rate B must be less than

$$B \leq \frac{1}{2t_r} \quad \text{...(3.40)}$$

which is sufficient to avoid significant intersymbol interference. This is based on a unipolar non-return no zero (UPNRZ) line code where two alternating bits constitute one cycle of the highest frequency component to be transmitted and the transmitted pulse condition lasts for a full bit period t_r.

3.4. Dispersion in Single Mode Fibers

The pulse broadening in single mode fibers is mainly because of intramodal or chromatic dispersion as only a single mode is allowed to propagate. Therefore, the bandwidth is limited by the finite spectral width of the source.

The **group delay τ_g or transit time** for a light pulse propagating along a unit length of single mode fiber, may be given as :

$$\boxed{\tau_g = \frac{1}{C}\frac{d\beta}{dk}}$$

where, C = Velocity of light in a vacuum

β = Propagation constant for a mode within the fiber

n_1 = Refractive index

k = Propagation constant for the mode in a vacuum

The total first order dispersion parameter or the chromatic dispersion of a single mode fiber, D_T is given by the derivative of specific group delay w.r.t. the vacuum length, l as :

$$D_T = \frac{d\tau_g}{d\lambda} \qquad \text{...(3.41)}$$

With material dispersion parameter, usually it is expressed in units of ps nm–1 km–1. When the variable λ is replaced by w, the total dispersion parameter will be :

$$D_T = \frac{-\omega}{\lambda}\left(\frac{d\tau_g}{dw}\right)$$

$$D_T = \frac{-\omega}{\lambda}\left(\frac{d^2\beta}{dw^2}\right) \qquad \text{...(3.42)}$$

When β varies nonlinearly with wavelength, the fiber exhibits intramodal dispersion. β may be expressed in terms of relative refractive index Δ and the normalized propagation constant b as :

$$b = \left[\frac{\left(\frac{\beta}{k}\right)^2 - n_2^2}{2n_1^2\Delta}\right] \qquad \text{...(3.43)}$$

$$\therefore \quad \beta = kn_1[1-2\Delta(1-b)]^{1/2} \qquad \text{...(3.44)}$$

The **rms pulse broadening** because of intramodal dispersion down a fiber of length L is given by the derivative of the group delay w.r.t. wavelength :

$$\text{Total rms pulse broadening} = \sigma_\lambda L \geq \left|\frac{d\tau_g}{d\lambda}\right|$$

$$= \frac{\sigma_\lambda L 2\pi}{c\lambda^2}\left(\frac{d^2\beta}{dk^2}\right)$$

where, σ_λ = Source rms spectral line width centred at a wavelength λ

The dependence of the pulse broadening on the fiber material's properties and the normalized propagation constant b. This gives rise to three inter-related effects which are as follows :

(1) **Material Dispersion Parameter D_M** defined by,

$$\frac{\lambda}{C}\left|\frac{d^2n}{d\lambda^2}\right| \quad \text{...(3.45)}$$

where, $n = n_1$ or n_2 for the core or cladding respectively.

(2) The **waveguide dispersion parameter D_W** defined by,

$$D_W = \left(\frac{n_1 - n_2}{\lambda C}\right) V \frac{d^2(Vb)}{dV^2} \quad \text{...(3.46)}$$

where, V = Normalized frequency for the fiber

Since normalized propagation constant b for a specific fiber is any dependent on V.

Then the normalized waveguide dispersion coefficient $\frac{Vd^2(Vb)}{dV^2}$ also depends on V.

(3) A **profile dispersion parameter D_P** which is proportional to $\frac{d\Delta}{d\lambda}$.

Therefore, the **total first order dispersion** D_T in a **practical single mode fiber** is given by :

$$D_T = D_M + D_W + D_P \text{ (ps nm}^{-1}\text{ km}^{-1}) \quad \text{...(3.47)}$$

where, D_M = Material dispersion

D_W = Waveguide dispersion

D_P = Profile dispersion component

But in **standard single mode fibers** the total dispersion tends to be dominated by the material dispersion of the fused silica.

3.5. Polarization Mode Dispersion

The effect of fiber birefringence on the polarization states of an optical signal are another source of pulse broadening. Particularly it is critical for high rate long haul transmission links. Birefringence can result from :

- Geometric irregularities of the fiber core.
- Internal stresses in the fiber.

The fundamental property of an optical signal is its polarization state. **Polarization** is electric field orientation of a light signal, which can vary significantly along the length of a fiber.

The varying birefringence along its length will cause each polarization mode to travel at a slightly different velocity. The resulting difference in propagation times between the two orthogonal polarization modes will result in pulse spreading. This is known as **polarization mode dispersion (PMD).**

3.5.1. Modal Birefringence

Single model fibers which are having nominal circular symmetry about the core axis allow the propagation of two nearly degenerate models with orthogonal polarizations. Therefore they are bimodal supporting HE_{11}^{x} and HE_{11}^{y} modes where the principle axes x and y are determined by the symmetry element of the fiber cross-section. Thus the fiber behaves as a birefringent medium due to the difference in effective refractive indices, and hence phase velocities for these two orthogonally polarized modes.

Therefore the modes have different propagation constants β_x and β_y. When the fiber cross-section is independent of the fiber length L in the Z-direction, then the model birefringence B_F for the fiber is given by :

$$B_F = \frac{\beta_x - \beta_y}{(2\pi/\lambda)} \quad \text{...(3.48)}$$

where, λ = Optical wavelength

Light polarized along one of the principle axes will retain its polarization for all L.

The fiber exhibit a linear retardation $\phi(z)$ because of the difference in phase velocities. It depends on the fiber length L in the z-direction and is given by : (assuming that the phase coherence of the two mode components is maintained)

$$\phi(z) = \left(\beta_x - \beta_y\right)L \quad \text{...(3.49)}$$

The phase coherence of the two mode components is achieved when,

The delay between the two transit times < Coherence time of the source ...(3.50)

Coherence time for the source = $\frac{1}{\delta f}$

where, δf = Uncorreleated source frequency width

The birefringent coherence is maintained over a length of fiber L_{bc} when :

$$L_{bc} \simeq \frac{C}{B_F\, \delta f}$$

$$L_{bc} = \frac{\lambda^2}{B_F\, \delta\lambda} \quad \text{...(3.51)}$$

where C = Velocity of light in a vacuum.

$\delta\lambda$ = Source line width

When phase coherence is maintained, it leads to a polarization state which is generally elliptical but which varies periodically along the fiber. This situation is shown in Fig. 3.6.

Here the incident linear polarization which is at 45° w.r.t. the x-axis becomes circular polarization at $\phi = \frac{\pi}{2}$ and linear again at $\phi = \pi$. The process continues through another circular polarization at $\phi = \frac{3\pi}{2}$ before returning to the initial linear polarization at $\phi = 2\pi$.

The characteristic length corresponding to this process is known as the **beat length,** which is given by

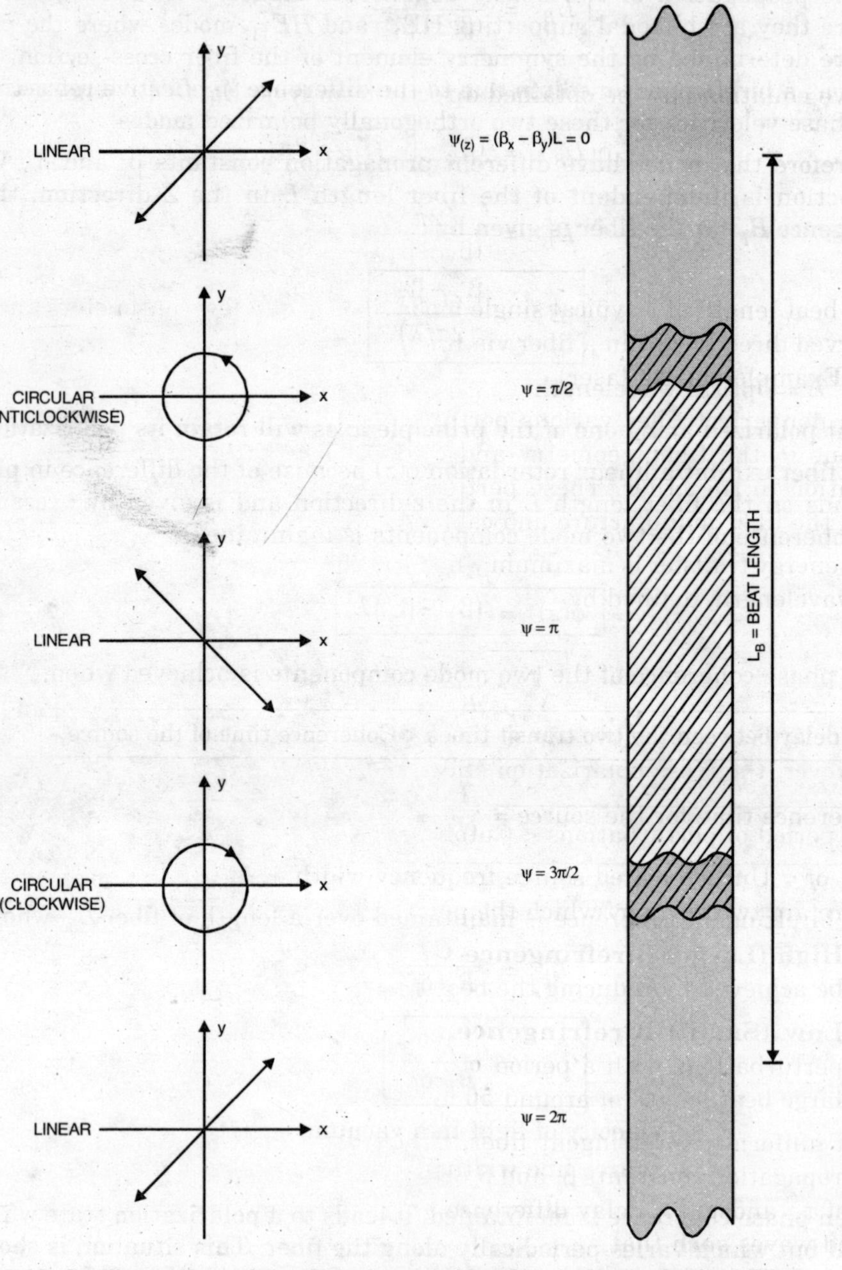

Fig. 3.16. An Illustration of the Beat Length in a Single Mode Optical Fiber.

$$L_B = \frac{\lambda}{B_F} \quad \text{...(3.52)}$$

Subtituting the value of B_F from Eq. (3.48) we get,

$$L_B = \frac{2\pi}{(\beta_x - \beta_y)} \quad ...(3.53)$$

Above equation may be obtained directly from from Eq. (3.49) where,

$$\phi(L_B) = (\beta_x - \beta_y)L_B = 2\pi$$

$$\therefore \quad L_B = \frac{2\pi}{(\beta_x - \beta_y)} \quad ...(3.54)$$

The beat length of a typical single mode fiber is of a few centimeters and its effect may be observed directly within a fiber via Rayleigh scattering with the use of a suitable visible source (Example He–Ne laser).

In a nonperfect fiber various perturbations along the fiber length such as strain or variations in the fiber geometry and composition lead to coupling of energy from one polarization to the other. These perturbations are difficult to avoid as they may easily occur in the fiber manufacture and cabling.

The energy transfer is maximum when the perturbations have a period Λ, corresponding to the wavelength, defined by

$$\Lambda = \frac{\lambda}{B_F} \quad ...(3.55)$$

However, the cross-polarization effect may be minimized, when

The period of perturbations < Cutoff period (Λ_c)

There are two ways by which the polarization maintaining fibers may be designed :

(1) High (Large) Birefringence : The maximization of the model birefringence may be achieved by reducing the beat length L_B to around 1 mm or less.

(2) Low (Small) Birefringence : The minimization of the polarization coupling perturbations with a period of Λ. This may be achieved by increasing Λ_c giving a large beat length of around 50 m or more.

In a uniformly birefringent fiber, the orthogonal fundamental modes have different phase propagation constants β_x and β_y. Hence the two modes exhibit different specific group delays of τ_{gx} and τ_{gy}. A delay difference $\delta\tau_g$ therefore occurs between the two orthogonally polarized waves such that,

$$\delta\tau_g = \tau_{gx} - \tau_{gy} \quad ...(3.56)$$

where, $\delta\tau_g$ = Polarization mode dispersion

Its range from significantly less than 1 ps km^{-1} in conventional single mode fibers to greater than 1 ns km^{-1} in high birefringence polarization maintaining fibers.

Since the two fundamental modes launched into single mode fiber have different group velocities, the output from the fiber length L will comprise two elements separated by a time interval $\delta\tau_g L$. For high birefringence fibers, the product $\delta\tau_g L$ provides a good estimate of pulse spreading in long fiber lengths. In this case the bandwidth B is given by

$$\boxed{B = \frac{0.9}{\delta\tau_g L}} \quad \text{...(3.57)}$$

But for short fiber lengths and fiber lengths longer than a characteristics coupling length L_c the

$$\boxed{\text{Pulse spreading} \propto (LL_c)^{1/2}} \quad \text{...(3.58)}$$

instead of simply L.

The maximum bit rate $B_{T\,(\text{max.})}$ for digital transmission in relation to polarization mode dispersion is given by

$$\boxed{B_{T\,(\text{max})} = \frac{B}{0.55}} \quad \text{...(3.59)}$$

SOLVED EXAMPLES

Example 3.1. A multimode graded index fiber exhibits a total pulse broadening of 1.2 ns over a distance of 30 K. Determine the pulse dispersion per unit length.

[PTU 2003-04]

Solution. Given that,

$$\text{Pulse broadening} = 1.2 \text{ ns}$$

$$\text{Distance} = 30 \text{ K}$$

Dispersion per unit length is given by

$$= \frac{\text{Pulse broadening}}{\text{Distance}}$$

$$= \frac{1.2 \times 10^{-3}}{30} = 0.4 \times 10^{-4} \text{ ns/m}$$

$$\boxed{\text{Dispersion} = 4 \times 10^{-5} \text{ ns/m}} \quad \textbf{Ans.}$$

Example 3.2. A multimode graded index fiber exhibits total pulse broadening of 0.1 μs over a distance of 10 km. Determine

(*a*) The maximum possible bandwidth on the link assuming no intersymbol interference.

(*b*) The pulse dispersion per unit length

(*c*) The bandwidth length product for the fiber.

Solution. Given that, $\tau = 0.1\ \mu s = 0.1 \times 10^{-6}$ sec

$L = 10$ km

(*a*) The maximum possible optical bandwidth is given by

$$B_{opt} = B_\tau = \frac{1}{2\tau} = \frac{1}{2 \times 0.1 \times 10^{-6}} = \frac{1}{0.2 \times 10^{-6}}$$

$$B_\tau = 5 \text{ MHz}$$

(*b*) The pulse dispersion per unit length is

$$\text{Dispersion} = \frac{0.1 \times 10^{-6}}{10} = 0.01\ \mu \text{ sec/km}$$

$$\text{Dispersion} = 0.01\ \mu\text{s/km} \quad \textbf{Ans.}$$

(*c*) The bandwidth length product is

$$B_{opt} \times L = 5 \text{ MHz} \times 10 \text{ km}$$

$$\boxed{B_{opt} \times L = 50 \text{ MHz km}} \quad \textbf{Ans.}$$

Example 3.3. A glass fiber is operating at a wavelength of 1.5 μm and its material dispersion parameter is 20 ps nm^{-1} km^{-1}. Determine the pulse broadening due to material dispersion within the fiber when light is launched from an injection laser source with a peak wavelength of 1.5 μm and an rms spectral width of 2 nm into a 30 km length of the fiber.

Solution. Given that, $M = 20$ ps nm^{-1} km^{-1}

$L = 30$ km

$\lambda = 1.5\ \mu m$

$\sigma_\lambda = 2$ nm

Pulse broadening due to material dispersion is given by

$$\sigma_m = \sigma_\lambda LM$$

$$= 2 \times 30 \times 20 = 1200$$

$$\sigma_m = 1200 \text{ ps} \quad \textbf{Ans.}$$

Example 3.4. A glass fiber exhibits material dispersion by $\left|\lambda^2 \left(\frac{d^2 n_1}{d\lambda^2}\right)\right|$ of 0.025. Determine the material dispersion parameter at a wavelength of 0.85 μm, and estimate the rms pulse broadening per km for a good LED source with an rms spectral width of 20 nm at this wavelength.

Solution. Given that, $\lambda = 0.85\ \mu m = 850$ nm

$$\left|\lambda^2 \left(\frac{d^2 n_1}{d\lambda^2}\right)\right| = 0.025$$

$$\sigma_\lambda = 20 \text{ nm}$$

The material dispersion parameter is given by

$$M = \frac{\lambda}{c}\left|\frac{d^2 n_1}{d\lambda^2}\right|$$

$$= \frac{1}{c\lambda}\left|\lambda^2\left(\frac{d^2 n_1}{d\lambda^2}\right)\right|$$

$$= \frac{1 \times 0.025}{2.998 \times 10^5 \times 850} \text{ s nm}^{-1} \text{ km}^{-1}$$

$$M = 98.1 \text{ ps nm}^{-1} \text{ km}^{-1} \quad \textbf{Ans.}$$

The rms pulse broadening due to material dispersion parameter M is given by

$$\sigma_m = \frac{\sigma_\lambda L}{c}\left|\lambda . \frac{d^2 n_1}{d\lambda^2}\right|$$

$$\sigma_m = \sigma_\lambda LM$$

Hence the rms pulse broadening per kilometer is given by

$$\sigma_m \text{ (1 km)} = 20 \times 1 \times 98.1 \times 10^{-12} = 1.96$$

$$\boxed{\sigma_m = 1.96 \text{ ns km}^{-1}} \quad \textbf{Ans.}$$

Example 3.5. The material dispersion in an optical fiber defined by $\left|\frac{d^2 n}{d\lambda^2}\right|$ is 4.0×10^{-2} μm^{-2}. Calculate the pulse broadening per kilometer due to material dispersion within the fiber when it is illuminated with an LED source with a peak wavelength of 0.9 μm and an rms spectral width of 45 nm.

Solution. Given that, $\left|\frac{d^2 n}{d\lambda^2}\right| = 4.0 \times 10^{-2}$

$$\lambda = 0.9 \text{ μm}$$

$$\sigma_\lambda = 45 \text{ nm}$$

The rms pulse broadening per km due to material dispersion is given by

$$\sigma_m = \frac{\sigma_\lambda . L}{C}\left|\lambda . \frac{d^2 n}{d\lambda^2}\right|$$

$$= \frac{45 \times 1}{3 \times 10^8}\left|900 \times \frac{4 \times 10^{-2}}{10^{-6}}\right|$$

$$= 15 \times 10^{-8} \times 900 \times 4 \times 10^4$$

$$\boxed{\sigma_m = 5.4 \text{ ns/km}} \quad \textbf{Ans.}$$

Example 3.6. Calculate the rms pulse broadening per kilometer for the fiber with the material dispersion parameter 95.0 ps nm^{-1} km^{-1}, when the optical source used is an injection laser with a relative spectral width $\frac{\sigma_\lambda}{\lambda}$ of 0.0012 at a wavelength of 0.85 μm.

Solution. Given that,

$$\lambda = 0.85 \ \mu\text{m}$$

$$M = 95.0 \text{ ps nm}^{-1} \text{ km}^{-1}$$

The rms spectral width can be calculated from the relative spectral width

i.e.

$$\sigma_\lambda = 0.0012\lambda$$

$$= 0.0012 \times 0.85 \times 10^{-6}$$

$$\sigma_\lambda = 1.02 \text{ nm}$$

The rms pulse broadening in terms of material dispersion parameter will be

$$\sigma_m = \sigma_\lambda LM$$

$$= 1.02 \times 1 \times 95.0 \times 10^{-12}$$

$$= 96.9 \times 10^{-12}$$

$$\boxed{\sigma_m = 0.096 \text{ ns km}^{-1}} \quad \textbf{Ans.}$$

Example 3.7. For a step index fiber whose NA = 0.275 and n_1 = 1.48, calculate how much a light pulse spread after travelling along 5 km.

Solution. Given that,

$$NA = 0.275$$

$$n_1 = 1.48$$

$$L = 5 \text{ km}$$

$$C = 3 \times 10^5 \text{ km}$$

Light pulse spread in single index fiber is given by

$$\delta T_s = \frac{L}{2n_1 C}(NA)^2$$

$$= \frac{5}{2 \times 1.48 \times 3 \times 10^5} \ (0.275)2$$

$$= \frac{5 \times 10^{-5} \times (0.275)^2}{8.88}$$

$$\boxed{\delta T_s = 423.8 \text{ ns}} \quad \textbf{Ans.}$$

Example 3.8. A 3 km optical link consists of multimode step index fiber with a core refractive index of 1.5 and a relative refractive index difference of 2%. Determine :

(*a*) The delay difference between the slowest and fastest modes at the fiber output.

(*b*) The rms pulse broadening due to intermodal dispersion on the link.

(c) The maximum bit rate that may be obtained without substantial errors on the link assuming only intermodal dispersion.

(d) The bandwidth-length product corresponding to (c).

Solution. Given that,

$$L = 3 \times 10^3 \text{ m}$$
$$n_1 = 1.5$$
$$D = 2\% = 0.02$$
$$C = 2.998 \times 10^8 \text{ m}$$

(a) The delay difference is given by

$$\delta T_s = \frac{Ln_1\Delta}{C}$$

$$= \frac{3 \times 10^3 \times 1.5 \times 0.02}{2.998 \times 10^8}$$

$$= \frac{90 \times 10^{-8}}{2.998}$$

$$\boxed{\delta T_s \cong 300 \text{ ns}} \quad \textbf{Ans.}$$

(b) The rms pulse broadening due to intermodal dispersion is given by

$$\sigma_s = \frac{Ln_1\Delta}{2\sqrt{3}C}$$

$$= \frac{3 \times 10^3 \times 1.5 \times 0.02}{2\sqrt{3} \times 2.998 \times 10^8}$$

$$= \frac{300 \times 10^{-12}}{2\sqrt{3}}$$

$$\boxed{\sigma_s = 86.7 \text{ ns}} \quad \textbf{Ans.}$$

(c) The maximum bit rate is given by

$$B_{\tau\,(\max)} = \frac{1}{2\tau} = \frac{1}{2\delta T_s}$$

$$= \frac{1}{2 \times 300 \times 10^{-9}}$$

$$B_{\tau\,(\max)} = 1.7 \text{ Mbits/sec} \quad \textbf{Ans.}$$

(d) Bandwidth length product,

$$B_{opt} \times L = 1.7 \times 3 = 5.1$$

$$\boxed{B_{opt} \times L = 5.1 \text{ MHz km}} \quad \textbf{Ans.}$$

Example 3.9. Calculate the rms pulse broadening per km due to intermodal dispersion for the multimode step index fiber of 6 km optical link with the 86.7 ns rms pulse broadening for an optimum near parabolic profile graded index fiber with the 1.5 core axis refractive index and 1% relative refractive index difference.

Solution. Given that,

$$\sigma_s\,(1\text{ km}) = 86.7 \text{ ns}$$

$$L = 6 \text{ km}$$

$$n_1 = 1.5$$

$$\Delta = 1\% = 0.01$$

The rms pulse broadening per km for the multimode step index fiber is given by

$$\frac{\sigma_s\,(1\text{ km})}{L} = \frac{86.7}{6} = 14.4 \text{ ns/km}$$

$$\frac{\sigma_s}{L} = 14.4 \text{ ns/km} \quad \textbf{Ans.}$$

The rms pulse broadening per km for the graded index fiber is

$$\sigma_g\,(1\text{ km}) = \frac{Ln_1\Delta^2}{20\sqrt{3}C}$$

$$= \frac{10^3 \times 1.5 \times (0.01)^2}{20\sqrt{3} \times 2.998 \times 10^8} = \frac{10^{-10}}{4\sqrt{3}}$$

$$\boxed{\sigma_g = 14.4 \text{ ps km}^{-1}} \quad \textbf{Ans.}$$

Example 3.10. A multimode optimum near parabolic graded index fiber has a material dispersion parameter of 30 ps nm^{-1} km^{-1} when used with good LED source of rms spectral width 25 nm. The fiber has a numerical aperture of 0.4 and a core axis refractive index of 1.48 and cladding of 1.47. Calculate the total rms pulse broadening per kilometer within the fiber assuming waveguide dispersion to be negligible. Also estimate the bandwidth length product for the fiber.

Solution. Given that,

$$M = 30 \text{ ps nm}^{-1} \text{ km}^{-1}$$

$$\sigma_\lambda = 25 \text{ nm}$$

$$\text{NA} = 0.4$$

$$n_1 = 1.48$$

$$n_2 = 1.47$$

The rms pulse broadening per km due to material dispersion is given by

$$\sigma_m = ML\sigma_\lambda = 30 \times 1 \times 25 = 750 \text{ ps/km}$$

The rms pulse broadening per km due to intermodal dispersion for the graded index fiber is given by

$$\sigma_s = \frac{Ln_1\Delta^2}{20\sqrt{3}C}$$

$$= \frac{10^3 \times 1.48 \times (0.01)^2}{20 \times 1.732 \times 3 \times 10^8}$$

$$= 14.24 \text{ ps/km}$$

Total rms pulse broadening per km is given by

$$\sigma_T = \sqrt{\sigma_m^2 + \sigma_s^2}$$

$$= \sqrt{(750)^2 + (14.24)^2}$$

$$\boxed{\sigma_T = 750.1 \text{ ps/km}} \quad \textbf{Ans.}$$

Example 3.11. A multimode step index fiber has a numerical aperture of 0.3 and a core refractive index of 1.45. The material dispersion parameter for the fiber is 250 ps nm^{-1} km^{-1} which makes material dispersion the totally dominating intramodal dispersion mechanism. Calculate :

(*a*) The total rms pulse broadening per km when the fiber is used with an LED source of the rms spectral width 50 nm.

(*b*) The corresponding bandwidth length product for the fiber.

Solution. Given that,

$$NA = 0.3$$

$$n_1 = 1.45$$

$$M = 240 \text{ ps nm}^{-1} \text{ km}^{-1}$$

$$\sigma_\lambda = 50 \text{ nm}$$

(*a*) The rms pulse broadening per km due to material dispersion is given by

$$\sigma_m \text{ (1 km)} = \sigma_\lambda LM$$

$$= 50 \times 1 \times 250$$

$$\sigma_m = 12.5 \text{ ns km}^{-1}$$

The rms pulse broadening per km due to intermodal dispersion for step index fiber is given by

$$\sigma_s = \frac{L(NA)^2}{4\sqrt{3}n_1 C} = \frac{10^3 \times (0.3)^2}{4\sqrt{3} \times 1.45 \times 2.998 \times 10^8} = \frac{10^{-5}}{0.207 \times 4\sqrt{3}}$$

$$\sigma_s = 29.9 \text{ ns km}^{-1}$$

The total rms pulse broadening per km for the multimode step index fiber is given by

where, ($\sigma_c \simeq \sigma_m$ and $\sigma_n = \sigma_s$)

$$\sigma_T = \sqrt{\sigma_m^2 + \sigma_s^2}$$

$$= \sqrt{(12.5)^2 + (29.9)^2}$$

$$= \sqrt{150.25 + 894.01}$$

$$= \sqrt{1050.26}$$

$$\boxed{\sigma_T = 32.4 \text{ ns km}^{-1}} \quad \textbf{Ans.}$$

(*b*) The bandwidth length product is given by

$$B_{opt} \times L = \frac{0.2}{\sigma_T} = \frac{0.2}{32.4 \times 10^{-9}} = 6.2 \times 10^6$$

$$\boxed{B_{opt} \times L = 6.2 \text{ MHz km}} \quad \textbf{Ans.}$$

Example 3.12. A single mode fiber maintains birefringent coherence over a length of 100 km when it is illuminated with a laser source of spectral width 1.5 nm and a peak wavelength of 1.32 μm. Find the beat length within the fiber.

[PTU 2003-04]

Solution. Given that,

$$L_{bc} = 100 \text{ km}$$
$$\lambda = 1.32 \text{ μm}$$
$$\delta f = 1.5 \text{ nm}$$

The fiber length L_{bc} is given by

$$L_{bc} = \frac{\lambda^2}{B_F . \delta\lambda}$$

$$100 \times 10^3 = \frac{(1.32 \times 10^{-6})^2}{B_F .(1.5 \times 10^{-9})}$$

$$\therefore \quad B_F = \frac{1.74 \times 10^{-12}}{15 \times 10^{-5}} = 0.116 \times 10^{-7}$$

and

$$B_F = \frac{\lambda}{L_B}$$

$$\Rightarrow \quad L_B = \frac{\lambda}{B_F} = \frac{1.32 \times 10^{-6}}{1.16 \times 10^{-8}} = 113.6 \times 10^3 \text{ m}$$

$$L_B = 113.6 \text{ km} \quad \textbf{Ans.}$$

Example 3.13. Two polarization maintaining fibers operating at a wavelength of 1.3 μm have beat lengths of 0.5 mm and 60 m. Determine the modal birefringence in each case.

Solution. Given that,

$$\lambda = 1.3 \text{ μm}$$
$$L_B = 0.5 \text{ mm, } 60 \text{ m}$$

The modal birefringence is given by

$$B_F = \frac{\lambda}{L_B}$$

For a beat length of 0.5 mm,

$$B_F = \frac{1.3 \times 10^{-6}}{0.5 \times 10^{-3}} = 2.6 \times 10^{-3}$$

$$\boxed{B_F = 2.6 \times 10^{-3}} \quad \textbf{Ans.}$$

For a beat length of 60 m,

$$B_F = \frac{1.3 \times 10^{-6}}{60} = 0.02166 \times 10^{-6}$$

$$\boxed{B_F = 2.17 \times 10^{-4}} \quad \textbf{Ans.}$$

Example 3.14. The beat length in a single mode optical fiber in 5 cm when light from an injection laser with a spectral line width of 1 nm and a peak wavelength of 0.5 μm is lauched into it. Determine the modal birefringence and estimate the coherence length in this case. Also calculate the difference between the propagation constants for the two orthogonal modes.

Solution. Given that,

$$\lambda = 0.5 \text{ μm}$$

$$L_B = 5 \text{ cm} = 0.05 \text{ m}$$

$$\delta\lambda = 1 \text{ nm}$$

The modal birefringence is given by

$$B_F = \frac{\lambda}{L_B}$$

$$B_F = \frac{0.5 \times 10^{-6}}{0.05} = 1 \times 10^{-5}$$

$$B_F = 1 \times 10^{-5} \quad \textbf{Ans.}$$

The coherence length is

$$L_{bc} \simeq \frac{\lambda^2}{B_F \delta\lambda} = \frac{(0.5 \times 10^{-6})^2}{1 \times 10^{-5} \times 10^{-9}} = \frac{25 \times 10^{-14}}{10^{-14}}$$

$$\boxed{L_{bc} = 25 \text{ m}} \quad \textbf{Ans.}$$

The difference between the propagation constant and for the two orthogonal modes is given by

$$\beta_x - \beta_y = \frac{2\pi}{L_B} = \frac{2 \times 3.14}{0.05} = 125.6$$

$$\boxed{\beta_x - \beta_y = 125.6} \quad \textbf{Ans.}$$

Example 3.15. The polarization mode dispersion in a uniformly birefringent single mode fiber is 600 ps km^{-1}. Determine the maximum bit rate that may be obtained on a 10 km repeaterless link assuming only polarization mode dispersion to occur.

Solution. Given that,

$$L = 10 \text{ km} = 10 \times 103 \text{ m}$$

$$\delta\tau_2 = 600 \text{ ps km}^{-1} = 600 \times 10^{-12} \text{ s km}^{-1}$$

The maximum bit rate is given by

$$B_{\tau\,(\max)} = \frac{B}{0.55}$$

where, $$B = \left(\frac{0.9}{\delta\tau_2 L}\right)$$

Therefore, $$B_\tau\,(\text{max}) = \frac{0.9}{0.55(\delta\tau_2 . L)}$$

$$= \frac{0.9}{0.55 \times 600 \times 10^{-12} \times 10 \times 10^3}$$

$$= \frac{0.9}{3.30 \times 10^{-6}} = 273 \times 10^3$$

$$\boxed{B_\tau\,(\text{max}) = 273 \text{ K bit/sec}}$$ **Ans.**

SUMMARY

1. Dispersion spreads the optical pulse as it travels along the fiber. Dispersion limits how fast information is transferred. Dispersion occurs when a pulse of light is spread out during transmission on the fiber.
2. Intramodal, or Chromatic, Dispersion occurs because light travels through different materials and different waveguide structures at different speeds. Material dispersion is dependent on the light wavelengths interaction with the refractive index of the core. Waveguide dispersion is a function of the size of the fiber's core relative to the wavelength of operation.
3. Intermodal, or Modal, dispersion occurs because each mode travels a different distance over the same time span.
4. The refractive index profile describes the value of the fiber's refractive index as a function of axial distance at any fiber diameter.
5. In step index fibers, the refractive index of the core is uniform and undergoes an abrupt change at the core-cladding boundary.
6. In graded index fibers, the refractive index of the core varies gradually as a function of radial distance from the fiber center.
7. Multimode step index profile fibers have a core of radius (α), and a constant refractive index n_1. A cladding of slightly lower refractive index n_2 surrounds the core.
8. The relative refractive index difference (Δ) is the difference in the core and cladding refractive index. The ability of the fiber to accept optical energy from a light source is related to Δ.
9. Single mode fiber cutoff wavelength is the smallest operating wavelength where single mode fibers propagate only the fundamental mode. At this wavelength, the 2nd-order mode becomes lossy and radiates out of the fiber core.
10. For dispersion, the maximum bit rate is given by

$$B_\tau\,(\text{max.}) = \frac{0.2}{\sigma} \text{ bit/sec}$$

11. The material dispersion parameter is given by

$$M = \frac{\lambda}{C}\left|\frac{d^2 n_1}{d\lambda^2}\right|$$

12. The dispersion results in an approximately Gaussian pulse shape. Total dispersion is given by

$$\Delta_t(\text{total}) = \sqrt{\Delta_t^2(ind) + \Delta_t^2(md) + \Delta_t^2(wg)}$$

13. In multimode step index fiber, the delay difference DTs between the extreme meridional ray and the axial ray is given by

$$\delta T_s = \frac{L.n_1}{c}.(\Delta)$$

14. The rms impulse response of a multimode step index fiber is given by

$$\sigma_s = \frac{L.(NA)^2}{4\sqrt{3}.n_1.c}$$

15. In multimode graded index fiber, the rms pulse broadening is given by

$$\sigma_g = \frac{\Delta}{D}(\sigma_s)$$

16. Indermodal dispersion is due to following two reasons :

- Material dispersion
- Waveguide dispersion.

17. Material dispersion can be determined by considering a plane wave propagating in an infinitely extended dielectric medium.

18. The group delay tmat due to material dispersion is expressed as

$$\tau_{\text{mat}} = \frac{L}{C}\left[n - \lambda\frac{dn(\lambda)}{d\lambda}\right]$$

19. The root mean square pulse broadening can be expressed as

$$\sigma = \left[\sigma^2_{\text{intermodal}} + \sigma^2_{\text{intramodal}}\right]^{1/2}$$

20. The cutoff wavelength for a single mode fiber is given by

$$\lambda_C \le \frac{2na\sqrt{n_1^2 - n_2^2}}{2.405}$$

21. Dispersion of the transmitted optical signal causes distortion for both digital and analog transmission along optical fibers.

22. Pulse broadening due to material dispersion results from the different group velocities of the various spectral components launched into the fiber from the optical source.

23. The rms pulse broadening due to material dispersion is given by

$$\sigma_m \simeq \frac{\sigma_\lambda L}{C}\left|\lambda.\frac{d^2 n_1}{d\lambda^2}\right|$$

REVIEW QUESTIONS

1. What do you understand by dispersion in optical fiber ? What is intersymbol interference or ISI ?
2. Briefly explain the reasons for pulse broadening due to material dispersion in optical fiber.
3. Derive an expression for the rms pulse broadening due to material dispersion in an optical fiber and define the material dispersion parameter.
4. Briefly explain the mechanism of intermodal dispersion in a multimode step index fiber.
5. Show that the total broadening of a light pulse δT_s due to intermodal dispersion in a multimode step index fiber may be given by

$$\delta T_s = \frac{L(NA)^2}{2n_1 c}$$

6. What do you understand by polarization mode dispersion (PMD) ? Derive an expression for bandwidth and maximum bit rate in digital transmission.
7. Explain dispersion mechanism with respect to single mode fibers indicating the dominating effects.
8. Briefly discuss how intramodal dispersion may be minimized within the single mode region.
9. What do you meant by mode coupling ? Explain the various irregularities in the fiber of its causes.
10. Derive an expression which gives the estimation of the rms impulse response of a multimode step index fiber.
11. A single mode step index fiber has a zero dispersion wavelength of 1.29 μm and exhibits total first order dispersion of 3.5 ps nm^{-1} km^{-1} at a wavelength of 1.32 μm. Determine the total first order dispersion in the fiber at a wavelength of 1.54 μm. **[PTU 2006]**
12. Determine the waveguide dispersion in a single mode step index fiber at a wavelength 1.34 μm when the fiber core radius and refractive index are 4.4 μm and 1.48 respectively.
13. The difference between the propagation constants for the two orthogonal modes in a single mode fiber is 250. It is illuminated with light of peak wavelength 1.55 μm from an injection laser source with a spectral line width of 0.8 μm. Calculate the coherence length within the fiber.
14. The material dispersion in an optical fiber defined by $\left|\frac{d^2 n_1}{d\lambda^2}\right|$ is 4.0×10^{-2} μm^{-2}. Determine the pulse broadening per kilometer due to material dispersion within the fiber when it is illuminated with an LED source with a peak wavelength of 0.9 μm and an rms spectral width of 45 nm.
15. An 8 km optical fiber link without repeaters uses multimode graded index fiber which has a bandwidth length product of 400 MHz im. Determine :

    (*a*) the total pulse broadening on the link.

    (*b*) the rms pulse broadening on the link.

16. Briefly explain
    (a) Intermodal dispersion.
    (b) Intramodal dispersion.
    (c) Chromatic dispersion.
17. Differentiate between intermodal and intramodal dispersion.
18. What do you meant by intramodal dispersion ? Show that the pulse spread for material dispersion is given as

$$\sigma_{mat} = \Delta_{mat} (\lambda) L\sigma_{\lambda}$$

19. What is intermodal dispersion ? Show that the delay difference between the extreme meridional ray and the axial ray in multimode step index fiber is given as

$$\delta T_{SI} = \left(\frac{Ln_1}{C}\right)\Delta$$

20. Describe the techniques employed and the fiber structures utilized to provide :
    (a) Dispersion shifted single mode fibers.
    (b) Dispersion flattened single node fibers.
21. In a single mode fibers, what do you meant by :
    (a) Modal birefringence.
    (b) The beat length.
22. Discuss dispersion mechanism with respect to single mode fibers indicating the dominating effects.
23. The light travels within a cladding faster than it does within a core. Why ?
24. What is the difference between conventional dispersion shifted (DSF) and dispersion flattered fibers ?
25. What is the pulse spread caused by material dispersion if $\Delta\lambda$ = 0.5 nm, L = 1 km and λ = 1550 nm ?
26. A multimode step index fiber gives a total pulse broadening of 95 ns over a 5 km length. Estimate the bandwidth length product for the fiber when a non-return to zero digital code is used.
27. A single mode step index fiber has a zero dispersion wavelength of 1.29 μm and exhibits total first order dispersion of 3.5 Ps nm^{-1} km^{-1} at a wavelength of 1.32 μm. Determine the total first order dispersion in the fiber at a wavelength of 1.54 μm.
28. Two polarization maintaining fibers operating at a wavelength of 1.3 μm have beat lengths of 0.7 mm and 80 m. Determine the modal birefringence in each case and comment on results.
29. Differentiate between intermodal and intramodal dispersion. **[PTU 2003-04]**
30. Explain modal birefringence and beat length in single mode fiber. What is polarization mode dispersion ?
31. Write a note on polarization maintaining fiber.
32. Explain in detail chromatic dispersion in single mode fiber. **[PTU 2007]**
33. What is group velocity dispersion. **[PTU 2008]**

❑❑❑

Optical Fiber Losses

INSIDE THIS CHAPTER

4.1. Introduction

We learned that propagation of light depends on the nature of light and the structure of the optical fiber. However that discussion did not describe how optical fibers affect system performance. In this case, system performance deals with signal loss and bandwidth.

Signal loss and system bandwidth describe the amount of data transmitted over a specified length of fiber. Many optical fiber properties increase signal loss and reduce system bandwidth. The most important properties that affect system performance are :

- **Fiber attenuation**
- **Dispersion.**

In addition to fiber attenuation and dispersion, other optical properties affect system performance. Fiber properties such as modal noise, pulse broadening and polarization can reduce the system performance. Modal noise, pulse broading and polarization are too complex to discuss at introductory level.

4.2. Attenuation Coefficient

Attenuation reduces the amount of optical power transmitted by the fiber. Attenuation controls the distance an optical signal can travel. Once the power of an optical pulse is reduced to a point where the receiver is unable to detect the pulse, an error occurs.

Attenuation is mainly a result of :

- **Light absorption**

- **Scattering**
- **Bending losses**

Attenuation is shown in Fig. 4.1.

Therefore attenuation is the loss of optical power as light travels along the fiber. **Signal attenuation** is defined as the ratio of optical input power (P_i) to the optical output power (P_0). **Optical input power** is the power injected into the fiber from an optical source. **Optical output power** is the power received at the fiber end or optical detector. The following equation defines signal attenuation as a unit of length :

$$\text{Attenuation } (\alpha) = \left(\frac{10}{L}\right) \log_{10}\left(\frac{P_i}{P_o}\right) \quad \text{...(4.1)}$$

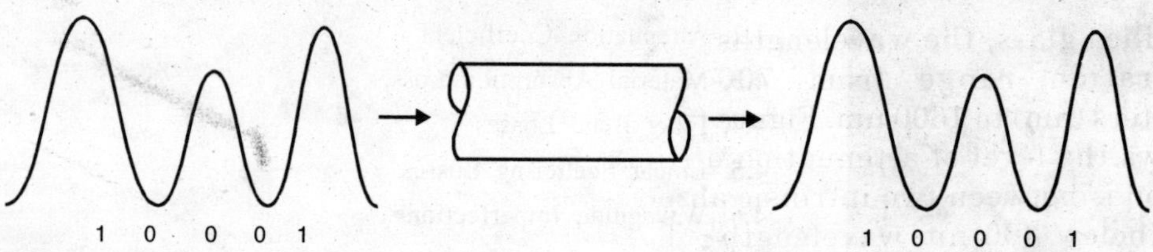

Fig. 4.1. Attenuation Reduced Power.

Since the length (L) is expressed in kilometers, therefore the **unit of attenuation is decibels/kilometers (dB/km).**

In optical fiber communication the attenuation is usually expressed in decibles per unit length (*i.e.* dB/km) as

$$\alpha_{dBL} = 10 \log_{10}\left(\frac{P_i}{P_o}\right) \quad \text{...(4.2)}$$

where, α_{dB} = Signal attenuation per unit length.

4.3. Material Absorption Losses

Absorption : Absorption is a major cause of signal loss in an optical fiber. Absorption is defined as the portion of attenuation resulting from the conversion of optical power into another energy form, such as heat. **Absorption in optical fibers is explained by three factors :**

- Imperfections in the atomic structure of the fiber material
- The intrinsic or basic fiber-material properties
- The extrinsic (presence of impurities) fiber-material properties

Imperfections in the atomic structure induce absorption by the presence of missing molecules or oxygen defects. Absorption is also induced by the diffusion of hydrogen molecules into the glass fiber. Since intrinsic and extrinsic material properties are the main cause of absorption, they are discussed further.

Material Absorption is a loss mechanism related to the material composition and the fabrication process for the fiber.

Intrinsic Absorption : Intrinsic absorption is caused by basic fiber-material properties. If an optical fiber were absolutely pure, with no imperfections or impurities, then all absorption would be intrinsic. Intrinsic absorption sets the minimal level of absorption. In fiber optics, silica (pure glass) fibers are used predominately. Silica fibers are used because of their low intrinsic material absorption at the wavelengths of operation.

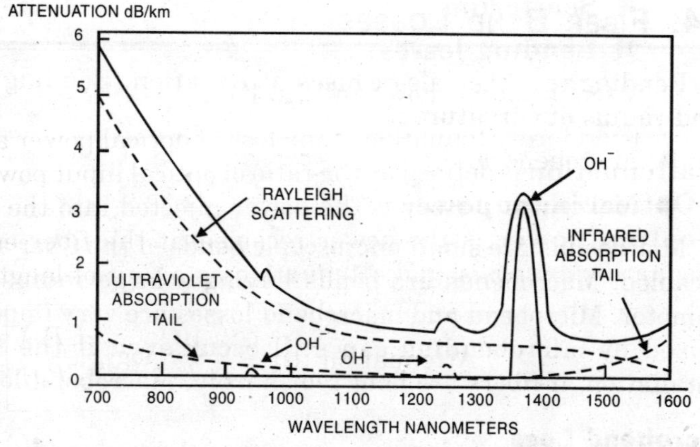

Fig. 4.2. Fiber Losses.

In silica glass, the wavelengths of operation range from 700 nanometers (nm) to 1600 nm. Figure 4.2 shows the level of attenuation at the wavelengths of operation. This wavelength of operation is between two intrinsic absorption regions. The first region is the ultraviolet region (below 400 nm wavelength). The second region is the infrared region (above 2000 nm wavelength).

Intrinsic absorption in the ultraviolet region is caused by electronic absorption bands. Basically, absorption occurs when a light particle (photon) interacts with an electron and excites it to a higher energy level. The tail of the ultraviolet absorption band is shown in Fig. 4.2.

The **main cause of intrinsic absorption** in the infrared region is the characteristic vibration frequency of atomic bonds. In silica glass, absorption is caused by the vibration of silicon-oxygen (Si-O) bonds.

Extrinsic Absorption : Extrinsic absorption is caused by impurities introduced into the fiber material. Trace metal impurities, such as iron, nickel, and chromium, are introduced into the fiber during fabrication. **Extrinsic absorption** is caused by the electronic transition of these metal ions from one energy level to another.

Extrinsic absorption also occurs when hydroxyl ions (OH^-) are introduced into the fiber. Water in silica glass forms a silicon-hydroxyl (Si-OH) bond. This bond has a fundamental absorption at 2700 nm. However, the harmonics or overtones of the fundamental absorption occur in the region of operation. These harmonics increase extrinsic absorption at 1383 nm, 1250 nm, and 950 nm. Figure 4.2 shows the presence of the three OH^- harmonics. The level of the OH harmonic absorption is also indicated.

These absorption peaks define three regions or windows of preferred operation. The first window is centered at 850 nm. The second window is centered at 1300 nm. The third window is centered at 1550 nm. Fiber optic systems operate at wavelengths defined by one of these windows.

The amount of water (OH^-) impurities present in a fiber should be less than a few parts per billion. Fiber attenuation caused by extrinsic absorption is affected by the level of impurities (OH^-) present in the fiber. If the amount of impurities in a fiber is reduced, then fiber attenuation is reduced.

4.4. Fiber Bend Losses

Bending the fiber also causes attenuation. Bending loss is classified according to the bend radius of curvature :

- Microbend loss.
- Macrobend loss.

Microbends are small microscopic bends of the fiber axis that occur mainly when a fiber is cabled. Macrobends are bends having a large radius of curvature relative to the fiber diameter. Microbend and macrobend losses are very important loss mechanisms. Fiber loss caused by microbending can still occur even if the fiber is cabled correctly. During installation, if fibers are bent too sharply, macrobend losses will occur.

Microbend Loss

Microbend losses are caused by small discontinuities or imperfections in the fiber. Uneven coating applications and improper cabling procedure increases microbend loss. External forces are also a source of microbends. An external force deforms the cabled jacket surrounding the fiber but causes only a small bend in the fiber. Microbends change the path that propagating modes take, as shown in Fig. 4.3. Microbend loss increases attenuation because low-order modes become coupled with high-order modes that are naturally lossy.

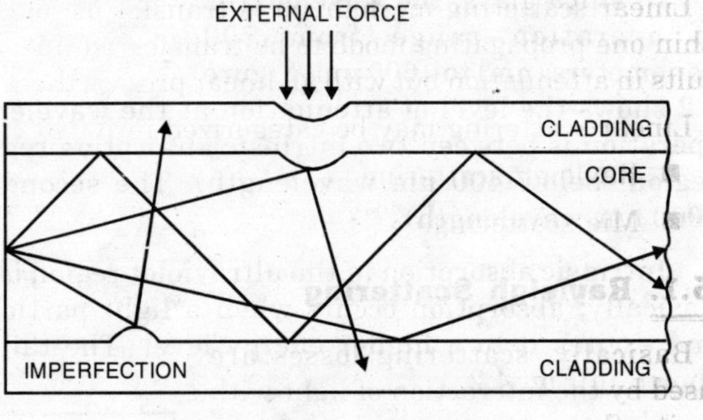

Fig. 4.3. Microbend Loss.

Macrobend Loss

Macrobend losses are observed when a fiber bend's radius of curvature is large compared to the fiber diameter. These bends become a great source of loss when the radius of curvature is less than several centimeters.

The loss can be represented by a radiation attenuation coefficient, which has the form

$$\boxed{\alpha_r = C_1 \exp(-C_2 R)} \quad \text{...(4.3)}$$

where C_1, C_2 = Constants which are independent of R.

R = Radius of curvature of the fiber bend.

Large bending losses occur in multimode fibers at a critical radius of curvature R_C, given by the equation

$$R_C \cong \frac{3n_1^2 \lambda}{4\pi\left(n_1^2 - n_2^2\right)^{1/2}} \quad \text{...(4.4)}$$

From the Eq. (4.4) it is clear that microbending losses may be reduced by

- Operating at the shortest wavelength.
- Designing fiber with large relative refractive index differences.

For a **single quasi guided mode**, bending loss is given by

$$R_{CS} \cong \frac{20\,\lambda}{(n_1 - n_2)^{1/2}}\left[2.748 - 0.996\,\frac{\lambda}{\lambda_C}\right]^{-3} \qquad ...(4.5)$$

where, λ_C = Cutoff wavelength for the single mode fiber.

4.5. Linear Scattering Losses

Linear scattering mechanism is transfer of some or all of the optical power contained within one propagating mode to be transferred linearly into a different mode. This process results in attenuation but with all linear process there is no change in frequency or scattering.

Linear scattering may be categorized into two main types :

- Rayleigh scattering.
- Mie scattering.

4.5.1. Rayleigh Scattering

Basically, scattering losses are caused by the interaction of light with density fluctuations within a fiber. Density changes are produced when optical fibers are manufactured. During manufacturing, regions of higher and lower molecular density areas, relative to the average density of the fiber, are created. Light traveling through the fiber interacts with the density areas as shown in Fig. 4.4. Light is then partially scattered in all directions.

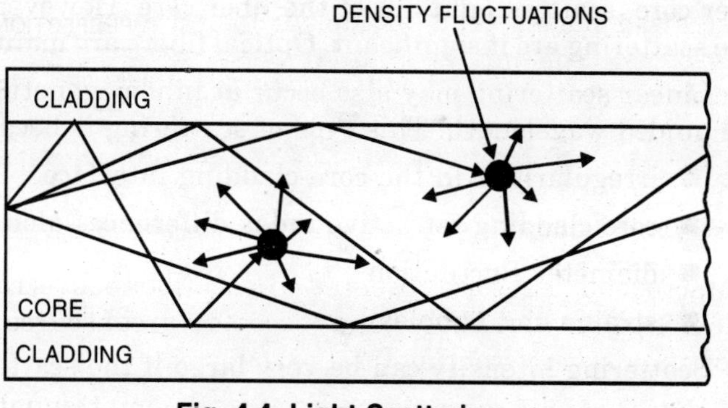

Fig. 4.4. Light Scattering.

In commercial fibers operating between 700 nm and 1600 nm wavelength, the main source of loss is called **Rayleigh scattering.** Rayleigh scattering is the main loss mechanism between the ultraviolet and infrared regions as shown in Fig. 4.4. Rayleigh scattering occurs when the size of the density fluctuation (fiber defect) is less than one-tenth of the operating wavelength of light. Loss caused by Rayleigh scattering is proportional to the fourth power of the wavelength ($1/\lambda^4$). As the wavelength increases, the loss caused by Rayleigh scattering decreases.

For a single component glass, it is given by

$$\gamma_R = \frac{8\pi^3}{3\pi^4}\, n^8 p^2 \beta_c\, KT_F \qquad ...(4.6)$$

where, γ_R = Rayleigh scattering coefficient
λ = Optical wavelength
n = Refractive index of medium
p = Average photoelastic coefficient
β_c = Isothermal compressibility
T_F = Fictive temperature
K = Boltzmann's constant

The **fictive temperature** T_F is defined as the temperature at which the glass can reach a state of thermal equilibrium and is closely related to the anneal temperature.

The Rayleigh scattering coefficient is related to the transmission loss factor (transmissivity) of the fiber £ following the relation :

$$£ = \exp(-\gamma_R L) \quad ...(4.7)$$

where, £ = Length of the fiber

It is clear from the equation (3.7) that the fundamental component of Rayleigh scattering is strongly reduced by operating at the longest possible wavelength.

4.5.2. Mie Scattering

If the size of the defect is greater than one-tenth of the wavelength of light, the scattering mechanism is called **Mie scattering.** Mie scattering, caused by these large defects in the fiber core, scatters light out of the fiber core. However, in commercial fibers, the effects of Mie scattering are insignificant. Optical fibers are manufactured with very few large defects.

Linear scattering may also occur at inhomogeneties and they are comparable in size to the guided wavelength. This type of scattering is because of fiber imperfections such as :

- irregularities in the core-cladding interface
- core cladding refractive index differences along the fiber length
- diameter fluctuation
- strains and bubbles.

Scattering intensity can be very large if the scattering inhomogeneities size is greater than $\frac{\lambda}{10}$.

Such inhomogeneities creates scattering in forward direction and is known as mie scattering, Mie scattering may cause significant losses. The inhomogeneities may be reduced by

- Removing imperfections because of glass manufacturing process.
- Coating of the fiber.
- Carefully controlled extrusion.
- Increasing the fiber guidance by increasing the relative refractive index difference.

4.6. Waveguide Imperfections

An ideal single mode fiber with a perfect cylindrical geometry guides the optical mode without energy leakage into the cladding layer. Imperfections at the core cladding interface can lead to additional losses which contribute to the net fiber loss. The physical process behind such loss is **Mie Scattering** and it occurs because of index inhomogeneities. All these variations can be kept below 1% and the resulting scattering loss is typically below 0.03 dB/km.

A major source of fiber loss in the form of cable is related to **random axial distortion** and it invariably occur during cabling when the fiber is pressed against a surface that is not perfectly smooth. This type of loss is known as **microbending losses.** Microbends cause an increase in the fiber loss for both multimode and single mode fibers and can result in an excessively large loss (~100 dB/km) if precautions are not taken to minimize them. For single mode fibers, microbending losses can be minimized by selecting V as close to the cut off value of 2.405.

Fiber bends constitute another source of scattering loss. In the mode, a part of mode energy is scattered into the cladding layer. The bending loss is given by

$$\text{Bending loss} \propto \exp\left(-\frac{R}{R_C}\right)$$

where, R = Radius of curvature of the fiber bend

$$R_C = \frac{a}{(n_1^2 - n_2^2)}$$

For single mode fibers, $R_C = 0.2 - 0.4$ μm typically and the bending loss is negligible (< 0.01 dB/km) for bend radius $R > 5$ mm. Since most macroscopic bends exceed $R = 5$ mm, **macrobending losses** are negligible practically.

SOLVED EXAMPLES

Example 4.1. After transversing a distance of 400 m in a fiber, an optical signal has lost 75% of its input power what is the loss in dB/km of the fiber.

Solution. Given, that, $L = 400 \text{ m} = 0.4 \text{ km}$

$$\frac{P_i}{P_o} = \frac{1}{0.25}$$

The loss in dB/km of the fiber is given by

$$\text{Loss} = \frac{10}{L}\log\left(\frac{P_i}{P_o}\right)$$

$$= \frac{10}{0.4}\log\left(\frac{P_i}{0.25\,P_i}\right)$$

$$= 25\log\left(\frac{1}{0.25}\right)$$

$$\boxed{\text{Loss} = 15 \text{ dB/km}} \quad \textbf{Ans.}$$

Example 4.2. An optical signal with an input power level of 1 mW to be launched into a fiber of attenuation loss 0.5 dB/km at λ = 0.82 μm. Calculate the power level in mW after it transverses a distance of 3 km.

Solution. Given that,

$$\lambda = 0.82 \text{ μm}$$
$$\alpha = 0.5 \text{ dB/km}$$
$$L = 3 \text{ km}$$
$$P_i = 1 \text{ mW}$$

Attenuation loss of optical fiber is given by

$$\alpha = \frac{10}{L}\log\left(\frac{P_i}{P_o}\right)$$

i.e.

$$P_o = P_i\left[10^{-\left(\frac{\alpha L}{10}\right)}\right]$$

$$= (1)\,P_i\left[10^{-\left(\frac{0.5\times 3}{10}\right)}\right] = \left[10^{-0.15}\right]$$

$$\boxed{P_o = 0.75 \text{ mW}} \quad \textbf{Ans.}$$

Example 4.3. The mean power launched in a fiber link is 1.5 mW and the fiber has an attenuation of 1.5 dB/km. Determine the maximum possible link length when the minimum mean optical power level at the detector is 2 μW.

Solution. Given that,

$$P_i = 1.5 \text{ mW}$$
$$P_o = 2 \text{ μW}$$
$$\alpha = 0.5 \text{ dB/km}$$

The signal attenuation is given by

$$\alpha = \frac{10}{L}\log_{10}\left(\frac{P_i}{P_o}\right)$$

$$\Rightarrow \quad L = \frac{10}{\alpha}\,.\log_{10}\left(\frac{P_i}{P_o}\right)$$

$$= \frac{10}{0.5}\log_{10}\left(\frac{1.5\times 10^{-3}}{2\times 10^{-6}}\right)$$

$$= \frac{10}{0.5}\log_{10}\left(\frac{1.5\times 10^{3}}{2}\right)$$

$$= 20\log_{10}(750)$$

$$L = 57.5\times 10^{3} \text{ m}$$

$$\boxed{L = 57.5 \text{ km}} \quad \textbf{Ans.}$$

Example 4.4. A 30 km long optical fiber has an attenuation of 0.8 dB/km at 1300 nm. If 200 μW of optical power is launched into the fiber, determine the optical output power P_{out}.

Solution. Given that,

$$\alpha = 0.8 \text{ dB/km}$$
$$L = 30 \text{ km}$$
$$\lambda = 1300 \text{ nm}$$
$$P_i = 200 \ \mu\text{W}$$

Attenuation as a unit of length is given by

$$\alpha = \left(\frac{10}{L}\right)\log_{10}\left(\frac{P_i}{P_o}\right)$$

$$0.8 = \left(\frac{10}{30}\right)\log_{10}\left(\frac{200\ \mu\text{W}}{P_o}\right)$$

$$2.4 = \log_{10}\left(\frac{200}{P_o}\right)$$

$$\boxed{P_o = 0.79\ \mu\text{W}} \quad \textbf{Ans.}$$

Example 4.5. The mean optical power at the fiber output is 3 μW when the mean optical power launched into an 8 km length of fiber is 120 μW. Determine :

(*a*) The overall signal attenuation or loss in decibels through the fiber.

(*b*) Signal attenuation per kilometer for the fiber.

(*c*) The overall signal attenuation for 10 km optical link using the same fiber with splices at 1 km intervals, each giving an attenuation of 1 dB.

(*d*) The numerical input/output power ratio in (C).

Solution. Given that,

$$L = 8 \text{ km},$$
$$P_i = 120\times10^{-6}\ \text{W}$$
$$P_o = 3\times10^{-6}\ \text{W}$$

(*a*) Signal attenuation in dB (α)

$$= 10\log_{10}\left(\frac{P_i}{P_o}\right)$$

$$= 10\log_{10}\left(\frac{120\times10^{-6}}{3\times10^{-6}}\right) = 10\log_{10}(40)$$

$$\boxed{\alpha = 16.0 \text{ dB}} \quad \textbf{Ans.}$$

(*b*) The signal attenuation per kilometer, *i.e.*,

$$\alpha_{dB}\,.\,L = 16 \text{ dB}$$

$$\alpha_{dB} = \frac{16}{L} = \frac{16.0}{8.0}$$

$$\boxed{\alpha_{dB} = 2.0 \text{ dB/km}} \quad \textbf{Ans.}$$

(*c*) The loss occurred along 10 km of the fiber is,

$$\alpha_{dB} . L = 2 \times 10 = 20 \text{ dB}$$

Given that the link also has nine splices (at 1 km intervals) each with an attenuation of 1 dB. Therefore the loss due to the splices is 9 dB.

Hence overall signal attenuation for the link is

$$\text{Signal attenuation} = 20 + 9$$

$$\boxed{\text{Signal attenuation} = 29 \text{ dB}} \quad \textbf{Ans.}$$

(*d*) The numerical value of input/output power ratio :

$$\frac{P_i}{P_o} = 10^{(\text{dB}/10)}$$

$$= 10^{(29/10)} = \left(10^{2.9}\right)$$

$$\boxed{\frac{P_i}{P_o} = 794.3} \quad \textbf{Ans.}$$

Example 4.6. The refractive index and the photoelastic coefficient for silica are 1.46 and 0.286 respectively if the silica has an estimated fictive temperature of 1400 K with an isothermal compressibility of 7 × 10^{-11} m^2 N^{-1}. Determine the theoretical attenuation in decibels per kilometers due to the fundamental Rayleigh scatterling in silica at optical wavelength of 0.63 μm and 1.30 μm. Boltzmann's constant is 1.38 × 10^{-23} JK^{-1}.

Solution. Given that,

$$T_F = 1400 \text{ K}$$

$$p = 0.206$$

$$n = 1.46$$

$$\beta_c = 7 \times 10^{-11}$$

The ray leight scattering coefficient is given by

$$\gamma_R = \frac{8\pi^3 n^8 p^2 \beta_o K\, T_F}{3\lambda^4}$$

$$= \frac{8(3.14)^3 \times (1.46)^8 \times (0.286)^2 \times 7 \times 10^{-11} \times 1.381 \times 10^{-23} \times 1400}{3 \times \lambda^4}$$

$$= 0.936$$

$$\text{Attenuation } (\alpha) = 10 \log_{10} \left(\frac{1}{£_{\text{km}}}\right)$$

$$= 10 \log_{10} \left(\frac{1}{0.936}\right)$$

$$= 10 \log_{10} (1.069)$$

$$\alpha = 0.3 \text{ dB/km} \quad \textbf{Ans.}$$

Example 4.7. A glass core optical fiber has an estimate fictive temperature of 758 K, isothermal compressibility of 8.4 × 10^{-11} m^2 N^{-1} and a photoelastic coefficient of 0.245. Determine the refractive index of the glass if the fiber has an attenuation resulting from Rayleigh scattering of 0.46 dB km^{-1} at a wavelength of 1 μm.

Solution. Given that,

$$\lambda = 1 \times 10^{-6} \text{ m}$$

$$p = 0.245$$

$$\beta_c = 8.4 \times 10^{-11}$$

$$T_F = 758$$

Rayleigh scattering coefficient is given by

$$\gamma_R = \frac{8\pi^3 n^8 p^2 \beta_c K T_F}{3\lambda^4}$$

$$= \frac{8(3.14)^3 n^8 (0.245)^2 \times (8.4 \times 10^{-11}) \times 1.381 \times 10^{-23} \times 758}{3 \times (1 \times 10^{-6})^4}$$

$$\gamma_R = \frac{1.895 \times 10^{-28}}{\lambda^4} \text{ m}^{-1}$$

At a wavelength of 0.63 μm :

$$\gamma_R = \frac{1.895 \times 10^{-28}}{(0.63 \times 10^{-6})^4} = \frac{1.895 \times 10^{-28}}{0.158 \times 10^{-24}}$$

$$= 1.199 \times 10^{-3} \text{ m}^{-1}$$

The transmission loss factor of fiber for one kilometer length is given by

$$£ = \exp(-\gamma_R . L)$$

$$= \exp\left(-1.199 \times 10^{-3} \times 10^3\right)$$

$$= 0.301.$$

Therefore the attenuation in dB per kilometer is

$$\alpha = 10 \log_{10} \left(\frac{1}{£_{km}}\right)$$

$$= 10 \log_{10} \left(\frac{1}{0.301}\right) = 10 \log_{10} (3.222)$$

$$\alpha = 5.2 \text{ dB/km} \quad \textbf{Ans.}$$

At a wavelength of 1.30 μm :

$$\gamma_R = \frac{1.895 \times 10^{-28}}{(1.3 \times 10^{-6})^4} = \frac{1.895 \times 10^{-28}}{2.856 \times 1\sigma^{24}}$$

$$= 0.664 \times 10^{-4}$$

Transmission loss factor is :

$$\pounds = \exp\left(-\gamma_R . L\right)$$

$$= \exp\left(0.664 \times 10^{-4} \times 10^{3}\right)$$

$$= 4.35 \times 10^{-6}\ (n^8)$$

$$\text{Attenuation } (\alpha) = 10 \log_{10}\left(\frac{1}{L_{\text{km}}}\right)$$

$$\Rightarrow \quad 0.46 = 10 \log_{10}\ (e^{\alpha} . L)$$

$$0.46 = 4.34\ (\alpha\ .\ L)$$

$$0.46 = 4.34 \times 4.35 \times 10^{-6} \times (n^8) \times 10^3$$

$$n^8 = \frac{0.46 \times 10^3}{4.34 \times 4.35}$$

$$n^8 = 24.365$$

$$n = (24.365)^{1/8}$$

$$n = 1.49 \qquad \textbf{Ans.}$$

SUMMARY

1. The most important properties of optical fiber that affects system performance are fiber attenuation and dispersion.
2. Attenuation controls the distance on optical signal can travel. It is because of light absorption, scattering and bending losses.
3. Signal attenuation as a unit of length is given by

$$\text{Attenuation } (\alpha) = \left(\frac{10}{L}\right) \log_{10}\left(\frac{P_i}{P_0}\right)$$

4. Absorption is defined as the portion of attenuation resulting from the conversion of optical power into another energy form such as heat.
5. The absorption of light may be intrinsic or extrinsic.
6. Intrinsic absorption is caused by basic fiber material properties in the near infrared region.
7. Extrinsic absorption is caused by impurities introduced into the fiber material. It also occurs when hydroxyl ions (OH^-) are introduced into the fiber.
8. Silica fibers are predominately used in fiber optic communications. They have low material absorption at the wavelengths of operation.
9. The wavelength of operation in fiber optics in between 700 nm and 1600 nm. The wavelength of operation is between the ultraviolet and infrared intrinsic absorption regions.
10. Scattering losses are caused by the interaction of light with density fluctuations within a fiber.

11. Rayleigh scattering is the main source of commercial fibers operating between 700 nm and 1600 nm.

12. Microbends are small microscopic bends of the fiber axis that occur mainly when a fiber is cabled. Macrobends are bends having a large radius of curvature relative to the fiber diameter.

13. Bending loss can be represented by a radiation attenuation coefficient and is given by

$$\alpha_r = C_1 \exp(-C_2 R)$$

14. Linear scattering loss may be categorized into

- Rayleigh scattering.
- Mie scattering.

15. The core-cladding interface, core cladding refractive index creates scattering in forward direction and is known as Mie scattering.

16. In single quasi guided mode the critical radius of curvature R_{CS} is given by

$$R_{CS} \simeq \frac{20\lambda}{(n_1 - n_2)^{1/2}}\left[2.748 - 0.996\frac{\lambda}{\lambda_C}\right]^{-3}$$

17. The main causes of scattering losses in glass are due to

- Microscopic variations in the material density.
- Composite fluctuations.
- Structural inhomogeneities.

18. The scattering loss at a wavelength, λ due to fluctuations is given by

$$\alpha_{\text{scat}} = \frac{8\pi^3}{3\lambda^4}(n^2 - 1)^2 K_B T_f B_T$$

19. Modal or speckle noise is observed in multimode fiber as fluctuations which have characteristics times longer than the resolution time of the detector.

REVIEW QUESTIONS

1. What do you meant by attenuation ? Discuss the different mechanism responsible for attenuation in optical fibers.
2. Briefly explain the various sources of extrinsic losses in optical fiber.
3. Explain absorption losses in optical fibers, comparing the intrinsic and extrinsic absorption mechanism.
4. Discuss the linear scattering losses in optical fibers with respect to
   (*a*) Rayleigh scattering
   (*b*) Mie scattering.
5. What do you understand by bending loss ? Discuss the parameters by which the microbending losses may be reduced.
6. List the major losses in optical fiber communication. How these losses can be minimized ?
7. Describe the phenomenon of modal noise in optical fibers and suggest how it may be avoided.

8. An optical signal at a specific wavelength has lost 55% of its power after travelling 3.5 km of fiber. What is the attenuation in dB/km of this fiber ?
9. An LED operating at 850 nm has a spectral width of 45 nm. What is the pulse spreading when a laser diode have a 2 nm spectral width is used ?
10. The mean power launched in a fiber link is 1.5 mW and the fiber has an attenuation of 0.5 dB/km. Determine the maximum possible link length when the minimum mean power optical power at the detector is 2 μW.
11. Silica has an estimated temperature of 1400 K with an isothermal compressibility of 7×10^{-11} m^2N^{-1}. The refractive index and the photoelastic coefficient for silica are 1.46 and 0.29 respectively. Determine the theoretical attenuation in dB/km due to fundamental Raylight scattering in silica at optical wavelengths of 0.7 μm.
12. An 11 km optical fiber link consisting of optimum near parabolic profile graded index fiber exhibits rms intermodal pulse broadening of 346 ps over its length. If the fiber has a relative refractive index difference of 1.5%, estimate the core axis refractive index. Hence determine the numerical aperture for the fiber. **[PTU 2008]**
13. Explain the difference between absorption and scattering processes as they determine the attentuation of optical power propagating in fibers. Explain the mechanism involved.
14. Explain intrinsic and extrinsic absorption in optical fiber material.
15. Explain the losses caused by linear and nonlinear scattering in optical fiber.
16. An optical signal at a specific wavelength has lost 55% of its power after travelling 3.5 km of fiber. What is the attenuation in dB/km of this fiber ?
17. Silica has an estimated temperature of 1400 K with an isothermal compressibility of 7×10^{-11} m^2N^{-1}. The refractive index and the photoelastic coefficient for silica are 1.46 and 0.206 respectively. Determine the theoretical attenuation in dB/km due to fundamental Raylight scattering in silica at optical wavelengths of 0.63 μm. **[PTU 2007]**
18. With the help of suitable diagrams explain in detail the basic attenuation mechanisms. On what factor these mechanisms depend. **[PTU 2007]**

❑❑❑

Nonlinear Optical Effects

Chapter 5

INSIDE THIS CHAPTER

5.1. INTRODUCTION

The design of a lightwave transmission system requires careful planning of factors such as :

- Fiber selection
- Choice and tuning of optoelectronic components
- Optical amplifier placement
- Path routing

The aim behind this planning is to create a network that meets the design criteria, is reliable, and is easy to operate and maintain. This chapter first gives a general overview of nonlinear processes in optical fibers. Above a certain optical power threshold the nonlinearities arise. The effect becomes negligible once the signal has become sufficiently attenuated after travelling a certain distance along the fiber. The nonlinear processes physically affect system performance. **These nonlinearities are :**

- Stimulated Raman scattering
- Stimulated Brillouin scattering
- Self phase modulation
- Cross phase modulation
- Four wave mixing

Mitigation of four wave mixing can be achieved by means of a special fiber design or by a chromatic dispersion compensation method.

5.2. General Overview of Nonlinearities

Optical nonlinearities can be classified into two general categories :

- Index related
- Scattering related.

Table 5.1 shows the summary of nonlinear effects in optical fibers.

TABLE 5.1.

S. No.	Nonlinearity Category	Single Channel	Multiple Channel
1.	Index related	Self phase modulation	• Cross phase modulation • Four wave mixing
2.	Scattering related	Stimulated brillouin scattering	Stimulated Raman scattering

The **first category** encompasses the nonlinear inelastic scattering processes. The **second category** of nonlinear effects arises from intensity dependent variations in the refractive index in a silica fiber, which is known as the **Kerr effect.**

5.3. Nonlinear Scattering

Optical waveguides do not always behave as linear channels where output optical power is directly proportional to the input optical power. Several nonlinear effects occur which causes scattering.

Nonlinear scattering is the transfer of optical power from one mode to be transferred in either the forward or backward direction to the same, or other modes at different frequency. There are two types of nonlinear scattering.

- Stimulated brillouin scattering
- Stimulated raman scattering.

5.3.1. Stimulated Brillouin Scattering (SBS)

The Stimulated Brillouin Scattering is nonlinear and in practical systems requires a power level of something above 3 mW for any serious effect to be observable. It also requires a long interaction length and a very narrow line width signal. In general the signal line width must be less than about 100 MHz (around 0.1 nm) for SBS. The effect in the forward direction is experienced as an increase in attenuation. This is rapid and adds noise to the signal. For narrow line width signals SBS imposes an upper limit on the usable transmit power.

In most systems SBS is not a major problem for the following reasons :

1. Direct modulation of the transmit laser's injection current produces a chirp and broadens the signal. This significantly reduces the impact of SBS.
2. The effect is less in 1300 nm systems than in 1550 nm systems due to the higher attenuation of the fiber.

3. Lasers capable of producing the necessary power level have become available and amplifiers are also a recent innovation.
4. At speeds of below 2.4 GHz it not necessary to use either very high power or very narrow line width lasers.
5. SBS effects decrease with increase in speed because of the signal broadening affect of the modulation.

In several cases where SBS could be a problem, the line width is often intentionally broadened. This can be done by using an additional RF modulation on the laser injection current, by using an external phase modulator or by using a self pulsating laser. The increasing line width may creates a problem against long distance transmission because it increases the effect of chromatic dispersion. However **SBS can be a major problem** in three situations :

1. In long distance systems where the span between amplifiers is long and the bit rate is low (below about 2.5 Gbps).
2. In WDM systems (up to about 10 Gbps) where the spectral width of the signal is very narrow.
3. In remote pumping of an erbium doped fiber amplifier (EDFA) through a separate fiber.

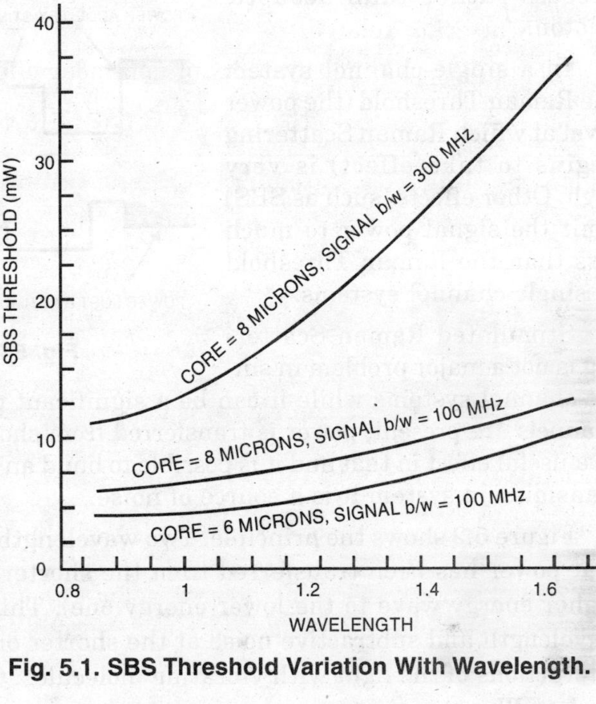

Fig. 5.1. SBS Threshold Variation With Wavelength.

Stimulated brillouin scattering (SBS) is the modulation of light through thermal molecular vibrations within the fiber. The scattered light appears as upper and lower sidebands which are separated from the incident light by the modulation frequency.

Brillouin scattering is significant above a threshold power density. Threshold power P_B is given by

$$P_B = 4.4 \times 10^{-3}\ d^2 \lambda^2\ \alpha_{dB} v \text{ Watts} \qquad ...(5.1)$$

where, d = Fiber core diameter in μm

λ = Operating wavelength in μm

α_{dB} = Fiber attenuation in dB

v = Source bandwidth in GHz.

P_B is the threshold optical power launched into a single mode optical fiber before SBs occurs.

5.3.2. Stimulated Raman Scattering

Stimulated raman scattering (SRS) is similar to stimulated brillouin scattering (SBS) except that a high frequency optical photon is generated in scattering process rather than acoustic photon.

In a single-channel system the Raman Threshold (the power level at which Raman Scattering begins to take effect) is very high. Other effects (such as SBS) limit the signal power to much less than the Raman Threshold in single-channel systems.

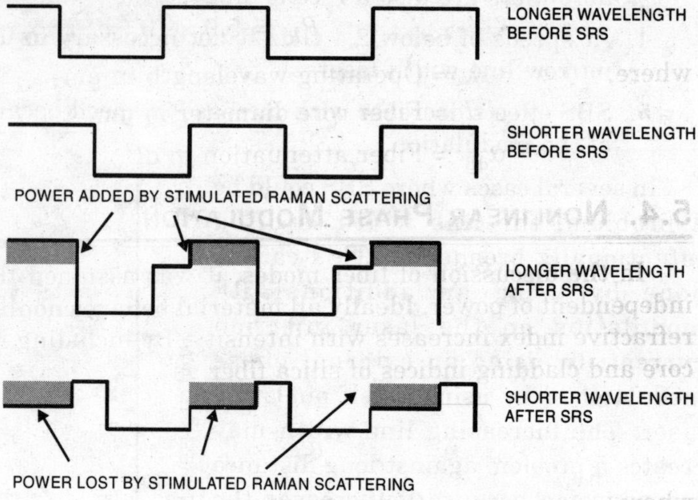

Fig. 5.2. Stimulated Raman Scattering.

Stimulated Raman Scattering is not a major problem in single-channel systems while it can be a significant problem in WDM systems. When multiple channels are present, power is transferred from shorter wavelengths to longer ones. This can be a useful effect in that and it is possible to build an optical amplifier based on SRS. But in the transmission system it is a source of noise.

Figure 5.2 shows the principle. Two wavelengths are shown before and after SRS. Notice that power has been transferred from the shorter wavelength to the longer one (from the higher energy wave to the lower energy one). This results in additive noise at the longer wavelength and subtractive noise at the shorter one. This power transfer is because of the interactions of the light with vibrating molecules. Optical power so transferred is known as **Stokes Wave.**

Important characteristics of SRS are :

1. The effect of SRS becomes higher as the signals are moved further and further apart. This is a problem because we would like to separate the signals as much as we can avoid four-wave mixing effects and when we do, we get SRS. SRS can take affect over about 40 THz below the higher frequency (shorter wavelength) involved *i.e.* it can extend over a range of wavelengths of about 300 nm longer than the shortest wavelength involved.

   The effect is maximized when the two frequencies are 13.2 THz apart.

2. SRS increases exponentially with increased power. At very high power it is possible that all of the signal power can be transferred to the Stokes Wave.

   One study concluded that to avoid SRS in a 10-channel WDM system with 1 nm channel spacing power levels need to be kept below 3 mw (per channel).

A SRS can occur in both forward and backward direction in optical fiber, and optical power threshold is of three orders of magnitude higher than brillouin threshold.

The threshold optical power (P_R) for SRS in a long single mode fiber is given by

$$P_R = 5.9 \times 10^{-2} d^2 \lambda \alpha_{dB} \quad ...(5.2)$$

where, λ = Operating wavelength in μm
d = Fiber core diameter in μm
α_{dB} = Fiber attenuation in dB.

5.4. NONLINEAR PHASE MODULATION

In the discussion of fiber modes, it was assumed that the refractive index of silica is independent of power. Ideally all material behaves nonlinearly at high intensities and their refractive index increases with intensity. By including nonlinear refraction, we modify the core and cladding indices of silica fiber as

$$n_j^1 = n_j + \bar{n}_2 \left(\frac{P}{A_{eff.}} \right) \quad ...(5.3)$$

where, n_j = Ordinary refractive index of the material
$\bar{n}_2$ = Nonlinear index coefficient
P = Optical power
$A_{eff.}$ = Effective mode area

The numerical value of $\bar{n}_2$ is about 2.6×10^{-20} m^2/W for silica fibers.

The nonlinearity in refractive index is known as the **Kerr nonlinearity.** This nonlinearity produces a carrier induced phase modulation of the propagating signal, which is called the **Kerr effect.**

5.4.1. Self Phase Modulation

By using first-order perturbation theory if we want to see that how fiber modes are affected by the nonlinear term, we find that mode shape does not change but the propagation constant becomes power dependent. It can be written as

$$\beta' = \beta + K_0 \bar{n}_2 \frac{P}{A_{eff.}}$$

$$= \beta + \gamma P \quad ...(5.4)$$

where, $$\boxed{\gamma = \frac{2\pi \bar{n}_2}{(A_{eff.} \lambda)}} \quad ...(5.5)$$

γ indicates the magnitude of nonlinear effect for SPM and it is an important nonlinear parameter with values ranging from 1 to 54 W^{-1}/km depending on the values of A_{eff} and the wavelength.

The term γ produces a nonlinear phase shift given by

$$\phi_{NL} = \int_0^L (\beta' - \beta)\, dz$$

$$= \int_0^L \gamma P(z)\, dz$$

$$\phi_{NL} = \gamma P_{\text{in}} L_{eff} \qquad ...(5.6)$$

where, $P(z) = P_{in} \exp(-\alpha z)$ accounts for fiber losses.

In Eqn. (5.6) P_{in} was assumed constant practically P_{in} makes ϕ_{NL} to vary with time.

In fact optical phase changes with time in exactly same manner as the optical signal. Since this nonlinear phase modulation is self induced, its associated nonlinear phenomenon is known as **self-phase modulation (SPM).** SPM leads to frequency chirping of optical pulses.

5.4.2. Cross Phase Modulation

The intensity dependence of refractive index in Eqn. (5.6) can also lead to another nonlinear phenomenon known as **cross-phase modulation (XPM).** It occurs when two or more optical channels are transmitted simultaneously inside an optical fiber using the WDM technique.

In such type of systems, the nonlinear phase shift for a specific channel not only depends on the power of that channel but also on the power of other channels.

Thus the **phase shift for the *j*th channel** becomes

$$\phi_j^{NL} = \gamma L_{eff}\left(P_j + 2\sum_{m \neq j} P_m\right) \qquad ...(5.7)$$

where,

- The sum extends over the number of channels.
- The factor of 2 indicates that XPM is twice as effective as SPM for the same amount of power.

The **total phase shift** depends on the powers in all channels and would vary from bit to bit depending on the bit pattern of the neighbouring channels. If we assume equal channel powers, in the worst case the phase shift in which all channels simultaneously carry 1 bits and all pulses overlap is given by

$$\phi_j^{NL} = \frac{\gamma}{\alpha}(2M-1)P_j \qquad ...(5.8)$$

It is difficult to estimate the impact of XPM on the performance of multichannel lightwave systems. It is because, we have assumed that XPM acts in isolation without dispersive effects and is valid only for CW optical beams.

5.5. Four-Wave Mixing

The power dependence of the refractive index as seen in Eqn. (5.8) has its origin in the third order nonlinear susceptibility denoted by χ. This nonlinear phenomenon is known as **four wave mixing (FWM)** and also originates from χ.

If three optical fields with carrier frequencies w_1, w_2 and w_4 present inside the fiber simultaneously, χ generates a fourth field with frequency w_4. w_4 is related to other frequencies by a rotation

$$w_4 = w_1 \pm w_2 \pm w_3 \qquad ...(5.9)$$

Frequency combinations of the form $w_4 = w_1 + w_2 - w_3$ are sometimes troublesome for multichannel communication systems because they can become nearly phase matched when channel wavelengths lie close to the zero dispersion wavelength. In fact the degenerate FWM process for which $w_1 = w_2$ is often the dominant process and impacts the system performance most.

In general for N wavelengths launched into a fiber, the number of generated mixing products M is

$$M = \frac{N^2}{2}(N-1) \qquad ...(5.10)$$

The efficiency of four wave mixing depends on :

- Fiber dispersion
- Channel spacing

Since the dispersion varies with length, the signal wave and the generated waves have different group velocities. This destroy the phase matching of the interactive waves and lowers the efficiency at which power is transferred to newly generated frequencies. The higher the group velocity mismatches and the wider the channel spacing, the lower the four wave mixing.

Nonlinear interaction constant *K* is defined by

$$K = \frac{32\pi^3 \chi_{1111}}{\eta_2 \lambda C}\left(\frac{L_{eff}}{A_{eff}}\right) \qquad ...(5.11)$$

where, χ_{1111} = Third order nonlinear susceptibility

η = Efficiency of fourwave mixing

L_{eff} = Effective length

A_{eff} = Effective cross-sectional area of the fiber

Fundamentally, FWM process can be known as **scattering process.** Here two photons of energies hw_1 and hw_2 are destroyed and their energy appears in the form of two new photons hw_3 and hw_4. Then the phase matching condition comes from the momentum conservation. The phase mismatch because of all four waves propagating in the same direction can be written as

$$\Delta = \beta(w_3) + \beta(w_4) - \beta(w_1) - \beta(w_2) \qquad ...(5.12)$$

where, $\beta(w)$ = Propagation constant for an optical field with frequency w.

In the degenerate core,

$$w_2 = w_1$$

$$w_3 = w_1 + \Omega$$

and $$w_3 = w_1 - \Omega$$

where, Ω = Channel spacing.

- FWM can also be useful in designing lightwave systems.
- Sometimes it is used for demultiplexing individual channels when time division multiplexing is used in the optical domain.
- It can also be used for wavelength conversion.
- Sometimes FWM in optical fibers is used for generating a spectrally inverted signal through the process of **optical phase conjugation.**

SOLVED EXAMPLE

Example 5.1. The threshold optical powers for stimulated Brillouin and Raman scattering in a long 8 μm core diameter single mode fiber are found to be 190 mW and 1.70 W respectively, when using an injection laser source with a bandwidth of 1 GHz. Calculate the operating wavelength of the laser and the attenuation in decibels per kilometer of the fiber at this wavelength.

Solution. Given that,

$$P_B = 190 \text{ mW}$$
$$P_R = 1.7 \text{ W}$$
$$2a = 8 \text{ μm} = d$$
$$V = BW = 1 \text{ GHz}$$

The optical power for stimulated Brillouin is given by

$$P_B = 4.4 \times 10^{-3}\, d^2\lambda^2\alpha_{dB}\, V$$

$$\Rightarrow \quad P_B = 4.4\times10^{-3}\times(8)^2\lambda^2\,.\,\alpha_{dB}\times1\text{ GHz}$$

$$\Rightarrow \quad 190\times10^{-3} = 0.2816\;\lambda^2.\alpha_{dB}$$

$$\Rightarrow \quad \lambda^2.\alpha_{dB} = 0.6747 \qquad ...(i)$$

The optical power for Raman scattering is given by

$$P_R = 5.9\times10^{-2}\, d^2\,\lambda\alpha_{dB}$$

$$\Rightarrow \quad 1.7 = 5.9\times10^{-2}\times(8)^2\,.\,\lambda\,.\,\alpha_{dB}$$

$$\lambda\alpha_B = 0.48 \qquad ...(ii)$$

For Eqns. (*i*) and (*ii*), we get

$$\lambda(0.45) = 0.6747$$

$$\boxed{\lambda = 1.49 \text{ μm}} \quad \textbf{Ans.}$$

Now, from Eqn. (*ii*), we get,

$$(1.49\text{ μm})\,.\,\alpha_{dB} = 0.45$$

$$\alpha_{dB} = \frac{0.45}{1.49\text{ μm}}$$

$$\boxed{\alpha_{dB} = 0.3 \text{ dB km}^{-1}} \quad \textbf{Ans.}$$

SUMMARY

1. There are two types of nonlinear scattering :
   - Stimulated Brillouin scattering.
   - Stimulated Ramon scattering.
2. Stimulated Brillouin scattering (SBS) may be regarded as the modulation of light through thermal molecular vibrations within the fiber.
3. Stimulated raman scattering (SRS) is similar to SBS except that a high frequency optical photon rather than an acoustic photon is generated in scattering process.
4. Optical nonlinearities can be classified as :
   - Index related
   - Scattering related
5. SBS effect decrease with increase in speed because of signal broadening affect of the modulation.
6. In Brillouin scattering, threshold power P_B is given by
$$P_B = 4.4 \times 10^{-3}\ d^2\lambda^2\ \alpha_{dB}\ v \text{ watts}$$
7. Threshold optical power PR for SRS is given by :
$$P_R = 5.9 \times 10^{-2}\ d^2\alpha_{dB}\ \lambda$$
8. Nonlinear phase modulation may be of two types :
   - Self phase modulation
   - Cross phase modulation
9. The efficiency of four wave mixing depends on :
   - Fiber dispersion
   - Channel spacing
10. FWM process is also known as scattering process.
11. Nonlinear interaction constant K is defined by
$$K = \frac{32\pi^3\chi_{1111}}{\eta_2 C\lambda}\left(\frac{L_{eff.}}{A_{eff.}}\right)$$

REVIEW QUESTIONS

1. Compare stimulated Brillouin and stimulated Raman scattering in optical fibers. How they may be avoided in optical fiber communication ?
2. A long single mode optical fiber has an attenuation of 0.5 dB/km when operating at a wavelength of 1.3 µm.. The fiber core diameter is 6 µm and the laser source bandwidth is 600 MHz. Compare the threshold optical powers for stimulated Brillouin and Raman scattering within the fiber at the specified wavelength.
3. What do you meant nonlinear effects ? Explain in detail the four wave mixing effect. **[PTU 2007]**
4. Briefly explain the various categories of optical nonlinearities.
5. What do you meant by stimulated Raman scattering ? What are its important characteristics ?
6. Discuss the Kerr nonlinearity and Kerr effect.
7. What is nonlinear phase modulation ? derive an expression for self phase modulation and cross phase modulation.
8. Briefly explain the various types of nonlinearities.
9. List the situations in which SBS can be a major problem.
10. What do you understand by self phase modulation ? Explain briefly.

❑❑❑

Optical Fibers : Manufacturing, Cables and Connectors

Chapter 6

INSIDE THIS CHAPTER

Manufacturing of fiber cables suitable for installation in an actual lightwave system invovles sophisticated technology with attention in many practical details. In this chapter, we shall discuss its various details.

6.1. Design Issues

Basically a step index fiber consists of a cylindrical core surrounded by a cladding layer and its index is slightly lower than the core. Core and cladding both use silica as the base material but the difference in the refractive indices is realized by doping the core, or the cladding, or both.

Dopants such as GeO_2 and P_2O_5 increase the refractive index of silica and suitable for the core. While dopants such as B_2O_3 and fluorine decrease the refractive index of silica and suitable for the cladding. The major design issues are related to the

- Refractive index profile
- Amount of dopants
- Core and cladding dimensions.

The diameter of the outermost cladding layer has the standard value of 125 μm for all communication grade fiber.

Matched clad and depressed clad are two types of single mode step index fibers.

- Matched cladding means that the fiber cladding is a single homogeneous layer of dielectric material.

- Depressed cladding means that the fiber cladding consists of two regions : an inner and outer cladding region.

6.2. Fabrication Methods

Basically fiber manufacturers use two methods to fabricate multimode and single mode glass fibers.

- Vapor phase oxidation
- Direct-melt process.

6.2.1. Vapor Phase Oxidation

In **vapor phase oxidation**, gaseous metal halide compounds, dopant material and oxygen are oxidized (burned) to form a white silica powder (SiCh). Manufacturers deposit the soot (sich) on the surface of a glass substrate (mandrel) or inside a hollow tube by one of the following **three methods :**

- Outside Vapor Phase Oxidation (OVPO)
- Inside Vapor Phase Oxidation (IVPO)
- Vapor Phase Axial Deposition (VAD)

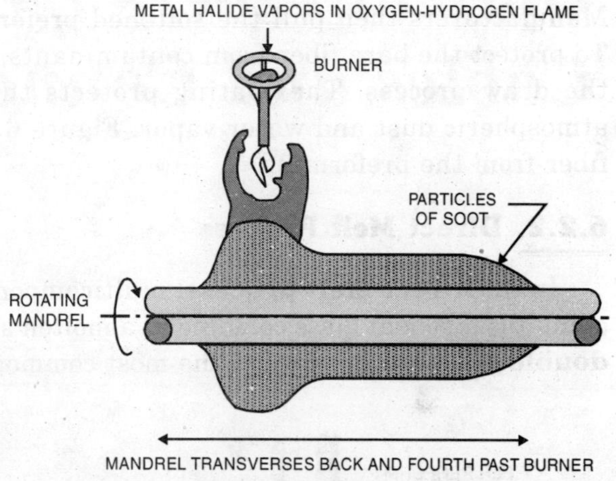

Fig. 6.1. OVPO Preform Preparation.

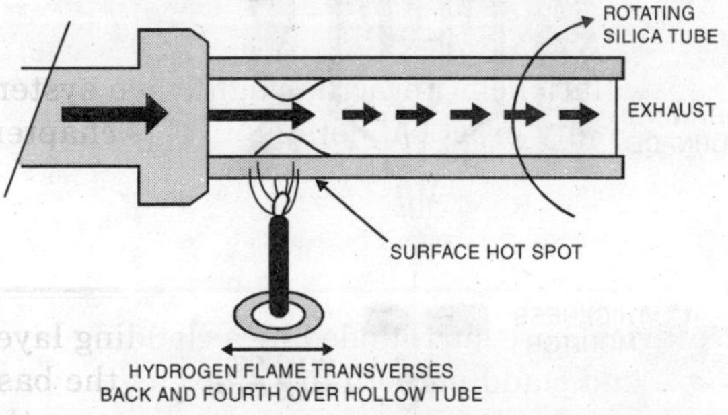

Fig. 6.2. IVPO Preform Preparation.

The soot forms the core and cladding material of the preform. The refractive index of each layer of soot is changed by varying the amount of dopant material being oxidized. Figures 6.1, 6.2 and 6.3 shows the different vapor phase oxidation preform preparation methods.

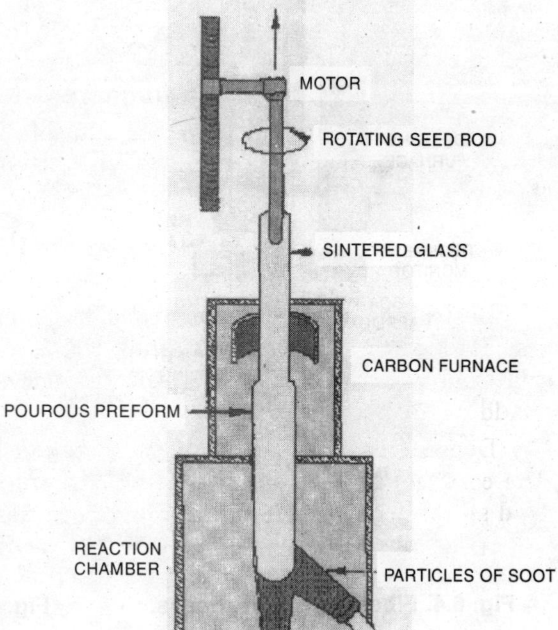

Fig. 6.3. VAD Preform Preparation.

During vapor phase oxidation, the mandrel or tube continuously moves from side to side and rotates while soot particles are deposited on the surface. This process forms cylindrical layers of soot on the surface of the mandrel or inside the hollow tube. This deposited material is transformed into a solid glass preform by heating the porous material (without melting). The solid preform is then drawn or pulled into an optical fiber by a process called fiber drawing.

Fiber Drawing

The fiber drawing process starts by feeding the glass preform into the drawing furnace. The drawing furnace softens the end of the preform to the melting point. Manufacturers then pull the softened preform into a thin glass filament (glass fiber). To protect the bare fiber from contaminants, manufacturers add an acrylate coating in the draw process. The coating protects the bare fiber from contaminants such as atmospheric dust and water vapor. Figure 6.4 shows the process of drawing an optical fiber from the preform.

6.2.2. Direct Melt Process

In the **direct-melt process**, multicomponent glass rods form the fiber structure. Rods of multicomponent glass combine in a molten state to form the fiber core and cladding. The **double-crucible method** is the most common direct-melt process.

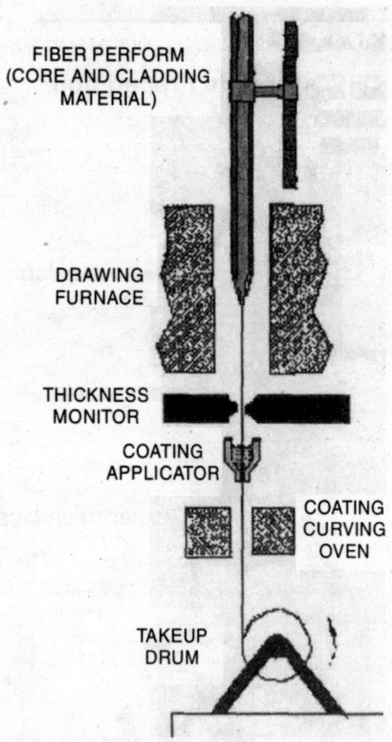

Fig. 6.4. Fiber Drawing Process.

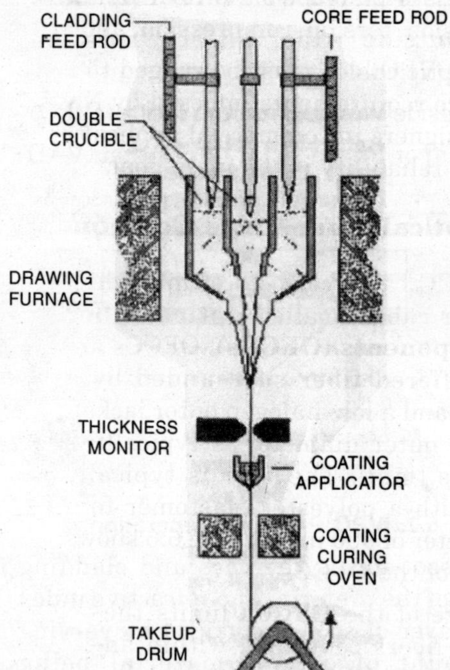

Fig. 6.5. Double-Crucible Fiber Drawing Process.

The double-crucible method combines the molten rods into a single perform using two concentric crucibles. Optical fibers are drawn from this molten glass using a similar fiber drawing process as in vapor phase oxidation. Figure 6.5 shows the double crucible drawing process.

6.3. OPTICAL CABLES

Optical fibers have small cross sectional areas. Without protection, optical fibers are fragile and can be broken. The optical cable structure protects optical fibers from environmental damage. Cable structure includes :

- Buffers
- Strength members
- Jackets

Many factors influence the design of fiber optic cables. The cable design relates to the cable's intended application. Properly designed optical cables perform the following functions :

- Protect optical fibers from damage and breakage during installation and over the fiber's lifetime.
- Provide stable fiber transmission characteristics compared with uncabled fibers. Stable transmission includes stable operation in extreme climate conditions.
- Maintain the physical integrity of the optical fiber by reducing the mechanical stresses placed on the fiber during installation and use. Static fatigue caused by tension, torsion, compression, and bending can reduce the lifetime of an optical fiber.

Fiber optic cables must be rugged to meet the optical, environmental, and mechanical performance requirements imposed by Navy systems. Cost is always a main consideration of cable designers in commercial applications. Cost is also considered in Navy applications, but system reliability is the main goal.

6.3.1. Optical Fiber Cable Component (OFCC) Cable

An OFCC cable consists of individual single fiber cables, called **optical fiber cable components (OFCCs).** OFCCs are a tight-buffered fiber surrounded by a rimid yarn and a low-halogen outer jacket. The OFCC outer diameter is typically 2 millimeters (mm). The fiber is typically buffered with a polyester elastomer to a total diameter of 900 μm. Figure 6.6 shows the design of the OFCCs.

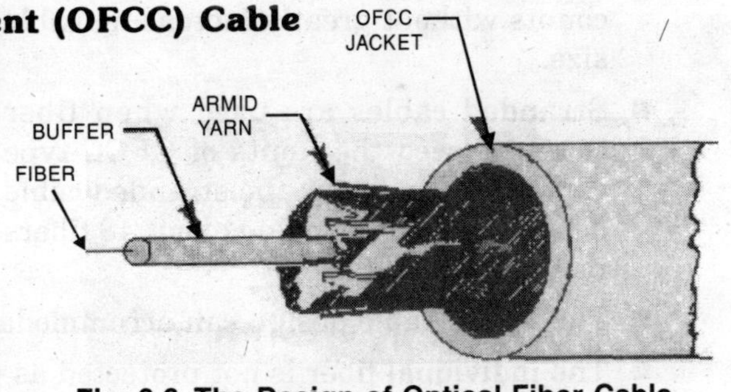

Fig. 6.6. The Design of Optical Fiber Cable Components (OFCCs).

The size of the OFCCs limits the amount of fibers contained within an OFCC cable. An OFCC cable generally contains less than 36 fibers (OFCCs). An OFCC cable of 0.5 inch cable outer diameter can accommodate about 12 fibers.

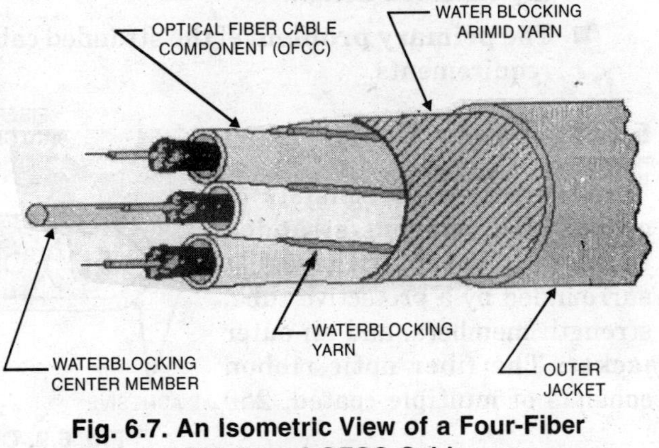

Fig. 6.7. An Isometric View of a Four-Fiber Shipboard OFCC Cable.

Figure 6.7 shows an isometric view of a four-fiber shipboard OFCC cable. In this multifiber cable design, the OFCCs surround a flexible central member in a helical manner. The central member may add to cable strength or only support the OFCCs.

- OFCC cables are easy to handle because each cable contains its own subcable.
- These OFCC subcables make it easy to reconfigure systems and handle individual fibers.
- Rugged OFCC cable design permits cable use in harsh environments, including Navy applications.
- OFCC-type cable is recommended for use in low-density (less than 24 fibers) Navy applications.

6.3.2. Stranded Cable

A stranded cable is a fiber optic cable consisting of buffered fibers stranded down the center of the cable surrounded by strength members and a protective jacket. Figure 6.8 shows a cross sectional view of the stranded cable. The fiber is typically buffered with a polyester elastomer to a total diameter of 900 µm. The **recommended use of** stranded cables is in medium-density (24 to 72 fibers) Navy applications.

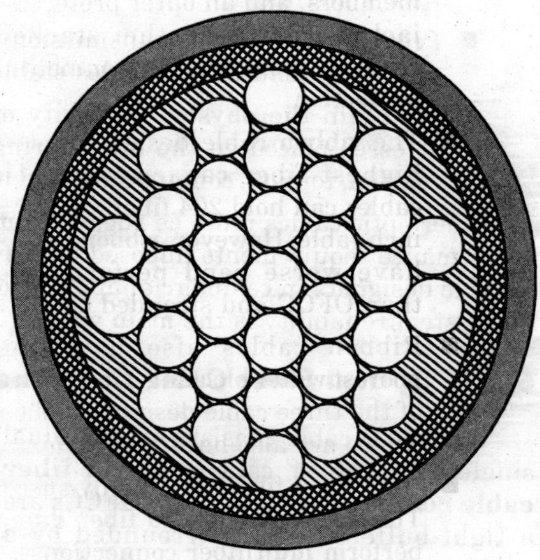

Fig. 6.8. Stranded Cable Design.

- Stranded cable designs increase fiber counts without greatly increasing cable size.
- Stranded cables are used when fiber counts exceed the limits of OFCC-type cables. For example, the stranded cable design can accommodate about 48 fibers in a 0.5-inch cable.
- The OFCC cable design can accommodate around 12 fibers.
- The individual fiber is not protected as well in the stranded design as in the OFCC design. For this reason more care is required in handling the individual fibers in the stranded design.
- The **primary problem** of the stranded cable design is in meeting the water blocking requirements.

6.3.3. Ribbon Cable

A ribbon cable consists of optical fiber ribbons stranded down the center of the cable surrounded by a protective tube, strength members, and an outer jacket. The fiber optic ribbon consists of multiple coated, 250

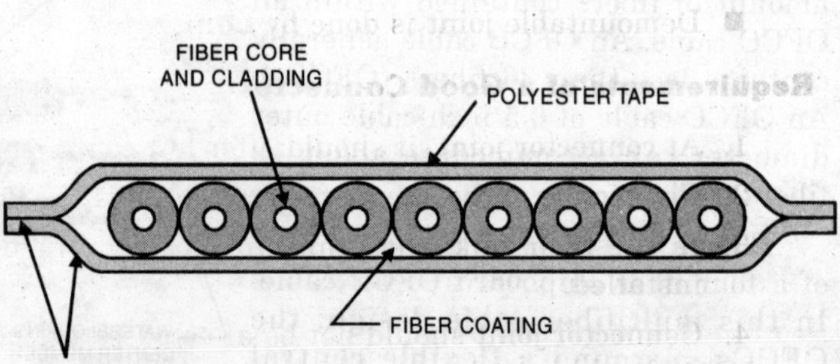

Fig. 6.9. Cross-section of a Fiber Optic Ribbon.

ocm diameter fibers sandwiched in a plastic material. Figure 6.9 shows a cross-sectional view of a 12-fiber ribbon.

Cable manufacturers stack these ribbons to form a rectangular cross-sectional array of fibers. **Stacked ribbons are the basic building blocks of the ribbon cable.**

Figure 6.10 shows this cross-sectional array of ribbons.

- Manufacturers introduce a controlled twist to the stacked ribbons to minimize fiber stress when the cable is bent.
- An inner plastic tube, strength members, and an outer protective jacket surround the stacked ribbons, providing environmental protection.
- The ribbon cable design has the highest fiber capacity. Ribbon cables can hold 204 fibers in a 0.5 inch cable. However, ribbon cables have worse bend performance than OFCC and stranded cables.
- Ribbon cables also have the poorest water blocking capabilities of the three cable designs. Ribbon cables are also hard to handle.

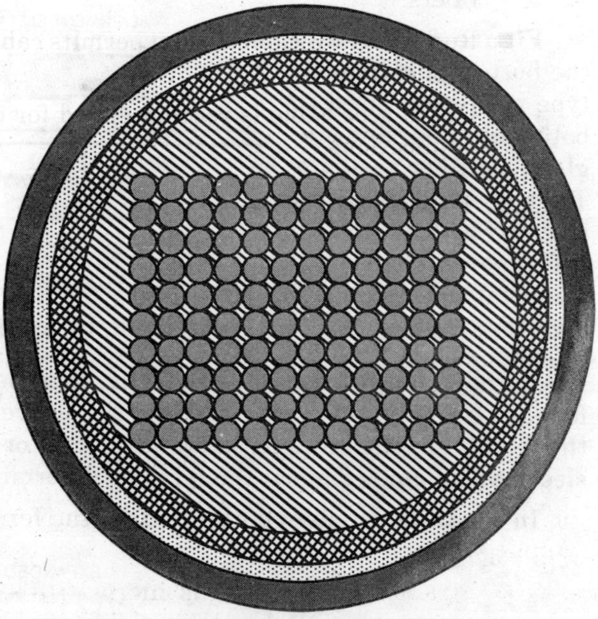

Fig. 6.10. Ribbon Cable Cross Sectional Array of Fibers.

- Individual fibers are highly susceptible to damage when separated from the ribbon. This susceptibility to fiber damage during fiber breakout makes it necessary to perform multifiber connections.

6.4. Fiber Connectors

Before connecting one fiber with the other fiber in the fiber optic communication link, one must decide whether the joint should be permanent or demountable. Based on this, we have two types of joints :

- A permanent joint is done by **splice**,
- Demountable joint is done by **connector**.

Requirements of a Good Connector

1. At connector joint, it should offer low coupling losses.
2. Connectors of the same type must be compatible from one manufacturer to another.
3. In the fiber link, the connector design should be simple so that it can be easily installed.
4. Connector joint should not be affected by temperature, dust and moisture. That is, it should have low environmental sensitivity.
5. It should be available at a lower cost and have a precision suitable to the application.

The coupling of light energy from one fiber to the other fiber using the connectors is based on two types of mechanism :

- Butt-joint alignment mechanism
- Expanded beam mechanism.

Figure 6.11 shows the butt-joint alignment type connectors used in both multimode and single mode fiber systems. These are straight sleeve (Fig. 6.11(*a*)) and the tapered sleeve (or) biconical sleeve connectors (Fig. 6.11(*b*)).

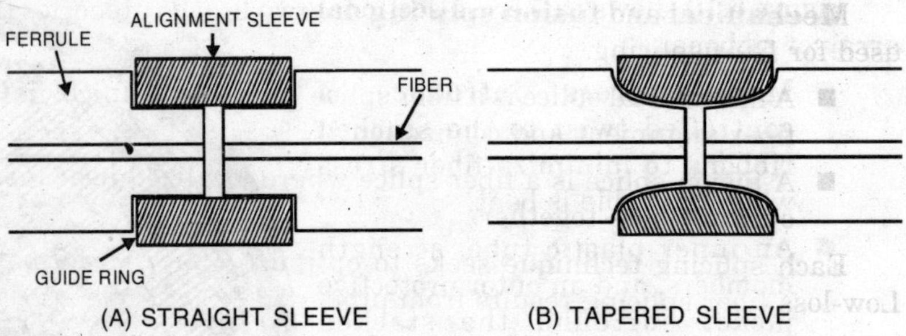

Fig. 6.11. Connectors Using Butt-Joint Alignment Designs.

In the **straight sleeve connector**, there is a metal, ceramic or molded plastic ferrule for each fiber and the ferrule fits into the sleeve. The fiber is epoxied into the drilled hole of the ferrule. In the straight sleeve connector or tapered sleeve connector the length of the sleeve and a guide ring on the ferrules determine the end separation of the fibers.

In the **tapered sleeve connector,** the ferrules and sleeves are tapered.

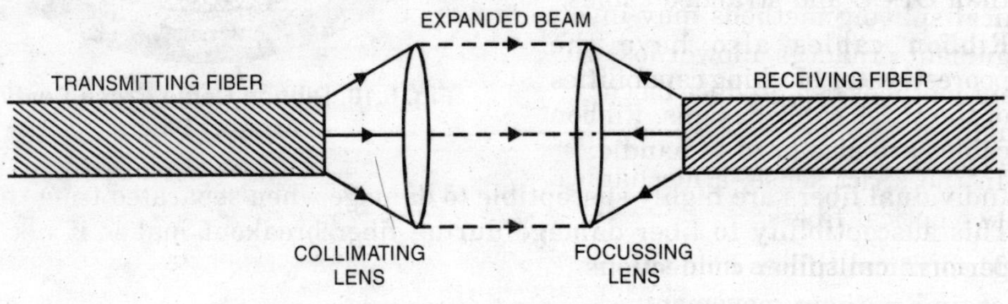

Fig. 6.12. Expanded Beam Connector.

Figure 6.12 shows the **expanded beam connector** employing collimating lens at the end of the transmitting fiber and focusing lens at the entrance end of the receiving fiber.

- The collimating lens collimating lens converts the light from the fiber into a parallel beam of light and the focusing lens converts the parallel beam of light into a focused beam of light on to the core of the receiving fiber.
- The fiber-to-lens distance is equal to the focal length of the lens. This expanded beam connector does not depend on lateral alignments and the optical processing elements can be easily inserted into the expanded beam between the fiber ends.
- The lenses are antireflection coated spherical micro lenses.
- To avoid losses due to fresnel reflection at the fiber-fiber joint, it is better to use an index matching fluid in the gap between the jointed fibers.
- When the index matching fluid has the same refractive index as the fiber core, Fresnel reflection losses are completely eliminated. But if there is any angular misalignment between fibers, there is an increased loss for the fibers with index matching fluid than for the fibers with air gap.

6.5. Fiber Optic Splices

A **fiber optic splice** is a permanent fiber joint whose purpose is to establish an optical connection between two individual optical fibers. Fiber optic splices also permit repair of optical fibers damaged during installation, accident, or stress. System designers generally require fiber splicing whenever repeated connection or disconnection is unnecessary or unwanted.

Mechanical and **fusion splicing** are two broad categories that describe the techniques used for fiber splicing.

- A mechanical splice is a fiber splice where mechanical fixtures and materials perform fiber alignment and connection.
- A fusion splice is a fiber splice where localized heat fuses or melts the ends of two optical fibers together.

Each splicing technique seeks to optimize splice performance and reduce splice loss. Low-loss fiber splicing results from proper fiber end preparation and alignment.

Fiber splice alignment can involve passive or active fiber core alignment. **Passive alignment** relies on precision reference surfaces, either grooves or cylindrical holes, to align fiber cores during splicing. **Active alignment** involves the use of light for accurate fiber alignment.

6.5.1. Mechanical Splices

Mechanical splicing involves using mechanical fixtures to align and connect optical fibers. Mechanical splicing methods may involve either passive or active core alignment. Active core alignment produces a lower loss splice than passive alignment. However, passive core alignment methods can produce mechanical splices with acceptable loss measurements even with single mode fibers.

- In the strict sense, a mechanical splice is a permanent connection made between two optical fibers.
- Mechanical splices hold the two optical fibers in alignment for an indefinite period of time without movement.
- The amount of splice loss is stable over time and unaffected by changes in environmental or mechanical conditions.
- If high splice loss results from assembling some mechanical splices, the splice can be reopened and the fibers realigned. Realignment includes wiping the fiber or ferrule end with a soft wipe, reinserting the fiber or ferrule in a new arrangement, and adding new refractive index material.
- The types of mechanical splices that exist for mechanical splicing include glass, plastic, metal, and ceramic tubes.
- Materials that assist mechanical splices in splicing fibers include transparent adhesives and index matching gels. **Transparent adhesives** are epoxy resins that seal mechanical splices and provide index matching between the connected fibers.

6.5.2. Glass or Ceramic Alignment Tube Splices

Mechanical splicing may involve the use of a glass or ceramic alignment tube, or capillary. The inner diameter of this glass or ceramic tube is slightly larger than the outer diameter of the fiber. A transparent adhesive, injected into the tube, bonds the two fibers together. The adhesive also provides index matching between the optical fibers. Figure 6.13 illustrates fiber alignment using a glass or ceramic tube.

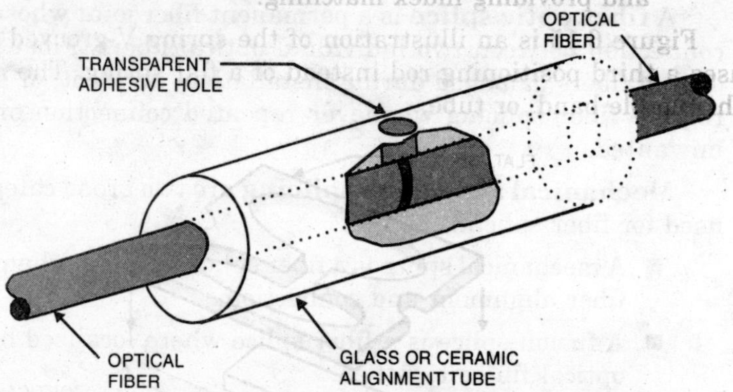

Fig. 6.13. A Glass or Ceramic Alignment Tube for Mechanical Splicing.

- This splicing technique relies on the inner diameter of the alignment tube. If the inner diameter is too large, splice loss will increase because of fiber misalignment.
- If the inner diameter is too small, it is impossible to insert the fiber into the tube.

6.5.3. V-Grooved Splices

Mechanical splices may also use either a **grooved substrate or positioning rods** to form suitable V-grooves for mechanical splicing. The basic V-grooved device relies on an open grooved substrate to perform fiber alignment. When inserting the fibers into the grooved substrate, the V-groove aligns the cladding surface of each fiber end. A transparent adhesive makes the splice permanent by securing the fiber ends to the grooved substrate. Figure 6.14 illustrates this type of open V-grooved splice.

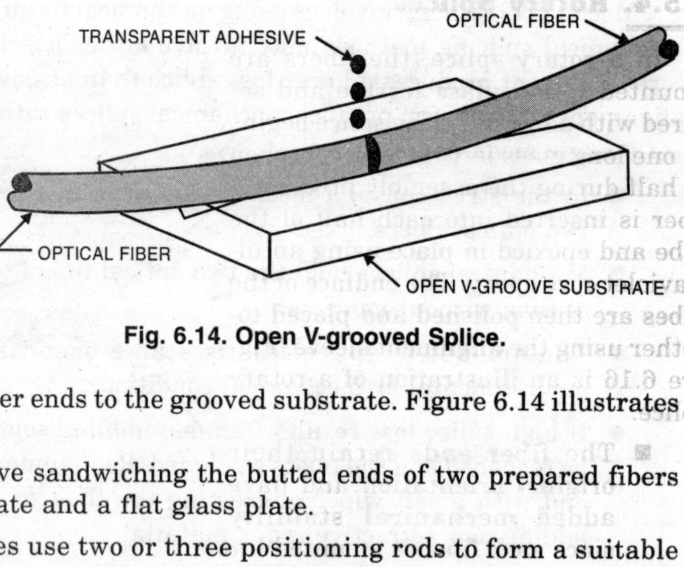

Fig. 6.14. Open V-grooved Splice.

- V-grooved splices may involve sandwiching the butted ends of two prepared fibers between a V-grooved substrate and a flat glass plate.
- Additional V-grooved devices use two or three positioning rods to form a suitable V-groove for splicing.
- The V-grooved device that uses two positioning rods is the spring V-grooved splice. This splice uses a groove formed by two rods positioned in a bracket to align the fiber ends.
- The diameter of the positioning rods permits the outer surface of each fiber end to extend above the groove formed by the rods. A flat spring presses the fiber ends into the groove maintaining fiber alignment.

- Transparent adhesive completes the assembly process by bonding the fiber ends and providing index matching.

Figure 6.15 is an illustration of the spring V-grooved splice. A variation of this splice uses a third positioning rod instead of a flat spring. The rods are held in place by a heat-shrinkable band, or tube.

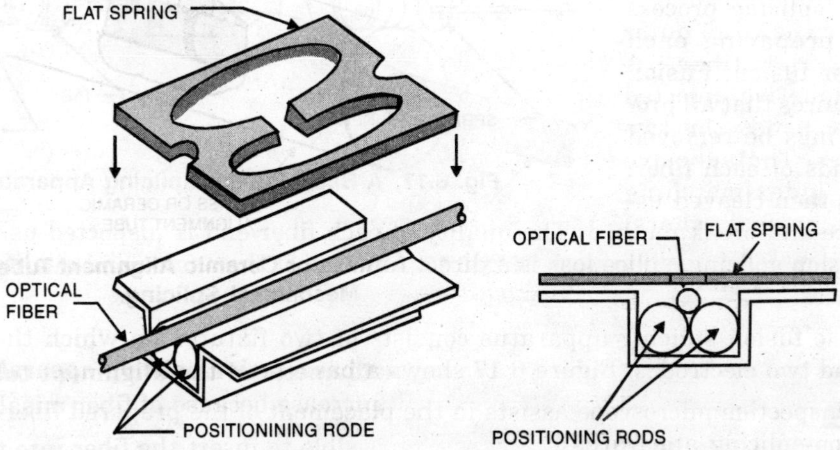

Fig. 6.15. Spring V-Grooved Mechanical Splice.

6.5.4. Rotary Splices

In a rotary splice, the fibers are mounted into a glass ferrule and secured with adhesives. The splice begins as one long glass ferrule that is broken in half during the assembly process. A fiber is inserted into each half of the tube and epoxied in place using an ultraviolet cure epoxy. The endface of the tubes are then polished and placed together using the alignment sleeve. Figure 6.16 is an illustration of a rotary splice.

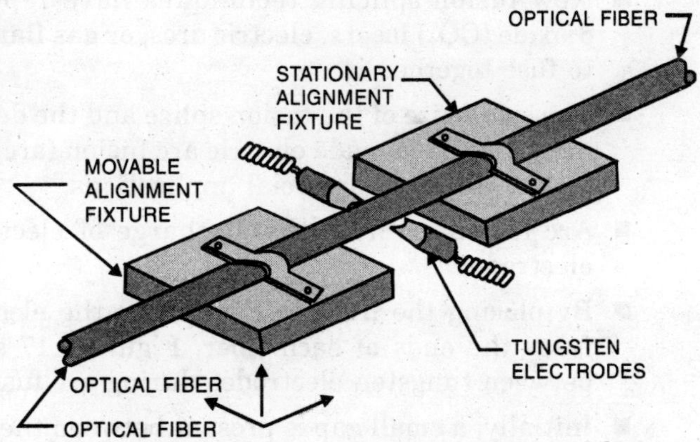

Fig. 6.16. Rotary Mechanical Splice.

- The fiber ends retain their original orientation and have added mechanical stability since each fiber is mounted into a glass ferrule and alignment sleeve.
- The rotary splice may use index matching gel within the alignment sleeve to produce low-loss splices.
- In shipboard applications, the Navy recommends using the rotary splice.
- The rotary splice is a low-loss mechanical splice that provides stable environmental and mechanical performance in the Navy environment.
- Stable performance means that splice loss does not vary significantly with changes in temperature or other environmental or mechanical conditions.

6.5.5. Fusion Splices

The process of fusion splicing involves using localized heat to melt or fuse the ends of two optical fibers together. The splicing process begins by preparing each fiber end for fusion. Fusion splicing requires that all protective coatings be removed from the ends of each fiber. The fiber is then cleaved using the score-and-break method. The quality of each fiber end is inspected using a microscope. In fusion splicing, splice loss is a direct function of the angles and quality of the two fiber-end faces.

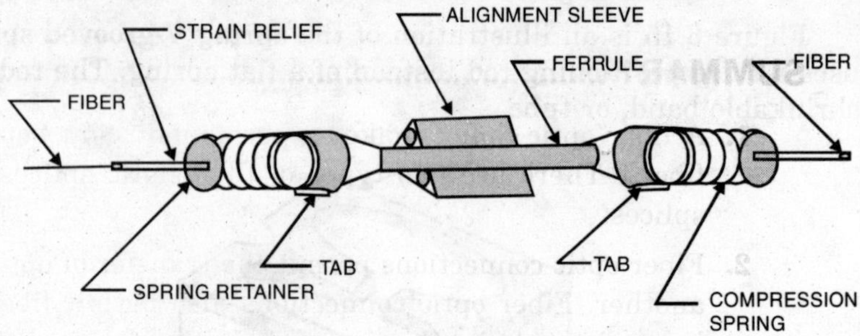

Fig. 6.17. A Basic Fusion Splicing Apparatus.

The basic fusion splicing apparatus consists of two fixtures on which the fibers are mounted and two electrodes. Figure 6.17 shows a basic fusion-splicing apparatus.

- An inspection microscope assists in the placement of the prepared fiber ends into a fusion-splicing apparatus.
- The fibers are placed into the apparatus, aligned, and then fused together. Initially, fusion splicing used nichrome wire as the heating element to melt or fuse fibers together.
- New fusion-splicing techniques have replaced the nichrome wire with carbon dioxide (CO_2) lasers, electric arcs, or gas flames to heat the fiber ends, causing them to fuse together.
- The small size of the fusion splice and the development of automated fusion-splicing machines have made electric arc fusion (arc fusion) one of the most popular splicing techniques in commercial applications.
- Arc fusion involves the discharge of electric current across a gap between two electrodes.
- By placing the fiber ends between the electrodes, the electric discharge melts or fuses the ends of each fiber. Figure 6.17 shows the placement of the fiber ends between tungsten electrodes during arc fusion.
- Initially, a small gap is present between the fiber ends. A short discharge of electric current is used to prepare the fiber ends for fusion.
- During this short discharge, known as **profusion**, the fiber ends are cleaned and rounded to eliminate any surface defects that remain from fiber cleaving.
- Surface defects can cause core distortions or bubble formations during fiber fusion.
- Fusion splicing yields typically vary between 25 and 75 percent depending on the strength and loss requirements for the splice and other factors.
- In general, fusion splicing takes a longer time to complete than mechanical splicing. Also, yields are typically lower making the total time per successful splice much longer for fusion splicing.
- Both the yield and splice time are determined to a large degree by the expertise of the fusion splice operator. Fusion splice operators must be highly trained to make

low-loss reliable fusion splices. For these reasons the fusion splice is **not recommended** for use in Navy shipboard applications.

SUMMARY

1. A fiber optic Splice makes a permanent joint between two fibers or two groups of fibers. There are two types of fiber optic splices—mechanical splices and fusion splices.
2. Fiber optic connections permit the transfer of optical power from one component to another. Fiber optic connections also permit fiber optic systems to be more than just point-to-point data communication links.
3. Fiber optic couplers distribute or combine optical signals between fibers. Couplers can distribute an optical signal from a single fiber into several fibers.
4. Fiber optic connection losses may affect system performance. Poor fiber end preparation and poor fiber alignment are the main causes of coupling loss.
5. Fiber-to-fiber connection loss is affected by intrinsic and extrinsic coupling losses. Intrinsic coupling losses are caused by inherent fiber characteristics. Extrinsic coupling losses are caused by jointing techniques.
6. The electrical power required to operate both injection lasers and LEDs is generally similar with typical currents levels of between 20 and 300 mA.
7. The speed of response of the two types of optical source is largely dictated by their respective radiative emission mechanism.
8. The finite spectral width of the optical source causes pulse broadening due to material dispersion on an optical fiber communication link.
9. When optical power is incident on the photodetector during the 0 bit period, then the system is said to exhibit a nonzero extinction ratio.
10. The extinction ratio ξ is usually defined as the ratio of the optical energy emitted in the 0 bit period to that emitted during the 1 bit period.
11. Intrinsic coupling losses are limited by reducing fiber mismatches between the connected fibers. This is done by procuring only fibers that meet stringent geometrical and optical specifications.
12. The three basic coupling errors that occur during fiber alignment are :
    - fiber separation (longitudinal misalignment)
    - lateral misalignment
    - angular misalignment
13. Fiber mismatches are a source of intrinsic coupling loss. As stated before, intrinsic coupling loss results from differences (mismatches) in the inherent fiber characteristics of the two connecting fibers.
14. Fiber mismatches are the result of inherent fiber characteristics and are independent of the fiber jointing techniques.

REVIEW QUESTIONS

1. Explain the three types of fiber alignment which may contribute to insertion loss at an optical fiber joint.
2. What do you meant by optical splices ? Briefly describe the different types of optical splicing used in optical communication.
3. With the help of suitable diagrams, explain the three common techniques used for the mechanical splicing of optical fibers.
4. Discuss the principle of operation of the two major types of fiber optic connectors.
5. Explain the function of optical switches. What are the advantages of optical switches over electrical switches ?
6. What do you understand by fiber connectors ? List the various requirements of a good conductor.
7. Briefly explain the various design issue involved in optical fiber manufacturing.
8. What do you meant by soot ? Discuss the two different methods to fabricate multi-mode and single mode glass fibers.

❑❑❑

Optical Transmitters

INSIDE THIS CHAPTER

The function of the optical transmitter is to convert an electrical input signal into the corresponding optical signal and then launch it into optical fiber serving as a communication channel. The **major component** of an optical transmitter is an optical source. Sometimes fiber optic communication systems use semiconductor optical sources such as light emitting diodes (LEDs) and semiconductor laser because of several advantages offered by them.

7.1. Basic Concepts

To understand the light generating mechanism within the major optical sources used in optical fiber communications we have to consider both the fundamental atomic concepts and the device structure. In this section, we elaborate on the basic principles which govern the operation of both these optical sources.

7.1.1. Absorption of Radiation

The light interaction with matter takes place in discrete packets of energy or quanta known as **photons**. The quantum theory suggests that atom exist only in certain discrete energy states such that atoms exist only in certain discrete energy states such that absorption and emission of light causes them to make a transition from one discrete energy to another.

If an atom is initially in a lower state 1, it can rise to a higher state 2 by absorbing a quantum of radiation (photon) of frequency f is given by

$$\boxed{E = E_2 - E_1}$$

$$hf = E_2 - E_1$$

$$\boxed{f = \frac{E_2 - E_1}{h}} \quad ...(7.1)$$

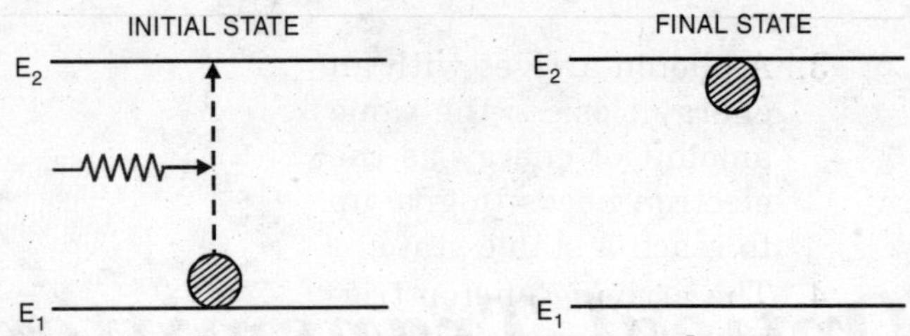

Fig. 7.1. Absorption.

where, $h = 6.626 \times 10^{-34}$ Js (Planck's Constant)

Figure 7.1 shows a two energy state or level atomic system where an atom is initially in the lower energy state E_1.

This process is known as **absorption of radiation.**

7.1.2. Spontaneous Emission

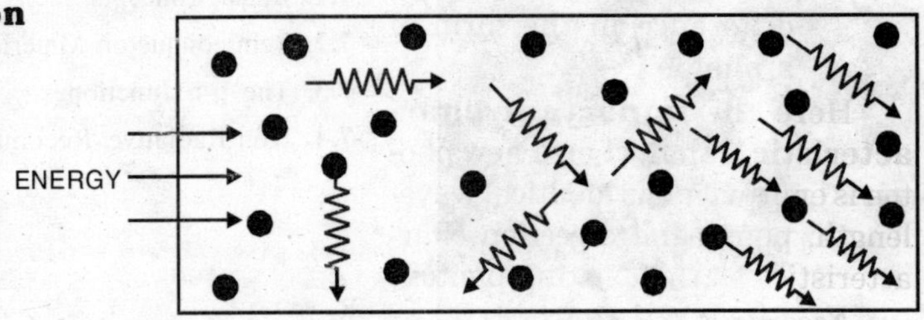

Fig. 7.2. Spontaneous Emission in Material.

If we take a material confine in some space and them bombard it with energy in a form that the material will absorb and then spontaneous emission of light occur. Materials capable of stimulated emission or distinguished by the fact that they have a high energy state *i.e.* **meta stable** because it can hold its high energy state for some time before decaying spontaneously. Figure 7.2 shows a material treated in this way.

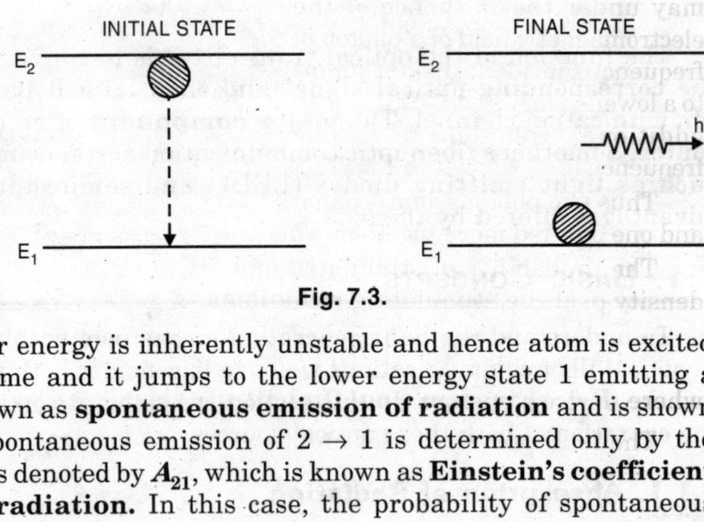

Fig. 7.3.

Excited atoms decay and emits photons randomly in all directions. Let us now consider an atom initially in higher (excited) state 2. Excited state with higher energy is inherently unstable and hence atom is excited state does not stay for longer time and it jumps to the lower energy state 1 emitting a photon of frequency *f*. This is known as **spontaneous emission of radiation** and is shown in Fig. 7.3. The probability of spontaneous emission of 2 → 1 is determined only by the properties of states 2 and 1. This is denoted by A_{21}, which is known as **Einstein's coefficient of spontaneous emission of radiation.** In this case, the probability of spontaneous emissions is independent of it.

7.1.3. Stimulated Emission

The process involved in stimulated emission is shown in Fig. 7.4.

1. An electron within at atom (or a molecule or an ion) starts in a low energy stable state often known as **ground state.**
2. Energy is supplied from outside and is absorbed by the atomic structure where upon the electron enters an excited (higher energy) state.

3. A photon arrives with an energy close to the same amount of energy as the electron needs to give up to reach a stable state.
4. The arriving photon triggers a resonance with the excited atom. Thus the excited electron leaves its excited state and transitions to a more stable state giving up the energy difference in the form of a photon.

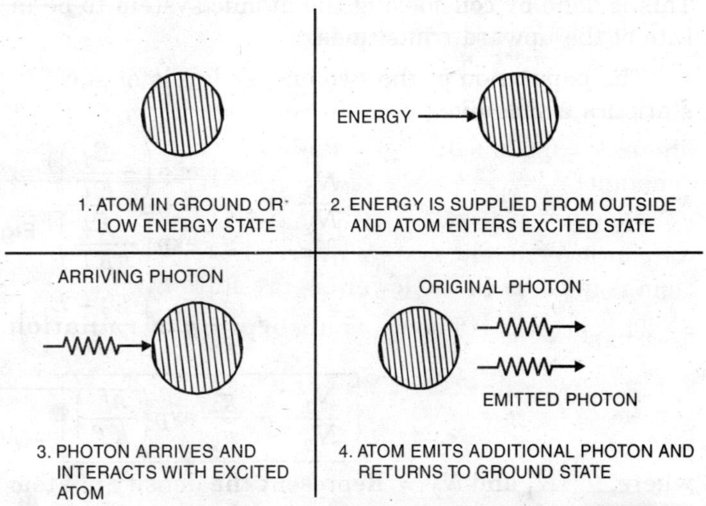

Fig. 7.4. Stimulated Emission.

Here the **important characteristic** is that when a new photon is emitted it has identical wavelength, phase and direction characteristics as the existing photon.

According to the Einstein, an atom in an excited energy state may under the influence of the electromagnetic field of a photon of frequency *f* incident upon it jumps to a lower energy state, emitting an additional photon of same frequency (*f*) as shown in Fig. 7.5.

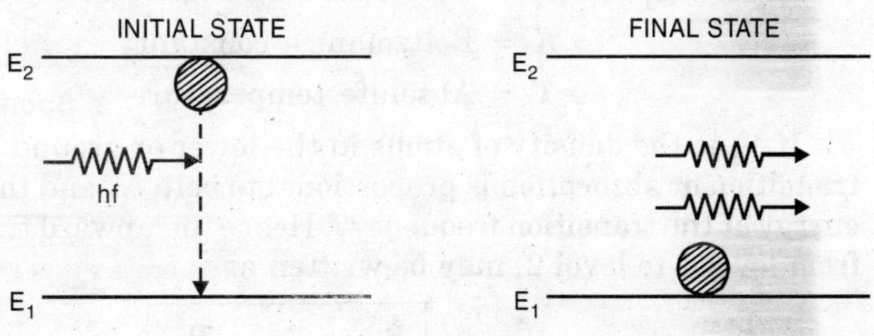

Fig. 7.5.

Thus two photons, one original and one emitted move together. This process is known as **stimulated emission of radiation.**

The probability of stimulated emission transition $2 \rightarrow 1$ is proportional to the energy density ρ_f of the stimulating radiation and is given by

$$\boxed{B_{21}\,\rho_f}$$

where B_{21} = **Einstein's coefficient of stimulated emission of radiation**

The total probability for an atom in state 2 to drop to the lower state 1 is therefore,

$$\boxed{P_{21} = A_{21} + B_{21}\rho_f}$$

Difference between the spontaneous and stimulated emission processes :

- In spontaneous emission the atom emits an electromagnetic waves which has no definite phase or directional relation with that emitted by another atom.
- In stimulated emission, the process is forced by the incident electromagnetic wave thus the emitted light by the atom is in phase with that of the incident electromagnetic wave. The emitted wave is in the same direction as that of the incident wave.

7.1.4. The Einstein Relations

In 1917, Einstein demonstrated that the rates of the three transition processes of absorption, spontaneous emission and stimulated emission were related mathematically.

This is done by considering the atomic system to be in thermal equilibrium such that the rate of the upward transitions.

The population of the two energy levels of such a system are described by Botlzmann statistics which give :

$$\frac{N_1}{N_2} = \frac{g_1 \exp\left(-\frac{E_1}{KT}\right)}{g_2 \exp\left(-\frac{E_2}{KT}\right)}$$

$$= \frac{g_1}{g_2} \exp\left(\frac{E_2 - E_1}{KT}\right) \quad ...(7.2)$$

$$\boxed{\frac{N_1}{N_2} = \frac{g_1}{g_2} \exp\left(\frac{hf}{KT}\right)} \quad ...(7.3)$$

where, N_1 and N_2 = Represent the density of atom in energy levels E_1 and E_2 respectively.

g_1 and g_2 = Corresponding degeneracies of the levels.

K = Boltzmann's constant

T = Absolute temperature

If N_1 is the density of atoms in the lower or ground energy state E_1, the rate of upward transition or absorption is proportional to both N_1 and the spectral density ρ_f of the radiation energy at the transition frequency f. Hence the upward transition rate R_{12} (electron transition from level 1 to level 2) may be written as

$$\boxed{R_{12} = N_1 \rho_f B_{12}} \quad ...(7.4)$$

where, B_{12} = Einstein coefficient of absorption

Similarly atoms in the higher or excited energy state can undergo electron transitions from level 2 to level 1 either spontaneously or through simulation by the radiation field. For spontaneous emission the average time an electron exists in the excited state before transition occurs is known as the **spontaneous lifetime τ_{21}.**

If N_2 is the density of atoms within the system with energy E^2 then the spontaneous emission rate is given by the product of N_2 and $\frac{1}{\tau_2}$. This may be written as $\boxed{N_2 A_{21}}$.

where, A_{21} = Einstein coefficient of spontaneous emission.

The rate of stimulated downward transition of an electron from level 2 to level 1 is given by $\boxed{N_2 \rho_f B_{21}}$.

where, B_{21} = Einstein coefficient of stimulated emission.

The **total transition rates** from level 2 to level 1, R_{21} is the sum of spontaneous and stimulated contributions. Hence,

$$\boxed{R_{21} = N_2 A_{21} + N_2 \rho_f B_{21}} \quad ...(7.5)$$

For a system in thermal equilibrium, the upward and downward transition rates must be equal and therefore,

$$R_{12} = R_{12}$$
$$N_1 \rho_f B_{12} = N_2 A_{21} + N_2 \rho_f B_{21}$$
$$\therefore \quad \rho_f = \frac{N_2 A_{21}}{N_1 B_{12} - N_2 B_{21}}$$

and
$$\rho_f = \frac{A_{21}/B_{21}}{\left(\frac{N_1 B_{12}}{N_2 B_{21}} - 1\right)} \qquad \text{...(7.6)}$$

Substituting the value of $\frac{N_1}{N_2}$ in above equation we get

$$\rho_f = \frac{A_{21}/B_{21}}{\left[\left(\frac{g_1 B_{12}}{g_2 B_{21}}\right)\exp\left(\frac{hf}{KT}\right)\right] - 1} \qquad \text{...(7.7)}$$

Since we have assumed that atomic system is in thermal equilibrium, it produces a radiation density which is identical to black body radiation. **According to Planks** the radiation spectral density for a black body radiating within a frequency range f to $f + d$ is given by

$$\rho_f = \frac{8\pi h f^3}{C^3} \frac{1}{\left[\exp\left(\frac{hf}{KT}\right) - 1\right]} \qquad \text{...(7.8)}$$

Comparing Equations (7.7) and (7.8) we get the **Einstein relation**

$$B_{12} = \left(\frac{g_2}{g_1}\right) B_{21} \qquad \text{...(7.9)}$$

and
$$\frac{A_{21}}{B_{21}} = \frac{8\pi h f^3}{C^3} \qquad \text{...(7.10)}$$

From Eqn. (7.9) it is observed that when the degeneracies of two levels are equal ($g_1 = g_2$), then the probabilities of absorption and stimulated emission are equal.

Therefore the ratio of the stimulated emission rate to the spontaneous emission rate is given by

$$\frac{\text{Stimulated Emission Rate}}{\text{Spontaneous Emission Rate}} = \frac{B_{21}\rho_f}{A_2} = \frac{1}{\left[\exp\left(\frac{hf}{KT}\right) - 1\right]} \qquad \text{...(7.11)}$$

7.1.5. Population Inversion

The condition of thermal equilibrium given by Boltzmann distribution defines that the lower energy level E_1 of the two level atomic system contains more atoms than the upper energy level E_2. This situation is normal for structures at room temperature is shown in Fig. 7.6.

To achieve optical amplification it is necessary to create a non-equilibrium

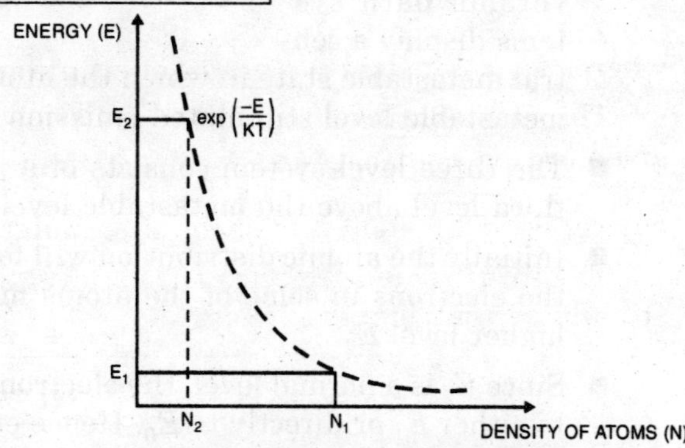

Fig. 7.6. Population in Two Energy Level Boltzmann Distribution.

distribution of atoms such that the population of the upper energy level is greater than that of lower energy level (*i.e.* $N_2 > N_1$). This condition is known as **population inversion** and is shown in Fig. 7.7.

The achieve population inversion it is necessary to excite atoms into the upper energy level E_2 and hence a non-equilibrium distribution is obtained. This process is achieved by using an external energy source and is known as **pumping**.

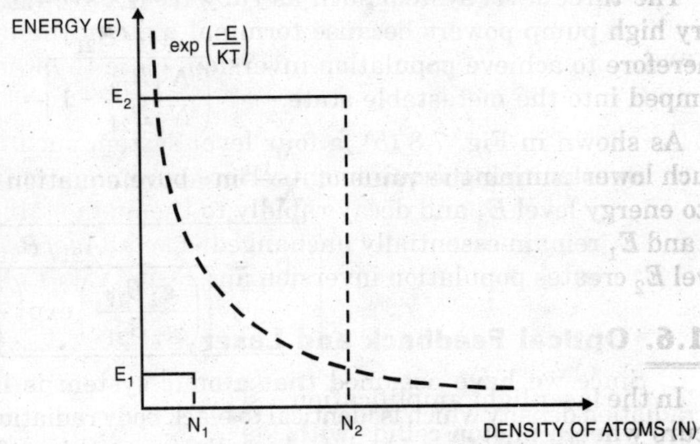

Fig. 7.7. Non-equilibrium Distribution Showing Population Inversion.

A common method used for pumping involves the application of intense radiation. In previous case atoms are excited into the higher energy state through stimulated absorption. When the two levels are equally degenerated (or not degenerated). Then

$$\boxed{B_{12} = B_{21}}$$

i.e. the two level system does not lend itself. Thus the probabilities of absorption and stimulated emission are not equal, providing at best equal population in the two levels.

Population inversion may be obtained in systems with three or four energy levels. The energy level diagrams for two such systems are shown in Fig. 7.8.

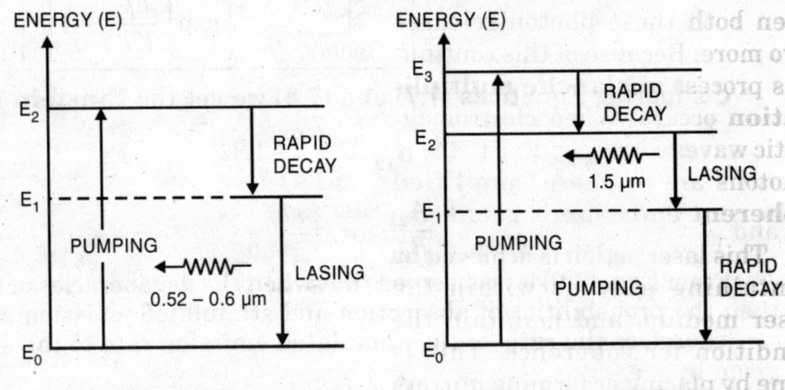

(A) THREE LEVEL SYSTEM—RUBY (CRYSTAL) LASER. (B) FOUR LEVEL SYSTEM—RUBY (CRYSTAL) LASER.

Fig. 7.8. Energy Level Diagram Showing Population Inversion and Lasing for Two Semiconductors Laser.

- For population inversion bath systems display a central metastable state in which the atoms spend an unusally long time and from this metastable level stimulated emission or lasing takes place.
- The three level system consists of a ground level E_0, a metastable level E_1 and a third level above the metastable level E_2.
- Initially the atomic distribution will follow Boltzmann's law. With suitable pumping the electrons in some of the atoms may be excited from the ground state into the higher level E_2.
- Since E_2 is a normal level, the electrons will rapidly decay by nonradiative processes to either E_1 or directly to E_0. Hence empty states will always be provided in E_2.
- The metastable level E_1 exhibits a much longer lifetime than E_2 which allows a large number of atoms to accumulate at E_1.

- Over a period of time the density of atoms in the metastable state N_1 increases above those in the ground state N_0 and a population inversion is obtained between these two levels. Then stimulated emission and hence lasing can occur creating radiative electron transitions between level E_1 and E_0.

The three level system such as ruby laser have a drawback that it generally requires very high pump powers because terminal state of the laser transition is the ground state. Therefore to achieve population inversion, more than half the ground state atoms must be pumped into the metastable state.

As shown in Fig. 7.8 (*b*), a four level system such as He-Ne laser is characterized by much lower pumping requirements. Here pumping excites the atoms from the ground state into energy level E_3 and decay rapidly to the meta stable level E_2. Since the populations of E_3 and E_1 remain essentially unchanged, a small increase in the number of atoms in energy level E_2 creates population inversion and lasing takes place between this level and level E_1.

7.1.6. Optical Feedback and Laser Oscillation

In the laser light amplification occurs when a photon collids with an atom in the excited energy state, it causes the stimulated emission of a second photon and then both these photon releases two more. Because of this continuous process **avalanche multiplication** occurs. When electromagnetic waves corresponding to these photons are in phase, **amplified coherent emission** is obtained.

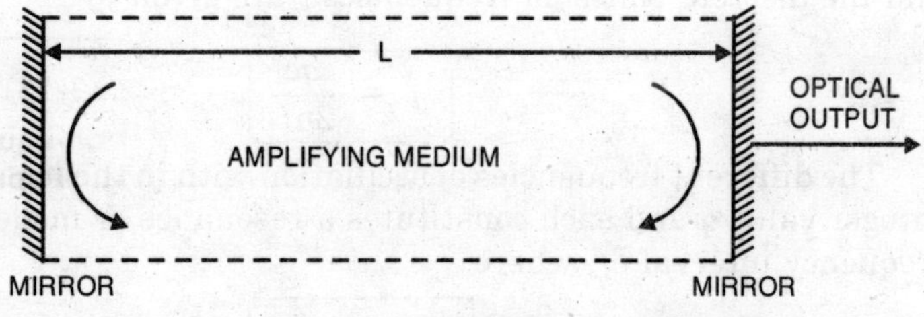

Fig. 7.9. Basic Laser Structure Incorporating Plane Mirrors.

This laser action is achieved by containing photons within the laser medium and maintain the condition for coherence. This is done by placing or forming mirrors at either end of the amplifying medium as shown in Fig. 7.9.

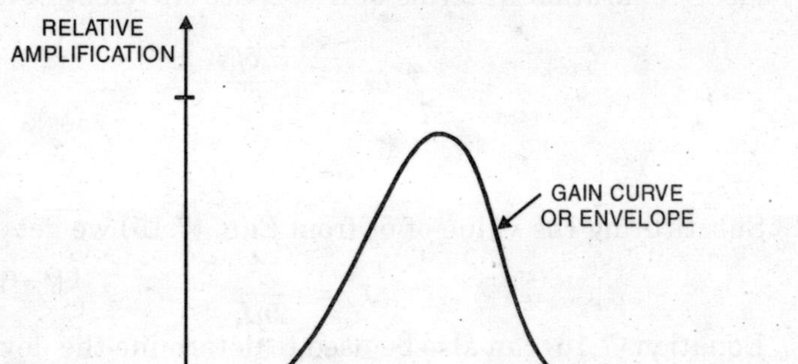

Fig. 7.10. Relative Amplification in the Laser Showing Gain Curve.

The optical cavity formed is more analogous to an oscillator than an amplifier because it provides positive feedback of the photons by reflection at the mirrors at either end of the cavity. Therefore the optical signal is fedback many times while receiving amplification is passes through the medium. Hence this structure acts as a **Fabry-Perot resonator**. Although the amplification of signal from a single pass through the medium is quite small, the net gain can be large after multiple passes. The losses in the amplifying medium result from factors such as :

- Absorption and scattering in the amplifying medium
- Absorption, scattering and diffraction at the mirrors
- Nonuseful transmission through the mirrors.

In the laser cavity oscillations occur over a small range of frequencies, hence the device is not a perfectly monochromatic source but emits over a narrow spectral spectral band. In this band the central frequency is obtained by the mean energy level difference of the stimulated emission transitions. Figure 7.10 shows the spectral emission from the device.

The structure forms a resonant cavity. When sufficient population inversion exists in the amplifying medium the radiation builds up and becomes established as standing waves between the mirrors. These standing waves exist only at frequencies for which the distance between the mirrors is an integral number of half wavelengths. If L is the optical spacing between the mirrors, the resonance condition along the axis of cavity is given by

$$L = \frac{\lambda q}{2n} \quad \text{...(7.12)}$$

where, λ = Emission wavelength.

q = An integer

n = Refractive index of the amplifying medium

and the discrete emission frequencies f are given by

$$f = \frac{qc}{2nL} \quad \text{...(7.13)}$$ where, c = Velocity of light

The different frequencies of oscillation with in the laser cavity are determined by various integer value q and each constitutes a resonance or mode. These modes are separated by a frequency interval δf, where

$$\delta f = \frac{c}{2nL} \quad \text{...(7.14)}$$

Mode separation in terms of free space wavelength is given by

$$\delta\lambda = \frac{\lambda . \delta f}{f} \qquad \left[\text{assuming, } \delta f << f \text{ and } f = \frac{c}{f}\right]$$

$$= \frac{\lambda^2}{c} . \delta f \quad \text{...(7.15)}$$

Substituting the value of δf from Eqn. (7.15) we get

$$\delta\lambda = \frac{\lambda^2}{2nL} \quad \text{...(7.16)}$$

Equation (7.16) can also be used to determine the device specral linewidth as a function of wavelength (in Hz).

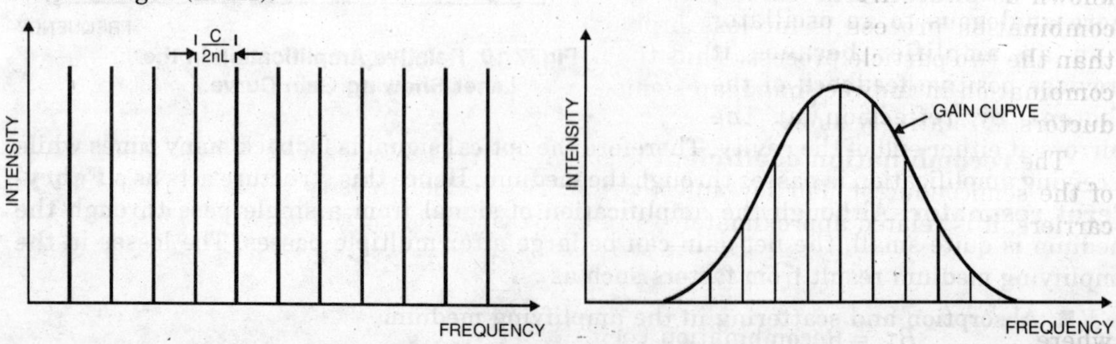

(A) MODES IN THE LASER CAVITY. (B) THE LONGITUDINAL MODES IN THE LASER OUTPUT.

Fig. 7.11.

When large number of modes are generated within the laser cavity, the spectral output from the device is defined by the **gain curve**. Thus laser emission will only include longitudinal modes contained within the spectral width of gain curve. Figure 7.11 shows the condition where several modes are prevent in laser output. Such type of device is known as **multimode.**

7.2. SEMICONDUCTOR MATERIALS

Direct Bandgap Semiconductors

The **direct bandgap semiconductors** are those in which electrons and holes on either side of the forbidden energy gap have the same value of crystal momentum and therefore direct recombination is possible. In direct bandgap semiconductors electroluminance is possible and this process is shown in Fig. 7.12.

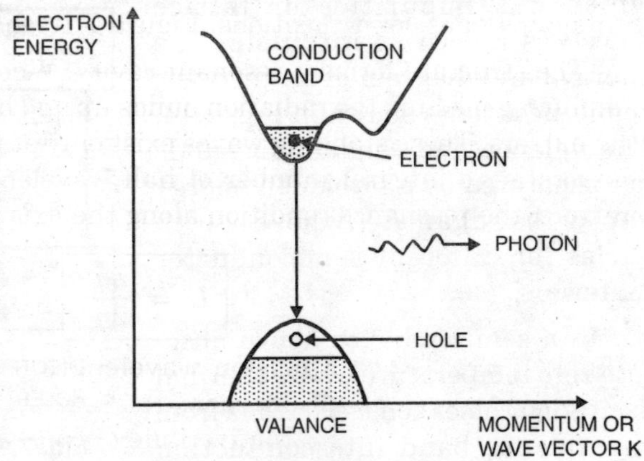

Fig. 7.12. Energy-Momentum Diagram of Direct Bandgap Semiconductor.

The maximum energy of valence band occurs at the same value of electron crystal momentum as the minimum energy of the condition band. Thus when electron hole recombination take place, the electrons momentum remains virtually constant and bandgap energy E_g may be emitted as light. This direct transition of an electron across the energy gap provides an efficient mechanism for photon emission and average time before the recombination, the minority carrier remains in a free state is short *i.e.* 10^{-8} to 10^{-10} sec.

Indirect Bandgap Semiconductors

In **indirect bandgap semiconductors** the maximum and minimum energies occur at different values of crystal momentum, as shown in Fig. 7.13.

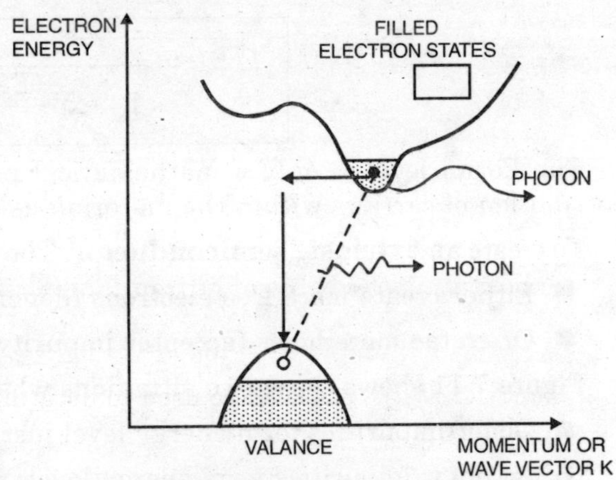

Fig. 7.13. Electron-Momentum Diagram of Indirect Bandgap Semiconductor.

Electron hole recombination occur when electron loses momentum such that the value of momentum occurs corresponds to the maximum energy of valence band. The conversion of momentum requires the emission or absorption of a third particle known as **photon**. The three particle recombination process is for less probable than the two particle process. Thus the recombination in indirect bandgap semiconductors is relatively slow.

The **recombination coefficient** is obtained from the measured absorption coefficient of the semiconductor. For low infected minority carrier density relative to the majority carriers, it is related approximately to radiative minority carrier lifetime τ_t by

$$\boxed{\tau_t = [B_\tau (N+P)]^{-1}} \quad ...(7.17)$$

where $B\tau$ = Recombination coefficient.

N and P = Majority carrier concentrations in the n and p type regions.

7.3. The p-n Junction

A perfect semiconductor crystal which has impurities or lattice defects is known as **intrinsic.** The energy band diagram of an intrinsic semiconductor is shown in Fig. 7.14. The valence and conduction bands are separated by a forbidden energy gap or bandgap E_g, whose width varies for different semiconductor materials.

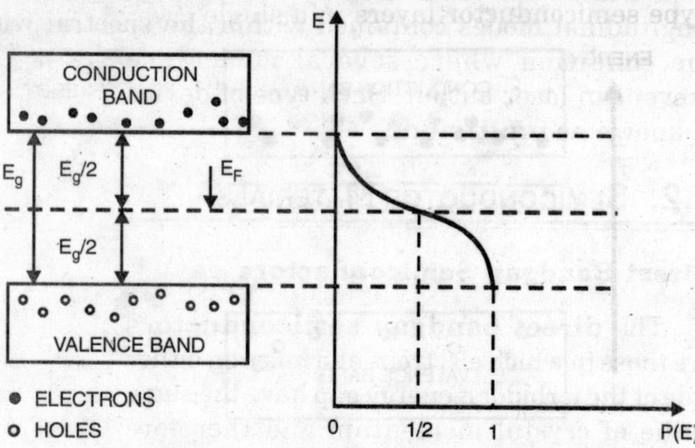

Fig. 7.14. (a) Energy Band Diagram of an Intrinsic Semiconductor, (b) The Fermi-Dirac Probability Distribution Corresponding to (a).

In a semiconductor above absolute zero temperature where thermal excitation raises some electrons from the valence band into conduction band, leaving empty hole states in the valance band. These thermally excited electrons in the conduction band and holes left in the valence band allow conduction through the material, and are known as **carriers.**

For a semiconductor in thermal equillibrium the energy level occupation is described by the **Fermi Dirac Distribution** function rather than **Boltzmann.** The probability $P(E)$ that an electron gains sufficient thermal energy at an absolute temperature T such that it will be found occupying a particular energy level E, is given by

$$P(E) = \frac{1}{1 + \exp\dfrac{(E - E_F)}{KT}} \qquad \text{...(7.18)}$$

The Fermi level is only a mathematical parameter but it gives an indication of the distribution of carriers within the material, as shown in Fig. 7.10(b).

To create an **extrinsic semiconductor.** The material is doped with impurity atoms which

- Either create more free electrons (donor impurity)
- Or cretae more holes (acceptor impurity)

Figure 7.11 shows these two situations where

- Donor impurities form energy level just below the conduction band.
- Acceptor impurities form energy levels just above the valence band.

When Donar Impurities are Added : Thermally excited electrons from the donor levels are raised into the conduction band to create an excess of negative charge carriers. The semiconductor is known as n-type and the majority carriers are electrons. For this carrier distribution, the fermi level is raised to a position above the centre of the bandgap, as shown in Fig. 7.15(a).

When Acceptor Impurities are Added : Thermally excited electrons are raised from the valance band to the acceptor impurity levels leaving an excess of positive charge carriers in the valance band and creating a p-type semiconductor where the majority carriers are holes. In this case Fermi level is lowered below the centre of the bandgap, as shown in Fig. 7.15(b).

Figure 7.16(a), shows the formation of p-n junction diode by creating adjoining p and n type semiconductor layers in a single crystal.

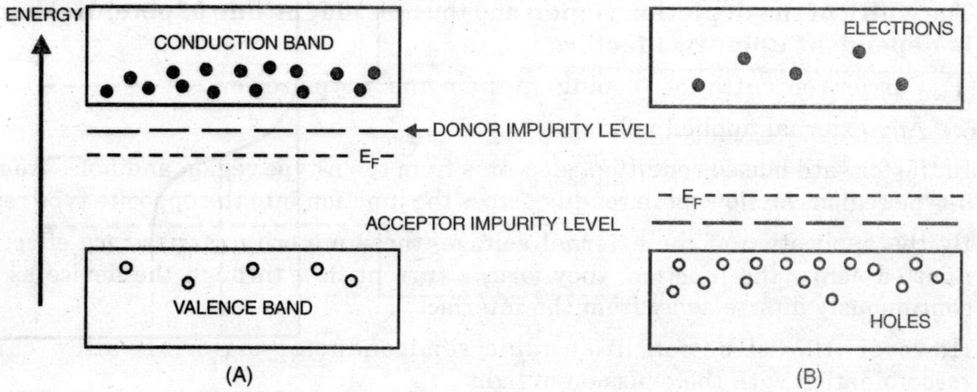

Fig. 7.15. Energy band diagrams : (a) n-type semiconductor, (b) p-type semiconductor.

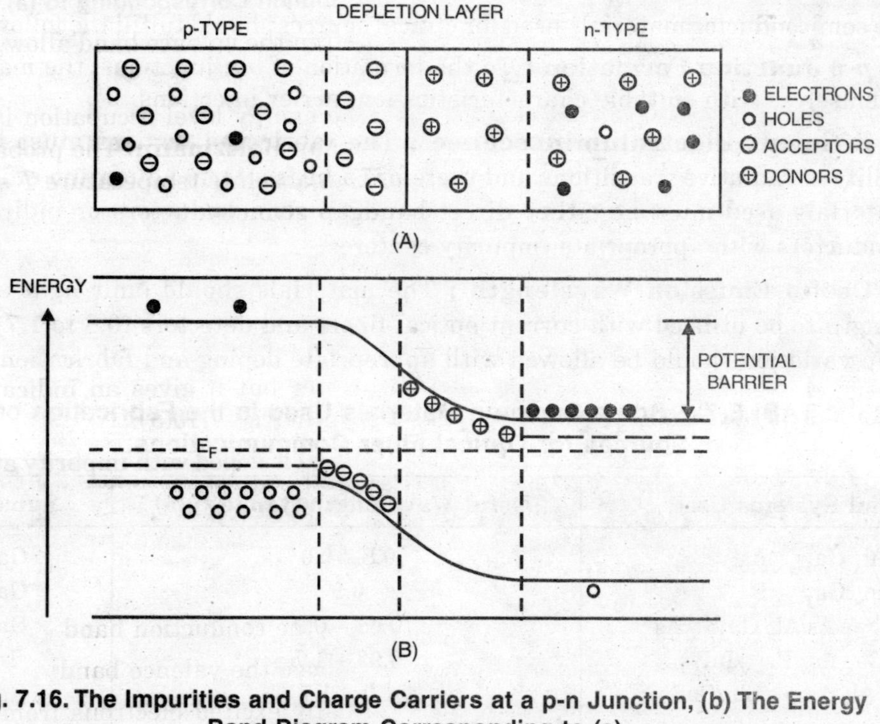

Fig. 7.16. The Impurities and Charge Carriers at a p-n Junction, (b) The Energy Band Diagram Corresponding to (a).

- A thin depletion region is formed at the junction through carrier recombination which effectively leaves it free of mobile charge carriers (both electrons and holes).
- This establishes a potential barrier between the p and n type regions which restricts the interdiffusion of majority carriers from their respective regions, as shown in Fig. 7.16(b).
- In the absence of an externally applied voltage, no current flows because potential barrier prevents the net flow of carriers from one region to another.

- In this equilibrium state of junction, the Fermi levels for the *p* and *n* type semiconductor is the same as shown in Fig. 7.16(*b*).
- The **width of the depletion region** and thus the **magnitude of potential barrier is dependent upon two factors :**

  (1) Carrier concentration (doping) in the *p* and *n* type region.

  (2) Any external applied voltage.
- In the forward biased condition, electrons from the n-type region and holes from the *p*-type region can flow more readily across the junction into the opposite type region.
- By the application of the external voltage, these minority carriers are effectively injected across the junction. they form a current flow through the device as they continuously diffuse away from the interface.
- However, this situation in suitable semiconductor materials allow carrier recombination with the emission of light.

Semiconductor Materials

The semiconductor materials used for optical sources should fulfill the following criteria.

(*a*) *p-n* Junction Formation : To the formation of *p-n* junctions, the materials must lend themselves with suitable characteristics for carrier injections.

(*b*) Efficient Electroluminescence : The fabricated devices must have a high probability of radiative transitions and therefore a high internal quantum efficiency. Thus the materials used must be either direct bandgap semiconductors or indirect bandgap semiconductors with appropriate impurity centeres.

(*c*) Useful Emission Wavelength : The materials should emit light at a suitable wavelength to be utilized with current optical fibers and detectors (0.8 to 1.7 μm). Ideally bandgap variation should be allowed with appropriate doping and fabrication.

TABLE 7.1. Some Common Materials Used in the Fabrication of Sources for Optical Fiber Communications.

Material Systems Used	Useful Wavelength Range (μm)	Substrate
$GaAs/Al_x Ga_{1-x} As$	0.8 – 0.9	GaAs
$GaAs/In_x Ga_{1-x} P$	0.9	GaAs
$Al_y Ga_{1-y} As/Al_x Ga_{1-x} As$	0.65 – 0.9	GaAs
$In_{1-x} Ga_x As_y P_{1-y}/InP$	0.92 – 1.7	InP
$Ga_{1-y} Al_y As_{1-x} Sb_x/GaSb$	1.0 – 1.7	GaSb

7.4. Non-Radiative Recombination

In this section the electroluminescent properties of forward biased *p-n* junction diode have been considered. In the semiconductor injection laser, stimulated emission by the recombination of the injected carriers is encouraged. **Thus injection laser has several advantages over other semiconductor sources.** These are as follows :

(*a*) High radiance due to the amplifying effect of stimulated emission. Generally injection lasers supply milliwatts of optical power output.

(*b*) Narrow bandwidth of the order of ≤ 1 nm which is used to minimize the effects of materials dispersion.
(*c*) Modulation capabilities can be extended to gigahertz range.
(*d*) Relative temporal coherence allows heterodyne (coherent) detection in high capacity systems but presently used in single mode systems.
(*e*) Good spatial coherence which allows the output to be focused by a lens into a spot which has a greater intensity than the dispersed in focused emission. This provides efficient coupling of optical output power into the fiber.

Efficiency : A number of ways are there to define the operational efficiency of the semiconductor laser :

The **differential external quantum efficiency** η_D is the ratio of the increase in photon output rate for a given increase in the number of injected electrons, and is given by

$$\eta_D = \frac{dP_e/hf}{dI/e}$$

$$\boxed{\eta_D \simeq \frac{dP_e}{dI(E_g)}} \quad ...(7.19)$$

where, P_e = Optical power emitted from device
I = Current
e = Charge on an electron
hf = Photon energy
E_g = Energy bandgap in eV.

η_D gives a measure of the rate of change of optical power output with current and hence defines the slope of the output characteristics in the lasing region for a particular device sometimes η_D is known as **slope quantum efficiency**. For a CW semiconductor laser, its value lies in the range of 40-60%.

Internal quantum effeciency (η_i) of the semiconductor laser is given by

$$\eta_i = \frac{\text{No. of photons produced in the laser cavity}}{\text{No. of injected electrons}} \quad ...(7.20)$$

Usually its value lies in the range of 50 to 100%. It is related to the differential external quantum efficiency by the given expression :

$$\boxed{\eta_D = \eta_i \frac{1}{\left\{\dfrac{2\bar{\alpha}L}{\ln\left(\dfrac{1}{r_1 r_2}\right)}\right\}}} \quad ...(7.21)$$

where, $\bar{\alpha}$ = Loss coefficient of the laser cavity
L = Length of laser cavity
r_1, r_2 = Cleaved mirror reflectivities.

Total efficiency (external quantum efficiency) η_T is defined as

$$\eta_T = \frac{\text{Total number of output photons}}{\text{Total number of injected electrons}}$$

$$= \frac{P_e/hf}{I/e}$$

$$\boxed{\eta_T \simeq \frac{P_e}{IE_g}} \qquad ...(7.22)$$

P_e changes linearly when the injection current I is greater threshold current Ith then :

$$\boxed{\eta_T \cong \eta_D\left(1 - \frac{I_{th}}{I}\right)} \qquad ...(7.23)$$

The **external power efficiency** of the device η_{ep} is converting electrical input to optical output is given by

$$\eta_{ep} = \frac{P_e}{P} \times 100$$

$$\boxed{\eta_{ep} = \frac{P_e}{IV} \times 100\%} \qquad ...(7.24)$$

where, $P = IV$ = d.c. electrical input power

Using Eqn. (7.24), **total efficiency** is given by

$$\boxed{\eta_{ep} = \eta_T\left(\frac{E_g}{V}\right) \times 100\%} \qquad ...(7.25)$$

SOLVED EXAMPLES

Example 7.1. An incandescent lamp operating at a temperature of 1000 K with an average operating wavelength is 0.5 μm. Determine the ratio of stimulated emission rate to the spontaneous emission rate.

Solution. Given that,

$$\lambda = 0.5 \text{ μm} = 0.5 \times 10^{-6} \text{ m}$$

$$T = 1000 \text{ K}$$

$$f = \frac{C}{\lambda} = \frac{3 \times 10^8}{0.5 \times 10^{-6}} = 6 \times 10^{14} \text{ Hz}$$

Ratio of stimulated emission rate and spontaneous emission rate is given by

$$= \frac{1}{\exp\left(\frac{hf}{KT}\right) - 1}$$

Substituting the values, we get

$$\frac{\text{Stimulated Emission Rate}}{\text{Spontaneous Emission Rate}} = \frac{1}{\exp\left[\dfrac{6.626\times10^{-34}\times6\times10^{14}}{1.38\times10^{-23}\times1000}\right]-1} = \exp[-28.8]$$

$$\boxed{\text{Ratio} = 3.1\times10^{-13}} \quad \textbf{Ans.}$$

Example 7.2. Determine the number of longitudinal modes and their frequency separation of a ruby laser contains a crystal length 4 cm with a refractive index of 1.78. The peak emission wavelength from the device is 0.55 μm.

Solution. Given that,

$$n = 1.78$$

$$L = 4 \text{ cm} = 4 \times 10^{-2} \text{ m}$$

$$\lambda = 0.55 \text{ μm} = 0.55 \times 10^{-6} \text{ m}$$

The number of longitudinal modes is given by

$$q = \frac{2nL}{\lambda} = \frac{2\times1.78\times0.04}{0.55\times10^{-6}}$$

$$\boxed{q = 2.6\times10^{5}} \quad \textbf{Ans.}$$

The frequency separation is given by

$$\delta f = \frac{C}{2nL} = \frac{3\times10^{8}}{2\times1.78\times0.04}$$

$$\boxed{\delta f = 2.1 \text{ GHz}} \quad \textbf{Ans.}$$

Example 7.3. A GaAs laser is operating at 850 nm wavelength and has a length of 500 μm. If its refractive index is 3.7, calculate the frequency and wavelength spacing.

Solution. Given that,

$$\lambda = 850 \text{ nm} = 850 \times 10^{-9} \text{ m}$$

$$L = 500 \text{ μm} = 500 \times 10^{-6} \text{ m}$$

$$n = 3.7.$$

The frequency spacing is given by

$$\delta f = \frac{C}{2Ln} = \frac{3\times10^{8}}{2\times3.7\times500\times10^{-6}}$$

$$= \frac{3\times10^{11}}{3.7} = 8.1\times10^{10} \text{ Hz}$$

$$\boxed{\delta f = 81 \text{ GHz}} \quad \textbf{Ans.}$$

The wavelength spacing is given by

$$\delta\lambda = \frac{\lambda^2}{2Ln}$$

$$= \frac{(850\times10^{-9})^2}{2\times500\times10^{-6}\times3.7} = \frac{(85)^2\times10^{-13}}{3.7} = 0.19\times10^{-9}$$

$$\boxed{\delta\lambda = 0.19 \text{ nm}} \quad \textbf{Ans.}$$

Example 7.4. Determine the external power efficiency of the device if the total efficiency of an injection laser with a GaAs active region is 18%. The voltage applied to the device is 2.5 V and the bandgap energy of GaAs is 1.43 eV.

Solution. Given that,

$$\eta_T = 18\% = 0.18$$

$$E_g = 1.43 \text{ eV}$$

$$V = 2.5 \text{ V}$$

The external power efficiency is given by

$$\eta_{ext.} = \eta_T\left(\frac{E_g}{V}\right)\times100\% = 0.18\times\left(\frac{1.43}{2.5}\right)\times100$$

$$\boxed{\eta_{ext.} \simeq 10\%} \quad \textbf{Ans.}$$

SUMMARY

1. The word LASER is an acronym for light amplification of stimulated emission of radiation.
2. Laser is a device to produce a strong monochromatic, collimated and highely coherent beam of light and depends on the phenomenon of **stimulated emission**.
3. Laser beam is highly monochromatic while the intensity of ordinary light sources decreases rapidly with distance.
4. The spatial and spectral characteristics of a semiconductor laser are strongly influenced by the properties of the function medium, such as doping and band tailing.
5. For laser action (lasing) to take place, there should be more electrons at higher energy conduction band than at the lower energy valance band. This situation is known as **population inversion**.
6. A semiconductor diode work like a laser if the following conditions are met
   - population inversion
   - stimulated emission.
   - positive feedback.
7. The external quantum efficiency is given by

$$\eta_{ext.} = \frac{\text{No. of photons escaping from laser's resonator}}{\text{No. of injected charge carriers}}$$

8. There are three types of efficiency of a laser diode :

- power efficiency (η_p)
- external quantum efficiency ($\eta_{ext.}$)
- internal quantum efficiency ($\eta_{int.}$)

9. Internal quantum efficiency is defined as

$$\eta_{int.} = \frac{\text{No. of photons radiated}}{\text{No. of injected charge carriers}}$$

10. There are three major components of laser :

- energy source
- active medium
- optical resonator

11. In a positive index waveguide, the central region has a higher refractive index than outer regions. In negative index waveguide, the central region of the active layer has a lower refractive index than the outer regions.

12. A heterojunction is an interface between two adjoining single crystal semiconductors with different bandgap.

13. Laser oscillations may occur in a direction transverse to the axis of cavity. This gives rise to resonant modes which are transverse to the direction of propagation.

14. Heterojunction may be classified as

- isotype (*n-n* or *p-p*)
- anisotype (*p-n*)

15. Stripe geometry provides optical confinement in the horizontal plane. The stripe is formed by the creation of high resistance areas on either side by means of proton bombardment or oxide isolation.

16. In the DFB optical gating is employed over the whole active region whereas in DBR gating is etched only near the cavity ends.

17. The dynamic behaviour of the injection laser is critical especially when it is used in high bit rate optical fiber communication systems.

18. The temperature dependence of the threshold current density *J*th is approximately exponential for most common structures and is given by

$$J_{th} \ \alpha \ \exp \frac{T}{T_0}$$

19. The relative intensity noise is defined by

$$\text{RIN} = \overline{\delta P_e^2(t)} = \int_0^\infty S_{RIN}(f).df$$

20. Mode hopping occurs in all single mode lasers and is a result of increase in temperature of the device junction. It alters the high output agains current characteristics of the laser.

21. The function of optical transmitter is to convert an electrical input signal into the corresponding optical signal and then launch it into optical fiber.

22. An optical source is the major component of an optical transmitter.

23. The frequency of an atom by absorbing a quantum of radiation (photon) is given by

$$f = \frac{E_1 - E_2}{h}$$

24. The population of the two energy levels of a system is described by Boltzmann statics, given by

$$\frac{N_1}{N_2} = \frac{g_1}{g_2} \exp\left(\frac{hf}{KT}\right)$$

25. The ratio of stimulated emission rate to the spontaneous emission rate is given by

$$\frac{\text{Stimulated emission rate}}{\text{Spontaneous emission rate}} = \frac{B_{21}\rho_f}{A_2} = \frac{1}{\exp\left(\frac{hf}{KT}\right) - 1}$$

26. In population inversion when the two levels are equally degenerated (or not degenerated), then

$$B_{12} = B_{21}$$

27. The resonance condition of a laser structure along the axis of cavity is given by

$$L = \frac{\lambda q}{2n}$$

28. The discrete emission frequency of laser oscillation is given by

$$f = \frac{qc}{2nL}$$

29. The direct semiconductors are in which electrons and holes on either side of the forbidden energy gap have the same value of crystal momentum.

30. The minority carrier lifetime τ_t of a semiconductor is given by

$$\tau_t = [B_\tau (N + P)]^{-1}$$

31. The differential external quantum efficiency, η_D is given by

$$\eta_D \cong \frac{dP_e}{dI(E_g)}$$

32. Total efficiency (external quantum efficiency) η_T is given by

$$\eta_T \simeq \frac{P_e}{IE_g}$$

33. The external power efficiency of the device η_{ep} is converting electrical input to optical output is given by

$$\eta_{ep} = \left(\frac{P_e}{IV}\right) \times 100\%$$

34. There are three types of efficiency of a laser diode :

- power efficiency (η_p)
- external quantum efficiency ($\eta_{ext.}$)
- internal quantum efficiency ($\eta_{int.}$)

35. Internal quantum efficiency is defined as

$$\eta_{int.} = \frac{\text{No. of photons radiated}}{\text{No. of injected charge carriers}}$$

REVIEW QUESTIONS

1. What are stimulated emission and spontaneous emission ? Explain the principle of laser action
2. What do you understand by population inversion ? **[PTU 2007]**
3. Differentiate between spontaneous and stimulated emission.
4. What do you meant by Einsteins relations ? Derive its expression also.
5. What do you understand by stripe geometry ? Explain the widely used stripe geometry for optical communication.
6. Discuss the various types of commonly used bandgap materials.
7. What do you meant by heterojunctions ? Discuss its various types also.
8. Describe the following terms :
   (*a*) mode hopping
   (*b*) frequency chirp
   (*c*) realibility
   (*d*) relaxation oscillator
   (*e*) short and coupled cavity lasers
   (*f*) laser modes.
9. What is frequency chirp ? **[PTU 2008]**
10. How does spontaneous and stimulated emission occur ? **[PTU 2008]**
11. Determine the external power efficiency of a GaAs planar LED. When the transmission factor of the GaAs and interface is 0.68 and internally generated optical power is 30% of the electrical power supplied. The refractive index of GaAs is 3.6.
12. Determine the slope efficiency of a laser diode operating at $\lambda = 1300$ nm if its external quantum efficiency of $\eta_{ext} = 0.1$.
13. A GaAs injection laser with a cavity of length 500 μm has a loss coefficient of 20 cm^{-1}. The measured differential external quantum efficiency of the device is 45%. Calculate the internai quantum efficiency of the laser. The refractive index of GaAs is 3.6.

❑❑❑

Optical Sources : Light Emitting Diode and Laser

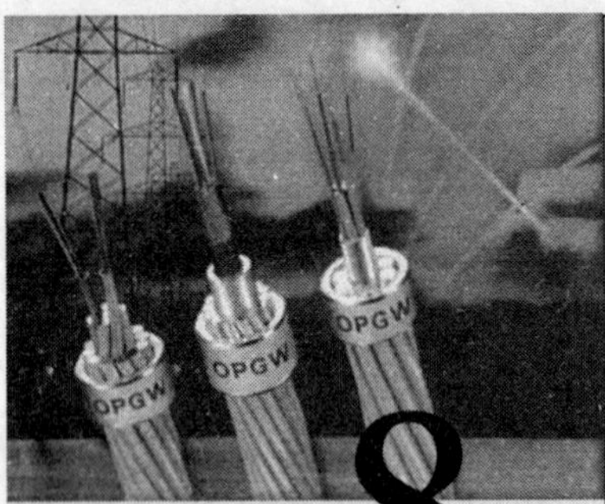

Chapter 8

INSIDE THIS CHAPTER

8.1. INTRODUCTION

Optical sources for fiber communication should satisfy some desirable properties in terms of intensity, radiation pattern, emission wavelength, spectral characteristics and response time. Semiconductor diode sources of the LED and LD types are a natural choice for fiber optic systems due to high efficiency, compact size, low power consumption and high reliability.

A light-emitting diode (LED) is a semiconductor diode that emits incoherent narrow spectrum light when electrically biased in the forward direction of the *p-n* junction. This effect is a form of **electroluminescence.**

An LED is usually a small area source, often with extra optics added to the chip that shapes its radiation pattern. The color of the emitted light depends on the composition and condition of the semiconducting material used, and can be infrared, visible, or near ultraviolet.

Direct bandgap materials such as GaAs are the most appropriate and efficient materials for generation of light. Indirect band materials (Ex. Si), if used will need photon assisted transitions to conserve the momentum. There are **several reasons** because of which **semiconductor light emitting diodes (LED) are preferred light sources** and these are :

1. They give adequate power.

2. They give adequately low noise.
3. The dimensional characteristics of LED's are compatible with those of optical fiber, which means that coupling of fibers with light source will be easier.
4. The light from LED can be varied or modulated by varying the applied voltage or current through the device which is essentially the information to be transmitted.

8.2. HETEROJUNCTIONS

A **heterojunction** is a junction in which the p and n type materials are different semiconductors.

A heterojunction is not unlike an ordinary *p-n* junction. The difference in bandgap energies creates a one-way barrier. Charge carriers (electrons or holes) are attracted over the barrier from the material of higher bandgap energy to the one of lower bandgap energy.

The **two basic LED configurations being used for fiber optics are :**

(1) Surface emitters (also known as burrus or front emitters)

(2) Edge emitters

- One major advantage of heterojunction LED is that it can be an edge emitter.
- In GaAlAs heterojunction the LED's light is created in a central region sandwiched between two regions that contain more aluminium. Because there is a heterojunction on either side of the active region, this structure is known as **double hetrostructure *DH*.**
- It's easy enough to construct a *p-n* junction that will emit light of the required wavelength.

As illustrated in Fig. 8.1, *p-n* junctions are necessarily very thin, flat and need to cover a relatively large area if they have to produce any meaningful amount of light. Further, light is spontaneously emitted in all directions and since the semiconductor material is transparent over the band of wavelengths produced, the light will disperse in all directions. It is very difficult to get any meaningful amount of light into a fiber from a regular *p-n* junction.

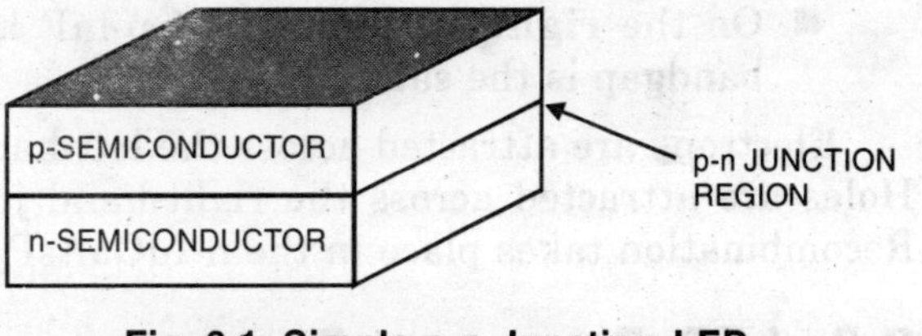

Fig. 8.1. Simple p-n Junction LED.

What is needed is a way of producing light in a more localized area, with a greater intensity and with some way of confining the light produced such that we can get it (or a lot of it), into a fiber. The heterojunction is the answer to this problem.

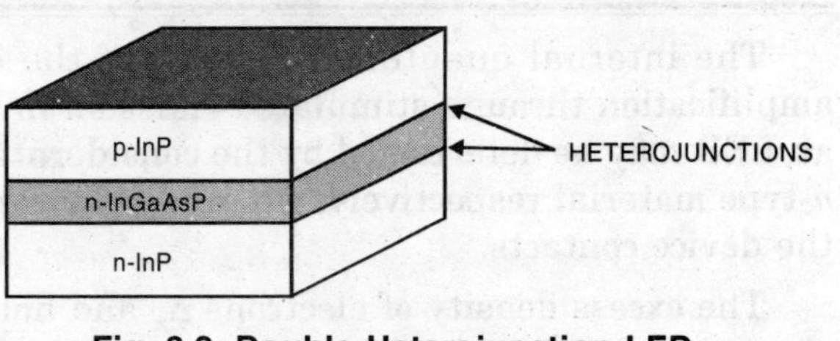

Fig. 8.2. Double Heterojunction LED.

Double Heterojunction : When a layer of material with a particular bandgap energy is sandwiched between layers of material with a higher energy bandgap a double heterojunction is formed. This is called a double heterojunction because there are two heterojunctions present—one on each side of the active material. The double heterojunction forms a barrier which restricts the region of electron-hole recombination to the lower bandgap material. This region is then called the **active region**. An energy diagram of a double heterojunction is shown in Fig. 8.3.

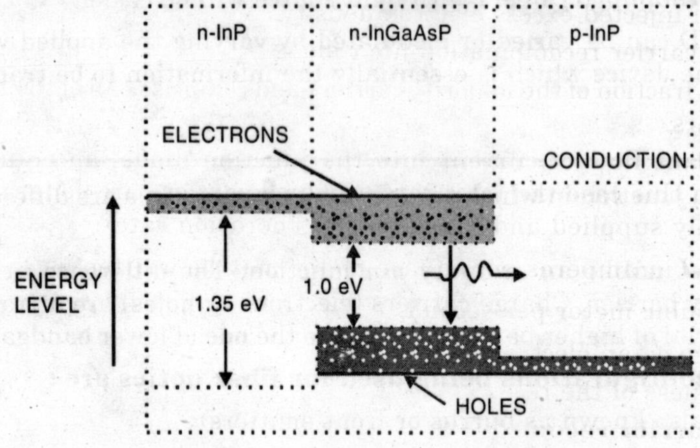

Fig. 8.3. Energy Bands in a Double Heterojunction.

The diagram shows the energy levels for the **three sections** of the double heterojunction.

- On the **left** is n-InP. The lower dotted line represents the energy level of the valence band in this material. The upper dotted line represents the lowest energy in the valence band for this material. Thus there is a bandgap difference of 1.35 eV.
- In the **middle section** of the diagram we see n-InGaAsP. Here the valence band is at a higher energy than the valence band of the adjacent n-InP. The conduction band is at a lower energy level.
- On the **right** we notice that p-lnP has higher energy levels than n-InP but the bandgap is the same.

Electrons are attracted across the left-hand junction from the n-lnP to the n-InGaAsP. Holes are attracted across the right-hand junction from the p-InP into the n-InGaAsP. Recombination takes place in the n-InGaAsP and spontaneous emission (or lasing) occurs.

8.3. LED Power and Efficiency

The internal quantum efficiency of the device is limited by the absense of optical amplification through stimulated emission in the LED. The power generated internally by an LED may be determined by the consideration of excess electrons and holes in the *p* and *n*-type material respectively, when it is forward biased and carrier injection takes place at the device contacts.

The excess density of electrons Δ_n and holes Δ_p is equal since the injected carriers are created and recombined in pairs such that charge neutrality is maintained within the structure.

In extrinsic materials one type of carrier will have much higher concentration than other type, example. In the *p*-type region the hole concentration will be much higher than the electron concentration.

Generally the **excess minority carrier density decays exponentially with time *t*** according to the relation

$$\Delta n = \Delta n(0) \exp\left(\frac{-t}{\tau}\right) \qquad ...(8.2)$$

where, $\Delta n(0)$ = Initial injected excess electron density.

t = Total carrier recombination life time.

Δn = Small fraction of the majority carriers and comprises all of the minority carriers.

When there is a constant flow of current into the junction diode, an equilibrium condition is established. In this case the total rate at which carriers are generated will be the sum of the externally supplied and the thermal generation rates.

The **current density *J*** in amperes per square meter may be written as

J/ed in electrons per cubic meter per second.

where, e = Charge on an electron.

d = Thickness of the recombination region.

A **rate equation for carrier recombination** in the LED :

$$\frac{d(\Delta n)}{dt} = \frac{J}{ed} - \frac{\Delta n}{\tau} (\text{m}^{-3}\text{s}^{-1}) \qquad ...(8.3)$$

The condition for equilibrium is obtained by setting the derivative in Eqn. (8.3) to zero. Hence the **steady state electron density** when a constant current is flowing into the junction region is :

$$\Delta n = \frac{J\tau}{ed} (\text{m}^{-3}) \qquad ...(8.4)$$

In steady state the total number of carrier recombinations per second or the **recombination rate r_t :**

$$r_t = \frac{J}{ed} = r_r + r_{nr} \ (\text{m}^{-3}) \qquad ...(8.5)$$

where r_r is the radiative recombination rate per unit volume and r_{nr} is the nonradiative recombination rate per unit volume.

From Eqn. (8.5) the **total number of recombinations per second R_t**:

$$R_t = \frac{i}{e} \qquad ...(8.6)$$

Internal Quantum Efficiency : The LED **internal quantum efficiency** η_{int} (the ratio of the radiative recombination rate to the total recombination rate) :

$$\eta_{int} = \frac{r_r}{r_t} = \frac{r_r}{r_r + r_{nr}} = \frac{R_r}{R_t} \qquad ...(8.7)$$

where R_r is the total number of radiative recombinations per second. Rearranging Eqn. (8.7) and substituting from Eqn. (8.6) gives :

$$R_r = \eta_{int} \frac{i}{e} \qquad ...(8.8)$$

The **optical power generated** internally by the LED, P_{int} is :

$$P_{int} = \eta_{int} \frac{i}{e} hf \quad \text{...(8.9)}$$

or

$$P_{int} = \eta_{int} \frac{hci}{e\lambda} \quad \text{...(8.10)}$$

For the exponential decay of excess carriers :

The **radiative minority carrier lifetime** is

$$\tau_r = \frac{\Delta n}{r_r}$$

The **nonradiative minority carrier lifetime** is

$$\tau_{nr} = \frac{\Delta n}{r_{nr}}.$$

Therefore from Eqn. (8.7) the **internal quantum efficiency** is :

$$\eta_{int} = \frac{1}{1+\left(\frac{r_{nr}}{r_r}\right)} = \frac{1}{\left[1+\left(\frac{\tau_r}{\tau_{nr}}\right)\right]} \quad \text{...(8.11)}$$

The **total recombination lifetime** τ can be written as $\tau = \frac{\Delta n}{r_t}$ which, using Eqn. (8.5), gives

$$\frac{1}{\tau} = \frac{1}{\tau_r} + \frac{1}{\tau_{nr}} \quad \text{...(8.12)}$$

Hence

$$\eta_{int} = \frac{\tau}{\tau_r} \quad \text{...(8.13)}$$

8.4. LED Structures

Mainly there are five major types of LED structure but only two have been widely used in optical fiber communication. The LED structures are :

- Planar LED
- Dome LED
- Surface emitter LED
- Edge emitter LED
- Superluminescent LED.

8.4.1. Planar LED

It is the **simplest of the structures** that are available and is fabricated by either liquid or vapour phase epitaxial processes over the whole surface of a GaAs substrate. This involves a *p*-type diffusion into the *n*-type substrate in order to create the junction as shown in Fig. 8.4.

(*i*) Forward current flow through the junction gives **Lambertian spontaneous emission** and the device emits light from all surfaces.

(*ii*) Only a limited amount of light escapes the structure due to total internal reflection, and therefore the radiance is low.

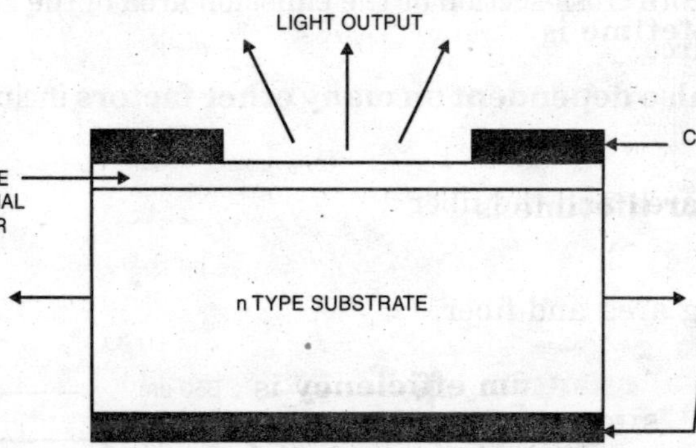

Fig. 8.4.

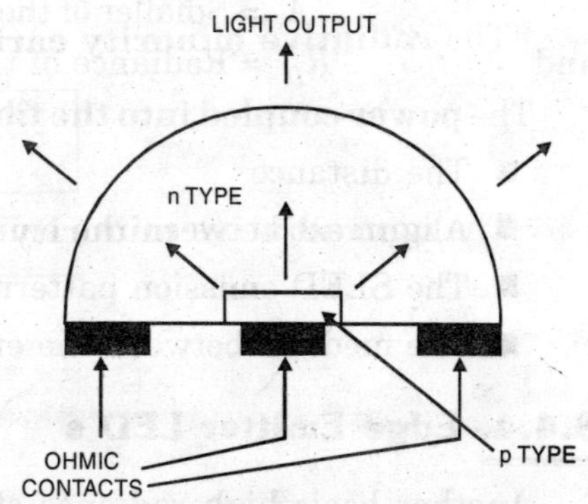

Fig. 8.5.

8.4.2. Dome LED

The structure of a typical dome LED is shown in Fig. 8.5.

(*i*) A hemisphere of *n*-type GaAs is formed around *p*-region.

(*ii*) The diameter of the dome is chosen to maximize the amount of internal emission reaching the surface within the critical angle of the GaAs air interface.

(*iii*) The device has a **higher external power efficiency** than the planar LED.

(*iv*) However, the geometry of the structure is such that the dome must be far larger than the active recombination area, which gives a greater effective emission area and thus reduces the radiance.

8.4.3. Surface Emitter LED's

A method for obtaining high radiance is to restrict the emission to a small active region within the device. The technique pioneered by **Burrus and Dawson** with homostructure devices was to use an etched well in a GaAs substrate in order to prevent heavy absorption of the emitted radiation and physically to accommodate the fiber. These structures have a **low thermal impedance** in the active region allowing high current densities and giving high radiation emission into the optical fiber.

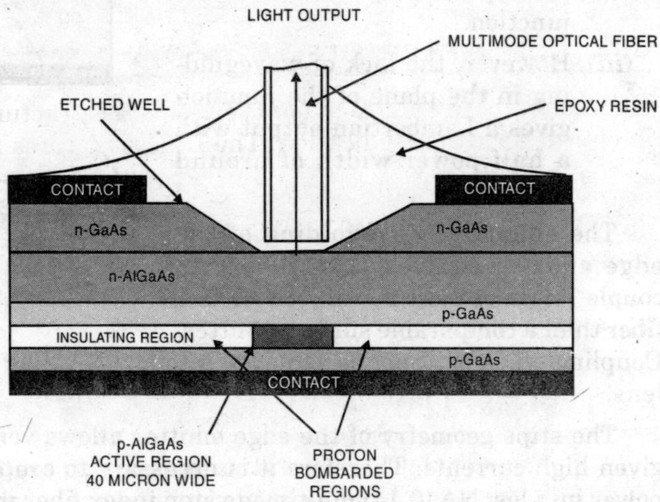

Fig. 8.6.

The structure of a high radiance etched well *DH* surface emitter for the 0.8 to 0.9 μm wavelength band is shown in Fig. 8.6.

The power coupled into a multimode step index fibre :

$$P_c = \pi(1-r)AR_D\,(NA)^2 \quad ...(8.14)$$

where r = Fresnel reflection coefficient at the fiber surface.

A = Smaller of the fiber core cross-section or the emission area of the source.

and R_D = Radiance of the source.

The **power coupled into the fiber is also dependent on many other factors** including :

- The distance
- Alignment between the emission area and the fiber
- The SLED emission pattern
- The medium between the emitting area and fiber.

8.4.4. Edge Emitter LED's

Another basic high radiance structure currently used in optical communications is the stripe geometry. This device has a similar geometry to a conventional contact stripe infection laser as shown in Fig. 8.7.

(*i*) A very thin active layer (50 to 100 μm) to reduce self absorption in the active layer.

(*ii*) The consequent waveguiding narrows the beam divergence to a half power width of around 30° in the plane perpendicular to the junction.

(*iii*) However, the lack of waveguiding in the plane of the junction gives a Lambertian output with a half power width of around 120°.

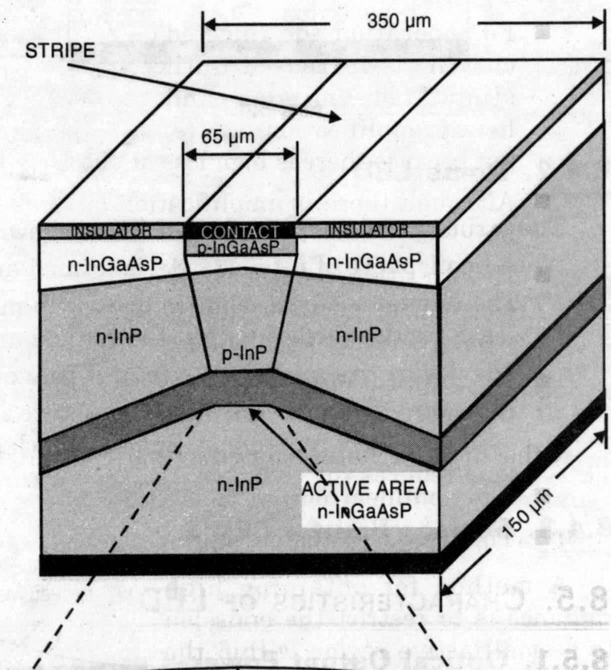

Fig. 8.7.

The enhanced waveguiding of the edge emitter enables it in theory to couple 7.5 times more power into low NA fiber than a comparable surface emitter. Coupling efficiency may be achieved into low NA fiber with surface emitters by the use of a lens.

The stipe geometry of the edge emitter allows very high carrier injection densities for given high currents. Therefore it is necessary to couple approaching a milliwatt of optical power into low NA (0.14) multimode step index fiber with edge emitting LED's operating at high drive currents *i.e.* 500 mA.

Edge emitters have a substantially **better modulation bandwidth** of the order of hundreds of megahertz than comparable surface emitting structures with the same drive level.

8.4.5. Super Luminescent LED's

A third device geometry which is already providing significants benefits over both SLED's and ELED's for communication applications is the superluminescent diode or SLD. This device gives **several advantages :**

1. A high output power.
2. A directional output beam.
3. A narrow spectral linewidth.

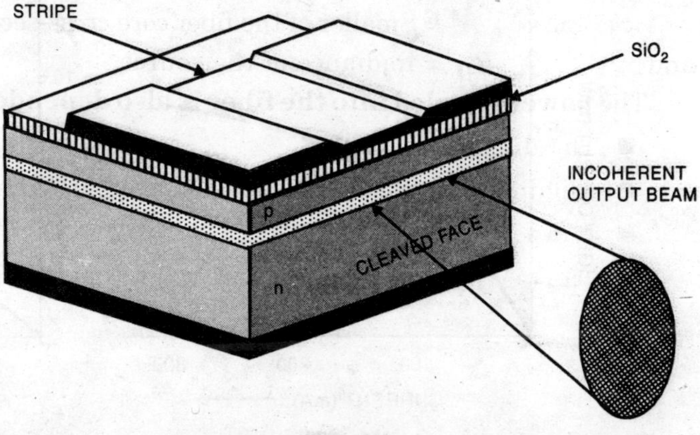

Fig. 8.8.

All of the which prove useful for coupling significant optical power levels into optical fiber to single mode fiber. Figure 8.8 shows the structures of the SLD.

- For operation the injected current is increased until stimulated emission and hence amplification occurs, but because there is high loss at one end of the device, no optical feedback takes place.
- Although there is amplification of the spontaneous emission, no laser oscillations builds up.
- However the operation in the current region for stimulated emission provides gain causing the device output to increase rapidly with increase in drive current due to what is effectively single pass amplification.
- High optical output power can therefore be obtained together with a narrowing of the spectral width which also results from stimulated emission.

Other **drawbacks associated with the SLD in comparison with conventional LED's** are :

- The nonlinear output characteristic
- The increased temperature dependence of the output power.

8.5. Characteristics of LED's

8.5.1. Optical Output Power–Current Characteristics

The ideal **light output power against current characteristic** for an LED is shown in Fig. 8.9.

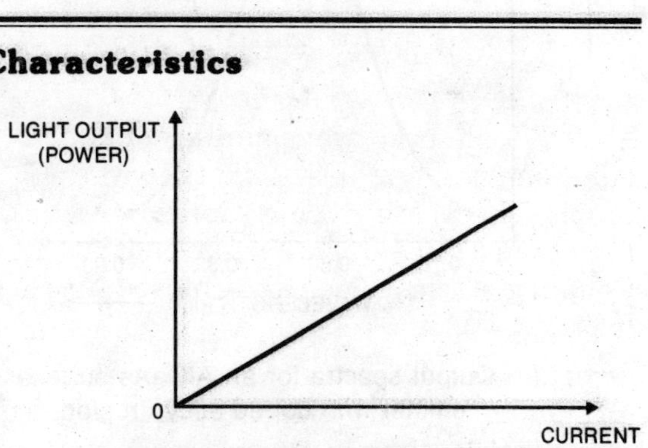

Fig. 8.9. An Ideal Light Output Against Current Characteristics an LED.

It is linear corresponding to the linear part of the injection laser optical power output characteristic before lasing occurs. Basically LED is a very linear device in comparison with the majority of injection lasers and hence it is more suitable for

analog transmission. Practically LED's exhibit nonlinear behaviour because it depends on configuration utilized.

Therefore it is necessary to use same form of **linearizing circuit technique to get the linear performance of the device** to allow its use in high quality analog transmission systems Fig. 8.10 (*a*) and (*b*) shows the light output against current characteristics for good surface and edge emitters respectively.

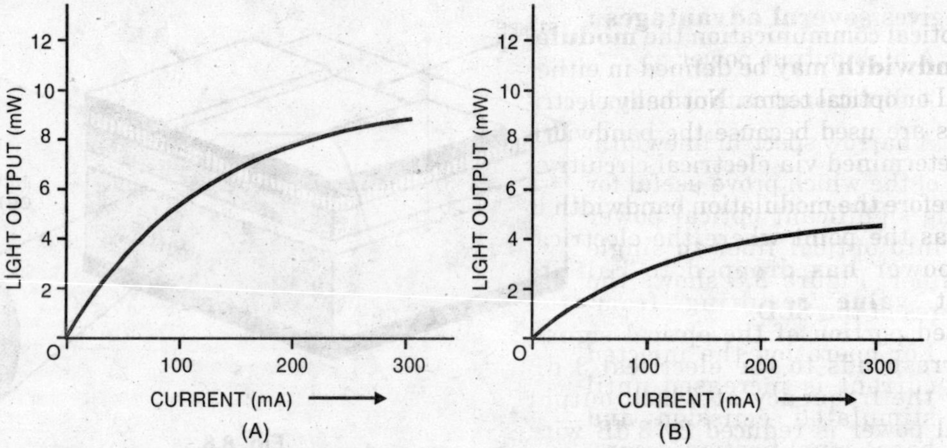

Fig. 8.10. Light Output Power Into Air Against d.c. Driven Current. (a) An AlGaAs Surface Emitter with 50 µm Diameter Dot Contact. (b) An AlGaAs Edge Emitter with a 65 µm Wide Stripe and 100 µm Length.

8.5.2. LED Output Spectrum

The spectral linewidth of an LED operating at room temperature in the 0.8 to 0.9 µm wavelength band is usually between 25 and 40 nm at the half maximum intensity points.

For materials with smaller bandgap energies operating in the 1.1 to 1.7 µm wavelength region the linewidth tends to increase to around 50 to 160 nm. **Two such types of output spectrums** are shown in Fig. 8.11.

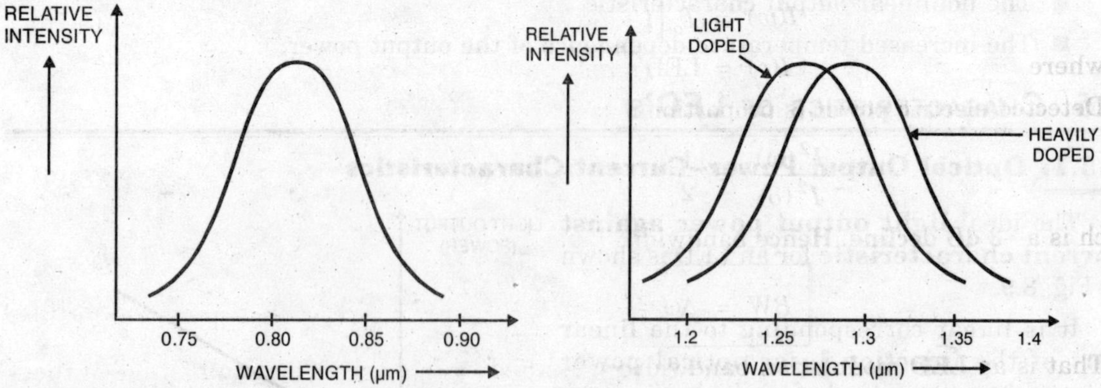

(a) Output spectra for an AlGaAs surface emitter with doped active region.

(b) Output spectra for an InGaAsP surface emitter showing both lightly doped and heavily doped cases.

Fig. 8.11. LED Output Spectra.

Figure 8.11(*b*) shows the increase in linewidth due to increased doping levels. There is a shift to lower peak emission wavelength through reduction in doping and hence the active layer composition must be adjusted if the same centre wavelength is to be maintained.

8.5.3. Modulation Response

In optical communication the **modulation bandwidth** may be defined in either electrical or optical terms. Normally electrical terms are used because the bandwidth can be determined via electrical circuitry.

Therefore the modulation bandwidth is defined as the point where the electrical signal power has dropped to half its constant value resulting from the modulated portion of the optical signal. This corresponds to the electrical 3 dB point or the frequency at which output electrical power is reduced by 3 dB with respect to the input electrical power as shown in Fig. 8.12.

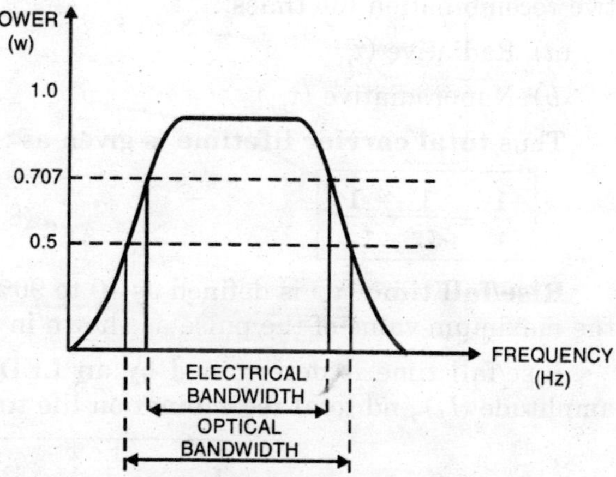

Fig. 8.12. Frequency Response of the Optical Source.

The **relationship between bandwidth and rise time** is given as

$$BW = \frac{0.35}{t_r} \quad ...(8.15)$$

There is a difference between the practical and theoretical value of *BW*. This discrepancy occurs because if the forward current is modulated at angular frequency (w), an LED's output light intensity $I(w)$ will vary as follows :

$$I(w) = I_o \left[1 + (\omega\tau)^2\right]^{-1/2} \quad ...(8.16)$$

where $I(o)$ = LED's light intensity at constant current.

Detected electric power is proportional to I^2. By taking,

$$\frac{I^2(\omega)}{I^2(o)} = \frac{1}{2} \quad ...(8.17)$$

which is a –3 dB decline. Hence bandwidth becomes,

$$BW = \Delta\omega = \frac{1}{\tau} \quad ...(8.18)$$

That is an LED modulation bandwidth is limited by the recombination lifetime of the charge carriers.

Another **important characteristics** of an LED is **power bandwidth product.** The product of LED's power output power and its modulation bandwidth is constant.

$$BW \times P = \text{Constant}$$

i.e., by increasing bandwidth, the output power will decrease and vice versa.

8.5.4. Life Time, Rise/Fall Time

Life time τ of the charge carriers is the time between the moment they excited and the moment they recombine. This is also known as **recombination life time** and it ranges from nanoseconds to milliseconds. Basically there are two recombination life times :

(*a*) Radiative (τ_r)

(*b*) Nanoradiative (τ_{nr}).

Thus **total carrier lifetime** is given as :

$$\frac{1}{\tau} = \frac{1}{\tau_r} + \frac{1}{\tau_{nr}} \quad ...(8.19)$$

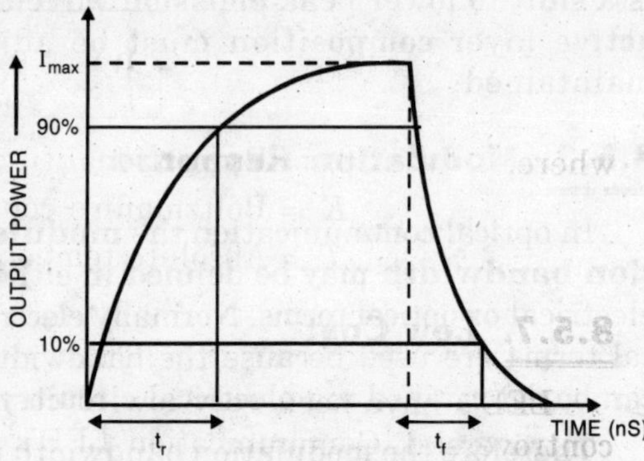

Fig. 8.13. Rise Time/Fall Time.

Rise/fall time (τ_r) is defined as 10 to 90% of the maximum value of the pulse as shown in Fig. 8.19.

Rise/fall time is determined by an LED's capacitance (C), input step current with amplitude (I_P) and total recombination life time (τ).

$$\tau_r = 2.2\left[\tau + \left(\frac{1.7 \times 10^{-4} \times T^o K \times C}{I_P}\right)\right] \quad ....(8.20)$$

where T = absolute temperature in K.

8.5.5. Light Output Temperature Dependence

The internal quantum efficiency of the LED's decreases exponentially with increasing temperature.

The edge-emitting device exhibits a greater temperature dependence than the surface emitter and the output of SLD is *strongly* dependent on the junction temperature.

Figure 8.14 shows the variation of optical output power with temperature for different LED's.

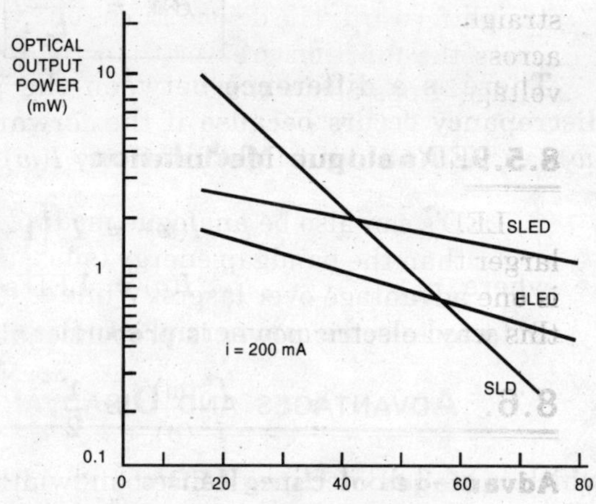

Fig. 8.14.

8.5.6. Reliability

LED's are generally not affected by the **catastrophic degradation.** They do **exhibit gradual gradation (defects).**

The optical output power $P_e(t)$ may be expressed as a function of the operating time t :

$$P_e(t) = P_{out} \exp(-\beta_r t) \quad ...(8.21)$$

where, P_{out} = Initial output power

and β_r = Degradation rate.

The **degradation rate** is characterized by the activation energy of homogeneous degradation E_a and is a function of temperature :

$$\beta_r = \beta_0 \exp\left[-\frac{E_a}{KT}\right] \quad ...(8.22)$$

where, β_0 = Proportionality constant,

K = Boltzmann's constant,

T = Absolute temperature of the emitting region.

8.5.7. Low Cost

LED's have been very low in cost compared to communication lasers. This is highly controversial. Communication LED's and lasers are not too different in their structures and are comparable in manufacturing cost. Connecting to single mode fiber (pigtailing) is significantly more costly than connecting to multimode fiber and since lasers are commonly used with single mode and LED's with multimode there is a cost difference here.

However whilst most LED's cant be used with single-mode fiber, lasers certainly can be used with multimode fiber. In early 1996, people began to use the lasers from CD-ROM players for short distance communications on MM fibre. These are about 1/10 the cost of standard communications LED's (simply because they are made in vast volume—several million per year).

8.5.8. Digital Modulation

LED's cannot produce pulses short enough to be used at gigabit speeds. However, systems using LED's operate quite well at speeds of up to around 300 Mbps. Digital modulation is straightforward. The device "turns on" when the forward voltage applied results in a potential across the junction greater than the bandgap energy required. It extinguishes when the voltage drops below that.

8.5.9. Analogue Modulation

LED's can also be analogue modulated quite simply by maintaining a forward bias just larger than the bandgap energy (since the device response is linear with current flow). This is one advantage over lasers. While lasers can be analogue modulated and are indeed used this way in some commercial situations, this is not an easy thing to do.

8.6. Advantages and Disadvantages of Using LED's

Advantages of Using LED's

- LED's produce more light per watt than do incandescent bulbs; this is useful in battery powered or energy saving devices.
- LED's can emit light of an intended color without the use of color filters that traditional lighting methods require.
- When used in applications where dimming is required, LED's do not change their color tint.
- LED's are ideal for use in applications that are subject to frequent on-off cycling.

- LED's, being solid state components, are difficult to damage with external shock. Fluorescent and incandescent bulbs are easily broken if dropped on the ground.
- LED's light up very quickly. A typical red indicator LED will achieve full brightness in microseconds.
- LED's used in communications devices can have even faster response times.
- LED's can be very small and are easily populated onto printed circuit boards.

Disadvantages of Using LED's

- LED's are currently more expensive, price per lumen, on an initial capital cost basis, than more conventional lighting technologies.
- LED performance largely depends on the ambient temperature of the operating environment. Driving an LED difficult in high ambient temperatures may result in overheating of the LED package, eventually leading to device failure.
- LED's must be supplied with the correct current. This can involve shunt resistors or regulated power supplies.
- LED's cannot be used in applications that need a sharply directive and collimated beam of light. LED's are not capable of providing directivity below a few degrees.

8.7. Introduction of Laser

The word LASER is an acronym for **light amplification by stimulated emission of radiation.** It is a device that produces optical radiation by the process that produces optical radiation. It is a device that produces optical radiation by the process of **stimulated emission.**

Laser light is different from the ordinary light source in following ways :

- Laser beam produces coherent light *i.e.* in same phase and of same frequency while ordinary light source such as incadescent lamp sources produces in coherent light.
- Laser beam is highly monochromatic while ordinary light sources spread over a wavelength range of 100 Å to 1000 Å.
- Laser beam is extremely intense while the intensity of ordinary light source decreases rapidly with distance.
- Laser beam does not diverge while ordinary light source are highly divergent *i.e.* the light travels long distances with spreading.

Many types of materials including gas, liquid and semiconductors can form the lasing medium. For optical fibre systems, the exclusively used laser sources are semiconductor laser diodes.

Semiconductor laser diodes were developed in the 1970s and they have found vast commercial applications in compact disc (CD) players. Semiconductor lasers are similar to other lasers such as conventional solid state and gas lasers in that the emitted radiation has spatial and temporal coherence. This means that laser radiation is highly monochromatic and it produces highly directional beams of light.

There are many reasons those **differentiate semiconductor lasers from other lasers :**

- In semiconductor lasers the quantum transitions are associated with the band properties of materials while in conventional lasers transitions occur between discrete energy levels.

- The semiconductor laser is very compact in size *i.e.* 0.1 mm long.
- The divergence of laser beam is considerably larger than in a conventional laser because of its very narrow active region.
- The spatial and spectral characteristics of a semiconductor laser are strongly influenced by the properties of junction medium *i.e.* doping and band tialing.

8.8. THE SEMICONDUCTOR INJECTION LASER

In this section the electroluminescent properties of forward biased *p-n* junction diode have been cansidered. In the semiconductor injection laser, stimulated emission by the recombination of the injected carriers is encouraged. Thus injection laser has **several advantages** over other semiconductor sources. These are as follows :

(*a*) High radiance due to the amplifying effect of stimulated emission. Generally injection lasers supply milliwatts of optical power output.

(*b*) Narrow bandwidth of the order of ≤ 1 nm which is used to minimize the effects of materials dispersion.

(*c*) Modulation capabilities can be extended to gigahertz range.

(*d*) Relative temporal coherence allows heterodyne (coherent) detection in high capacity systems but presently used in single mode systems.

(*e*) Good spatial coherence which allows the output to be focused by a lens into a spot which has a greater intensity than the dispersed infocused emission. This provides efficient coupling of optical output power into the fiber.

These advantages together with the compatibility of injection laser with optical fibers led to the early developments of the device in 1960s. Early injection lasers had the form of a **Fabry Perot cavity.** Often it is fabricated in gallium arsenide which was the major III-V compound semiconductor with electroluminescent properties at the appropriate wavelength for first generation systems.

Fabry-Perot Lasers

The Fabry-Perot laser is conceptually like an LED with a pair of end mirrors. The mirrors are required to create the right conditions for lasing to occur but practically it is somewhat more complex. The Fabry-Perot laser gets its name from the fact that its cavity acts as a Fabry-Perot resonator.

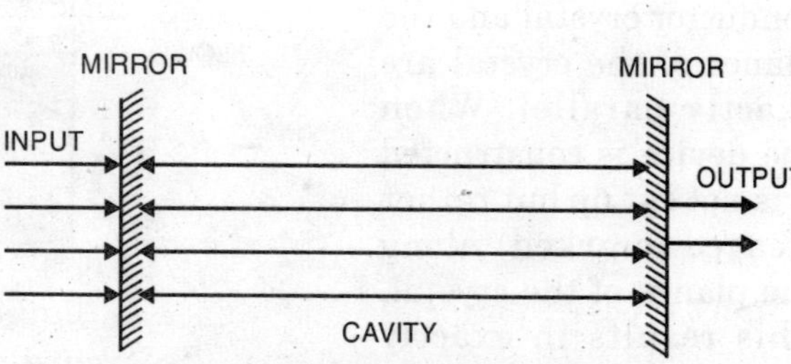

Fig. 8.15. Fabry-Perot filter.

The **principle of the Fabry-Perot filter** is illustrated in Fig. 8.15. When we put two mirrors opposite to each another they form a resonant cavity. Light will bounce between the two mirrors. When the distance between the mirrors is an integral multiple of half wavelengths, the light will reinforce itself. Those wavelengths which are not resonant undergo destructive interference and are reflected away.

In this principle the light is emitted within the cavity itself rather than arriving from outside. In some sense every laser cavity is a Fabry-Perot cavity. But when the cavity is very long compared to the wavelength involved we get a very large number of resonant wavelengths all of which are very close together. Thus the important filtering characteristics of the Fabry-Perot cavity are lost.

Wavelengths produced are related to the distance between the mirrors by the following formula :

$$C_l = \frac{\lambda x}{2n} \qquad ...(8.23)$$

where λ = Wavelength

Cl = Length of the cavity

x = An arbitrary integer : 1, 2, 3, 4...

n = Refractive index of active medium

This is a very simple equation and the only variable in the equation is the refractive index of the gain medium (dielectric) in the cavity.

Construction of a Fabry-Perot Laser

In its basic form an FP laser is like an edge-emitting LED with mirrors on the ends of the cavity. In an ideal laser there is no problem of guiding the light at all. Lasing takes place only between the mirrors and the light produced is exactly positioned but it is not very simple.

These devices are made from a single semiconductor crystal and the planes of the crystal are exactly parallel. When the device is constructed it is not cut up but rather cleaved (cracked) along the planes of the crystal. This results in exactly parallel mirrors at each end of the cavity. The interface of the semiconductor medium (refractive index usually around 3.5) and air (RI around 1.1) forms a mirror.

The active layer is very thin and the refractive index difference between the material of the

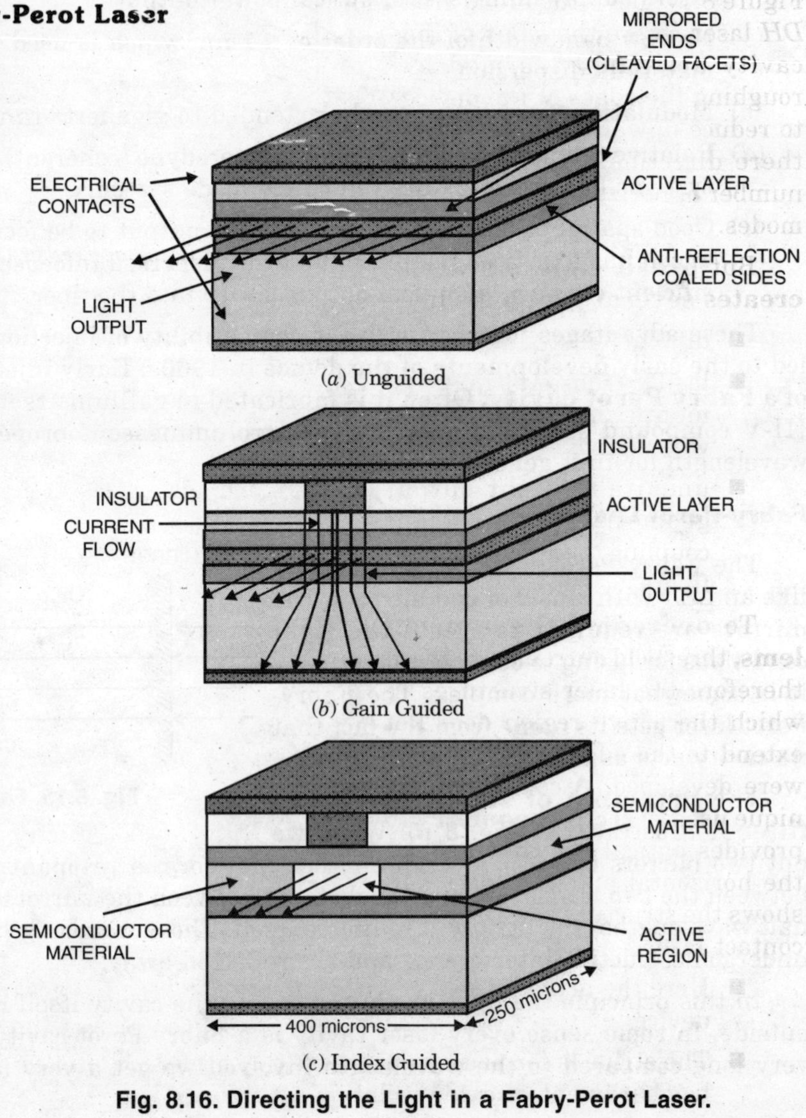

Fig. 8.16. Directing the Light in a Fabry-Perot Laser.

active layer and the surrounding material is not good. Thus lasing is not obtained in the vertical (transverse) mode. We can get lasing in the lateral mode but this is minimized by either coating the sides with an anti-reflection material or just making sure the sides are rough.

8.9. Stripe Geometry

The *DH* laser structure provides optical confinement in the vertical direction through the refractive index step at the hetero function interfaces but lasing takes place across the whole width of the device. Figure 8.17 shows the broad area *DH* laser where the sides of the cavity are simply formed by roughing the edges of the device to reduce unwanted emission in there directions and limit the number of horizontal transverse modes.

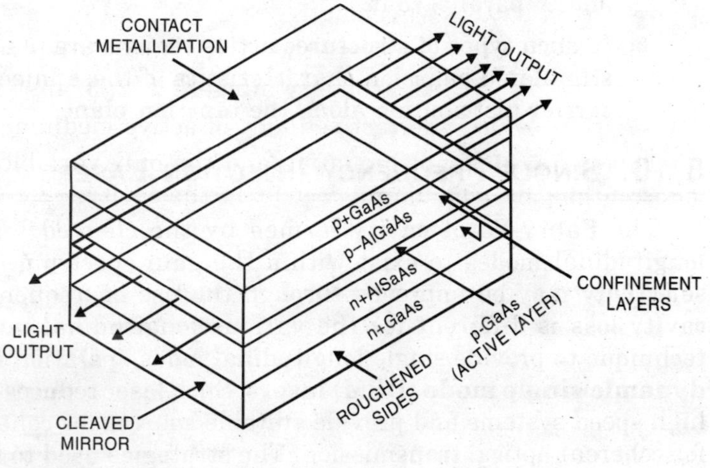

Fig. 8.17. A Broad Area GaAs/AlGaAs DH Injection Laser.

But **broad emission area creates several problems** like :

- difficult heat sinking.
- lasing from multiple filaments in the relatively wide active area.
- unsuitable light output geometry for efficient coupling to the cylinderical fibers.

To **overcome these problems,** threshold current is reduced therefore the laser structures in which the active region does not extend to the edges of the device were developed. A common technique used in **stripe geometry** to provides optical containtment in the horizontal plane. Figure 8.18 shows the structure of a DH stripe contact laser.

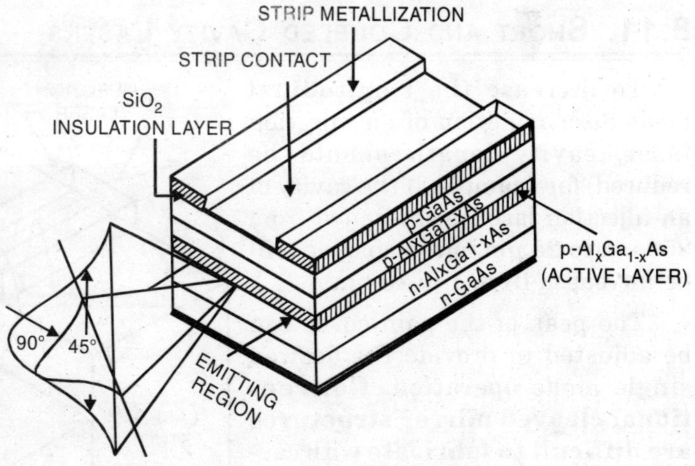

Fig. 8.18. Schematic Representation of an Oxide Stripe AlGaAs DH Injection Laser.

- Here the active region is within the stripe because major current flow through the device.
- The stripe is formed by the creation of high resistance area on either side by proton bombardment or oxide isolation techniques.

- Therefore the stripe acts as a guiding mechanism which overcomes the major problems of broad area device.
- The stripe contact device also gives single transverse (in direction parallel to junction plane) mode operation while broad area device tends to allow multimode operation in this horizontal plane.
- Typically the output beam divergence is 45° perpendicular to the plane of the junction and 9° parallel to it.
- In such types of structures active regions are planar and continuous. Therefore the stimulated emission characteristics of these injection lasers are determined by the carrier distribution along the function plane.

8.10. Single Frequency Injection Lasers

The Fabry-Perot cavity formed by the cleaved laser mirrors may allows several longitudinal modes to exist within the gain spectrum of the device. Longitudinal mode selectivity may be improved through the use of frequency selective feedback so that the cavity loss is different for the various longitudinal modes. Devices which employ this technique to provide single longitudinal mods operation is known as **single frequency** or **dynamic single mode (DSM) lasers**. Such laser reduces fiber intermodal dispersion within high speed systems and provide suitable sources for coherent optical transmission sources for coherent optical transmission. The strategies used to achieve single frequency laser are

(1) Short cavity resonator.

(2) Coupled cavity resonator.

(3) Distributed feedback.

8.11. Short and Coupled Cavity Lasers

To increase the longitudinal mode discrimination of an injection laser, cavity length should be reduced, for example, if the cavity of an injection laser is shortened from 250 μm to 25 μm, mode spacing will be increaesd from 1 to 10 nm.

The peak of the gain curve can be adjusted to provide the desired single mode operation. **Conventional cleaved mirror structures are difficult to fabricate** with cavity length less than 50 μm, therefore configurations employing resonators *i.e.* either microcleaved or etched have been utilized. Such resonators form a short cavity with length 10 μm–20 μm in a direction normal to the active region provide stable single frequency resonator.

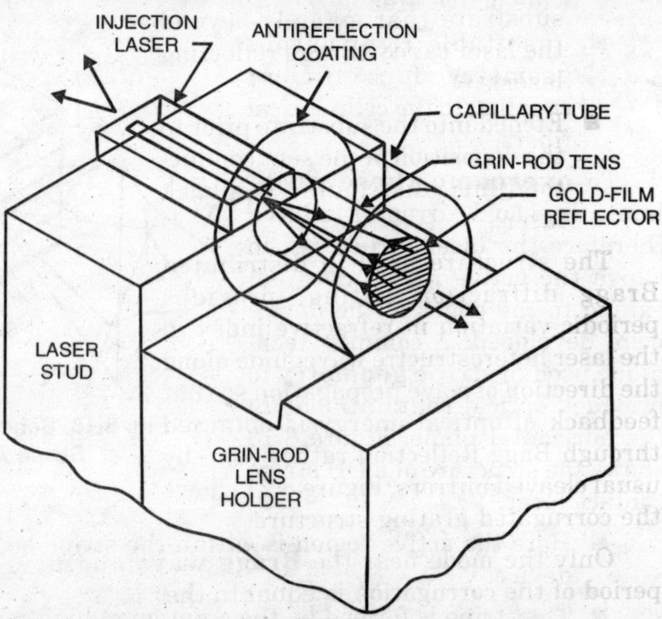

Fig. 8.19. Short External Cavity Laser Using GRIN Rod Lens.

Multiple element resonators or resonators with distributed reflectors also give a loss mechanism. Mode selectivity in such a coupled cavity laser is obtained when the longitudinal modes of each Fabry-Perot cavity coincides. Figure 8.19 shows an example of three mirror resonator which uses a **graded index (GRIN)** rod lens to enhance the coupling to an external mirror.

Figure 8.20 shows the alternative approach in which two active laser sections are separated by a gap of approximately a single wavelength. When the gap is obtained by recleaving a finished laser chip into two partially attached segments it yields the **cleaved coupled cavity (C^3)** laser. This four mirror resonantor device has provided dynamic single mode operation with side mode suppression ratio of several thousands. The single frequency emission of C^3 device can be tuned discretely over a range of 26 nm by varying the current through one section.

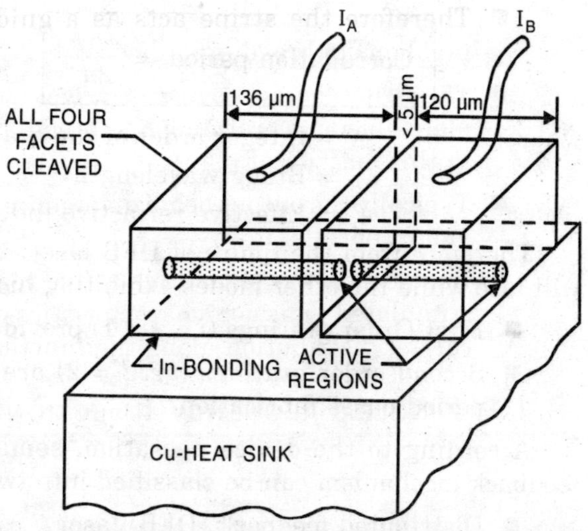

Fig. 8.20. Cleaved Coupled Cavity Laser.

8.12. Distributed Feedback Lasers (DFB)

An elegant approach used to pick out a single longitudinal mode is to use a diffraction grating. The grating can be :

- Separate from the substrate.
- Etched into the portion of the substrate that extends beyond the laser cavity (Bragg reflecting laser).
- Etched into the substrate prior to the deposition of the semiconductor films (distributed feedback laser).

The structure used in distributed Bragg diffraction gating, provides periodic variation in refractive index in the laser heterostructre waveguide along the direction of wave propagation so that feedback of optical energy is obtained through Bagg Reflection rather than by usual cleaved mirrors. Figure 8.21 shows the corrugated grating structure.

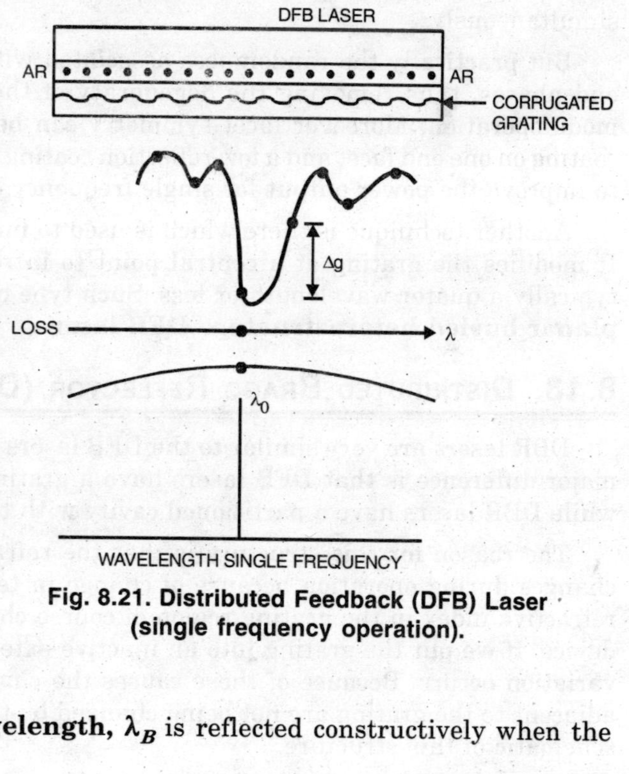

Fig. 8.21. Distributed Feedback (DFB) Laser (single frequency operation).

Only the mode near the **Bragg wavelength, λ_B** is reflected constructively when the period of the corrugation is equal to the

$$\text{Corrugation period} = \frac{l\lambda_B}{2n_e} \qquad ...(8.24)$$

where, λ = Integer order of the grating

λ_B = Bragg wavelength

n_e = Effective refractive index of the waveguide

Therefore from the figure of DFB laser, it may be observed that this particular mode will lase while the other modes exhibiting higher losses are suppressed from oscillation.

- First Order gratings (*i.e.* $l = 1$) provide the strongest coupling within the device.
- Second order gratings (*i.e.* $l = 2$) are sometimes used because their large spatial period eases fabrication.

According to the device operation, semiconductor lasers employing the distributed feedback mechanism can be classified into two broad categories :

- Distributed feedback (DFB) laser
- Distributed Bragg Reflector (DBR) laser.

In the DFB laser the optical grating is usually applied over the entire active region which is pumped.

When considering a DFB laser with both end facets antireflection (AR) coated, then two modes located symmetrically on either side of the Bragg wavelength will experience the same lowest threshold gain within an ideal symmetrical structure and will therefore lase simultaneously.

But practically the randomness associated with the cleaving process creates different end phases, thus removing the degeneracy of the modal gain and providing only single mode operation. Moreover facet symmetry can be increased by placing a high refraction coating on one end facet and a low reflection coating on the other (**known as hi-lo structure)** to improve the power output for single frequency operation.

Another technique is there which is used to improve the performance of the DFB laser. It modifies the grating at a central point to introduce an additional optical phase shift, typically a quater wavelength or less. Such type of structure is kown as **double channel planar buried heterostructure DFB laser.**

8.13. Distributed Bragg Reflector (DBR) Lasers

DBR lasers are very similar to the DFB lasers as described in the previous section. The major difference is that DFB lasers have a grating within the active region of the cavity while DBR lasers have a partitioned cavity with the grating in a region that is not active.

The reason for this structure is that the refractive index within the cavity of a laser changes during operation because of change in temperature and electron flux. Change of refractive index in the grating region of course changes the operational wavelength of the device. If we put the grating into an inactive extension of the cavity then less wavelength variation occurs. Because of these causes the characteristics of the material immediately adjacent to the grating are not being changed by the laser's operation. Figure 8.22 shows a schematic of this structure.

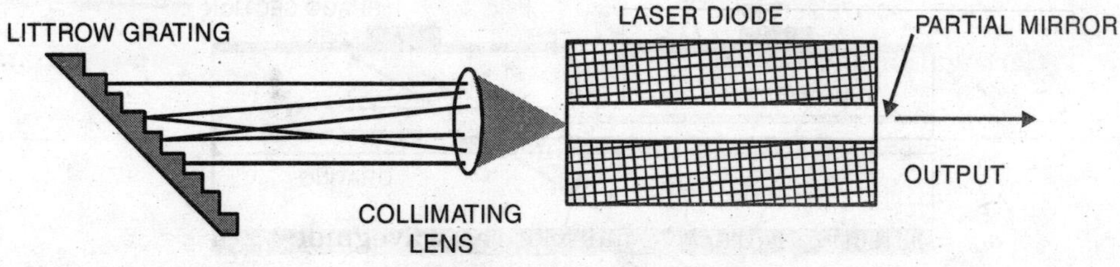

Fig. 8.22. DBR Laser-Schematic.

As in the DFB laser no end mirrors required as the grating can provide sufficient reflection. However end mirrors are always used in an asymmetric configuration. Typically this is 30% mirror on the back facet and either just a cleaved facet or a 4% mirror on the front (output) facet.

Typically DBR lasers produce a single line only with a line width of around .0001 nm. The major problem is that there can be significant absorption in the inactive region near the grating. This causes a loss in efficiency.

8.13.1. Tunable DBR Lasers

There are several situations where it is required to change (tune) the laser's wavelength.

- Some WDM systems required to tune the laser very quickly within a few tens of nanoseconds to a particular wavelength (or channel). A number of proposals for optical LAN and MAN networks require this ability.
- In other WDM situations we may want to establish a path through a wide area network by using wavelength routing. This would be appropriate in a wide area telecommunication backbone network where tuning is used to configure and re-configure routes through the network. In this case a tuning time of around a second would be quite acceptable.
- Still other WDM situations might require the setting up of "calls" across the network (like telephone calls). Set up time in this case can be perhaps a hundred microseconds or so.
- In early fixed channel WDM systems there were two main problems with laser wavelengths :
  (*i*) When lasers are made it is extremely difficult and costly to specify the wavelength exactly. Randomness determines the required wavelengths and rejects the rest but this is expensive.
  (*ii*) Over the time the materials of which the laser is made deteriorate a bit and the wavelength changes. This is fine in single-channel systems but anathema to WDM systems as wavelengths will start to run into one another.

In both of these cases a laser that can be tuned slowly over a relatively narrow range that can make a big improvement. So different systems require different kinds of tunability *i.e.* range and speed.

Figure 8.23 shows two electronically tunable DBR lasers. One is a two-section device and the other has three sections.

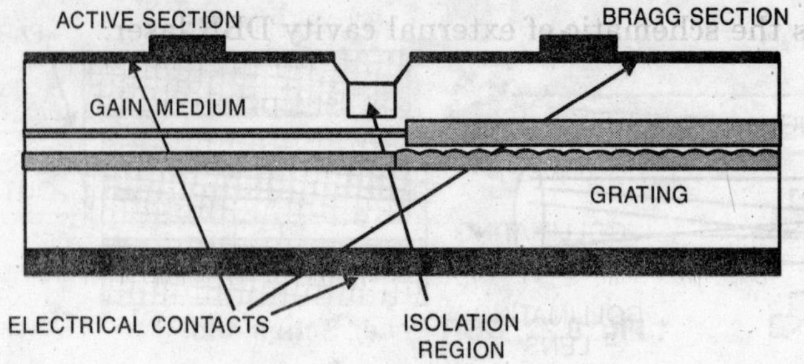

(a) Tunable 2-section DBR Laser.

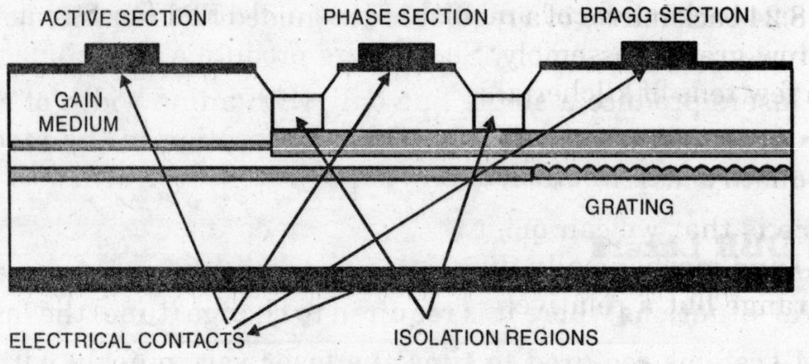

(b) Tunable 3-section DBR Laser.

Fig. 8.23.

The tunable 2-section device has a discontinuous tuning characteristic over a range of a bit less than 10 nm while the three-section device can be continuously tuned over the same range.

Operation (of the 3-section device) is as follows :

- Current in the left-hand section of the device flows through the gain medium and here lasing (amplification) takes place.
- In the centre and rightmost sections the electron flux within the cavity causes the cavity to change in refractive index within these sections.
- The change in RI causes the wavelength of the light emitted to change also.
- The right-hand section has the grating inside it (or rather immediately adjacent to it) and this is used for broad tuning.
- Fine tuning is accomplished by varying the current in the "phase section" (middle part) of the laser cavity.

A disadvantage of the 3-section device is that it requires relatively complex electronics for control.

8.13.2. External Cavity DBR Lasers

An interesting form of DBR laser can be constructed where the resonant cavity extends outside the laser chip itself. Since the length of cavity is extended thus the possible number of longitudinal modes increases exponentially. The major reason for the external cavity is that we can use a much stronger (more selective) grating.

Figure 8.24 shows the schematic of external cavity DBR laser.

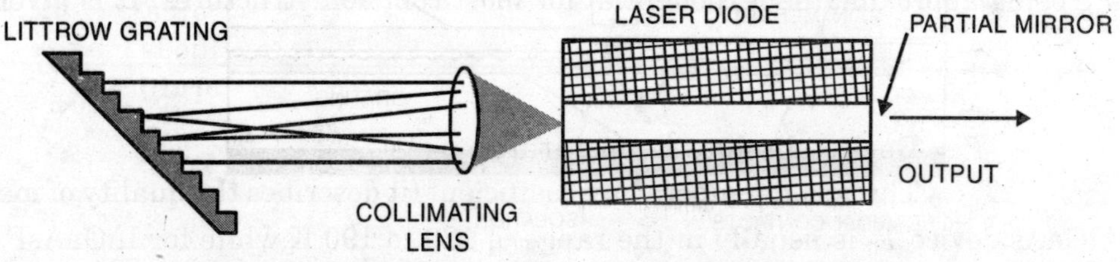

Fig. 8.24. External Cavity DBR Laser Schematic.

As shown in Fig. 8.24 back mirror of a regular index-guided FP laser is removed and replaced with a lens and Littrow grating assembly. Such lasers produce a single line only with a very narrow line width (a few tens of kilohertz).

Here the **major problem** is that these devices are significantly more expensive than other types of high quality laser.

A **useful feature** is that we can mount the grating on a moveable platform (such as a piezoelectric crystal) and mechanically tune the device by tilting the grating. This provides a very wide tuning range but a relatively slow tuning speed.

8.14. Injection Laser Characteristics

The injection laser is used for optical fiber communication thus its characteristics should be studied which may affect its efficient operation. Some major operating characteristics of the device which generally apply to all the various materials and structures are discussed here.

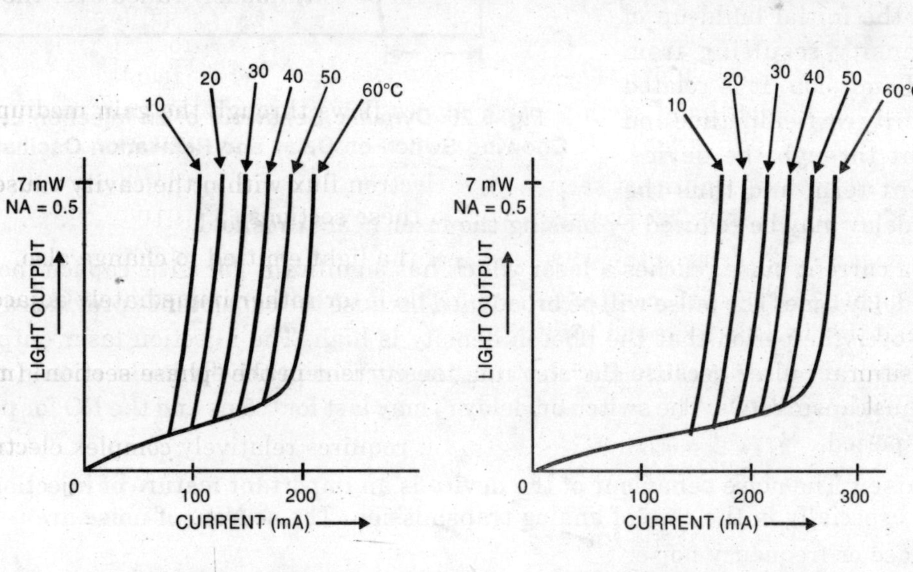

Fig. 8.25. Variation in Threshold Current With Temperature for Gain Guided Injection Lasers.

(1) Threshold Current Temperature Dependence : Generally the threshold current tends to increase with temperature, the temperature dependence of the threshold current density J_{th} being approximately exponential for most common structures. It is given by

$$J_{th} \;\alpha\; \exp\left(\frac{T}{T_0}\right) \qquad \text{...(8.25)}$$

where, T = Device absolute temperature.

T_0 = Threshold temperature coefficient (it describes the quality of material)

For AlGaAs device T_0 is usually in the range of 120 to 190 K while for InGaAsP device it is between 40 and 75 K. The increase in threshold current with temperature for AlGaAs devices can be accounted for with reasonable accuracy by consideration of increasing energy spread of electrons and holes injected into the conduction and valence bands.

(2) Dynamic Response : When injection laser is used in high bit rate (wideband) optical fiber communication systems, its dynamic behaviour is critical. The application of a current step to the device gives **switch on delay**, sometimes followed by high frequency damped oscillations known as **relaxation oscillations (RO).** These transient phenomenon occur when the electron and photon populations within the structure come into equilibrium, as shown in Fig. 8.32.

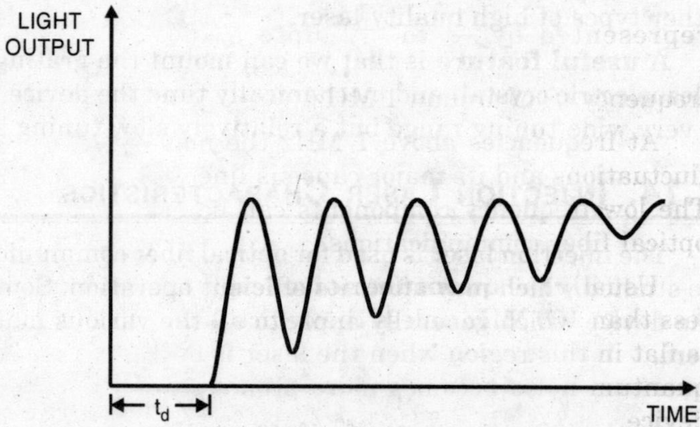

Fig. 8.26. Dynamic Behaviour of an Injection Laser Showing Switch on Delay and Relaxation Oscillations.

The **switch on delay** is caused by the initial build up of photon density resulting from stimulated emission. It is related to the minority carrier lifetime and the current through the device. The current term and thus the switch on delay may be reduced by biasing the laser near threshold.

When a current pulse reaches a laser which has significant parasitic capacitance after the initial delay time. The pulse will be broadened because the capacitance provides a source of current over the period that the photon density is high. The injection laser output can consist of several pulses because the electron density is repetitively build up and quickly reduced, thus causing ROs. The switch on delay t_d may last for 0.5 ns and the RO for perhaps twice that period.

(3) Noise : The noise behaviour of the device is an important feature of injection laser operation, especially in the case of analog transmission. The sources of noise are :

(*a*) Phase or frequency noise.

(*b*) Instabilities in operation

- kinks in the light output against current characteristics.
- self pulsation.

(*c*) Reflection of light back into the device.

(*d*) Mode partition noise.

(*b*), (*c*) and (*d*) types of noise can be reduced by using mode stabilized devices and optical isolators. **Phase noise** is an intrinsic property of all laser types, it results from the discrite and random spontaneous or stimulated transitions which cause intensity fluctuations in optical emission. Each event causes a sudden jump in the phase of the electromagnetic field generated by the device. Figure 8.27 shows the spectral density of phase or frequency noise characteristic represented by $\frac{1}{f}$ to $\frac{1}{f^2}$ upto a frequency (*f*) of around 1 MHz.

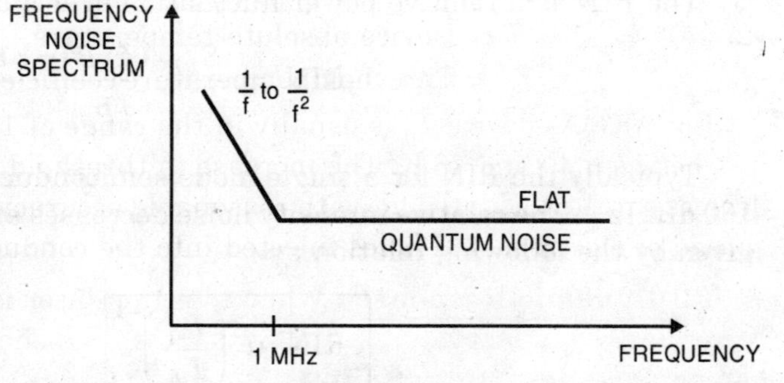

Fig. 8.27. Spectral Characteristics Showing Injection Laser Phase Noise.

At frequencies above 1 MHz the noise spectrum is flat or white because of quantum fluctuations and its major cause is line width broadening within the semiconductor lasers. The low frequency components can be easily tracked and is not a major problem within optical fiber communications.

Usually the **quantum noise** levels are low for injection lasers operating at frequencies less than 100 MHz unless the device is biased within 10% of threshold. The noise spectrum is flat in this region when the laser is operating above threshold *i.e.* in wideband systems, quantum noise becomes more pronounced. This occurs mainly in the case of multimode device.

Amplitude or intensity fluctuations of the output from semiconductor injection lasers also leads to **optical intensity noise**. These fluctuations occur because of temperature variations or spontaneous emission contained in the laser output. The random intensity fluctuations create a noise source known as **relative intensity noise** (RIN) which is defined in terms of

- Mean square power fluctuations $\overline{\delta P_e^2}$.
- Mean optical power squared $\left(\overline{P}_e\right)^2$.

The relative intensity noise (RIN) is given by

$$\boxed{\text{RIN} = \frac{\overline{\delta P_e^2}}{\left(\overline{P}_e\right)^2}} \qquad \text{...(8.26)}$$

RIN is measured in dB Hz^{-1}. The power fluctuation is written as

$$\overline{\delta P_e^2(t)} = \int_0^\infty S_{RIN}(f).df \qquad \text{...(8.27)}$$

where $S_{RIN}(f)$ = Power spectral density of the relative intensity noise.

$S_{RIN}(w)$ and $S_{RIN}(f)$ is related by :

$$S_{RIN}(f) = 2\pi\, S_{RIN}(w) \quad ...(8.28)$$

where, $w = 2\pi f$

The RIN as a relative power fluctuation over a bandwidth B which is defined as 1 Hz :

$$\text{RIN} = \frac{S_{RIN}(f)\, B\,(=1\text{ Hz})}{\left(\overline{P}_e\right)^2} \quad ...(8.29)$$

Typically the RIN for a single mode semiconductor laser would lie in the range 130 to 160 dB Hz^{-1}. The relative intensity noise decreases as the injection current level I increases, given by the following relation :

$$\text{RIN}\ \alpha \left(\frac{I_{-1}}{I_{th}}\right)^{-3} \quad ...(8.30)$$

where, I_{th} = Laser threshold current.

When an optical field at a frequency f is incident with power $P_0(t)$ on a photodetector whose quantum efficiency is α then the output photocurrent $I_p(t)$ is :

$$I_p(t) = \frac{\eta e\, P_0(t)}{hf} \quad ...(8.31)$$

where, e = Charge on an electron

h = Planck's constant.

Therefore an optical power fluctuation $\delta P_0(t)$ will cause a fluctuating current component :

$$\delta I_p(t) = \frac{\eta e\, P_0(t)}{hf}$$

which exhibits a mean square value :

$$\overline{i^2(t)} = \delta\, I_p^2(t) = \frac{\eta^2 e^2}{(hf)^2}\, \delta\, P_o^2(t) \quad ...(8.32)$$

If we consider the fluctuation in the incident optical power at the detector to result from RIN in the laser emission and transporting P_e for P_o, then mean square noise current in the output of the detector $\overline{i_{RIN}^2}$ due to these fluctuations is

$$\overline{i_{RIN}^2} = \frac{\eta^2 e^2}{(hf)^2}\,(RIN)\left(\overline{P}_e\right)^2 B \quad ...(8.33)$$

Optical feedback from unwanted external reflections can also affect the intensity and frequency stability of semiconductor laser. This effect is reduced with multimode lasers because the reflections are distributed among many fiber modes.

Mode partition noise occurs in single mode devices as a result of the residual side modes in the laser output spectrum. It also occurs in multimode semiconductor lasers when modes are not well stabilized.

(4) Realibility : Device realibility is a major problem with injection lasers some progress has been made because in early days the device lifetimes were very short (a few hours). The degradation behaviour is separated into two major processes :

(*a*) *Catastrophic Degradation :* It is because of the mechanical damage of the mirror facts and leads to partial or complete laser failure. It is caused by the average flux density within the structure at the fact and may be limited by using the device in a pulsed mode. But its operation may restrict the operation and lifetime of CW devices.

(*b*) *Gradual Degradation :* Its mechanism may be separated into two categories

- defect formation in the active region.
- degradation of the current confining functions.

Normally there degradations are characterized by an increase in the threshold current for the laser which is often accompained by a decrease in its external quantum efficiency.

Defect formation in the active region can be promoted by the high density of recombining holes within the device. Internal damage may be caused by energy released, which results the presence of strain and thermal gradients by these nonradiative carrier recombination processes. If nonradiative electron hole recombination occurs at the damaged surface of a laser where it has been roughened, this accelerates the diffusion of the point defects into the active region of the device. Therefore the emission characteristics of the active region gradually deteriorate through the accumulation of point defects until the device is no loger useful. These defect structure are known as **dark spot defects (DSDs).**

Degradation of the current confining junctions occur in many index guided laser structures and they utilize current restriction layers so that most of the injected current will flow through the active region. The current flowing outside the active region in BH lasers in known as **leakage current**. Mode degradation increases leakage current. Therefore thershold current is required to be increased *i.e.* quantum effeciency is lowered with ageing.

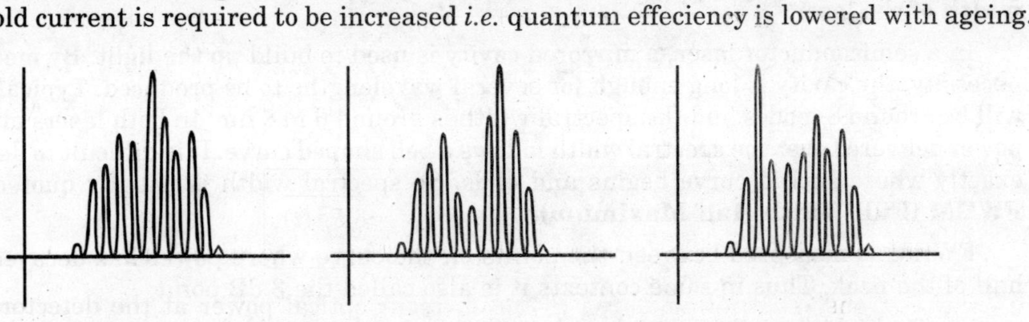

Fig. 8.28. Typical Mode Hopping Behaviour in an Unguided FP Laser.

(5) Mode Hopping : Hole burning triggers mode hopping. When the strong and dominant mode decreases other modes are able to increase and become more dominant. Thus the laser produces light in one mode for a very short duration of time and then it hops to another mode, and then to another and then to another. The whole range of resonant modes within the gain spectrum may be covered. This happens very quickly (a few tens of picoseconds per hop), thus the effect of the laser producing a band of wavelengths is achieved.

It is important to note that when a single mode dominates, the others are usually not suppressed entirely but they are strongly attenuated. When the signal is sent on a dispersive medium, mode hopping can become an additional source of noise. This is because each mode is at a different wavelength and each wavelength will travel at a different speed within the fiber. Not only the pulse will disperse but the dispersion will be irregular and random in nature. Figure 8.28 shows the typical mode hopping behaviour in an unguided FP laser.

(6) Frequency Chirp : When the power is applied to a laser, there is an abrupt change in the carrier (electron and hole) flux density in the cavity caused by the lasing operation itself. This charge carrier density is one factor that affects the refractive index and the temperature in the cavity increases quite rapidly. This temperature increase is too localized to affect the length of the cavity immediately but it does contribute to changing the refractive index of the material in the active region (within the cavity). These changes in the RI of the cavity produce a rapid change in the centre wavelength of the signal produced. In the case of semiconductor lasers a downward chirp is produced. The wavelength shifts to a longer wavelength than it was immediately at the start of the pulse. It is not a major problem in short distance single-channel transmissions but in long distance applications and in WDM systems chirp can be a very serious problem. This is due to the fact that it broadens the spectral width of the signal and interacts with other aspects of the transmission system to produce distortion. The chirp problem is the main reason so that people use external modulators for transmission rates in excess of 1 Gbps.

8.15. Communication Laser Characteristics

Researchers have developed a multitude (perhaps a hundred or so) of different types of communication lasers. The important features of communication lasers are as follows:

(1) Spectral Width

It is a fact that most simple semiconductor lasers do not produce a single wavelength of light. They produce a range of wavelengths. This range of wavelengths is called the **spectral width** of the laser.

In a semiconductor laser, a mirrored cavity is used to build up the light. By mechanical necessity, the cavity is long enough for several wavelengths to be produced. Typically there will be around 8 modes and the spectral width is around 6 to 8 nm. In both lasers and LEDs power delivered over the spectral width follows a bell shaped curve. It is difficult to determine exactly where such a curve begins and ends. So spectral width is usually quoted as the **FWHM (Full Width Half Maximum).**

FWHM is measured between the points on the curve where power has decayed to one half of the peak. Thus in some contexts it is also called the 3 dB point.

Spectral width is very important because :

(*i*) The wider the spectrum of the light source, the more dispersion the signal will suffer when travelling on the fibre.

(*ii*) In a Wavelength Division Multiplexing (WDM) system it is desirable to pack the channels as closely together as possible in order to maximize the number of channels. The narrower the spectral width, the more channels we can have.

(*iii*) We can't use frequency or phase modulation techniques or coherent detection methods unless the linewidth (expressed as occupied bandwidth) is significantly less than the bandwidth of the modulating signal (100 to 1 is a good ratio).

(*iv*) Very narrow spectral width signals are subject to a number of nonlinear effects which are generally thought to be undesirable.

(2) Linewidth

Instead of producing a continuous range of wavelengths over their spectral width, semiconductor lasers produce a series of lines at a number of discrete wavelengths. Lines themselves vary in width (in different types of lasers) very significantly. The linewidth is inversely proportional to the coherence length of the laser.

(3) Coherence Length and Coherence Time

A particular laser line is emitted at a very specific wavelength corresponding to one mode (light path) in the laser's cavity. Over the time this wavelength varies somewhat around a center wavelength (the amount of variation is the linewidth). If a sample emission is taken from one line of a laser at a particular time is of exactly the same wavelength and phase as another sample taken at a later time then the laser is said to be coherent over that time.

The length of time that coherence is maintained is called the **coherence time**. The length that the signal could travel in a vacuum during that time is called the **coherence length.**

The coherence time for an LED is of the order of half a picosecond and its coherence length is around 15 microns. A simple laser may have a coherence time of perhaps half of a nanosecond and a coherence length of perhaps 15 centimetres. A very good quality narrow linewidth laser could have a coherence time of perhaps a microsecond and a coherence length of up to 200 metres.

It is relatively obvious that the range of variation of the signal (the linewidth) should be inversely related to the coherence length (and time).

$$\text{Length}_{\text{coherence}} = c \times \text{Time}_{\text{coherence}} = \frac{\lambda^2}{\Delta\lambda} \quad ...(8.34)$$

where $\Delta\lambda$ = Linewidth and λ = Centre wavelength.

(4) Power

The signal is attenuated as it travels on the fiber and thus the higher the signal power we use, the further you can go without needing to regenerate it. In addition, theory tells us that in an optical receiver of a given type we need a certain fixed minimum amount of power per bit transmitted. If we have a working system and want to double the bit rate we must double the power (or double the receiver sensitivity). But transmitters have limits to their power and receivers have limits to their sensitivity. We can get a higher bit rate by reducing the attenuation (by shortening the distance between stations) thereby increasing the signal power at the receiver.

In some systems signal power, more than fiber capacity is the limiting factor.

(5) Operating Wavelength (or Range)

Lasers must be able to operate on wavelengths appropriate to the system being designed.

The operating wavelength of a laser depends on :

- the materials used for lasing (in exactly the same way as the wavelength of an LED depends on the same materials)
- on the geometry of the laser cavity.

(6) Frequency (Wavelength) Stability

In a single-channel system using incoherent detection, a bit of instability (wander) in the laser wavelength doesn't matter too much. However, if the system is using WDM techniques, each laser must keep within its allocated band.

Fabry-Perot lasers vary an enormous. 4 nm per degree Celsius of temperature variation. Most of the single-mode lasers are significantly better than this, but temperature control is critical. When most lasers are modulated (turning them on and off) they produce a **chirp** at the beginning of each pulse (this is a transient frequency shift of as much as several gigahertz). This is a problem in WDM systems and ones using coherent receivers. Chirp is caused by two factors :

1. The depletion of carriers within the active region (cavity) of the laser immediately after inversion is achieved. This causes a shift in RI of the cavity and therefore a shift in the resonant wavelength.
2. By instantaneous heating effects. The cavity itself heats up after all, the energy (light) produced is infrared and this also changes the RI of the material in the cavity. Different types of semiconductor laser react differently to these two effects.

(7) Switching Time and Modulation

A fundamental operational characteristic of any laser in which modulation techniques are possible and how fast they can operate. In general, all lasers can be modulated by 0 K (on/off keying) and some by FSK (frequency shift keying). Other modulation techniques require an external modulator to be placed into the light beam after it is generated.

(8) Tuning Range and Speed

In many proposed WDM systems, transmitters, and/or receivers need to be switched between different wavelengths (channels). There are many techniques for doing this. In general, the faster the device can switch, the narrower will be the range of channels over which it can switch.

Another point is that tunable lasers are seldom capable of continuous tuning over an unbroken range of wavelengths. When they are tuned they **jump** from one wavelength to another (corresponding to the resonance modes of the laser cavity).

SOLVED EXAMPLES

Example 8.1. A double heterojunction InGaAsP LED operating at 1310 nm has radiative and non-radiative recombination times of 30 and 100 ns respectively. The injected current is 40 mA. Calculate :

(*a*) Bulk recombination life time.

(*b*) Internal quantum efficiency.

(*c*) Internal power level.

Solution. Given that,

$$\lambda = 1310 \text{ nm} = 1310 \times 10^{-9} \text{ m}$$

$$\tau_r = 30 \text{ n sec}$$

$$\tau_{nr} = 100 \text{ n sec}$$

$$i = 40 \text{ mA}$$

(*a*) Bulk recombination life time τ is given by

$$\frac{1}{\tau} = \frac{1}{\tau_r} + \frac{1}{\tau_{nr}}$$

$$= \frac{1}{30} + \frac{1}{100} = \frac{130}{3000}$$

$$\boxed{\tau = 23.07\, n \text{ sec.}} \quad \textbf{Ans.}$$

(*b*) Internal quantum efficiency is given by

$$\eta_{\text{int}} = \frac{\tau}{\tau_r}$$

$$= \frac{23.07}{30}$$

$$\boxed{\eta_{\text{int}} = 0.769} \quad \textbf{Ans.}$$

(c) Internal power level is given by

$$P_{\text{int}} = \eta_{\text{int}} \frac{hci}{e\lambda}$$

$$= \frac{0.769 \times 6.62 \times 10^{-34} \times 3 \times 10^8 \times 0.04}{1.6 \times 10^{-19} \times 1.31 \times 10^{-6}}$$

$$\boxed{P_{\text{int}} = 2.913 \text{ mW}} \quad \textbf{Ans.}$$

Example 8.2. Determine the power radiated by an LED if its quantum efficiency is 3% and the weak wavelength is 670 nm.

Solution. Given that,

$$\lambda = 670 \text{ nm} = 670 \times 10^{-9} \text{ m}$$

$$h = 6.62 \times 10^{-34}$$

$$C = 3 \times 10^8$$

Typically values of I for LED's are in the range of 50 to 150 mA. Hence taking $i = 50$ mA

Radiated power is given by

$$P(\text{mW}) = [\eta_{\text{int}}\, E_P(eV)]\, I(mA)$$

$$= \left[\eta_{\text{int}} \cdot \frac{hc}{\lambda}\right] I$$

$$= \frac{3}{100} \times \frac{6.62 \times 10^{-34} \times 3 \times 10^8}{670 \times 10^{-9}} \times 50 \times 10^{-3}$$

$$= \frac{1}{100} \times 1.86 \times 50 = 0.93 \times 10^{-3} \text{ W}$$

$$\boxed{P = 0.93 \text{ mW}} \quad \textbf{Ans.}$$

Example 8.3. A LED emits the light having a peak wavelength of 890 nm have radiative recombination time of 100 ns. If the bulk recombination life time is 130 ns and drive current is 14 mA. Determine the non-radiative recombination time.

Solution. Given that,

$$\lambda = 890 \text{ nm} = 890 \times 10^{-9} \text{ m}$$

$$\tau_r = 100 \text{ ns}$$

$$\tau = 130 \text{ ns}$$

$$i = 14 \text{ mA}$$

The bulk recombination life time τ is given by

$$\frac{1}{\tau} = \frac{1}{\tau_r} + \frac{1}{\tau_{nr}}$$

$$\frac{1}{130} = \frac{1}{100} - \frac{1}{\tau_{nr}}$$

$$\frac{1}{\tau_{nr}} = \frac{1}{100} - \frac{1}{130} = \frac{30}{13000}$$

$$\tau_{nr} = \frac{1300}{3}$$

$$\boxed{\tau_{nr} = 433 \text{ ns}} \quad \textbf{Ans.}$$

Example 8.4. Determine the power internally generated within the device when the peak emission wavelength is 0.87 μm at a drive current of 40 mA. The internal quantum efficiency of the device is 0.625.

Solution. Given that,

$$\lambda = 0.87 \text{ μm} = 0.87 \times 10^{-6} \text{ m}$$

$$\eta_{\text{int}} = 0.625$$

$$i = 40 \text{ mA}$$

The internally generated power is given by

$$P_{\text{int}} = \eta_{\text{int}} \cdot \frac{hci}{e\lambda}$$

$$= \frac{0.625 \times 6.62 \times 10^{-34} \times 3 \times 10^{8} \times 40 \times 10^{-3}}{1.6 \times 10^{-19} \times 0.87 \times 10^{-6}}$$

$$= 35.6 \times 10^{-3} \text{ W}$$

$$\boxed{P_{\text{int}} = 35.6 \text{ mW}} \quad \textbf{Ans.}$$

Example 8.5. A planar LED is fabricated from GaAs which has a refractive index of 3.6. Calculate the optical power emitted into air as a % of the internal optical power for the device if the transmission factor at the crystal air interface is 0.68.

Solution. Given that,

$$F = 0.68$$

$$nx = 3.6$$

Refractive index (n) for air = 1.

The optical power emitted is given by

$$P_e \simeq \frac{P_{int}\, Fn^2}{4Fnx^2}$$

$$\simeq \frac{P_{int} \times 0.68 \times 1}{4 \times (3.6)^2} \cong 0.013\, P_{int.}$$

Thus emitted power is only 1.3% of the optical power generated internality.

Example 8.6. The light output from the GaAs is coupled into a step index fiber with a numerical aperture of 0.2, a core refractive index of 1.4 and a diameter larger than the diameter of the device. Determine :

(*a*) Coupling efficiency into the fiber.

(*b*) Optical loss relative to the power emitted from the LED.

Solution. Given that, $NA = 0.2$

$n = 1.4$

(*a*) Coupling efficiency is given by

$$\eta_c = (NA)^2$$

$$= (0.2)^2 = 0.04$$

$$\boxed{\eta_c = 0.04} \quad \textbf{Ans.}$$

(*b*) The optical loss is determined by

$$\text{Loss} = -\log_{10} \frac{P_c}{P_e}$$

$$= -\log_{10} (\eta_c)$$

$$= -\log_{10} (0.04)$$

$$\boxed{\text{Loss} = 14.0 \text{ dB}} \quad \textbf{Ans.}$$

Example 8.7. A *DH* surface emitter has an emission area diameter of 50 μm and is but joined to an 80 μm core step index fiber with a numerical aperture of 0.15. The radiance of device is 30 W Sr^{-1} cm^{-2} at a operating drive current of 40 mA. Determine the optical power coupled into the fiber if the Fresnel reflection coefficient at the index matched fiber surface is 0.01.

Solution. Given that, $r = 0.01$

$NA = 0.15$

$R_D = 30 \text{ W sr}^{-1} \text{ cm}^{-2}$

$i = 40$ mA

$$\text{Radius, } (R) = \frac{50}{2} \,\mu\text{m} = 25 \times 10^{-6} \text{ m} = 25 \times 10^{-4} \text{ cm.}$$

Optical power coupled into the fiber is given by

$$P_c = \pi(1-r)A\,R_D\,(NA)^2$$

and the emission area (*A*) is calculated

$$A = \pi(25\times10^{-4})^2$$

$$A = 1.96 \times 10^{-5}\ \text{cm}^2.$$

Therefore,

$$P_c = 3.14\ (1 - 0.01)\ 1.96 \times 10^{-5} \times 30 \times (0.15)^2$$

$$= 41.1 \times 10^{-6}\ \text{W}$$

$$\boxed{P_c = 41.1\ \mu\text{W}}\quad \textbf{Ans.}$$

Example 8.8. A surface emitting LED launches 150 μM of optical power into a millimode step index fiber. Calculate the overall power conversion efficiency if the 25 mA forward current is flowing the device and the corresponding forward voltage across the diode is 2.5 V.

Solution. Given that, P_c = 150 μW = 150 × 10^{-6} W

Forward current = 25 mA

Forward Voltage drop = 2.5 V

The overall power conversion efficiency of may be given by

$$\eta_{pc} = \frac{P_c}{P}$$

and

$$P = \text{Forward current} \times \text{Forward voltage drop}$$

$$= 25 \times 10^{-3} \times 2.5 = 6.25 \times 10^{-2}\ \text{W}$$

Therefore,

$$\eta_{pc} = \frac{150\times10^{-6}}{6.25\times10^{-2}}$$

$$\eta_{pc} = 24 \times 10^{-4}$$

$$\boxed{\eta_{pc} = 0.24\%}\quad \textbf{Ans.}$$

Example 8.9. When a constant dc drive current is applied to the device, the optical output power is 300 μW. If the minority carrier recombination life time for an LED is 5 ns, calculate the optical output power when the device is modulated with an rms drive current corresponding to dc drive current at 20 MHz frequency.

Solution. Given that, P_{dc} = 300 μW = 300 × 10^{-6} W

f = 20 MHz = 20 × 10^6 Hz

τ_i = 5 ns = 5 × 10^{-9} s

The optical output power $P_e(W)$ of the device is given by

$$P_e = P_{dc}\left[1+(\omega\tau_i)^2\right]^{-1/2}$$

$$= 300 \times 10^{-6} \left[1 + \left(2\pi \times 20 \times 10^{6} \times 5 \times 10^{9}\right)^{2}\right]^{-1/2}$$

$$= 300 \times 10^{-6} \left[1 + \left(2 \times 3.14 \times 10^{8} \times 10^{-9}\right)^{2}\right]^{-1/2}$$

$$= 300 \times 10^{-6} \, [1.39]^{-1/2} = 254.2 \times 10^{-6} \text{ W}$$

$$\boxed{P_e = 254.2 \, \mu\text{W}} \quad \textbf{Ans.}$$

Example 8.10. If the average operating wavelength of a laser is 0.5 µm with 1000 K temperature. The refractive index of laser is 1.38 with 1.3 × 10^4 longitudinal modes. Determine the optical spacing between mirrors.

Solution. Given that, $\lambda = 0.5 \, \mu\text{m} = 0.5 \times 10^{-6}$ m

$n = 1.38$

$T = 1000$ K

$q = 1.3 \times 10^{4}$

Optical spacing between the mirrors 'L' is given by

$$L = \frac{\lambda q}{2n}$$

Substituting the values, we get

$$L = \frac{0.5 \times 10^{-6} \times 1.3 \times 10^{4}}{2 \times 1.38} = \frac{6.5 \times 10^{-3}}{2.76}$$

$$\boxed{L = 2.4 \times 10^{-8} \text{ m}} \quad \textbf{Ans.}$$

Example 8.11. Calculate the optical gain at threshold of a ruby laser diode having the following values.

$$\boldsymbol{R_1 = R_2 = 0.32}$$

$$\boldsymbol{\bar{\alpha} = 10 \text{ cm}}$$

$$\boldsymbol{L = 500 \, \mu\text{m}}$$

Solution. The optical gain per unit in laser diode is given by

$$\bar{g}_{th} = \bar{\alpha} + \frac{1}{2L} \ln \left[\frac{1}{R_1 R_2}\right]$$

Substituting the values we get

$$= 10 + \frac{1}{2 \times 500 \times 10^{-6}} \ln \left[\frac{1}{0.32 \times 0.32}\right]$$

$$= 10 + 1000 \ln \frac{1}{(0.32)^2}$$

$$\boxed{\bar{g}_{th} = 33.7 \text{ cm}^{-1}} \quad \textbf{Ans.}$$

Example 8.12. Calculate the radiative minority carrier lifetimes in GaAs and Si when the minority carriers are electrons injected into the P region which has a hole concentration of 10^{+18} cm^{-3}. The injected electron density is small compared to the majority carrier density.

Solution. The minority carrier life time τ_r is given as

$$\tau_r = [B_r(N+P)]^{-1}$$

But in the P region, $P >> N$. Therefore,

For GaAs, $$\tau_r = [B_r.N]^{-1} = \left[7.21\times10^{-10}\times10^{18}\right]^{-1} = 1.39\times10^{-9} \text{ sec.}$$

$$\boxed{\tau_r(\text{GaAs}) = 1.39 \text{ ns}}$$ **Ans.**

For Si, $$\tau_t = \left[1.79\times10^{-15}\times10^{18}\right]^{-1} = 5.58\times10^{-4}$$

$$\boxed{\tau_r(\text{Si}) = 0.56 \text{ ms}}$$ **Ans.**

Example 8.13. A GaAs injection laser has an optical cavity length 250 μm and width 100 μm. Determine the threshold density at normal operating temperature the gain factor $\bar{\beta}$ is 21 × 10^{-3} A cm^{-3} and the loss coefficient is 10 cm^{-1}. Given that the reflectivity for normal incidence of plane wave is 0.32.

Solution. Given that,

$$\bar{\beta} = 21\times10^{-3} \text{ A cm}^{-3}$$
$$\bar{\alpha} = 10 \text{ cm}^{-1}$$
$$L = 250 \text{ μm} = 250\times10^{-6} \text{ m}$$
$$r = 0.32$$

The threshold current is given by

$$J_{th} = \frac{1}{\bar{\beta}}\left[\bar{\alpha} + \frac{1}{L}\ln\frac{1}{r}\right]$$

Substituting the values, we get

$$J_{th} = \frac{1}{21\times10^{-3}}\left[10 + \frac{1}{250\times10^{-6}}\ln\frac{1}{0.32}\right]$$

$$= \frac{1}{21\times10^{-3}}\left[10 + 4\times10^{-3}\ln\left(\frac{100}{32}\right)\right] = 2.65\times10^{3} \text{ A cm}^{-2}$$

$$\boxed{J_{th} = 2.65\times10^{3} \text{ A cm}^{-2}}$$ **Ans.**

Example 8.14. Calculate the ratio of threshold current densities at 20°C and 80°C for a AlGaAs injection laser with T_0 = 160 K.

Solution. The threshold current density is given by

$$J_{th} \;\alpha\; \exp\left(\frac{T}{T_0}\right)$$

For the AlGaAs device

$$J_{th}\text{ (at 20° C)} \;\; \alpha \;\; \exp\left(\frac{293}{160}\right) = 6.24$$

$$J_{th}\text{ (at 80° C)} \;\; \alpha \;\; \exp\left(\frac{353}{160}\right) = 9.08$$

Hence the ratio of current densities :

$$\frac{J_{th}\text{(80° C)}}{J_{th}\text{(20° C)}} = \frac{9.08}{6.24}$$

$$\boxed{\text{Ratio} = 1.46} \quad \textbf{Ans.}$$

SUMMARY

1. Light emitting diodes (LED's) are semiconductor *p-n* junction operating under proper forward biased conditions and are capable of emitting external spontaneous radiations in the visible range (370 nm to 770 nm).
2. LED's are special diodes that emits light when connected in a circuit. They are frequently used as "pilot" light in electronic appliances.
3. In a regular diode the electron hole recombinations release energy in the thermal rather than the variable portion of the spectrum.
4. The internal quantum efficiency is given by

$$\eta_{int} = \frac{R_r}{R_r + R_{nr}}$$

5. The bulk recombination life time τ is given by

$$\frac{1}{\tau} = \frac{1}{\tau_r} + \frac{1}{\tau_{nr}}$$

6. LED's that emits different colours are made of different semiconductor materials. The semiconductor materials that are used for the active layer of an optical source must have a direct bandgap.
7. A heterojunction is a junction in which the *p* and *n*-type materials are different semiconductors.
8. Basically two LED's configurations are used for fiber optics
   - Surface emitters.
   - Edge emitters.
9. A single heterostructure (SH) is sometimes used for for window for surface emitters.
10. In double heterostructure (ΔH) the upper and lower confinement layers are so designed because they have a lower index of refraction and thus confine the light by reflection at the interface.

11. The two different types of LED emission pattern are
    - Surface emitting LED.
    - Edge emitting LED.
12. The light power coupled into a step index fiber having a numerical aperture (NA) can be given by

$$P_{in} = P_o(NA)^2$$

$$P_{in} = P_o \sin^2 QA$$

13. The rise/fall time of an LED is given by

$$t_r = 2.2\left[\tau + \left(\frac{1.7\times10^{-4}\times T^{o}K\times C}{I_P}\right)\right]$$

14. In electronics the general relationship between bandwidth and rise time is given by

$$BW = \frac{0.35}{t_r}$$

15. The external power efficiency of an LED is given by

$$\eta_{ep} = \frac{P_e}{P}\times 100\%$$

16. There are five major types of LED structure :
    - Surface emitter LED
    - Edge emitter LED
    - Superluminescent LED
    - Planar LED
    - Dome LED
17. The power coupled P_c into a multimode step index fiber may be given by

$$P_c = \pi(1-r)\,AR_D\,(NA)^2$$

18. Rapid degradation in LED's is similar to that in a injection lasers because of growth of dislocations and precipitate type defects in the active region.
19. The optical output power $P_e(t)$ may be expressed as a function of operating t and is given by

$$P_e(t) = P_{out}\exp(-\beta_\tau\,.t)$$

20. To transmit informations via an optical fiber communication system modulation of light with the information signal is required. The modulation may be intensity, frequency phase or polarization (direction) with either digital or analog signals.
21. The LED's have several drawbacks in comparison with injection lasers :
    - Low optical power coupled into a fiber.
    - Low modulation bandwidth.
    - Harmonic distortion.

22. The planar LED is fabricated by either liquid or vapour phase pitoxial processes over the whole surface of a GaAs substrate.

23. In optical communication the modulation bandwidth may be defined in either electrical or optical terms.

24. The quantity of emitted photons are reduced by three loss mechanisms :

- Absorption within the LED material.
- Fresnel loss.
- Critical angle loss.

25. The relation between Fresnel loss and the reflection coefficient is given as

$$R = \left[\frac{\bar{n}_2 - \bar{n}_1}{\bar{n}_2 + \bar{n}_1}\right]^2$$

26. Light emitting diodes are very realiable and long lasting if it operates within the power, voltage, current and temperature limits specified by the manufacturer.

27. The word LASER is an acronym for light amplification of stimulated emission of radiation.

28. Laser is a device to produce a strong monochromatic, collimated and highely coherent beam of light and depends on the phenomenon of **stimulated emission.**

29. Laser beam is highly monochromatic while the intensity of ordinary light sources decreases rapidly with distance.

30. The spatial and spectral characteristics of a semiconductor laser are strongly influenced by the properties of the function medium, such as doping and band tailing.

31. For laser action (lasing) to take place, there should be more electrons at higher energy conduction band than at the lower energy valance band. This situation is known as **population inversion**.

32. A semiconductor diode work like a laser if the following conditions are met

- population inversion
- stimulated emission.
- positive feedback.

33. There are three major components of laser :

- energy source
- active medium
- optical resonator

34. There are two types of index guided LDs

- positive index
- negative index

35. In a positive index waveguide, the central region has a higher refractive index than outer regions. In negative index waveguide, the central region of the active layer has a lower refractive index than the outer regions.

36. Index guided lasers can be made using following constructions

- buried heterostructure (BH)
- a selectively diffused construction

- a varying thickness structure
- a bent layer configuration.

37. The quantum well fabrication technique makes lasing action more efficient because it modifies the density of energy levels available for electrons and holes.

38. Quantum well laser diodes are available with the following structures

- single quantum well (SQW)
- multiple quantum well (MQW)
- graded index separate confinement heterostructures (GRINSCH).

39. In distributed feedback laser diode word **distributed** mean reflection takes place not a one point but many points dispersed along the active region and word **feedback** means to returns stimulated photons to an active medium.

40. Axial modes is the most important characteristic of laser. There can be classified into two categries :

- lateral (transverse) mode
- longitudinal mode

41. Basically there are two techniques for the fabrication of gain guided laser structure :

- proton isolated stripe
- *p-n* junction isolated stripes

42. Laser oscillations may occur in a direction transverse to the axis of cavity. This gives rise to resonant modes which are transverse to the direction of propagation.

43. Stripe geometry provides optical confinement in the horizontal plane. The stripe is formed by the creation of high resistance areas on either side by means of proton bombardment or oxide isolation.

44. The strategies used to achieve single frequency laser are :

- short cavity resonator
- coupled cavity resonator
- distributed feedback laser.

45. Semiconductors lasers using distributed feedback mechanism may be classified as

- distributed feedback (DFB) laser
- distributed Bragg reflector (DBR) laser

46. In the DFB optical gating is employed over the whole active region whereas in DBR gating is etched only near the cavity ends.

47. The dynamic behaviour of the injection laser is critical especially when it is used in high bit rate optical fiber communication systems.

48. The temperature dependence of the threshold current density *Jth* is approximately exponential for most common structures and is given by

$$J_{th} \;\alpha\; \exp \frac{T}{T_0}$$

49. The relative intensity noise is defined by

$$\text{RIN} = \overline{\delta P_e^2(t)} = \int_0^\infty S_{RIN}(f).df$$

50. Mode hopping occurs in all single mode lasers and is a result of increase in temperature of the device junction. It alters the high output agains current characteristics of the laser.

51. The degradation behaviour of injection lasers may be separated into two major process

- catastrophic degradation
- gradual degradation

REVIEW QUESTIONS

1. Discuss the light source materials that are used in manufacturing LED's. Also explain their advantages also.
2. Briefly explain, how does light radiation occur in a semiconductor diode ?
3. What are the major advantages of a hetero structured LED over a homostructured one ?
4. What do you understand by the term external quantum efficiency and internal quantum efficiency ?
5. Explain necessary requirements of an optical sources used for the purpose of optical communication.
6. Draw the electric circuit of an LED and explain the function of each component.
7. List the suitable schemes for improvement in coupling efficiency.
8. Briefly describe the modulation bandwidth of an LED. Draw the frequency response of the optical source showing the electrical and optical bandwidth.
9. Differentiate between a surface emitting LED (SLED) and an edge emitting LED (ELED).
10. What do you understand by double heterostructure ? Draw the schematic diagram of an edge-emitting DH LED.
11. What is an LED ? List their advantages and disadvantages also.
12. List the various characteristics of LEDs and explain them in detail.
13. Discuss the LED emission pattern for surface emitting LED and edge emitting LED.
14. Explain the various techniques used to couple light from an LED in an optical fiber. **[PTU 2005]**
15. What do you understand by 'FWHM'. How does it apply to an LED's characteristics ?
16. Describe the mechanism of emission of light from light emitting diode.
17. List some materials which are used in fabrication of IREDs and visible LED's.
18. On what factors does the absorption rate for a transition of an atom from a lower energy to higher energy state depend ?
19. Describe with the aid of suitable diagram the mechanism giving the emission of light from an LED. Discuss the properties of LED in relation to its use as an optical source for communication.
20. Outline the common LED structures for optical fiber communication discussing their relative merits and drawbacks. In particular compare surface and edge emitting devices.
21. Discuss the lens coupling to LEDs to optical fibers and outline the various techniques employed.
22. What are direct bandgap and indirect bandgap semiconductors ? Give atleast two examples of each. Which of these are suitable for fabricating LEDs and Why ?
23. Describe the mechanism of emission of light from light emitting diode. What are the types of structure of light emitting diode ? Explain any one of them. **[PTU 2005]**

24. Explain the efficiency and modulation capability of light emitting diode.

25. What do you understand by intensity modulation of an optical source ? Explain the use of direct intensity modulation of semiconductor optical source.

26. What do you understand by injection luminescense ? **[PTU 2007]**

27. With the help of energy bandgap diagram, explain the transition process in light emitting diode.

28. What do you meant by confinement layer and how it help in confining and guiding the light within the device ?

29. The power generated internally within a double heterojunction LED is 28.4 mW at a drive current of 60 mA. Determine the peak emission wavelength from the device when the radiotive and non radiotive recombination lifetimes of the minority carriers in the active region are equal.

30. Determine the optical modulation bandwidth for the LED if the device emits 840 μW of optical power into air when modulated at a frequency of 150 MHz.

31. Determine the electrical modulation bandwidth for an LED with a carrier recombination lifetime of 8 ns. The frequency response of the device may be assumed to be gaussian.

32. Find the external power efficiency of GaAs planar LED having refractive index of 3.6 with transmission factor of GaAs air interface 0.70. The internally generated optical power is 30% of the electric power supplied.

33. The radiative and nonradiative recombination lifetimes of the minority carriers in the active region of a double heterojunction LED are 60 ns and 100 ns respectively. Determine the total carrier recombination lifetime and power internally generated within the device when the peak emission wavelength is 0.87 μm at a drive current of 40 mA.

34. Determine the external power efficiency of a GaAs planar LED. When the transmission factor of the GaAs and interface is 0.68 and internally generated optical power is 30% of the electrical power supplied. The refractive index of GaAs is 3.6.

35. For a LED, determine the fraction of injected charges which produce photons if 2 mW of optical power and radiated with a drive current of 50 mA at 1.3 μm.

36. The forward current through a GaAsP red LED emitting at 670 nm wavelength is 30 mA. If the internal quantum efficiency of GaAsP is 0.1, what is the optical power generated by the LED.

37. What are stimulated emission and spontaneous emission ? Explain the principle of laser action.

38. Briefly explain the principle of laser action. What are the necessary actions required for lasing action to take place ?

39. What do you understand by population inversion ? **[PTU 2007]**

40. Differentiate between spontaneous and stimulated emission.

41. What do you meant by einsteins relations ? Derive its expression also.

42. What is population inversion ? Explain the mechanism of population inversion in a three level or four level system.

43. Why indirect bandgap material is used for manufacturing LEDs. **[PTU 2008]**

44. What do you meant by heterojunctions ? Discuss its various types also.

45. What are stimulated emission and spontaneous emission ? Explain the principle of laser action.

46. What is population inversion ? Explain the mechanism of population inversion in a three level or four level system.

47. Explain the necessity of carrier confinement and optical confinement in semiconductor laser. How these confinements are achieved ? Explain the structure of any one double heterojunction semiconductor laser.

48. What is meant by optical and electrical confinement in a LASER ? How is it achieved ? Explain.

49. Draw and discuss the block diagram of optical receiver. What are the various sources of noise in the receiver ?

50. Briefly describe the various types of noise due to spontaneous fluctuations in optical fiber communication system.

51. Derive the Bose-Einstein Distribution to describe the random statistics of light emitted in black body radiation.

52. Explain various noise and disturbances associated with signal detection in optical receiver.

53. What is an optical front end of a reciever ? What is the major factor involved in the design of an optical front end ?

54. List two of the commonest types of receiver circuit.

55. Discuss the working of transimpedance preamplifier. What are its advantages over high impedance FET pre-amplifier.

56. Draw and discuss the schematic showing major elements of an optical fiber receiver.

57. Describe the possible sources of noise in optical fiber receivers. Briefly discuss what do you meant by quantum noise. Consider this phenomenon with respect to :

(*a*) Digital signalling. (*b*) Analog transmission.

58. What trade-off does the designer of an optical front end have to consider when choosing a type of pre-amp ?

59. Draw the equivalent circuit of a receiver and derive an expression for its input admittance.

60. What are the sources of receiver sensitivity degradation ?

61. What do you understand by photodiode ? Discuss their noise sources also.

62. Derive an expression of SNR for the *p-n* and *p-i-n* photodiode receiver.

63. What is avalanche photodiode receiver ? Determine the factor on which avalanche photodiode receiver depends.

64. Draw and discuss the digital fiber optic receiver. List the different types of receiver structures also.

65. Discuss the importance of load resistance on both thermal noise and post detection bandwidth in optical fiber communication receivers.

66. Explain the following mechanism associated with noise in optical fiber communication fiber communication.

(*a*) Quantum shot noise

(*b*) Avalanche excess noise

(*c*) fiber mode partition noise

(*d*) Thermal noise

67. Discuss the three main front end amplifiers configuration currently adopted for optical fiber communication with circuit diagram, comment on their merits and drawbacks.

68. Discuss the expression for the SNR in an APD receiver given by

$$\frac{S}{N} = \frac{M^2 I_P^2}{2eB(I_P + I_d)M^{2+x} + \dfrac{4KTBF_n}{R_L}}$$

with respect to various sources of noise present in the receiver. How may this this expression be modified to give the optimum avalanche multiplication factor ?

69. Explain the necessity of preamplifier in an optical receiver. Mention the types of preamplifier used and explain the working of any one of them.

70. A silicon *p-i-n* photodiode has a quantum efficiency of 65% at a wavelength of 0.8 μm. Determine :

(*a*) the mean photocurrent when the detector is illuminated at a wavelength of 0.8 μm with 5 μW of optical power.

(*b*) The rms quantum noise current in a post detection bandwidth of 20 MHz.

(*c*) The SNR in dB.

71. An analog optical fiber communication system requires an SNR of 40 dB at the detector with a post detection bandwidth of 30 MHz. Calculate the minimum optical power required at the detector if it is operating at a wavelength of 0.9 μm with a quantum efficiency of 70%.

72. A digital optical fiber link employing ideal binary signalling at a rate of 50 M bits s^{-1} operates at a wavelength of 1.3 μm. The detector is a germanium photodiode which has a quantum efficiency of 45% at this wavelength. An alarm is activated at the receiver when the bit error rate drops below 10^{-5}. Calculate the theoretical minimum optical power required at the photodiode in order to keep the alarm inactivated.

73. The germanium photodiode incorporated into an optical fiber receiver working at a wavelength of 1.55 μm has a dark current of 500 nA at the operating temperature. When the incident optical power at this wavelength is 10^{-6} W and the responsivity of the device is 0.6 A/W, shot noise dominates in the receiver. Determine the SNR in dB at the receiver when the post detection bandwidth is 100 MHz.

74. A silicon RAPD has a quantum efficiency of 95% at a wavelength of 0.9 μm, has an excess avalanche noise factor of M0.3 and a capacitance of 2 pF. It may be assumed that the post detection bandwidth is 25 MHz, and that the dark current in the device is negligible at the operating temperature of 290 K. Determine the minimum incident optical power which can yield an SNR of 23 dB.

75. A germanium APD with (x = 1) operates at a wavelength of 1.35 μm where its responsivity is 0.45 A/W. The dark current is 200 nA at the operating temperature of 250 K and the device capacitance is 3 pF. Determine the maximum possible SNR when the incident optical power is 8×10^{-7} W and the post detection bandwidth without equalization is 560 MHz.

76. What is meant by excess avalanche noise factor F(M) ? Give two possible ways of exproding this factor in analytical terms. Comment briefly on their relative merits.

❑❑❑

Optical Receivers

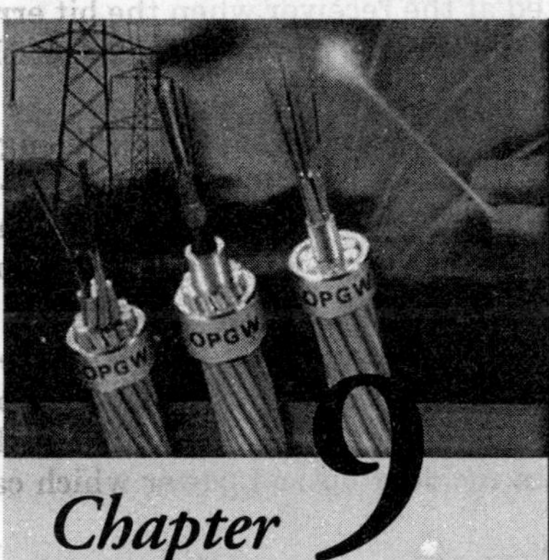

INSIDE THIS CHAPTER

9.1. Introduction

In optical fiber communication systems, optical signals that reach fiber optic receivers are generally attenuated and distorted, as shown in Fig. 9.1.

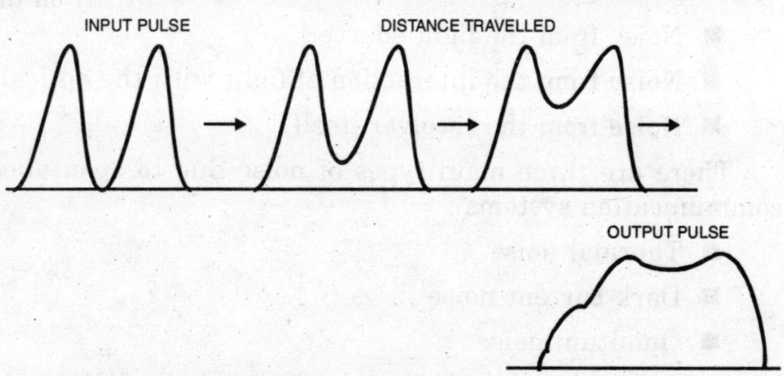

Fig. 9.1. Attenuated and Distorted Optical Signals.

The optical fiber receiver must convert the input and amplify the resulting electrical signal without distorting it to a point. An optical receiver consist of :

- Photodetector
- An amplifier
- Signal Processing Circuitry

In most fiber optic system, the **optical detector** is a PIN photodiode or APD. Receiver performance varies depending on the type of detector used.

The **amplifier** is generally described as having two stages :

- Pre-amplifier
- Post-amplifier.

The **Preamplifier** is defined as the first stage of amplification following the optical detector. The **postamplifier** is defined as the remaining stages of amplification required to raise the detector's electrical signal to a level suitable for further signal processing. The preamplifier is the domain contributor of the electrical noise in the receiver. Because of this, its design has a significant influence in determining the sensitivity of the receiver.

The **signal processing circuitry** processes the amplifies signal into a form suitable for interfacing circuitry. For digital receivers, this circuitry may include low pass filters and comparators. For analog receivers this circuitry may also include low pass filters.

The key operational parameters used to define receiver performance are :

- Receiver Sensitivity
- Bandwidth
- Dynamic Range

One goal in designing fiber optic receivers is to optimize receiver sensitivity. To increase sensitivity, receiver noise resulting from signal dependent shot noise and thermal noise must be kept at a minimum.

While designing receiver it is useful to regard the limit on the performance of the system set by the signal to noise ratio (SNR) at the receiver. Therefore it is necessary to outline noise sources within optical fiber systems.

9.2. Noise Sources in Optical Fiber Communication

Noise is defined as any spurious or undesired distrubance that mask the received signal in a communication system. There are many sources of noise in fiber optic systems, they are :

- Noise from the light source.
- Noise from the interaction of light with the optical fiber.
- Noise from the receiver itself.

There are three main types of noise due to spontaneous fluctuations in optical fiber communication systems :

- Thermal noise
- Dark current noise
- Quantum noise

Noise introduced by the receiver is either signal dependent or signal independent. Signal dependent noise results from the random generations of electrons by the incident optical power. Signal independent noise is independent of the incident optical power level.

9.2.1. Thermal Noise

This noise results from the random motion of electrons in a conducting medium. Thermal noise arises from

- Photodetector
- Load resistor.

Amplifier noise also contributes to thermal noise. A reduction in thermal noise is possible by increasing the value of the load resistor. However, increasing the value of load resistor to reduce thermal noise reduces the receiver bandwidth. In APDs, the thermal noise is unaffected by the internal carrier multiplication.

The thermal noise current it in a resistor R may be expressed by its mean square value and is given by

$$i_t^2 = \frac{4KTB}{R} \qquad ...(9.1)$$

where,
K = Boltzmann's constant
T = Absolute temperature
B = Post detection bandwidth of the system.

9.2.2. Dark Current Noise

Dark current noise results from dark current that continues to flow in the photodiode when there is no incident light. Dark current noise is independent of the optical signal. The dark current gives random fluctuations about the average particle flow of the photocurrent the dark current noise $\overline{i_d^2}$ is given by

$$\overline{i_d^2} = 2eBI_d \qquad ...(9.2)$$

where,
e = Charge on an electron
I_d = Dark current
B = Bandwidth

9.2.3. Quantum Noise

Quantum noise results from the random generation of electrons by the incident optical radiation. The energy of quantum or photon is given by

$$E = hf \qquad ...(9.3)$$

The quantum behaviour of electromagnetic radiation is considered at optical frequencies since $hf > KT$ and quantum fluctuations dominate over thermal fluctuations.

The detection of light by a photodiode is a **discrete process** because electron hole pair generation results from the absorption of a photon and the signal emerging from the detector is dictated by the statistics of photon arrivals. This statistic follows a discrete probability distribution which is independent of the number of photons.

The probability $P(z)$ of detecting z photons intime period τ when it is expected on average to detect z_m photons follows the **Poisson distribution.**

$$P(z) = \frac{z_m^2 \exp(-z_m)}{z!} \qquad ...(9.4)$$

where, z_m = Variance of the probability distribution.

The electron rate r_e generated by incident photons is

$$r_e = \frac{\eta P_0}{hf} \qquad ...(9.5)$$

The number of electrons generated in time τ is equal to the average number of photons detected over this time period z_m. Therefore,

$$z_m = \frac{\eta P_0 \tau}{hf} \quad ...(9.6)$$

Independent atoms emit incoherent light and therefore there is no phase relationship between the emitted photons. This property gives exponential intensity distribution for incoherent light which if averaged over the Poisson distribution gives

$$P(z) = \frac{z_m^2}{(1 + z_m)^{z+1}} \quad ...(9.7)$$

Equation (9.7) is identical to **Bose-Einstein distribution**, which is used to describe the random statistics of light emitted in black body radiation.

9.2.4. Digital Signalling Quantum Noise

It is possible to calculate a fundamental lower limit to the energy in digital optical fiber systems. In the absence of light and neglecting dark current, no current will flow. Thus an error can occur if light pulse is present and no electron hole pairs are generated. When a light pulse is present, the probability of no pairs being generated is given by

$$P\left(\frac{0}{1}\right) = \exp(-z_m) \quad ...(9.8)$$

If $P(e)$ is the system error probability in the receiver described $P(0/1)$, therefore

$$P(e) = \exp(-z_m) \quad ...(9.9)$$

Equation (9.9) represents an **absolute receiver sensitivity** and allows the determination of a fundamental limit in digital optical communications. This is the **minimum pulse energy (E_{min})** required to maintain a given bit error rate (BER) and is known as **quantum limit**.

The above analysis assumes that the photodetector emits no electron-hole pair in the absence of illumination. In this sense it is considered perfect.

9.2.5. Analog Transmission Quantum Noise

In analog optical fiber systems quantum noise means shot noise and it has Poisson statics. The shot noise current is on the photocurrent I_p is given by

$$\overline{i_s^2} = 2eBI_p \quad ...(9.10)$$

By neglecting other noise sources, the SNR at the receiver may be written as

$$\frac{S}{N} = \frac{I_p^2}{\overline{i_s^2}} \quad ...(9.11)$$

Substituting the value of $\overline{i_s^2}$ we get

$$\frac{S}{N} = \frac{I_p}{2eB} \quad ...(9.12)$$

The SNR in terms of incident optical power P_0 given by

$$\frac{S}{N} = \frac{\eta P_0 e}{hf\ 2eB}$$

$$\boxed{\frac{S}{N} = \frac{\eta P_0}{2hfB}} \quad ...(9.13)$$

The incident optical power required at the receiver can be calculated by eqn. (9.13).

9.3. Receiver Noise

Figure 9.2 shows the block diagram of the front end of an optical receiver and the various noise sources associated with it.

The basic requirements for a fiber optic communication receiver are

- Wide dynamic range
- Excellent receiver sensitivity
- Fast acquisition time
- Wide range of bit rates.

All these parameters are important for point to point datalinks, local area a networks (LANs), terrestrial communication and many more applications.

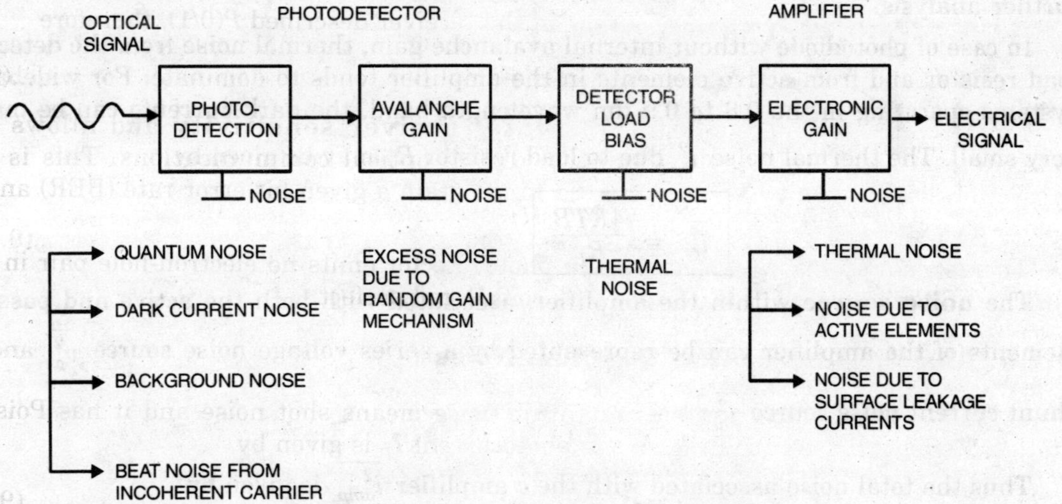

Fig. 9.2. Block Diagram of the Front End of an Optical Receiver.

The **sensitivity** is an important parameter of receiver because the repeater spacing is directly dependent upon it. The repeater spacing *i.e.* the spacing between the two repeaters may be increased if the receiver sensitivity and dynamic range are high. The receiver sensitivity itself depend on the amount of noise introduced by photodetector/photodiode.

A receiver should be capable of accepting power levels over a wide range of values of the input. The difference in decibel between maximum to minimum power range which a receiver can accept or detect is known as **dynamic range.** For a good receiver the dynamic range should be high.

The majority of noise sources shown in Fig. 9.2 apply to both main types of optical detector (p-i-n and avalanche photodiode). The noise generated from background radiation and beat noise generated from the various spectral components are negligible in the both type of optical fiber receiver.

9.3.1. p-n and p-i-n Photodiode Receiver

Basically in photodiodes there are two main sources of noise

- Dark Current Noise
- Quantum Noise

Both of which may be regarded as shot noise on the photocurrent. The total shotnoise i_{Ts}^2 is given by

$$\overline{i_{Ts}^2} = 2eB\left(I_p + I_d\right) \quad ...(9.14)$$

But the noise due to background radiation should also be considered, thus the above equation may be written as

$$\overline{i_{Ts}^2} = 2eB\left(I_p + I_d + I_b\right) \quad ...(9.15)$$

Since I_b is usually negligible, the expression given in Eqn. (9.15) will be used in the further analysis.

In case of photodiode without internal avalanche gain, thermal noise from the detector load resistor and from active elements in the amplifier tends to dominate. For wideband systems operating in the 0.8 to 0.9 μm wavelength band, the dark currents can be made very small. The thermal noise i_t^2 due to load resistor R_L may be given by

$$\overline{i_t^2} = \frac{4KTB}{R_L} \quad ...(9.16)$$

The **noise source** within the amplifier, associated with both the active and passive elements of the amplifier can be represented by a series voltage noise source $\overline{v_a^2}$ and a shunt current noise source $\overline{i_a^2}$.

Thus the total noise associated with the c amplifier $\overline{i_{amp}^2}$ is given by

$$\overline{i_{amp}^2} = \int \left(i_d^2 + v_a^2 \left|y\right|^2\right) df \quad ...(9.17)$$

where, y = Shunt admittance (combines the shunt capacitances and resistances)
f = Frequency

An equilibrium circuit for the front end of the receiver is shown in Fig. 9.3.

The SNR for the p-n and p-i-n photodiode receiver may be obtained by summing all the noise contributions and is given by

$$\frac{S}{N} = \frac{I_p^2}{2eB\,(I_p + I_d) + \dfrac{4KTB}{R_L} + \overline{i_{amp}^2}} \qquad \text{...(9.18)}$$

- The thermal noise contribution may be reduced by inreasing the value of load resistor R_L and this reduction may be limited by bandwidth consideration.
- The amplifier noise $\overline{i_{amp}^2}$ may be reduced with low detector and amplifier capacitance.

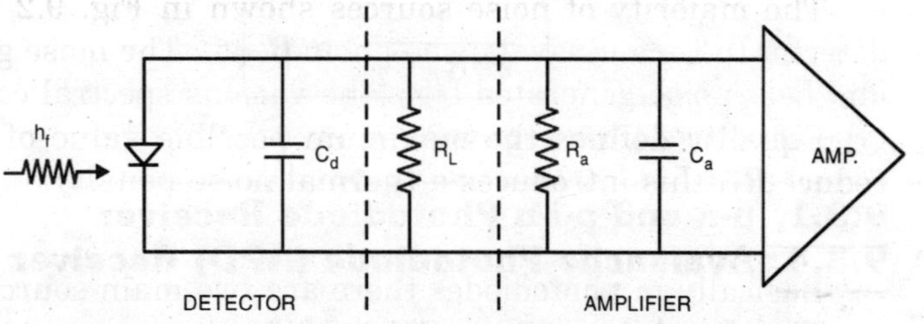

cd = Capacitance of detector

ca = Effective input capacitance

Ra = Resistance of the amplifier

Fig. 9.3. The Equivalent Circuit For the Front End of an Optical Fiber Receiver.

The noise figure F_n for the amplifier may be obtained when amplifier noise $\overline{i_{amp}^2}$ is referred to the load resistor R_L. This allows $\overline{i_{amp}^2}$ to combined with the thermal noise from the load resistor $\overline{i_t^2}$ to give

$$\overline{i_t^2} + \overline{i_{amp}^2} = \frac{4KTBF_n}{R_L} \qquad \text{...(9.19)}$$

Therefore, from the Eqn. (9.18) the expression for SNR may be rewritten as

$$\frac{S}{N} = \frac{I_p^2}{2eB\,(I_p + I_d) + \dfrac{4KTBF_n}{R_L}}$$

Thus if the noise figure F_n for the amplifier is known, SNR can be determined.

9.3.2. Receiver Capacitance

From the equivalent circuit shown in Fig. 9.3, the total capacitance for the front end of an optical receiver c_T is given by

$$c_T = c_d + c_a \qquad \text{...(9.20)}$$

where, c_d = Detector capacitance

c_a = Amplifier input capacitance.

The total capacitance can also be minimized from the bandwidth penalty which occurs due to time constant of c_T and the load resistor R_L.

9.3.3. Bandwdith

The reciprocal of the time constant $2\pi R_L c_T$ must be greater than or equal to the post detection bandwidth B,

$$\frac{1}{2\pi R_L C_T} \geq B \qquad ...(9.21)$$

Equality defines the maximum possible value of B. To increase B it is necessary to reduce R_L, this introduces a thermal noise penalty.

9.3.4. Avalanche Photodiode (APD) Receiver

The signal current into the amplifier is increased by the internal gain mechanism in an APD receiver. This improves the SNR because the load resistance and amplifier noise remain unaffected.

The dark current and quantum noise are increased by the multiplication process, it become a limiting factor because the random gain mechanism introduces excess noise into the receiver. If the photocurrent is increased by a factor M (**mean avalanche multiplication factor**), then the shot noise is also increased by an excess noise factor M^x such that the total shot noise $\overline{i_{SA}^2}$ is given by

$$\overline{i_{SA}^2} = \left(2eB\,I_p + I_d\right) M^{2+x} \qquad ...(9.22)$$

where, x = 0.3 to 0.5 (for silicon APDs)
= 0.7 to 1.0 (for germanium or III-V allow APDs)

The **SNR for the avalanche photodiode** may be obtained by summing the combined noise contribution from the load resistor and the amplifier, the SNR for the APD is given by

$$\frac{S}{N} = \frac{M^2 I_p^2}{2eB\,(I_p + I_d)\,M^{x+2} + \dfrac{4KTB\,F_n}{R_L}} \qquad ...(9.23)$$

The relative significance of the combined thermal and amplifier noise term is reduced due to the avalanche multiplication of the shot noise term. Thus the Eqn. (9.23) may be rewritten as

$$\frac{S}{N} = \frac{I_p^2}{2eB\,(I_p + I_d)\,M^x + \dfrac{4KTB\,F_n}{R_L} M^{-2}} \qquad ...(9.24)$$

In the denominator the first term increases with increasing M whereas the second term decreases.

1. For **low value of *M*,** when the signal level is increased the combined thermal and amplifier noise term dominates, the total noise power is virtually unaffected and it improves SNR.
2. When ***M* is large,** the thermal and amplifier noise term becomes insignificant and the SNR decreases with increasing M at the rate of M^2.

Therefore an optimum value of multiplication factor M_{OP} exists which maximizes the SNR. It is given by

$$\frac{2eB(I_p + I_d)\,M_{op}^x}{\left(\dfrac{4KTB\,F_n}{R_L}\right) M_{op}^{-2}} = \frac{2}{x} \qquad ...(9.25)$$

and therefore

$$M_{op}^{x+2} = \frac{4KT\,F_n}{x\,e\,R_L\,(I_p + I_d)} \qquad ...(9.26)$$

9.3.5. Excess Avalanche Noise Factor

The value of excess noise factor is dependent upon :

- Detector material
- The shape of the electric field profile within the device.
- Whether the avalanche is initiated by holes or electrons.

Sometimes it is represented as $F(M)$. Let us consider one approximation for excess noise factor,

$$F(M) = M^x \qquad ...(9.27)$$

The resulting noise is assumed to be white with a Gaussian distribution. Another relationship is given by

$$F(M) = M\left[1-(1-K)\left(\frac{M-1}{M}\right)^2\right] \qquad ...(9.28)$$

where injected electrons are the only carriers and K is the ratio of the ionization coefficients of holes and electrons.

If the only carries are injected holes, then

$$F(M) = M\left[1+\left(\frac{1-K}{K}\right)\left[\frac{M-1}{M}\right]^2\right] \qquad ...(9.29)$$

when K is small, the best performance is achieved. The value of K for silicon APDs 0.02 and 0.10 where as for germanium and III-V alloy APDs K is between 0.3 and 1.0.

9.4. Receiver Structures

An optical receiver is treated as optical to electrical convertor. The receiver is designed with the aim to achieve maximum sensitivity for a given bit error rate (BER). Figure 9.4 shows the block diagram of receiver.

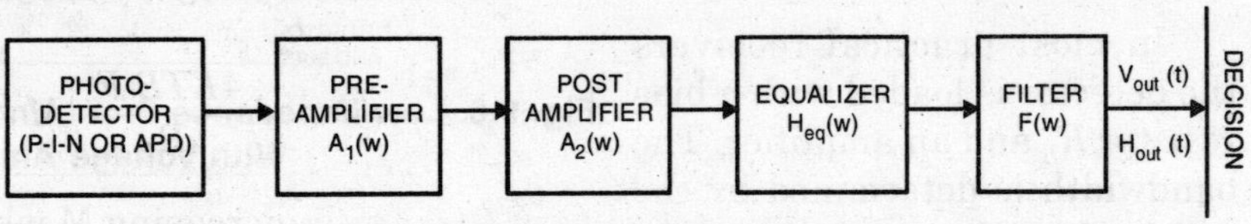

Fig. 9.4. Block Diagram of a Digital Fiber Optic Receiver.

The signal voltage as the output of the linear channel is

$$V_s(w) = Z_T(w)\,.\,I_s(w) \qquad ...(9.30)$$

where, $Z_T(w)$ = Total system transfer function including equalizer.

$I_s(w)$ = Fourier transform of $i_s(t)$.

$i_s(t)$ = Current produced due to electrons which are generated within the photodiode or detector.

Figure 9.5 shows the full equivalent circuit for the digital optical fiber receiver, where optical detector is represented as a current source i_{det}.

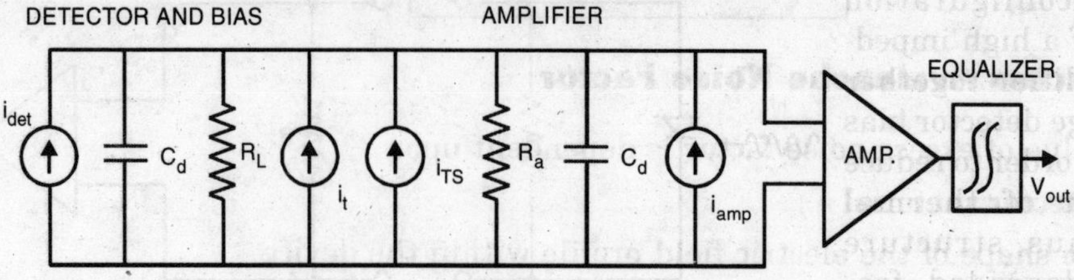

Fig. 9.5. A Full Equivalent Circuit for a Digital Optical Fiber Receiver Including the Various Noise Sources.

The noise sources (i_t, i_{TS} and i_{amp}), the immediately following amplifier and equalizer are also shown in Fig. 9.5 Equalization compensates for distortion of the signal due to the combined transmitter, medium and receiver characteristics.

The **equalizer** is a **frequency shaping filter** which has a frequency response *i.e.* the inverse of the overall system frequency response. The equalizer may also apply in wideband systems to boost the high frequency components to correct the overall amplitude of the frequency response. Basically there are three receiver structures :

1. Low impedance front end.
2. High impedance front end.
3. The trans-impedance front end.

9.4.1. Low Impedance Front End

Figure 9.6 shows the simplest and most common receiver structure with voltage amplifier.

To make suitable design choices, it is required to considered both bandwidth and noise.

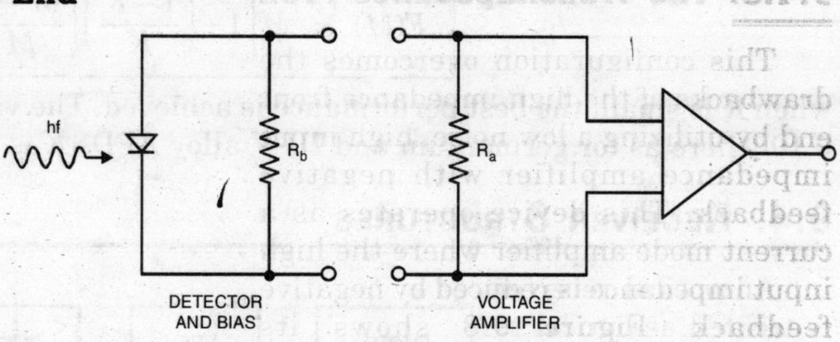

Fig. 9.6. Low Impedance Front End Optical Fiber Receiver With Voltage Amplifier.

In most practical receivers the detector is loaded with a bias resistor R_b and an amplifier. The bandwidth is determined by the passive impedance which appears across the detector terminals. The resistance R_L may be modified to incorporate the parallel resistance of the detector bias R_b and the amplifier input resistance R_a. The **total modified load resistance R_{TL}** is given

$$R_{TL} = \frac{R_b R_a}{R_b + R_a} \quad ...(9.31)$$

To achieve an optimum bandwidth both R_b and R_a must be minimized. This gives a low impedance front end design for the receiver amplifier. But this design allows thermal noise to dominate within the receiver and it may limit its sensitivity. Therefore this structure demands a trade off between bandwidth and sensitivity.

9.4.2. High Impedance (Integrating) Front End

This configuration consists of a high impedance amplifier together with a large detector bias resistor in order to reduce the effect of thermal noise. Thus structure gives a degraded frequency response. The detector output is effectively integrated over a large time constant and must be restored by differentiation. At a later stage, this may be performed by the correct equalization and it is shown in Fig. 9.7.

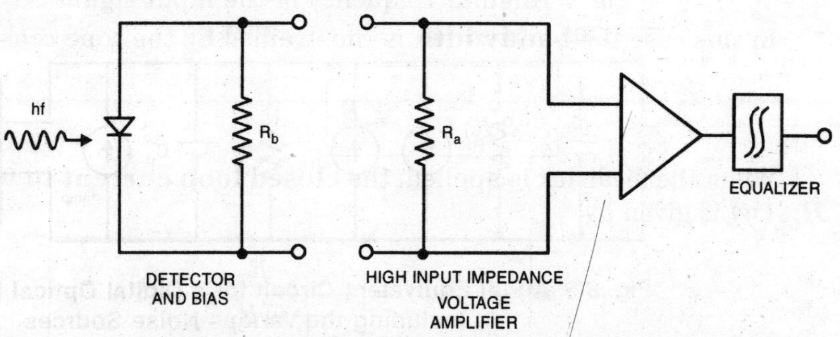

Fig. 9.7. High Impedance Integrating Front End Optical Fiber Receiver With Equalized Voltage Amplifier.

Therefore **this structure gives a significant improvement in sensitivity over the low impedance front end design but it creates a heavy demand for equalization.**

There is a **problem of limited dynamic range**, which results from the attenuation of the low frequency signal components by the equalization process and it causes the amplifier to saturate at high signal levels. The reduction in dynamic range is dependent upon the amount of integration and subsequent equalization employed.

9.4.3. The Transimpedance Front End

This configuration overcomes the drawbacks of the high impedance front end by utilizing a low noise, high input impedance amplifier with negative feedback. This device operates as a current mode amplifier where the high input impedance is reduced by negative feedback. Figure 9.8 shows its equivalent circuit.

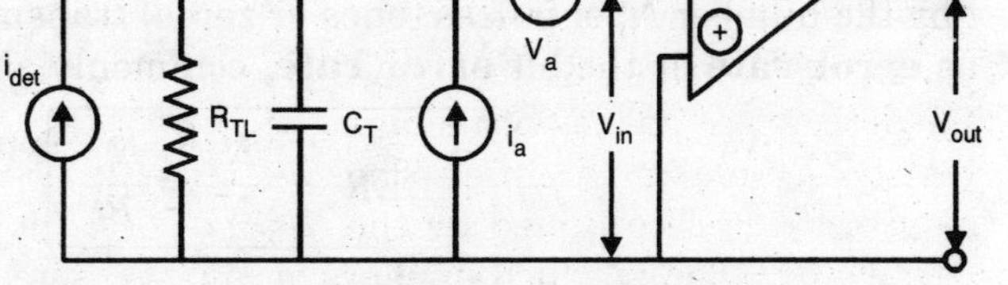

Fig. 9.8. An Equivalent Circuit for Trans-Impedance Front End Optical Receiver.

In this equivalent circuit the parallel resistances and capacitances are combined into R_{TL} and C_T respectively. The **open loop current to voltage transfer function $H_{OL}(w)$** for this configuration may be written

$$H_{OL}(w) = -G\frac{V_{in}}{i_{det}}$$

$$H_{OL}(w) = -G\frac{R_{TL}\dfrac{1}{jwC_T}}{R_{TL}+\dfrac{1}{jwC_T}} = \frac{-G.R_{TL}}{1+jwR_{TL}C_T}\ (\text{VA}^{-1})$$

where, G = Open loop voltage gain of amplifier

w = Angular frequency of the input signal

In this case the **bandwidth** is constrained by the time constant and is given by

$$\frac{1}{2\pi R_L C_T} \geq B \quad ...(9.32)$$

When the feedback is applied, the **closed loop current to voltage transfer function** $\boldsymbol{H_{CL}(w)}$ is given by

$$H_{CL}(w) = \frac{-R_f}{1+\left(\frac{jwR_f C_T}{G}\right)} \text{ (VA}^{-1}) \quad ...(9.33)$$

where, R_f = Feedback resistor.

In this case, the **permitted electric bandwidth (*B*)** may be written as

$$B \leq \frac{G}{2\pi R_f C_T} \quad ...(9.34)$$

By comparing the Eqns. (9.32) and (9.34), it is clear that trans-impedance or feedback amplifier provides a much greater bandwidth than the amplifiers without feedback.

Therefore the **advantages of transimpedance front end optical fiber receivers** are :

- It provides a far greater bandwidth without equalization than the high impedance front end.
- The trans-impedance configuration has great dynamic range over the front end.

9.5. Bit Error Rate

There are several ways to measure the rate of error occurred in digital data stream. A simple approach is to divide the number N_e of errors occurring over a certain time interval t by the number N_t of pulses (ones or zeros) transmitted during this interval. This is known as **error rate** or the bit **error rate,** commonly abbreviated BER. Therefore,

$$\text{BER} = \frac{N_e}{N_t} = \frac{N_e}{Bt}$$

where, $B = \frac{1}{T_0}$, is the bit rate

The error rate is expressed by a number, example such as 10^{-9}, which states that on average one error occurs for every billion pulses sent.

For optical fiber communication, typical error rates range from 10^{-9} to 10^{-12}. This error rate depends on the signal to noise ratio at the receiver.

9.6. Receiver Sensitivity

Optical communication systems use a BER value to specify the performance requirements for a particular transmission link application. Example : SONET/SDH Networks specify that the BER must be 10^{-10} or lower.

To achieve a desired BER at a given data rate, a specific minimum average optical power level must arrive at the photodetector. The value of this minimum power level is known as the **receiver sensitivity.** There are **two ways to define the receiver sensitivity :**

(1) Average optical power (P_{ave}) in dBm incident on the photodetector.

(2) Optical modulation amplitude (DMA) given in terms of peak to peak current at the photodetector output.

The receiver sensitivity is defined as :

$$P_{sensitivity} = \frac{7.37}{M}\left[5.6\times10^{-19}\ M F(M) B + 9.10\times10^{-12}\ B^{1/2}\right]$$

where $F(M)$ = Photodiode noise figure

M = Gain of the photodiode

B = Bit rate

9.7. Minimum Received Power

Suppose we have an ideal photodetector which has unity quantum efficiency and which produces no dark current *i.e.* no electron hole pairs are generated in the absence of an optical pulse. Given this condition, it is possible to find the minimum received optical power required for a specific bit error rate performance in a digital system. This minimum received power level is known as the **quantum limit,** since all system parameters are assumed ideal and the performance is limited only by the photodetection statistics.

Let us assume that an optical pulse of energy E falls on the photodetector in a time interval t. By the receiver it is interpreted as a 0 pulse if no electron-hole pairs are generated with the pulse present.

The probability that n = 0 electrons are emitted in a time interval t is

$$P_r(0) = e^{-\overline{N}} \qquad ...(9.35)$$

where, $\overline{N}$ = Average number of electron-hole pairs

$$\overline{N} = \frac{\eta E}{hv} \qquad ...(9.36)$$

SOLVED EXAMPLES

Example 9.1. An InGaAs pin photodiode has the following parameters at a wavelength of 1100 nm :

I_D = 4 nA, η = 0.90, R_L = 1000 Ω and the surface leakage current is negligible. The incident optical power is 300 nW and the receiver bandwidth is 20 MHz. Find the value of quantum noise current, dark current and thermal noise current.

Solution. Given that,

$$I_D = 4 \text{ nA}$$
$$\eta = 0.90$$
$$R_L = 1000\ \Omega$$
$$\lambda = 1100 \text{ nm}$$
$$P_0 = 300 \text{ nW}$$
$$B = 20 \text{ MHz}$$

Quantum noise is given by

$$I_Q = \sqrt{2eI_PB}$$

where,
$$I_P = R.P_0 = \frac{\eta e P_0}{hv} = \frac{\eta \lambda e P_0}{hc}$$

$$= \frac{0.90 \times 1100 \times 10^{-9} \times 1.6 \times 10^{-19} \times 300 \times 10^{-19}}{6.626 \times 10^{-34} \times 3 \times 10^8}$$

$$I_P = \frac{4752 \times 10^{-35}}{19.86 \times 10^{-26}} = 2.4 \times 10^{-7} \text{ A}$$

Therefore,
$$I_Q = \sqrt{2 \times 1.6 \times 10^{-19} \times 2.4 \times 10^{-7} \times 20 \times 10^6}$$

$$= \sqrt{64 \times 2.4 \times 10^{-20}}$$

$$= 12.4 \times 10^{-10} \text{ amp.}$$

$$\boxed{I_Q = 1.24 \text{ nA}} \quad \textbf{Ans.}$$

Dark current,
$$i_D = \sqrt{2eBI_d}$$

$$= \sqrt{2 \times 1.6 \times 10^{-19} \times 20 \times 10^6 \times 4 \times 10^{-9}}$$

$$= \sqrt{2.56 \times 10^{-20}}$$

$$i_D = 1.6 \times 10^{-9} \text{ Amp}$$

$$\boxed{i_D = 0.16 \text{ nA}} \quad \textbf{Ans.}$$

Thermal noise current,

$$i_t = \sqrt{\frac{4KTB}{R_L}}$$

$$= \sqrt{\frac{4 \times 1.38 \times 10^{-23} \times 300 \times 20 \times 10^6}{1000\,\Omega}}$$

$$= \sqrt{323 \times 10^{-18}} = 18 \times 10^{-9} \text{ Amp.}$$

$$\boxed{i_t = 18 \text{ nA}} \quad \textbf{Ans.}$$

Example 9.2. A digital optical fiber communication system operating at a wavelength of 1 μm requires a maximum bit error rate of 10^{-9}. Determine :

(*a*) The theoretical quantum limit at the receiver in terms of the quantum efficiency of the detector and the energy of an incident photon.

(*b*) The minimum incident optical power required at the detector in order to achieve the above bit rate when the system is employing ideal binary signalling at 10 Mbit/sec and assuming the detector is ideal.

Solution. Given that, $\lambda = 1\ \mu m = 10^{-6}$ m

$$BER = 10^{-9}$$

(*a*) The probability of error is given by

$$P(e) = \exp(-Z_m)$$

$$10^{-9} = \exp(-Zm)$$

$$\therefore \quad Z_m = 20.7$$

(Z_m = average number of photons detected in a time period τ for a BER of 10^{-9})

and
$$Z_m = \frac{\eta P_0 \tau}{hf} = 20.7$$

The minimum pulse energy = Quantum limit = E_{min}.

$$E_{min} = P_0\tau = \frac{20.7\, hf}{\eta}$$

Therefore the quantum limit at the receiver to maintain a maximum BER of 10^{-9} is

$$\boxed{\text{Quantum limit} = \frac{20.7\, hf}{\eta} \quad \textbf{Ans.}}$$

(*b*) The minimum pulse energy :

$$P_0\tau = \frac{20.7\, hf}{\eta}$$

Therefore the average received optical power,

$$P_0 = \frac{20.7\, hf}{\eta\,\tau}$$

For ideal binary signalling there are an equal number of ones and zeros. Thus the average received power may be considered to arrive over two bit periods, and

$$P_0\,(\text{binary}) = \frac{20.7\, hf}{2\tau\eta} = \frac{20.7\, hf\, B_T}{2\eta}$$

(η = 1, for ideal detector)

$$P_0\,(\text{binary}) = \frac{20.7 \times 6.626 \times 10^{-34} \times 3 \times 10^{14} \times 10^{7}}{2}$$

$$= \frac{6.21 \times 6.626 \times 10^{-13}}{2}$$

$$= 20.6 \times 10^{-12}\ \text{W}$$

$$= 20.6\ \text{pW}.$$

When the reference level is one watt :

$$P_0 = 10 \log_{10} (P_0)$$

$$= 10\log_{10}(2.06\times10^{-11})$$

$$= 3.14 - 110$$

$$P_0 = -106.9 \text{ dBW}$$

When the reference level is one milliwatt :

$$P_0 = 10\log_{10}(2.06\times10^{-8})$$

$$= 3.14 - 80$$

$$\boxed{P_0 = -76.9 \text{ dBm}}$$

Therefore the required minimum incident optical power is 20.6 pW or –76.9 dBm.

Example 9.3. An optical fiber system operating at a wavelength of 1 μm has a post detection bandwidth of 5 MHz. Assuming an ideal detector and considering only quantum noise on the signal, determine the incident optical power necessary to achieve on SNR of 50 dB at the receiver.

Solution. Given that,

$$\eta = 1 \text{ (ideal detector)}$$

$$\text{SNR} = 50 \text{ dB}$$

$$\lambda = 1 \text{ μm}$$

$$B = 5 \text{ MHz}$$

The SNR is given by,

$$\frac{S}{N} = \frac{\eta P_0}{2hfB}$$

$$\frac{S}{N} = 50 \text{ dB} \Rightarrow 10\log_{10}\left(\frac{S}{N}\right) = 50$$

$$\frac{S}{N} = 10^5$$

Therefore,

$$P_0 = \frac{2hfB\left(\frac{S}{N}\right)}{\eta}$$

$$= \frac{2\times6.62\times10^{-34}\times3\times10^{14}\times5\times10^6\times10^5}{1}$$

$$= 198.6\times10^{-9}$$

$$P_0 = 198.6 \text{ nW}$$

In dB,

$$P_0 = 10\log_{10}(198.6\times10^{-6})$$

$$= -40 + 2.98$$

$$\boxed{P_0 = -37.02 \text{ dBm.}}\quad \textbf{Ans.}$$

Example 9.4. A silicon p-i-n photodiode incorporated into an optical receiver has a quantum efficiency of 60% when operating at a wavelength of 0.9 μm. The dark current in the device at this operating point is 3 nA and total resistance is 4 kΩ. The incident optical power at this wavelength is 200 nW and the post detection bandwidth of the receiver is 5 MHz at a temperature of 20°C. Calculate

(*a*) Total shot noise

(*b*) Thermal noise

Solution. Given that,

$$\eta = 60\% = 0.60$$
$$\lambda = 0.9\ \mu m$$
$$I_d = 3nA$$
$$P_0 = 200\ nW$$
$$T = 293$$
$$R_L = 4\ k\Omega$$
$$B = 5\ MHz$$

(*a*) Total shot noise is given by

$$\overline{i_{Ts}^2} = 2eB\,(I_d + I_P)$$

where,

$$I_P = \frac{\eta P_0 e}{hf} = \frac{\eta P_0 e\lambda}{hc}$$
$$= \frac{0.6 \times 200 \times 10^{-9} \times 1.6 \times 10^{-19} \times 0.9 \times 10^{-6}}{6.626 \times 10^{-34} \times 3 \times 10^{8}}$$
$$= \frac{172.8 \times 10^{-34}}{19.878 \times 10^{-26}}$$
$$I_P = 87.1\ nA$$

Therefore, Total Shot Noise :

$$\overline{i_{Ts}^2} = 2 \times 1.6 \times 10^{-19} \times 5 \times 10^{6}\left[(3 + 87.1) \times 10^{-9}\right]$$
$$= 16 \times 10^{-13} \times 90.1 \times 10^{-9}$$
$$\boxed{\overline{i_{Ts}^2} = 1.44 \times 10^{-19}} \quad \textbf{Ans.}$$

The rms shot noise current is

$$\boxed{\left(\overline{i_{Ts}^2}\right)^{1/2} = 3.79 \times 10^{-10}\ A} \quad \textbf{Ans.}$$

(*b*) Thermal noise is

$$\overline{i_t^2} = \frac{4KTB}{R_L}$$
$$= \frac{4 \times 1.38 \times 10^{-23} \times 293 \times 5 \times 10^{6}}{4 \times 10^{3}} = \frac{8086.8 \times 10^{-7}}{4 \times 10^{3}}$$
$$\boxed{\overline{i_t^2} = 2.02 \times 10^{-17}} \quad \textbf{Ans.}$$

The rms thermal noise current is

$$\left(\overline{i_t^2}\right)^{1/2} = \left(2.02\times10^{-17}\right)^{1/2}$$

$$\left(\overline{i_t^2}\right)^{1/2} = 4.49\times10^{-9}\text{ A} \qquad \textbf{Ans.}$$

Example 9.5. A photodiode has a capacitance of 8 pF. Calculate :

(*a*) Maximum load resistance which allows an 6 MHz post detection bandwidth.

(*b*) The bandwidth penalty with the same load resistance when the following amplifier also has an input capacitance of 4 pF.

Solution. Given that,

$$B = 6\text{ MHz}$$

$$C_d = 8\text{ pF}$$

$$C_a = 4\text{pF}$$

(*a*) The maximum bandwidth is given by

$$B = \frac{1}{2\pi R_L C_d}$$

$$R_L(\max) = \frac{1}{2\pi C_d B}$$

$$= \frac{1}{2\times3.14\times8\times10^{-12}\times6\times10^{6}}$$

$$= \frac{1\times10^{6}}{96\times3.14} = 3.32\times10^{3}$$

$$\boxed{R_L(\max) = 3.32\text{ k}\Omega} \qquad \textbf{Ans.}$$

(*b*) Also considering the amplifier capacitance, the maximum bandwidth.

$$B = \frac{1}{2\pi R_L(C_d + C_a)}$$

$$= \frac{1}{2\times3.14\times3.32\times10^{3}\times12\times10^{-12}}$$

$$= \frac{1\times10^{9}}{250.1952} \simeq 4\times10^{6}\text{ Hz}$$

$$\boxed{B = 4\text{ MHz.}} \qquad \textbf{Ans.}$$

Example 9.6. The receiver has an amplifier with a noise figure of 3 dB. Determine the SNR at the output of the receiver under the following conditions :

$$I_P = 87.1\times10^{-9}\text{ A}$$

$$\overline{i_{Ts}^2} = 1.44\times10^{-19}\text{ A}^2$$

$$\overline{i_t^2} = 2.02\times10^{-17}\text{ A}^2$$

Solution. Given that, $F_n = 3$ dB

$$\therefore \quad 10 \log_{10} (F_n) = 3$$

$$F_n = 2$$

The SNR is given by

$$\frac{S}{N} = \frac{I_P^2}{2eB(I_P + I_d) + \frac{4KTB\,F_n}{R_L}}$$

$$= \frac{I_P^2}{\overline{i_{Ts}^2} + \left(\overline{i_t^2} \times F_n\right)}$$

$$= \frac{(87.1 \times 10^{-9})^2}{(1.44 \times 10^{-19}) + (2.02 \times 10^{-17} \times 2)}$$

$$= \frac{7586.41 \times 10^{-18}}{405.44 \times 10^{-19}}$$

$$\boxed{\frac{S}{N} = 1.87 \times 10^2}$$

SNR is dB :

$$\text{SNR} = 10 \log_{10} (1.87 \times 10^2)$$

$$\boxed{\text{SNR (in dB)} = 22.72 \text{ dB}}$$ **Ans.**

Example 9.7. A good silicon APD (x = 0.3) has a capacitance of 5 pF, negligible dark current and is operating with a post detection bandwidth of 50 MHz. When the photocurrent before gain is 10^{-7} A and temperature is 18 °C. Determine the maximum SNR improvement between M = 1 and $M = M_{OP}$ assuming all operating condition are maintained.

Solution. Given that,

$$C_d = 5\text{pF} = 5 \times 10^{-12} \text{ F}$$

$$B = 50 \text{ MHz} = 50 \times 10^6 \text{ Hz}$$

$$T = 18°\text{C} = 291 \text{ K}$$

$$I_P = 10^{-7}$$

The maximum value of load resistance is

$$R_L = \frac{1}{2\pi C_d B}$$

$$= \frac{1}{2 \times 3.14 \times 5 \times 10^{-12} \times 50 \times 10^6}$$

$$= \frac{1}{1570 \times 10^{-6}}$$

$$R_L = 636.5\ \Omega$$

When $M = 1$: The SNR is given by,

$$\text{SNR} = \frac{I_P^2}{2eBI_P + \dfrac{4KTB}{R_L}}$$

where $I_d = 0$ and $F_n = 1$

Therefore,

$$\frac{S}{N} = \frac{(10^{-7})^2}{(2 \times 1.6 \times 10^{-19} \times 50 \times 10^6 \times 10^{-7}) + \dfrac{4 \times 1.38 \times 10^{-34} \times 291 \times 50 \times 10^6}{636.5}}$$

$$= \frac{10^{-14}}{1.602 \times 10^{-18} + \dfrac{80316 \times 10^{-26}}{636.5}}$$

$$= \frac{10^{-14}}{1.602 \times 10^{-18} + 1.263 \times 10^{-15}}$$

$$= \frac{10^{-14}}{1264.602 \times 10^{-18}} = 0.0791 \times 10^2$$

$$\boxed{\frac{S}{N} = 7.91}$$

The SNR in dB :

$$\frac{S}{N} = 10 \log_{10} (7.91)$$

$$\boxed{\frac{S}{N} = 8.98 \text{ dB}} \quad \textbf{Ans.}$$

When $M = M_{OP}$ and $x = 0.3$:

$$M_{OP}^{2+x} = \frac{4KT}{x\, e\, R_L\, I_P}$$

where, $I_d = 0$ and $F_n = 1$

$$\therefore \quad M_{OP}^{2.3} = \frac{4 \times 1.381 \times 10^{-23} \times 291}{0.3 \times 1.6 \times 10^{-19} \times 636.5 \times 10^{-7}}$$

$$= \frac{1607.484 \times 10^{-23}}{305.52 \times 10^{-26}}$$

and

$$M_{OP} = (5.255 \times 10^3)^{1/2.3} = (5.255 \times 10^3)$$

$$M_{OP} = 41.54$$

The SNR at M_{OP} is given by

$$\frac{S}{N} = \frac{M^2 I_P^2}{2eBI_P M^{2.3} + \dfrac{4KTB}{R_L}}$$

$$= \frac{(41.54)^2 \times (10^{-7})^2}{[2 \times 1.6 \times 10^{-19} \times 50 \times 10^6 \times 10^{-7} \times (41.54)^{2.3}] + \left[\frac{4 \times 1.38 \times 10^{-23} \times 291 \times 50 \times 10^6}{636.5}\right]}$$

$$= \frac{1725.57 \times 10^{-14}}{1.602 \times 10^{-18} \times (41.34)^{2.3}] + 1.263 \times 10^{-15}} = 1780$$

$$\boxed{\frac{S}{N} = 1.78 \times 10^3}$$

The SNR in dB is

$$\frac{S}{N} = 10 \log_{10} (1.78 \times 10^3)$$

$$\boxed{\frac{S}{N} = 32.50 \text{ dB}} \quad \textbf{Ans.}$$

Therefore the SNR improvement over $M = 1$ is (32.5 dB – 8.98 dB) = 23.5 dB.

Example 9.8. A germanium APD (with $x = 1$) is incorporated into an optical fiber receiver with a 10 kΩ load resistance. When operated at a temperature of 120 K, the required minimum photocurrent to give a SNR of 35 dB at the output of the receiver is found to be a factor of 10 greater than the dark current. If the noise figure of the following amplifier at this temperature is 1 dB and the post detection bandwidth is 10 MHz, determine :

(*a*) The minimum photocurrent

(*b*) Optimum avalanche multiplication factor.

Solution. Given that,

$$x = 1$$

$$\text{SNR} = 35 \text{ dB} = 3.16 \times 10^3$$

$$T = 120 \text{ K}$$

$$F_n = 1 \text{ dB} = 1.26$$

$$B = 10 \text{ MHz} = 10^6 \text{ Hz}$$

$$R_L = 10 \text{ k}\Omega = 10^4 \ \Omega$$

(*a*) The SNR is given by (with $x = 1$ and $M = M_{OP}$)

$$\frac{S}{N} = \frac{M_{OP}^2 I_P^2}{2eB(I_P + I_d) M_{OP}^3 + \frac{4KTB}{R_L}} \quad ...(i)$$

where,
$$M_{OP}^3 = \frac{4KT F_n}{e R_L (I_P + I_d)}$$

$$M_{OP} = \left[\frac{4KT F_n}{e R_L (I_P + I_d)}\right]^{1/3} \quad ...(ii)$$

Substituting the value of M_{OP} in eqn. (*i*) we get

$$\frac{S}{N} = \frac{\left[\dfrac{4KT\,F_n}{eR_L\,(I_P + I_d)}\right]^{2/3} I_P^2}{\dfrac{8\,KTB\,F_n}{R_L} + \dfrac{4KTB\,F_n}{R_L}}$$

Since, $I_d = 0.1\,I_P$, the SNR will be

$$\frac{S}{N} = \frac{\left(\dfrac{4KT\,F_n}{1.1\,eR_L}\right)^{2/3} I_P^{4/3}}{\dfrac{12\,KTB\,F_n}{R_L}}$$

Therefore the minimum photo current, I_P is

$$I_P^{4/3} = \left(\frac{S}{N}\right)\frac{\left(\dfrac{12\,KTB\,F_n}{R_L}\right)}{\left(\dfrac{4\,KT\,F_n}{1.1\,e\,R_L}\right)^{2/3}}$$

$$= \frac{\left[\dfrac{3.16\times10^{3}\times12\times1.38\times10^{-23}\times120\times10^{7}\times1.26}{10^{4}}\right]}{\left(\dfrac{4\times1.38\times10^{-23}\times120\times1.26}{1.1\times1.6\times10^{-19}\times10^{4}}\right)^{2/3}}$$

$$= \frac{7912.2355\times10^{-17}}{\left[\dfrac{0.34.624\times10^{-6}}{1.76}\right]^{2/3}} = \frac{7912.2355\times10^{-17}}{\left[474.22\times10^{-6}\right]^{2/3}}$$

$$I_P^{2/3} = \frac{7912.24\times10^{-17}}{2.82\times10^{-4}}$$

$$I_P = \left[2805.76\times10^{-13}\right]^{3/4}$$

$$\boxed{I_P = 6.87\times10^{-8}\text{ A}}$$ **Ans.**

(*b*) Therefore the optimum avalanche multiplication factor is

$$M_{OP} = \left[\frac{4\times1.38\times10^{-23}\times120\times1.26}{1.6\times10^{-19}\times10^{3}\times1.1\times6.87\times10^{-8}}\right]^{1/3}$$

$$= \left[\frac{834.624\times10^{-23}}{12.20\times10^{-24}}\right]^{1/3} = (684.12)^{1/3}$$

$$\boxed{M_{OP} = 8.84}$$ **Ans.**

Example 9.9. A high input impedance amplifier which in employed in an optical fiber receiver has an effective input resistance of 4 MΩ which is matched to a detector bias resistor of the same value. Determine :

(*a*) The maximum bandwidth that may be obtained without equalization if the total capacitance C_T is 6 pF.

(*b*) The mean square thermal noise current per unit bandwidth generated by this high input impedance amplifier configuration when it is operating at a temperature of 300 K.

Solution. Given that,

$$R_a = 4\ \text{M}\Omega = 4 \times 10^6\ \Omega$$
$$R_b = 4\ \text{M}\Omega = 4 \times 10^6\ \Omega$$
$$C_T = 6\ \text{pF} = 6 \times 10^{-12}\ \text{F}$$
$$T = 300\ \text{K}$$

(*a*) The maximum bandwidth is given by

$$B = \frac{1}{2\pi R_{TL} C_T}$$

where,

$$R_{TL} = \frac{R_a R_b}{R_a + R_b}$$

$$R_{TL} = \frac{(4 \times 10^6)^2}{8 \times 10^6} = 2\ \text{M}\Omega$$

Therefore,

$$B = \frac{1}{2 \times 3.14 \times 2 \times 10^6 \times 6 \times 10^{-12}}$$

$$= \frac{10^6}{75.36}$$

$$B = 1.33 \times 10^4\ \text{Hz} \qquad \textbf{Ans.}$$

(*b*) The mean square thermal noise current per unit bandwidth is given by

$$\overline{i_t^2} = \frac{4KT}{R_{TL}}$$

$$= \frac{4 \times 1.38 \times 10^{-23} \times 300}{2 \times 10^6}$$

$$\boxed{\overline{i_t^2} = 8.28 \times 10^{-27}\ \text{A}^2\text{Hz}^{-1}} \qquad \textbf{Ans.}$$

Example 9.10. A silicon photodiode has a responsivity of 0.5 AW^{-1} at a wavelength of 0.85 μm. Determine the minimum incident optical power required at the photodiode at this wavelength in order to maintain a bit error rate of 10^{-7}. When utilizing ideal binary signalling at a rate of 35 M-bits S^{-1}.

Solution. Given that,

$$R = 0.5\ \text{AW}^{-1}$$
$$\lambda = 0.5\ \mu\text{m}$$

$$P(e) = \text{BER} = 10^{-7}$$

$$r = 35 \text{ M bits S}^{-1}$$

Bit error rate is given by $P_e = e^{(-z_m)}$

$$10^{-7} = \left(e^{-z_m}\right)$$

$$\Rightarrow \quad z_m = 16.12$$

where, $$z_m = \frac{\eta P_0 z}{hf} = \frac{P_0 z}{e} \cdot \frac{\eta e}{h f}$$

$$\Rightarrow \quad P_0 . z = 51.6 \times 10^{-19}$$

$$P_0 = \frac{51.6 \times 10^{-19} \times 35 \times 10^6}{2}$$

$$P_0 = 9.03 \times 10^{-11}$$

P_0 in dB, $$= 10 \log_{10} \left(\frac{P_0}{P_r}\right)$$

$$P_0 = 10 \log_{10} (P_0) \quad \text{[For reference power level of 1 mW]}$$

$$= 10 \log_{10} (9.03 \times 10^{-8})$$

$$= -70.4 \text{ dB}$$

$$\boxed{P_0 = -70.4 \text{ dB}} \quad \textbf{Ans.}$$

SUMMARY

1. The receiver in an intensity modulated/direct modulation (IM/DD) optical fiber communication system consists of a photodetector and an amplifier.
2. Noise is defined by any spurious or undesired disturbances that mask the received signal in a communication system.
3. The three types of noise due to spontaneous fluctuations in optical fiber communication systems are :
   - Thermal noise.
   - Dark current noise.
   - Quantum noise.
4. Thermal noise current it in a resistor R is given by

$$i_t^2 = \frac{4KTB}{R}$$

5. The dark current noise is given by

$$\bar{i}_d^2 = 2eBI_d$$

6. The Bose-Einstein distribution is used to describe the random statistics of light emitted in black body radiation, and is given by

$$P(z) = \frac{z_m^2}{(1 + z_m)^{2+1}}$$

7. **The basic requirements for a fiber optic communication receiver are :**

- **Wide dynamic range.**
- **Excellent receiver sensitivity.**
- **Fast acquisition time.**
- **Wide range of bit rates.**

8. **The incident optical power required at the receiver in analog transmission quantum noise is given by**

$$\frac{S}{N} = \frac{\eta P_0}{2hfB}$$

9. **The dynamic range is defined by the difference in decibels between maximum to minimum power range which a receiver can accept or detect.**

10. **In photodiodes the two main sources of noise are :**

- **Dark current noise.**
- **Quantum noise.**

11. **The total shot noise in a photodiode receiver is given by**

$$\bar{i}_{TS}^2 = 2eB(I_P + I_d + I_b)$$

12. **The SNR for the *p-n* and *p-i-n* photodiode receiver is given by**

$$\frac{S}{N} = \frac{I_P^2}{2eB(I_P + I_d) + \frac{4KTB}{R_L} + i_{amp}^2}$$

13. **The three receiver structures are :**

- **Low impedance front end.**
- **High impedance front end.**
- **Trans-impedance front end.**

14. **The trans-impedance front end configuration overcomes the drawbacks of the high impedance front end by utilizing a low noise, high input impedance amplifier with negative feedback.**

15. **An optical receiver consist of**

- **Photodetector.**
- **An amplifier.**
- **Signal processing circuitry.**

16. **The key parameters used to define receiver performance are :**

- **Receiver sensitivity.**
- **Bandwidth.**
- **Dynamic range.**

REVIEW QUESTIONS

1. Draw and discuss the block diagram of optical receiver. What are the various sources of noise in the receiver ? **[PTU 2005]**
2. Briefly describe the various types of noise due to spontaneous fluctuations in optical fiber communication system.
3. Derive the Bose-Einstein Distribution to describe the random statistics of light emitted in black body radiation.
4. Explain various noise and disturbances associated with signal detection in optical receiver.
5. What is an optical front end of a reciever ? What is the major factor involved in the design of an optical front end ?
6. List two of the commonest types of receiver circuit. **[PTU 2003]**
7. Discuss the working of transimpedance preamplifier. What are its advantages over high impedance FET pre-amplifier ?
8. Draw and discuss the schematic showing major elements of an optical fiber receiver.
9. Describe the possible sources of noise in optical fiber receivers. Briefly discuss what do you meant by quantum noise. Consider this phenomenon with respect to :

   (*a*) Digital signalling. (*b*) Analog transmission.
10. What trade-off does the designer of an optical front end have to consider when choosing a type of pre-amp ?
11. Draw the equivalent circuit of a receiver and derive an expression for its input admittance.
12. What are the sources of receiver sensitivity degradation ?
13. What do you understand by photodiode ? Discuss their noise sources also.
14. Derive an expression of SNR for the *p-n* and *p-i-n* photodiode receiver. **[PTU 2005]**
15. What is avalanche photodiode receiver ? Determine the factor on which avalanche photodiode receiver depends.
16. Draw and discuss the digital fiber optic receiver. List the different types of receiver structures also.
17. Discuss the importance of load resistance on both thermal noise and post detection bandwidth in optical fiber communication receivers.
18. Explain the following mechanism associated with noise in optical fiber communication fiber communication.

    (*a*) Quantum shot noise

    (*b*) Avalanche excess noise

    (*c*) fiber mode partition noise

    (*d*) Thermal noise
19. Discuss the three main front end amplifiers configuration currently adopted for optical fiber communication with circuit diagram, comment on their merits and drawbacks.
20. Discuss the expression for the SNR in an APD receiver given by

$$\frac{S}{N} = \frac{M^2 I_P^2}{2eB(I_P + I_d)M^{2+x} + \dfrac{4KTBF_n}{R_L}}$$

with respect to various sources of noise present in the receiver. How may this this expression be modified to give the optimum avalanche multiplication factor ?

21. Explain the necessity of preamplifier in an optical receiver. Mention the types of preamplifier used and explain the working of any one of them.

22. A silicon *p-i-n* photodiode has a quantum efficiency of 65% at a wavelength of 0.8 μm. Determine :

(*a*) the mean photocurrent when the detector is illuminated at a wavelength of 0.8 μm with 5 μW of optical power.

(*b*) The rms quantum noise current in a post detection bandwidth of 20 MHz.

(*c*) The SNR in dB.

23. An analog optical fiber communication system requires an SNR of 40 dB at the detector with a post detection bandwidth of 30 MHz. Calculate the minimum optical power required at the detector if it is operating at a wavelength of 0.9 μm with a quantum efficiency of 70%.

24. A digital optical fiber link employing ideal binary signalling at a rate of 50 M bits s^{-1} operates at a wavelength of 1.3 μm. The detector is a germanium photodiode which has a quantum efficiency of 45% at this wavelength. An alarm is activated at the receiver when the bit error rate drops below 10^{-5}. Calculate the theoretical minimum optical power required at the photodiode in order to keep the alarm inactivated.

25. The germanium photodiode incorporated into an optical fiber receiver working at a wavelength of 1.55 μm has a dark current of 500 nA at the operating temperature. When the incident optical power at this wavelength is 10^{-6} W and the responsivity of the device is 0.6 A/W, shot noise dominates in the receiver. Determine the SNR in dB at the receiver when the post detection bandwidth is 100 MHz.

26. A silicon RAPD has a quantum efficiency of 95% at a wavelength of 0.9 μm, has an excess avalanche noise factor of M0.3 and a capacitance of 2 pF. It may be assumed that the post detection bandwidth is 25 MHz, and that the dark current in the device is negligible at the operating temperature of 290 K. Determine the minimum incident optical power which can yield an SNR of 23 dB.

27. A germanium APD with (x = 1) operates at a wavelength of 1.35 μm where its responsivity is 0.45 A/W. The dark current is 200 nA at the operating temperature of 250 K and the device capacitance is 3 pF. Determine the maximum possible SNR when the incident optical power is 8×10^{-7} W and the post detection bandwidth without equalization is 560 MHz.

28. What is meant by excess avalanche noise factor F(M) ? Give two possible ways of exproding this factor in analytical terms. Comment briefly on their relative merits.

❑❑❑

Optical Detectors

Chapter 10

INSIDE THIS CHAPTER

10.1. INTRODUCTION

A transducer is a device that converts input energy of one form into output energy of another. An optical detector is a transducer that converts an optical signal into an electrical signal. This is done by generating an electrical current proportional to the intensity of incident optical radiation. The relationship between the input optical radiation and the output electrical current is given by the **detector responsivity.**

The predominant types of light detector used in communications systems rely on the principle of ionization in a semiconductor material. There are a number of different kinds of devices but they can all be viewed as variations on a central principle rather than as devices that involve radically different principles.

Photoconductors are semiconductors devices that can convert optical signals into electrical signals. Basically there are **three processes for a general photoconductors :**

(*i*) Generation of carrier by incident light.

(*ii*) Carrier transport and/or multiplication by whatever current gain mechanism may be present.

(*iii*) Interaction of current with external circuitry to provide the output signal.

Properties of Optical Detector

Fiber optic communications systems require that optical detectors meet specific performance and compatibility requirements. Many of the requirements are similar to those of an optical source. Fiber optic systems require that optical detectors :

(*i*) Be compatible in size to low-loss optical fibers to allow for efficient coupling and easy packaging.

(*ii*) Have a high sensitivity at the operating wavelength of the optical source.

(*iii*) Have a sufficiently short response time (sufficiently wide bandwidth) to handle the system's data rate.

(*iv*) Contribute low amounts of noise to the system.

(*v*) Maintain stable operation in changing environmental conditions, such as temperature.

(*vi*) It must be of low cost.

(*vii*) Low bias voltage.

(*viii*) High reliability.

(*ix*) High fidelity.

(*x*) Large electrical response to the received output signal.

(*xi*) Dark current, leakage current and shunt conductance should be low.

Optical detectors that meet many of these requirements and are suitable for fiber optic systems are semiconductor photodiodes. The principal optical detectors used in fiber optic systems include semiconductor positive-intrinsic-negative (PIN) photodiodes and avalanche photodiodes (APDs).

10.2. IMPORTANT PARAMETERS OF PHOTODETECTORS

When discussing photodetectors some important parameters are there :

(1) Detector Responsivity : Responsivity is the ratio of the optical detector's output photocurrent in amperes to the incident optical power in watts.

The responsivity of a detector is a function of the wavelength of the incident light and the efficiency of the device in responding to that wavelength. For a particular material, only photons of certain wavelengths will generate a photocurrent when they are absorbed. Additionally, the detector material absorbs some wavelengths better than others. These two properties cause the wavelength dependence in the detector responsivity.

Responsivity is a useful parameter for characterizing detector performance because it relates the photocurrent generated to the incident optical power.

(2) Spectral Response Range : This is the range of wavelengths over which the device will operate.

(3) Response Time : This is a measure of how quickly the detector can respond to variations in the input light intensity. Response time can be affected by dark current, noise, linearity, back reflection and edge effect as shown in Fig. 10.1.

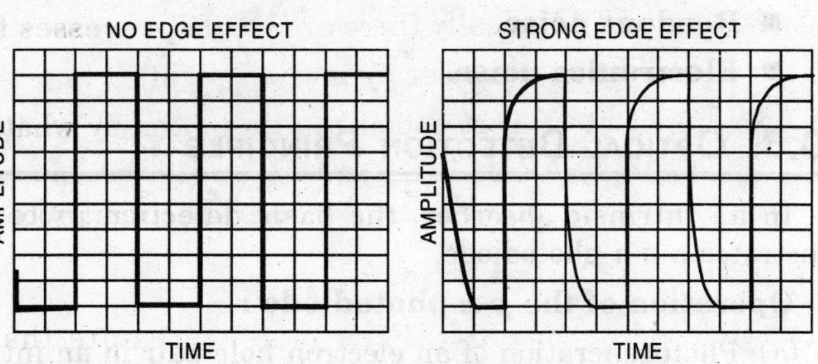

Fig. 10.1. Edge Effect.

Edge effect results from the fact that detectors only provide fast response in their center region. The outer region of the detector has a higher responsivity than the center region, which can cause problems when aligning the fiber to the detector. The higher responsivity may fool one into thinking they have aligned the fiber to the center region. Because response is much slower at the edge, this misalignment will reduce the response time of the detector.

(4) Quantum Efficiency : Ratio of primary electron-hole pairs created by incident photons to the photons incident on the detector material.

Detective quantum efficiency is given by :

$$DQE = \left[\frac{\dfrac{S}{N_{out}}}{\dfrac{S}{N_{in}}}\right]$$

So $$\frac{S}{N_{out}} = \frac{S}{N_{in}} * DQE^{1/2}$$

Poisson statistics $$\frac{S}{N_{in}} = N^{1/2}$$

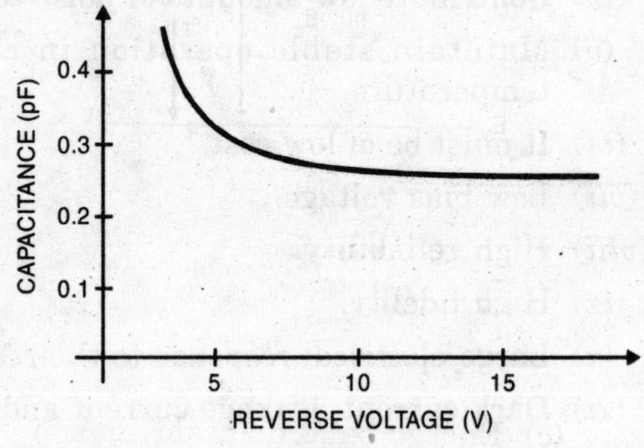

Fig. 10.2. C-V Curve.

(5) Capacitance : Dependent upon the active area of the device and the reverse voltage across the device. This relationship is illustrated in Fig. 10.2.

(6) Noise Characteristics : The level of noise produced in the device is critical to its operation at low levels of input light. There are several types of noise and these are :

- **Photon statistics limited noise** (photon arrival rate, by Poisson statistis, goes as qrt (n), n = number of photons).
- **Sky noise** caused by sky transparency changes and/or seeing/scintillation.
- **Johnson noise-thermal agitation** of electrons in detector (dark noise) reduced by cooling detector.
- **Readout noise.**
- **Electronics noise.**

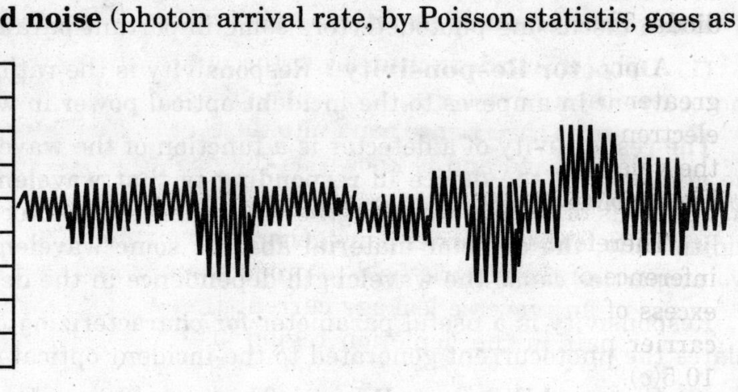

Fig. 10.3.

10.3. Optical Detection Principle

In an intrinsic absorber, the basic detection process is shown in Fig. 10.4, which illustrates a *p-n* photodiode.

Operation of the *p-n* photodiode :

(*a*) Photogeneration of an electron hole pair in an intrinsic semiconductor.

(*b*) Reverse bias *p-n* junction showing carrier drift in depletion region.

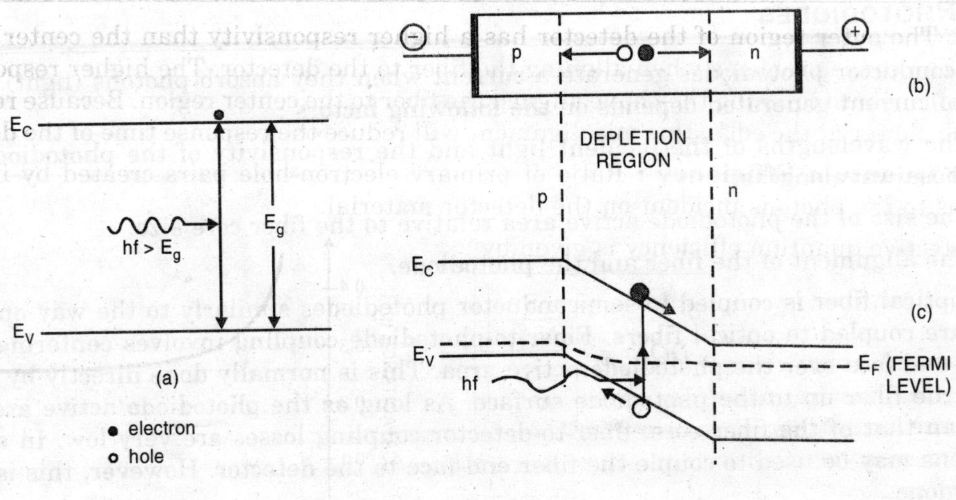

Fig. 10.4.

(*c*) Energy band diagram showin photo generation and separation of an electron hole pair.

This device is reverse biased and because of developed electric field mobile carriers more to their respective majority sides. Therefore on either side of junction a depletion region or layer is created. This barrier stops the majority carriers to cross the junction in the opposite direction to the field. However, the field accelerates minority carriers from both sides to the opposite side of the junction, forming the reverse leakage current of the diode. Thus intrinsic conditions are created in the depletion region.

A photon incident in or near the depletion region of this device which has an energy greater than or equal to the bandgap energy E_g of the fabricating material will excite on electron from the valence band into the conduction band. This process leaves an empty hole in the valence band and is known as the photogeneration of an electron hole pair as shown in Fig. 10.5(*a*).

Therefore, the carrier pairs generated near the junction are separated and swept under the inference of the electric field to produce a displacement by current in the external circuit in excess of any reverse leakage current [Fig. 10.5(*b*)]. Photo-generation and separation of a carrier pair in the depletion region of this reverse biased *p-n* junction is shown in Fig. 10.5(*c*).

The depletion region should be sufficiently thick to allow a large fraction of the incident light to be absorbed in order to achieve maximum carrier-pair generation. However, since long carrier drift times in the depletion region restrict the speed of operation of the photodiode, it is required to limit its width. Thus there is a trade off between the number of photons absorbed *i.e.* sensitivity and the speed of response.

Types of optical detectors :

The optical detectors are of three types :

(*i*) Photoemissive types (phototubes and photomultiplier tubes).

(*ii*) Photodiodes (*pn*-junction photodiodes *p-i-n* photodiodes and avalanche photodiodes).

(*iii*) Photoconductors.

10.4. Photodiodes

Semiconductor photodiodes generate a current when they absorb photons (light). The amount of current generated depends on the following factors :

- The wavelengths of the incident light and the responsivity of the photodiode at those wavelengths.
- The size of the photodiode active area relative to the fiber core size.
- The alignment of the fiber and the photodiode.

The optical fiber is coupled to semiconductor photodiodes similarly to the way optical sources are coupled to optical fibers. Fiber-to-photodiode coupling involves centering the flat fiber-end face over the photodiode active area. This is normally done directly by butt coupling the fiber up to the photodiode surface. As long as the photodiode active area is larger than that of the fiber core, fiber-to-detector coupling losses are very low. In some cases a lens may be used to couple the fiber end-face to the detector. However, this is not typically done.

There are four different types of photodiodes :

(*i*) *p-n* photodiodes.

(*ii*) *p-i-n* photodiodes.

(*iii*) Avalanche photodiodes (APD).

(*iv*) Schottky barrier photodiode.

10.5. p-n Photodiodes

Figure 10.5 shows a reverse biased *p-n* photodiode with both depletion and diffusion regions.

The principle involved in a *p-n* diode is simply the principle of the LED in reverse. That is light is absorbed at a *p-n* junction rather than emitted.

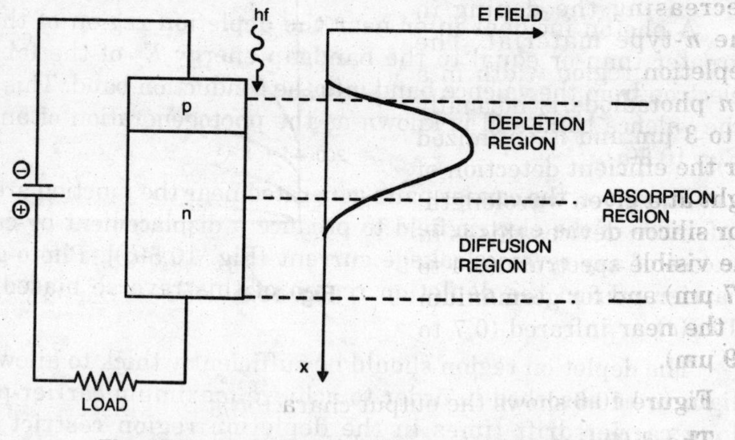

Fig. 10.5. *p-n* Photodiode Showing Depletion and Diffusion Regions.

- A reverse biased *p-n* junction passes no current unless something happens to promote electrons from the valence band to the conduction band within the depletion zone. At ordinary room temperatures there is a small amount of this due to the action of heat and thus there is a very small current.
- Photons of sufficient energy can be absorbed and cause the promotion of an electron from the valence band to the conduction band (and of course the simultaneous generation of a hole in the valence band).

- The free electron and hole created by the photon absorption now are attracted to their opposite charges at either side of the junction and current flows. The big problem here is that the depletion zone in a *p-n* junction is extremely thin.

Most light passes through without being absorbed in the junction, instead it is absorbed in the doped material at either side of the junction. Ultimately many of the electron-hole pairs created outside the junction do end up being attracted to the junction and creating current. However the process is quite slow and *p-n* junction devices are not fast enough for current communications applications.

As shown in Fig. 10.5, photons may be absorbed in both the depletion and diffusion regions. The absorption regions position and width depends upon :

- the energy of the incident photons
- the material from which the photodiode is fabricated.

The absorption region may extend completely throughout the device in the case of the weak absorption of photons. Therefore the electron hole pairs are generated in both the depletion and diffusion regions.

In the depletion region the carrier pair separate and drift under the influence of the electric field, whereas in the outside region the hole diffuses towards the depletion region in order to be collected. The diffusion process is very slow as compared to drift and thus limits the response of the photodiode.

Since the photons are absorbed in the depletion region, it should be made as long as possible by decreasing the doping in the *n*-type material. The depletion region width in a *p-n* photodiode is normally 1 to 3 μm and is optimized for the efficient detection of light at a given wavelength. For silicon devices this is in the visible spectrum (0.4 to 0.7 μm) and for germanium in the near infrared (0.7 to 0.9 μm).

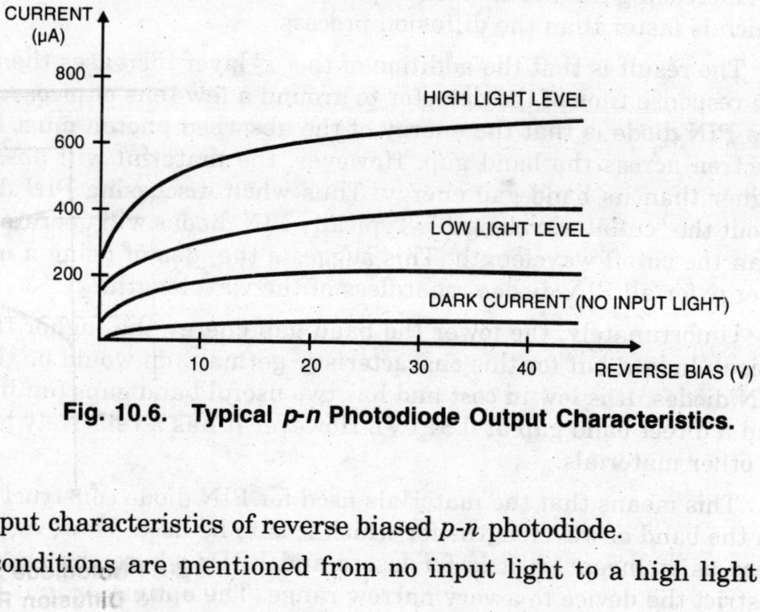

Fig. 10.6. Typical *p-n* Photodiode Output Characteristics.

Figure 10.6 shows the output characteristics of reverse biased *p-n* photodiode.

The different operating conditions are mentioned from no input light to a high light level.

10.6. p-i-n Photodiode

The problem created by the extreme thinness of a *p-n* junction is simplified by making it thicker. The junction is extended by the addition of a very lightly doped layer called the **intrinsic zone** between the *p* and *n* doped zones. Thus the device is called a *p-i-n* diode rather than a *p-n* diode. This is illustrated in Fig. 10.7.

The wide intrinsic (*i*) layer has only a very small amount of dopant and acts as a very wide depletion layer. There are a number of improvements here :

(*i*) It increases the chances of an entering photon being absorbed because the volume of absorbent material is significantly increased.

(*ii*) Because it makes the junction wider, it reduces the capacitance across the junction. The lower the capacitance of the junction the faster the device response.

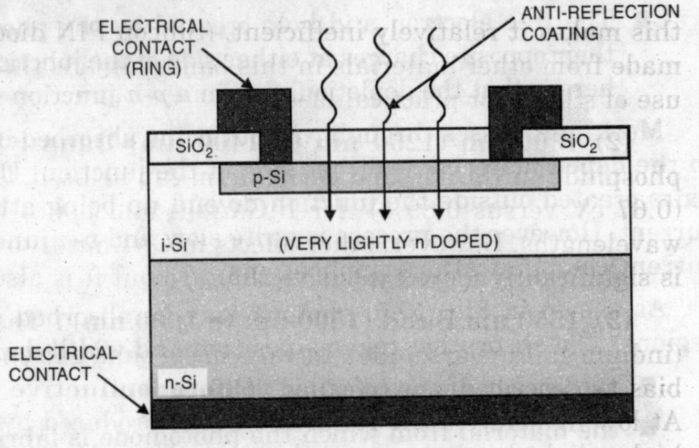

Fig. 10.7. Typical Silicon *p-i-n* Diode Schematic.

(*iii*) There are two ways of current carriage across the junction :

- Diffusion
- Drift.

Increasing the width of the depletion layer favours current carriage by the drift process which is faster than the diffusion process.

The result is that the addition of the "*i*" layer increases the responsivity and decreases the response time of the detector to around a few tens of picoseconds. The key to operation of a PIN diode is that the energy of the absorbed photon must be sufficient to promote an electron across the band gap. However, the material will absorb photons of any energy higher than its band gap energy. Thus when discussing PIN diodes it is common to talk about the "cutoff wavelength". Typically PIN diodes will operate at any wavelength shorter than the cutoff wavelength. This suggests the idea of using a material with low band gap energy for all PIN diodes regardless of the wavelength.

Unfortunately, the lower the band gap energy the higher the "dark current" (thermal noise). Indeed but for this characteristic germanium would be the material of choice for all PIN diodes. It is low in cost and has two useful band gaps (an indirect band gap at 0.67 eV and a direct band gap at 0.81 eV). However it has a relatively high dark current compared to other materials.

This means that the materials used for PIN diode construction are different depending on the band of wavelengths for which it is to be used. However, this restriction is nowhere near as stringent as it is for lasers and LEDs where the characteristics of the material restrict the device to a very narrow range. The optimal way is to choose a material with a band gap energy slightly lower than the energy of the longest wavelength we want to detect.

10.6.1. Types of Materials Used

Typical materials used in the three communication wavelength "windows" are as follows :

(1) 500-1000 nm Band : Silicon PIN diodes operate over a range of 500 to 1120 nm as silicon has a band gap energy of 1.11 eV. Since silicon technology is very low cost, silicon is the material of choice in this band. However, silicon is an indirect band gap material and this makes it relatively inefficient. (Silicon PIN diodes are not as sensitive as PIN diodes made from other materials in this band.) This is the same characteristic that prevents the use of silicon for practical lasers.

(2) 1300 nm (1250 nm to 1400 nm) Band : In this band indium gallium arsenide phosphide (InGaAsP) and germanium can be used. Germanium has lower band gap energy (0.67 eV versus 0.89 eV for InGaAsP) and hence it can theoretically be used at longer wavelengths. However, other effects in Ge limit it to wavelengths below 1400 nm. InGaAsP is significantly more expensive than Ge but it is also significantly more efficient.

(3) 1550 nm Band (1500 nm to 1600 nm) : The material used here is usually InGaAs (indium gallium arsenide). InGaAs has a band gap energy of 0.77 eV. Operation with reverse bias as described above (called **"Photoconductive Mode") has a significant problem.** At low light levels the random current produced by ambient heat is a source of noise. At higher light levels however this is not a problem and the device offers the advantages of much higher speeds (than the alternative mode of operation) and linear response characteristics over a wide range.

The higher speed characteristics are the result of lowered capacitance caused by the widening of the depletion layer in the presence of reverse bias. If the device is operated without an externally applied current the natural potentials of the *p-i-n* junction will cause electrons and holes to migrate across the junction anyway. This is called "Photovoltaic Mode". Thus we get a small voltage (around 15 V) developed across the device.

The **advantage** of this is that there is much less "dark current" caused by ambient heat. Thus in this mode the device is more sensitive but the output requires immediate amplification because of the low voltage levels produced.

The **efficiency** of *p-i-n* diodes at long wavelengths can be improved by the use of hetero-structures. For example using GaAIAs for the *p*-layer and GaAs for the *i* and *n*-layers. Incident light must pass through the *p*-layer before it enters the *i*-layer. If it is absorbed by the *p*-layer then the energy is lost and does not contribute to the current produced. By using a material with a high band gap energy for the *p*-layer (higher than the energy of the incident photons) we can prevent incident long wavelength light from being absorbed. This means that there is less light lost before it reaches the *i*-layer where we want it to be absorbed.

10.6.2. Response Time

There are several factors that influence the response time of a photodiode and its output circuitry (see Fig. 10.8). The most important of these are :

- the thickness of the detector active area
- the detector *RC* time constant

The detector thickness is related to the amount of time required for the electrons generated to flow out of the detector active area. This time is referred to as the **electron transit time.** The thicker the detector active area, the longer the transit time will be.

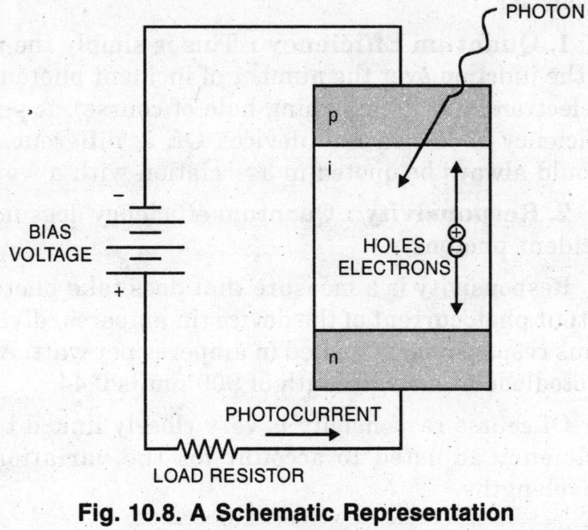

Fig. 10.8. A Schematic Representation of a Photodiode.

The capacitance (C) of the photodiode and the resistance (R) of the load form the RC time constant. The capacitance of the photo detector must be kept small to prevent the RC time constant from limiting the response time. The photodiode capacitance consists mainly of the junction capacitance and any capacitance relating to packaging. The RC time constant is given by

$$\tau_{RC} = RC \quad \text{...(10.1)}$$

- Trade-offs between fast transit times and low capacitance are necessary for high-speed response. However, any change in photodiode parameters to optimize the transit time and capacitance can also affect responsivity, dark current, and coupling efficiency.
- A fast transit time requires a thin detector active area, while low capacitance and high responsivity require a thick active region.
- The diameter of the detector active area can also be minimized. This reduces the detector dark current and minimizes junction capacitance.
- However, a minimum limit on this active area exists to provide efficient fiber-to-detector coupling.

Linearity

Reverse-biased photodetectors are highly linear devices. Detector linearity means that the output electrical current (photocurrent) of the photodiode is linearly proportional to the input optical power.

Reverse-biased photodetectors remain linear over an extended range (6 decades or more) of photocurrent before saturation occurs. Output saturation occurs at input optical power levels typically greater than 1 milliwatt (mW). Because fiber optic communication systems operate at low optical power levels, detector saturation is generally not a problem.

10.6.3. Measures of Efficiency in p-i-n Diode

There are two common measures quoted when the efficiency of PIN photodetectors is discussed.

1. Quantum Efficiency : This is simply the ratio of the number of electrons collected at the junction over the number of incident photons. In an ideal situation 1 photon releases 1 electron (and its matching hole of course). A perfect quantum efficiency is therefore an efficiency of "1". In real devices QE is different at each operating wavelength and so it should always be quoted in association with a wavelength.

2. Responsivity : Quantum efficiency does not take account of the energy level of the incident photons.

Responsivity is a measure that does take photon energies into account. It is simply the output photocurrent of the device (in amperes) divided by the input optical power (in watts). Thus responsivity is quoted in amperes per watt. A typical value of responsivity for a silicon photodiode at a wavelength of 900 nm is 0.44.

Of course responsivity is very closely linked to quantum efficiency. It is just quantum efficiency adjusted to account for the variation in energy level implied by different wavelengths.

10.7. Avalanche Photodiodes (APDs)

APDs amplify the signal during the detection process. They use a similar principle to that of "photomultiplier" tubes used in nuclear radiation detection. In the photomultiplier tube :

1. A single photon acting on the device releases a single electron.
2. This electron is accelerated through an electric field until it strikes a target material.
3. This collision with the target causes "impact ionisation" which releases multiple electrons.
4. These electrons are then themselves accelerated through the field until they strike another target.
5. This releases more electrons and the process is repeated until the electrons finally hit a collector element.

Thus, through several stages, one photon has resulted in a current of many electrons. APDs are of course different from photomultiplier tubes. Photomultiplier tubes are vacuum tubes with metallic targets arranged in stages down the length of the tube. APDs use the same principle but multiplication takes place within the semiconductor material itself. This process in APDs typically results in an internal amplification of between 10 and 100 times. Avalanche photodiode is shown in Fig. 10.9.

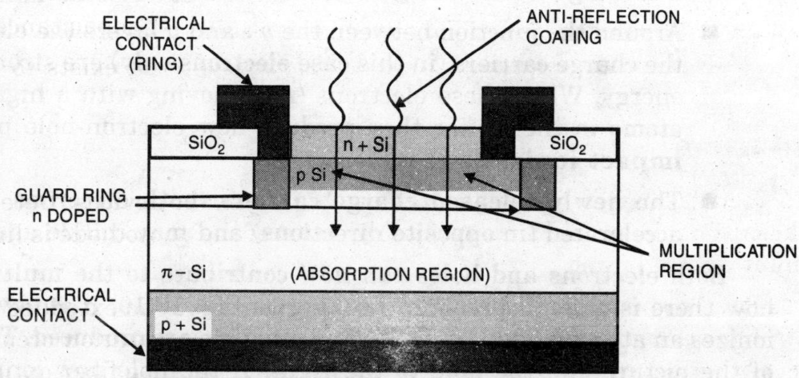

Fig. 10.9. Avalanche Photodiode (APD).

In its basic form an APD is just a *p-i-n* diode with a very high reverse bias. A reverse bias of 50 volts is typical for these devices compared with the regular *p-i-n* diodes used in the photoconductive mode which is reverse biased to only around 3 volts (or less). In the past some APDs on the market have required reverse bias of several hundred volts although recently lower voltages have been achieved.

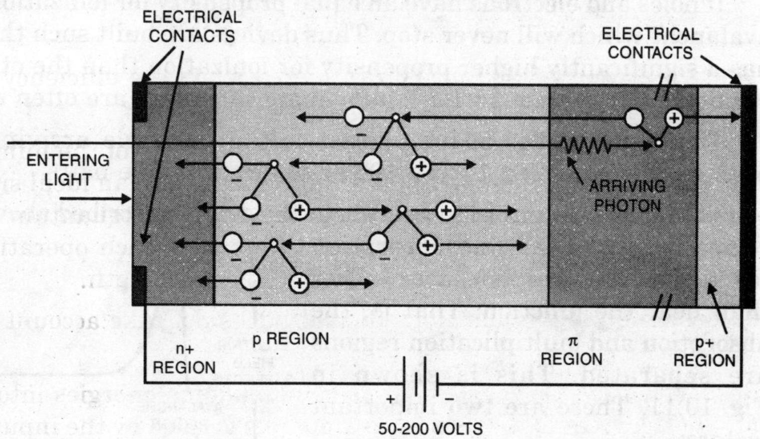

Fig. 10.10. Avalanche (Amplification) Process. The *p*-Region has been Enlarged to Show Avalanche Process.

The main structural difference between an APD and a *p-i-n* diode is that the *i* layer (which is lightly *n*-doped in a *p-i-n* structure) is lightly *p*-doped and renamed the layer. It is typically thicker than an *i*-zone and the device is carefully designed to ensure a uniform electric field across the whole layer. The guard ring shown in the figure serves to prevent unwanted interactions around the edges of the multiplication region.

The device operation is as follows :

- Arriving photons generally pass straight through the *n*+-*p* junction (because it is very thin) and are absorbed in the *n* layer. This absorption produces a free electron in the conduction band and a hole in the valence band.
- The electric potential across the *n* layer is sufficient to attract the electrons towards one contact and the holes towards the other. In the figure electrons are attracted towards the *n*+ layer at the top of the device because being reverse biased it carries the positive charge. The potential gradient across the *n* layer is not sufficient for the charge carriers to gain enough energy for multiplication to take place.
- Around the junction between the *n*+ and *p* layers the electric field is so intense that the charge carriers (in this case electrons only) are strongly accelerated and pick up energy. When these electrons (now moving with a high energy) collide with other atoms in the lattice they produce new electron-hole pairs. This process is called **impact ionization.**
- The newly released charge carriers (both electrons and holes) are themselves accelerated (in opposite directions) and may collide again.

Both electrons and holes can now contribute to the multiplication process. However now there is a small problem. Looking at Fig. 10.10, it is obvious that when an electron ionizes an atom an additional electron and hole are produced. The electron moves to the left of the picture and the hole to the right. If the hole now ionizes an atom it releases an electron (and a hole) and the electron moves to the left and starts again.

If holes and electrons have an equal propensity for ionization we can get an uncontrolled avalanche which will never stop. Thus devices are built such that one of the charge carriers has a significantly higher propensity for ionization than the other. In silicon, electrons are the dominant carrier. In III-V alloy materials holes are often employed.

The result of the above process is that a single arriving photon can result in the production of between 10 and 100 or so electron-hole pairs.

The important things to note about the device described above are that the multiplication region is very small and absorption takes place within the *n* layer rather than near the junction. That is, the absorption and multiplication regions are separated. This is shown in Fig. 10.11. There are two important factors:

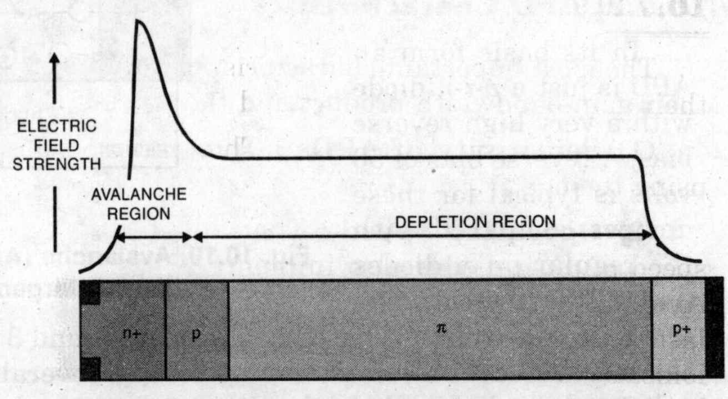

Fig. 10.11. Electric Field Strengths in an APD.

1. The strength of the required electric field is extremely high (Of the order of 10^6 volts/ metre.) In the presence of such a strong field, imperfections in the multiplication region (such

as lattice mismatches, impurities and even variations in dopant concentration) can produce small areas of uncontrolled multiplication called **microplazmas**. To control this effect the multiplication region needs to be small.

This is the reason for the guard ring mentioned above. Around the edges of the multiplication region we tend to get irregularities and imperfections in the material. Without the guard ring these serve as sites for microplazmas. In addition, to create an electric field of the required strength we need an applied bias voltage that increases with the thickness of the multiplication region. (Double the thickness of the region - double the required voltage.) High voltages (indeed voltages above about 12 volts) are expensive and difficult to handle in semiconductor devices and so we aim to minimize the applied voltage.

2. For the above reason the junction region is very thin and therefore it can't absorb many incident photons. The *n* layer is the absorption layer.
3. Many APDs are designed so that the depletion layer extends through the entire *p*-region to the *n* region boundary.

10.7.1. Types of Materials Used

As with *p-i-n* diodes different materials are usually employed for each of the three important wavelength bands:

(1) 800 nm to 1 Micron Band : In this band silicon is usually employed although germanium will also work reasonably well. Germanium devices however produce higher noise levels than silicon ones. As mentioned above for *p-i-n* structures, silicon has a relatively high band gap energy and is therefore used only for wavelengths shorter than about 1 micron. In practice, short wavelengths are only used for very short distance (less than 500 metres) communications. Attenuation over such short distances is generally not great enough to require the sensitivity of an APD (or to justify its cost).

(2) 1310 nm Band : This is important as it is the band used by most existing long distance communications systems. Germanium APDs are used extensively but III-V semiconductor alloys are increasing in usage because of the high noise levels in germanium.

(3) 1550 nm Band : III-V APDs are used widely in the 1550 nm band. The most common materials system in use is InGaAs/lnP where the majority charge carrier is hole rather than electron.

10.7.2. APD Characteristics

The most important characteristics of APDs are their sensitivity, their operating speed, their gain-bandwidth product and the level of noise.

(1) Sensitivity of APDs : The extreme sensitivity of APDs is the major reason for using them.

(2) Operating Speed : The same factors (such as device capacitance) that limit the speed of *p-i-n* diodes also influence APDs. However, with APDs there is another factor: **Avalanche Buildup Time**. Because both carriers can create ionisation an avalanche can last quite a long time. This is caused by the to and fro movement of electrons and holes as ionisations occur. Provided the propensity for ionisation in the minority carrier is relatively low, the avalanche will slow and stop but this takes some time. Avalanche Buildup Time thus limits the maximum speed of the APD.

(3) Gain-Bandwidth Product : The accepted measure of goodness of a photodetector is the gain-bandwidth product. This is usually expressed as a gain figure in dB multiplied by the detector bandwidth (the fastest speed that can be detected) in GHz. A very good current APD might have a gain-bandwidth product of 150 GHz.

(4) Noise : APDs are inherently noisy as the multiplier effect applies to all free electrons including those made free by ambient heat. This is especially a problem in longer wavelength devices where the band gap energy is low. In the design of devices great care is taken to make sure that the electrical potential gradient is sufficient for multiplication (ionisation) to occur. Greater levels of bias can cause spontaneous ionisation events.

In many long-distance, wide-area applications where sensitivity is more important than other factors, the APD has been commonly used. However as speed increases and amplified systems are introduced, the noise produced by APDs becomes a limitation. In these systems people use *p-i-n* detectors with preamplifiers as we get much higher sensitivity than an APD and much lower noise.

10.7.3. Multiplication Factor (*M*)

The multiplication factor *M* is a measure of the internal gain provided by the APD. It is a ratio of the total output current to the primary current, is given by

$$M = \frac{I}{I_P} \quad \text{...(10.2)}$$

where, I = Total output current at operating voltage (*i.e.* where carrier multiplication occurs)

I_P = Initial or primary photocurrent (*i.e.* before carrier multiplication occurs)

10.7.4. The Photomultiplication Factor (M_{Ph})

It is defined as the multiplied photocurrent at low voltages where no carrier multiplication takes place, is given by

$$M_{Ph} = \frac{I_{Ph}}{I_{Pho}} \quad \text{...(10.3)}$$

$$M_{Ph} = 1 + P + P^2 + ... + P^{\infty}$$

or $$M_{Ph} = \frac{1}{(1-P)} \quad \text{...(10.4)}$$

where, P = Probability that an accelerating electron will create an EHP.

In the above equation

First Term : It is one and gives the injected carrier

Second Term : Probability that the injected carrier generates another carrier.

Third Term : It is the probability that the injected carrier generates a second carrier

'*P*' is a very sensitive function of voltage and it is given by

$$P = \left(\frac{v}{v_B}\right)^n \quad \text{...(10.5)}$$

Hence photomultiplication factor, M_{Ph} is given by :

$$M_{Ph} = \left[\frac{1}{1 - \left(\frac{v}{v_B} \right)^n} \right] \qquad ...(10.6)$$

where, v_B = Breakdown voltage

v = Applied voltage

n = Experimentally determined constant

10.8. Hetero-Interface Photodetectors

Of available materials, APDs made with silicon have the best response, the highest gain and the lowest noise. However, silicon can't detect at wavelengths longer than about 1 micron because of its band gap energy. It can't absorb light at longer wavelengths.

The idea of a hetero junction APD is to replace the *p*-layer silicon material with some material that will absorb light in the long wavelength bands. In GaAs is one such material. The electric fields inside the device are organized such that the detection takes place in the In GaAs material but multiplication is allowed only in the silicon *i*-layer (this minimizes noise). Illumination can be from either side although illumination through the substrate material (silicon) is generally preferred.

The big problem is how to match the lattices of the Si *i*-layer and the In GaAs material. This has been done using a process called "wafer fusion". The resultant device has been called a Silicon Hetero Interface Photodetector (SHIP). This structure has been proven to provide a much higher gain-bandwidth product with lower noise than existing APDs. A product of 350 GHz has been reported. At the present time these devices are in research but commercial products are expected very soon.

10.9. Travelling Wave Photodetectors

When we try to make a *p-i-n* detector operate at extremely high speeds a serious problem arises. The maximum response frequency is determined by the time it takes for electrons and holes to drift/diffuse across the *i*-layer. (Both drift and diffusion contributes to the migration of a charge carrier across the *i*-layer.) Drift and diffusion are slow and it takes a finite amount of time for the device to respond to a pulse of light. To make the device go faster you have to reduce the thickness of the *i*-layer. But reducing the thickness of the *i*-layer increases capacitance effects between the *p*-layer and the *n*-layer. This increased capacitance acts to slow the device response. To counter the increased

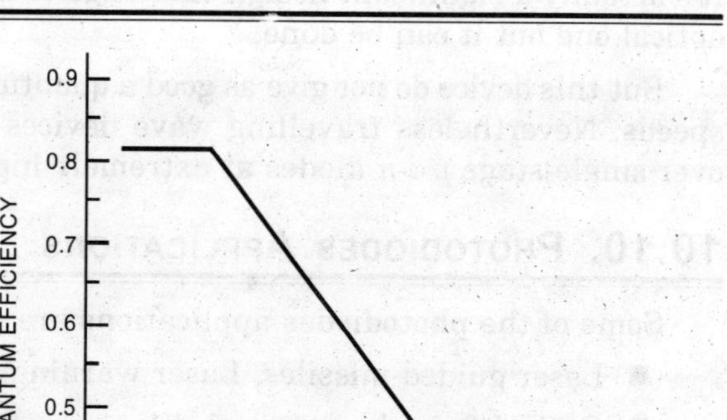

Fig. 10.12. Variation of Quantum Efficiency with Device Response.

capacitance you have to reduce the surface area of the device! So basically as the device gets faster it also must get much smaller. As it becomes smaller it produces less current.

As shown in Fig. 10.12, the above effect causes a serious reduction in the quantum efficiency (QE) of *p-i-n* detectors at speeds above about 10 Gbps. The "natural" loss of efficiency at higher speeds comes about because a detector requires a certain number of photons to reliably detect a bit. When the speed is doubled the number of photons required remains the same. Thus to get the same output level from a detector every time the speed is doubled then the detector halves in its sensitivity.

The principle of a travelling wave detector is illustrated in Fig. 10.13. A number of *p-i-n* detectors are integrated into an optical waveguide such that light not absorbed in one will travel on to the next. If the outputs were just joined together you wouldn't get a lot of improvement because the outputs come at different times depending on when light arrives at the device. In the TW configuration, light entering the waveguide on the left of the device encounters the *p-i-n* detectors one after another. Each detector would normally have a simple integrated preamplifier in this configuration. Output of the detectors is placed on an electrical waveguide on which the electrical signal travels in the same direction as the optical signal.

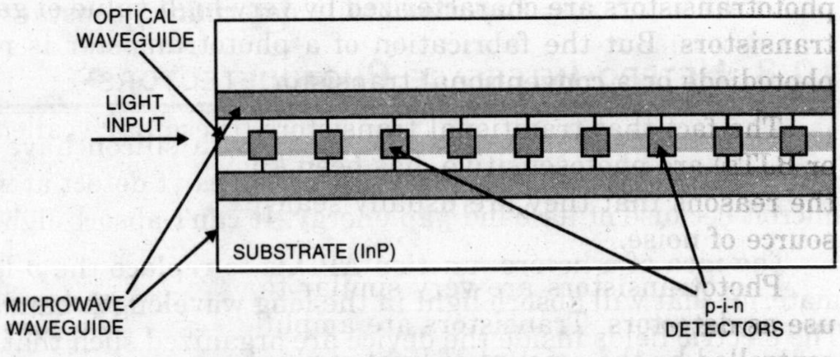

Fig. 10.13. Travelling Wave Photodetector—Principle.

The whole idea here is to match the speed of light propagation in the optical waveguide to the speed of electrical propagation in the electric waveguide. Thus the outputs of all the detectors "add-up" in phase as each detector places its output signal onto the waveguide. It's actually a significant design challenge to match the electronic propagation speed to the optical one but it can be done.

But this device do not give as good a quantum efficiency as a regular *p-i-n* diode at slower speeds. Nevertheless travelling wave devices give a very significant improvement in QE over single-stage *p-i-n* diodes at extremely high speeds (double or triple).

10.10. Photodiodes Applications

Some of the photodiodes applications are :

- Laser guided missiles, Laser warning, Laser range finders
- Optical free air communication
- Automotive anti collision optical radar
- Laser alignment and control systems
- Spectral analysis (medical)
- Two colour sensor (combined with IR detector chip)
- Monitoring of Hg lamps for sterilization
- Film processing

- Flame monitoring
- Scintillator read out
- Spectral monitoring of Earth ozone layer (environmental)
- Space applications (solar sensors, star sensors)

10.11. PHOTOTRANSISTORS

A phototransistor is a photodetector employing bipolar or unipolar transistors. The phototransistors are characterized by very high value of gains as compared to the ordinary transistors. But the fabrication of a phototransistor is rather more complicated than a photodiode or a conventional transistor.

The fact that traditional transistors (these days called "Bipolar Junction Transistors" or BJTs) are photosensitive, has been known ever since they were invented. This is one of the reasons that they are usually sealed in light tight cans—because unwanted light is a source of noise.

Phototransistors are very similar to ordinary BJTs except that they are designed for use as detectors. Transistors are amplifiers and in a phototransistor the amplifier gain is controlled by the amount of light striking the device.

Light arriving in the device is absorbed and creates charge carrier pairs in many different places. The action is different depending on where the charge carrier pairs are created. However, in an intentional phototransistor, absorbtion is intended to take place in the depletion zone between the emitter and the base. Creation of charge carriers here causes a current to flow in the E-B circuit which is then amplified by the transistor action of the device. These are much lower in noise and have a higher output than APDs, but are significantly less responsive than either APDs or *p-i-n* diodes.

The hetero-junction phototransistors may be used because it permits low emitter-base and collector-base junction capacitances together with low base resistance. This is done through low emitter and collector doping levels coupled with heavy doping of the base, and allows large current gain. The potential barrier developed by the hetero-junction at the emitter-base junction effectively eliminates holes injection from the base when the junction is forward biased. This gives good emitter base injection efficiency. The optical gain G_0 of the device is given by

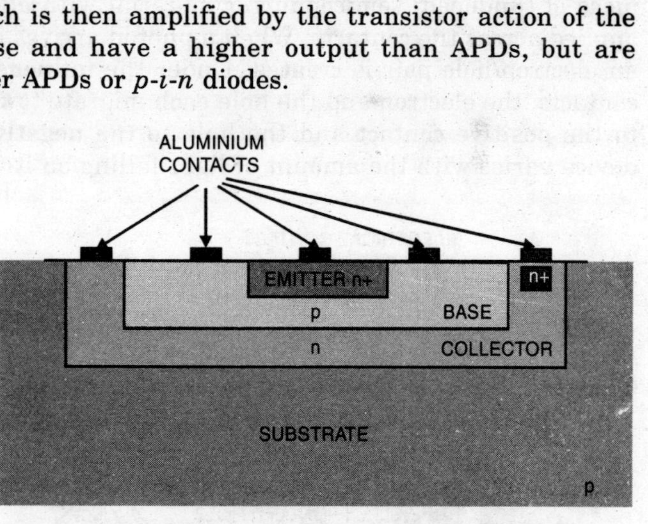

Fig. 10.14. Bipolar Junction Transistor (BJT).

$$G_0 \cong \eta h_{FE} = \frac{hf}{e}\frac{I_C}{P_0} \qquad ...(10.7)$$

where η = Quantum efficiency of base-collector photodiode

h_{FE} = Common emitter current gain

I_C = Collector current

P_0 = Incident optical power

e = Electronic charge

hf = Photon energy

The hetero-junction phototransistor is capable of operating over the 0.9 to 1.3 μm wavelength band giving optical gains in excess of one hundred.

The **major problem** with phototransistors is materials. We would like to use silicon or gallium arsenide but these have too great a band gap energy and are limited to detecting wavelengths shorter than about 1 micron. Germanium is usable in the 1300 nm band but while it is easily possible to build transistors with smaller band gap materials (such as InP) there is no established technology to do so. This takes away much of the potential cost advantage in the 1550 region.

The major use of phototransistors is in non-communication applications using visible (or near visible) light. Alarm systems (light beam detection) and remote controls for TV sets and automobiles are among the most common uses. Phototransistors are occasionally built as part of an integrated circuit. In this configuration they are referred to as **Integrated Preamplifier Detectors (IPDs).**

10.12. Photoconductors

A photoconductor is a device which is used to detect optical signals.

Photoconductors are the simplest conceivable optical detector. This device consists of a piece of (undoped) semiconductor material with electrical contacts attached. A voltage is applied across the contacts. When a photon arrives in the semiconductor it is absorbed and an electron/hole pair is created. Under the influence of the electric field between the two contacts, the electron and the hole each migrate toward one of the contacts *i.e.* the electron to the positive contact and the hole to the negative contact. Thus the resistance of the device varies with the amount of light falling on it.

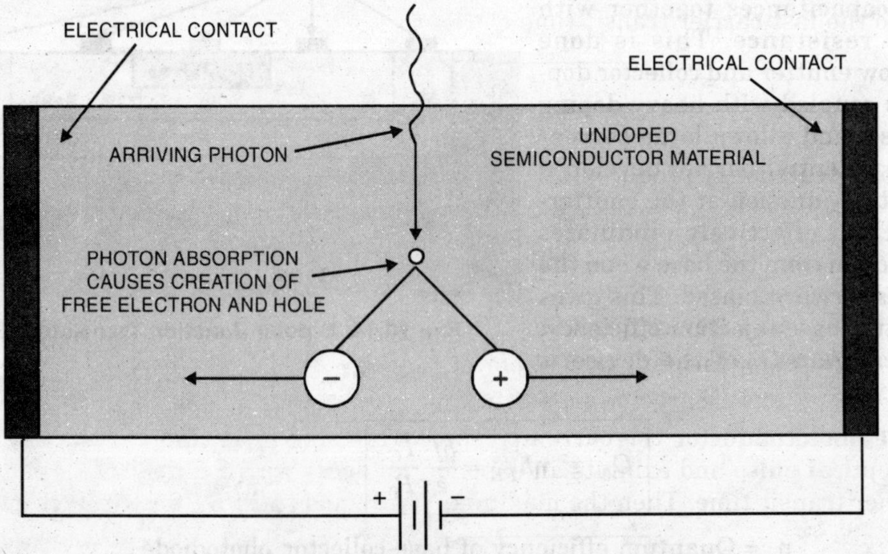

Fig. 10.15. Photoconductive Detector—Principle.

Practical photoconductive detectors are usually structured as shown in Fig. 10.15.

- A flat piece of semiconductor has metal contacts plated onto it.
- The contact areas are **inter digitized** so that there is only a short distance between the contacts. This is done because the charge carriers drift quite slowly through the semiconductor material and this limits the speed of the device.
- An interesting point is that the device exhibits gain. This is because electron drift is much faster than hole drift.
- Electrons tend to arrive at the positive contact before a hole arrives at the other side. This then causes the negative electrode to emit an electron into the semiconductor.
- This electron in turn may arrive at the positive electrode before the hole arrives at the negative electrode and yet another electron is emitted into the material.

This type of detector is quite useful at longer wavelengths (10 to 30 microns) but has not been used seriously for communication applications.

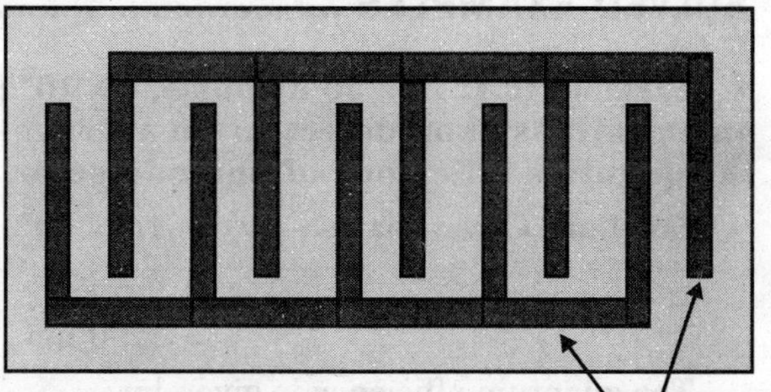

Fig. 10.16. Practical Photoconductive Detector—Schematic.

Since the fast electrons are collected at the anode, the corresponding holes are still proceeding across the channel. This creates an absence of electrons and hence a net positive charge in the channel region. The excess charge is immediately compensated by the injection of further electrons from the cathode into the channel. Thus further electrons may be generated from the absorption of a single photon. This factor creates a **photoconductive gain, *G*.**

Photoconductive gain is defined as the ratio of the slow carrier transit time (or life time) t_s to the fast carrier transit time t_f. Hence

$$G = \frac{t_s}{t_f} \quad ...(10.8)$$

The photocurrent I_P produced by the photoconductor is written as,

$$I_P = \frac{\eta P_0 e}{hf} G$$

where, η = Device quantum efficiency

P_0 = Incident optical power

e = Charge on an electron

In the photoconductor the current response remains for a time t_s after the end of an incident optical pulse and exhibits an exponential decay with a time constnat equal to this slow carrier transit time. Then the maximum 3 dB bandwidth B_m for the device is given by

$$B_m = \frac{1}{2\pi f_s} \quad ...(10.9)$$

Thus the gain bandwidth product of a photoconductor will be

$$\boxed{G \cdot B_m = \frac{1}{2\pi t_f}} \qquad \text{...(10.10)}$$

When t_f is fixed, photoconductive gain can only be obtained at the expense of the maximum bandwidth permitted by the device. Therefore there is a trade off between gain and speed of response.

The major difference between a photoconductor and photomultiplier tube is that an electron is excited into the conduction band rather than being removed. Thus, the minimum photon energy constraint is changed from ϕ to E_g for an intrinsic photoconductor. The photoconductor that is frequently used for measuring the intensity of light is CdS.

SOLVED EXAMPLES

Example 10.1. In a 50 ns pulse, 6×10^6 photons at a wavelength of 1200 nm fall on an InGaAs photodetector. On average 4.2×10^6 EHPs are generated. Calculate the quantum efficiency of photodetector.

Solution. Given that,

$$r_e = 4.2 \times 10^6$$

$$r_P = 6 \times 10^6$$

$$\lambda = 1200 \text{ nm}$$

The quantum efficiency is given by

$$\eta = \frac{\text{No. of electrons collected } (r_e)}{\text{No. of incident photons } (r_P)}$$

$$= \frac{4.2 \times 10^6}{6 \times 10^6} = 0.70$$

$$\boxed{\eta = 70\%} \quad \textbf{Ans.}$$

Example 10.2. The responsivity of photodiode is 0.85 A/W and the input power saturation is 1.5 mW. What is the photocurrent if the incident light power is 1 mW ?

Solution. Given that,

$$R = 0.85 \text{ A/W}$$

$$P_0 = 1 \text{ mW}$$

Responsivity of a photodiode is given by

$$R = \frac{I_P}{P_0}$$

$$\Rightarrow \quad I_P = R \times P_0$$

$$= 0.85 \times 1 \text{ mW} = 0.85 \text{ mA}$$

$$I_P = 0.85 \text{ mA} \quad \textbf{Ans.}$$

Example 10.3. Determine the responsivity of an InGaAs photodiode if its quantum efficiency is equal to 60%. The energy gap of InGaAs is 0.75 eV.

Solution. Given that, $E = 0.75 \text{ eV} = 0.75 \times 1.6 \times 10^{-19}$ V

$C = 3 \times 10^{8}$ m

$\eta = 60\% = 0.60$

The photon energy, $E = hf = \dfrac{hc}{\lambda}$

$$\lambda = \frac{hc}{E} = \frac{6.62 \times 10^{-34} \times 3 \times 10^{8}}{0.75 \times 1.6 \times 10^{-19}}$$

$$= \frac{19.86 \times 10^{-26}}{1.2 \times 10^{-19}}$$

$$\lambda = 1664 \times 10^{-9} \text{ m} = 1664 \text{ nm}$$

Hence, responsivity, $R = \left(\dfrac{\eta}{1248}\right) \lambda \text{ (nm)}$

$$= \frac{0.60 \times 1664}{1248}$$

$$\boxed{R = 0.8 \text{ A/W}} \quad \textbf{Ans.}$$

Example 10.4. When 3×10^{12} photons each with a wavelength of 0.65 μm are incident on a photodiode, on average 1.5×10^{12} electrons are collected at the terminals of the device. Determine the quantum efficiency and the responsivity of the photodiode at 0.65 μm. **[PTU, Dec. 2007]**

Solution. Given that, $\lambda = 0.65$ μm

$r_e = 1.5 \times 10^{12}$

$r_P = 3 \times 10^{12}$

Quantum efficiency, $\eta = \dfrac{r_e}{r_P}$

$$= \frac{1.5 \times 10^{12}}{3 \times 10^{12}} = 0.50$$

$$\eta = 50\% \quad \textbf{Ans.}$$

Responsivity, $R = \dfrac{\eta e \lambda}{hc}$

$$= \frac{0.5 \times 1.6 \times 10^{-19} \times 0.65 \times 10^{-6}}{6.626 \times 10^{-34} \times 3 \times 10^{8}}$$

$$= \frac{0.52 \times 10^{-25}}{19.878 \times 10^{-26}} = 0.261 \text{ A/W}$$

$$\boxed{R = 0.261 \text{ A/W}} \quad \textbf{Ans.}$$

Example 10.5. A photodiode has a quantum efficiency of 65% when photons of energy 1.5 × 10^{-19} J are incident upon it. Calculate

(*a*) The operating wavelength of photodiode.

(*b*) The incident optical power required to obtain a photocurrent of 1.5 μA.

Solution. Given that,

$$E = 1.5 \times 10^{-19} \text{ J}$$
$$\eta = 65\% = 0.65$$
$$e = 1.6 \times 10^{-19}$$
$$I_P = 1.5 \text{ μA} = 1.5 \times 10^{-6} \text{ A}$$

(*a*) The photon energy,

$$E = hf$$
$$E = \frac{hc}{\lambda}$$
$$= \frac{6.626 \times 10^{-34} \times 3 \times 10^{8}}{1.5 \times 10^{-19}}$$
$$= \frac{19.878 \times 10^{-26}}{1.5 \times 10^{-19}}$$
$$\boxed{\lambda = 1.32 \text{ μm}} \quad \textbf{Ans.}$$

(*b*) Responsivity is given by,

$$R = \frac{\eta e}{hf}$$
$$= \frac{0.65 \times 1.6 \times 10^{-19}}{1.5 \times 10^{-16}}$$
$$= \frac{1.04 \times 10^{-3}}{1.5}$$
$$R = 0.694 \text{ A/W}$$

and

$$R = \frac{I_P}{P_0} \Rightarrow P_0 = \frac{I_P}{R}$$
$$= \frac{1.5 \times 10^{-6}}{0.694} = 2.16 \times 10^{-6}$$
$$\boxed{P_0 = 2.16 \text{ μW}} \quad \textbf{Ans.}$$

Example 10.6. GaAs has a bandgap energy of 1.43 eV at 300 K. Determine the wavelength above which an intrinsic photodetector fabricated from this material will cease to operate.

Solution. Given that,

$$E_g = 1.43 \text{ eV} = 1.43 \times 1.6 \times 10^{-19} \text{ V}$$
$$c = 3 \times 10^{8}$$

The long wavelength cutoff is given by

$$\lambda_c = \frac{hc}{E_g}$$

$$= \frac{6.626 \times 10^{-34} \times 3 \times 10^8}{1.43 \times 1.6 \times 10^{-19}}$$

$$= \frac{19.878 \times 10^{-26}}{2.288 \times 10^{-19}} = 0.867 \times 10^{-6} \text{ m}$$

$$\boxed{\lambda_c = 0.867 \text{ μm}} \quad \textbf{Ans.}$$

Therefore, the GaAs photodetector will cease to operate above 0.87 μm.

Example 10.7. A p-i-n photodiode has a quantum efficiency of 50% at a wavelength of 0.9 μm. Calculate :

(i) its responsivity at 0.9 μm.

(ii) the received optical power if the mean photocurrent is 10^{-6} A.

(iii) the corresponding number of received photons at this wavelength.

Solution. Given that,

$$\eta = 50\% = 0.50$$

$$\lambda = 0.9 \text{ μm} = 900 \text{ nm}$$

$$I_P = 10^{-6} \text{ Amp}$$

(a) Responsivity,

$$R = \frac{\eta q}{h\nu}$$

$$= \left(\frac{\eta}{1248}\right)(\lambda \text{ nm})$$

$$= \frac{0.50}{1248} \times 900 = 0.36 \text{ A/W}$$

$$R = 0.36 \text{ A/W} \quad \textbf{Ans.}$$

(b)

$$R = \frac{I_P}{P_0}$$

$$\Rightarrow \quad P_0 = \frac{I_P}{R}$$

$$= \frac{10^{-6}}{0.36} = 2.77 \times 10^{-6} \text{ W}$$

$$\boxed{R = 2.77 \times 10^{-6} \text{ W}} \quad \textbf{Ans.}$$

(c) Received optical power is given by

$$P_0 = n\frac{hc}{\lambda}$$

$$2.77 \times 10^{-6} = n \times \frac{6.62 \times 10^{-34} \times 3 \times 10^{8}}{0.9 \times 10^{-6}}$$

$$n = \frac{24.93 \times 10^{-13}}{19.86 \times 10^{-26}}$$

$$\boxed{n = 1.255 \times 10^{14}} \quad \textbf{Ans.}$$

Example 10.8. The quantum efficiency of a particular silicon RAPD is 80% for the detection of radiation at a wavelength of 0.9 μm. When the incident optical power is 0.5 μW, the output current from the device is 12 μA. Determine the multiplication factor of the photodiode under these conditions.

Solution. Given that,

$$\eta = 80\% = 0.80$$

$$\lambda = 0.9\ \mu m = 0.9 \times 10^{-6}\ m$$

$$P_0 = 0.5\ \mu W = 0.5 \times 10^{-6}\ W$$

$$I = 12\ \mu A = 12 \times 10^{-6}\ A$$

The responsivity is given by,

$$R = \frac{\eta e \lambda}{hc}$$

$$= \frac{0.80 \times 1.6 \times 10^{-19} \times 0.9 \times 10^{-6}}{6.626 \times 10^{-34} \times 3 \times 10^{8}}$$

$$R = \frac{1.152 \times 10^{-25}}{19.878 \times 10^{-26}} = 0.581\ A/W$$

and the photocurrent,

$$I_P = P_0 R$$

$$= 0.5 \times 10^{-6} \times 0.581$$

$$= 2.405 \times 10^{-6}$$

$$I_P = 0.241\ \mu A$$

Multiplication factor,

$$M = \frac{I}{I_P}$$

$$= \frac{12 \times 10^{-6}}{0.291 \times 10^{-6}}$$

$$\boxed{M = 41.24} \quad \textbf{Ans.}$$

Example 10.9. A given silicon avalanche photodiode has a quantum efficiency of 65% at a wavelength of 850 nm. Suppose 0.5 μW of optical power produces a multiplied photo current of 10 μA. Calculate the responsivity and multiplication factor.

Solution. Given that,

$$\eta = 65\% = 0.65$$

$$\lambda = 850\ nm = 850 \times 10^{-9}\ m$$

$$I = 10\ \mu A = 10 \times 10^{-6}\ A$$

$$P_0 = 0.5\ \mu W = 0.5 \times 10^{-6}\ W$$

$$\text{Responsivity, } R = \frac{\eta q}{hc} = \frac{\eta q \lambda}{hc}$$

$$= \frac{0.65 \times 1.6 \times 10^{-19} \times 850 \times 10^{-9}}{6.626 \times 10^{-34} \times 3 \times 10^{8}}$$

$$= \frac{884 \times 10^{-28}}{19.875 \times 10^{-26}} = 44.5 \times 10^{-2}$$

$$\boxed{R = 0.445} \quad \textbf{Ans.}$$

$$\text{Multiplication factor, } M = \frac{I}{I_P}$$

and

$$I_P = RP_0$$

$\therefore$

$$M = \frac{I}{R.P_0}$$

$$= \frac{10 \times 10^{-6}}{0.445 \times 0.5 \times 10^{-6}} = \frac{10}{0.2225}$$

$$\boxed{M = 45.45} \quad \textbf{Ans.}$$

Example 10.10. A germanium p-i-n photodiode with active dimensions of 75 × 50 μm has a quantum efficiency of 55% when operating at a wavelength of 1.3 μm. At this wavelength the measured dark current is 8 nA. Calculate :

(*a*) The noise equivalent power

(*b*) Specific detectivity

It may be assumed that dark current is the dominant noise source.

Solution. Given that,

$$\lambda = 1.3\ \mu m = 1.3 \times 10^{-6}\ m$$

$$A = 75 \times 50\ \mu m = 75 \times 10^{-6} \times 50 \times 10^{-6}$$

$$\eta = 55\% = 0.55$$

$$I_d = 8\ nA = 8 \times 10^{-9}\ A$$

(*a*) The noise equivalent power is given by

$$NEP = \frac{hc(2eI_d)^{1/2}}{\eta e \lambda}$$

$$= \frac{6.626 \times 10^{-34} \times 3 \times 10^{8} \times (2 \times 1.6 \times 10^{-19} \times 8 \times 10^{-9})^{1/2}}{0.55 \times 1.6 \times 10^{-19} \times 1.3 \times 10^{-6}}$$

$$= \frac{19.878 \times 10^{-26} \times (25.6 \times 10^{-28})^{1/2}}{1.144 \times 10^{-25}}$$

$$NEP = 8.78 \times 10^{-14}\ W \quad \textbf{Ans.}$$

(*b*) Specific detectivity, $D^* = \dfrac{A^{1/2}}{NEP}$

$$= \frac{(75\times 50\times 10^{-12})^{1/2}}{8.78\times 10^{-14}} = \frac{(3750\times 10^{-12})^{1/2}}{8.78\times 10^{-14}}$$

$$= \frac{6.1\times 10^{-5}}{8.78\times 10^{-14}}$$

$$\boxed{D^* = 6.95\times 10^{8}}\ \textbf{Ans.}$$

Example 10.11. The phototransistors has a collector current of 15 mA when the incident optical power at a wavelength of 1.26 μm is 125 μW. Calculate :

(*a*) The optical gain of the device

(*b*) The common emitter current gain if the quantum efficiency of the base collector photodiode at a wavelength of 1.26 μm is 60%.

Solution. Given that,
I_C = 15 mA
λ = 1.26 μm
P_0 = 125 μW
η = 60% = 0.60

(*a*) The optical gain is given by

$$G_0 = \frac{hc}{\lambda e}\left(\frac{I_c}{P_0}\right)$$

$$= \frac{6.626\times 10^{-34}\times 3\times 10^{8}\times 15\times 10^{-3}}{1.26\times 10^{-6}\times 1.6\times 10^{-19}\times 125\times 10^{-6}}$$

$$= \frac{298.17\times 10^{-29}}{251\times 10^{-31}} = 1.181\times 10^{2}$$

$$\boxed{G_0 = 118.1}\ \textbf{Ans.}$$

(*b*) The common emitter current gain is

$$\eta_{FE} = \frac{G_0}{\eta}$$

$$= \frac{118.1}{0.6} = 196.8$$

$$\boxed{\eta_{FE} = 196.8}\ \textbf{Ans.}$$

Hence a common emitter current gain of 197 gives an optical gain of 118.

Example 10.12. In an phtoconductive detector the electron transit time is 5 ps. Determine the maximum 3 dB bandwidth permitted by the device when its photoconductive gain is 70.

Solution. Given that, $t_f = 5 \text{ ps} = 5 \times 10^{-12}$ s

$G = 70$

The maximum 3 dB bandwidth permitted by the photoconductor is given by

$$B_m = \frac{1}{2\pi t_f G}$$

$$= \frac{1}{2 \times 3.14 \times 5 \times 10^{-12} \times 70}$$

$$= \frac{1}{21.98 \times 10^{-12}}$$

$$= 4.547 \times 10^8 \text{ Hz}$$

$\therefore$ $\boxed{B_m = 454.7 \text{ MHz}}$ **Ans.**

Example 10.13. When 10^{11} photons per second each with an energy of 1.28×10^{-19} J are incident on an ideal photodiode, what is the output photocurrent ?

Solution. Given that, $r_P = 10^{11}$

$h_f = 1.28 \times 10^{19}$

The incident optical power is given by

$$P_0 = \frac{10^{11}}{1.28 \times 10^{19}}$$

$$= \frac{10^{-8}}{1.28} = 0.78 \times 10^{-7}$$

$$P_0 = 7.8 \ \mu\text{W}$$

Therefore, output photo current,

$$I_P = \frac{\eta P_0 e}{hf}$$

$$= \frac{1 \times 7.8 \times 10^{-6} \times 1.6 \times 10^{-19}}{1.28 \times 10^{19}}$$

$$= \frac{12.48 \times 10^{-44}}{1.28} = 9.75 \times 10^{-44} \text{ Amp}$$

$\boxed{I_P = 9.75 \times 10^{-44} \text{ amp}}$ **Ans.**

Example 10.14. A typical photodiode has a responsivity of 0.40 A/W for a He-Ne laser (λ = 632.8 mm). The active area of the photodiode is 2 mm². What will be the output photocurrent if the incident flux is 100 μW/² mm² ?

Solution. Given that,

$R = 0.40$ A/W

$$\text{Incident flux} = 100\ \mu W/mm^2$$
$$\text{Active area} = 2\ mm^2$$

Incident power will be,

$$P_0 = 100 \times 2\ \mu W = 200\ \mu W$$

Since, Responsivity, $R = \frac{I_P}{P_0}$

$$\Rightarrow \quad I_P = R \cdot P_0$$
$$= 0.40 \times 200 \times 10^{-6}$$
$$= 80 \times 10^{-6}\ \text{Amp}$$
$$\boxed{I_P = 0.08\ \text{mA}} \quad \textbf{Ans.}$$

Example 10.15. An APD has quantum efficiency of 50% at 1.3 µm. When illuminated with optical power of 0.4 µm at this wavelength, it produces an output photocurrent of 8 µA, after avalanche gain. Calculate the multiplication factor of the diode.

Solution. Given that,

$$\eta = 50\% = 0.05$$
$$\lambda = 1.3\ \mu m$$
$$P_0 = 0.4\ \mu m$$
$$I = 8\ \mu A$$

Now Responsivity, $R = \frac{\eta e \lambda}{hc}$

$$= \frac{0.50 \times 1.6 \times 10^{-19} \times 1.3 \times 10^{-6}}{6.626 \times 10^{-34} \times 3 \times 10^{8}}$$

$$R = \frac{1.04 \times 10^{-25}}{19.86 \times 10^{-26}} = 0.524\ \text{A/W}$$

Therefore, the output photocurrent,

$$I_P = P_0 \cdot R$$
$$= 0.4 \times 10^{-6} \times 0.524$$
$$I_P = 0.2096\ \mu A$$

Now, multiplication factor,

$$M = \frac{I}{I_P}$$

$$= \frac{8 \times 10^{-6}}{0.2096 \times 10^{-6}} = 38.17$$

$$\boxed{M = 38.17} \quad \textbf{Ans.}$$

Example 10.16. The electron transit time in a photoconductive detector is 4.5 ps. Determine the maximum 3 dB bandwidth permitted by the device when its photoconductive gain is 80.

Solution. Given that,

$$t_f = 4.5 \text{ ps}$$
$$G = 80$$

Maximum 3 dB bandwidth permitted is

$$B_m = \frac{1}{2\pi t_f G}$$

$$= \frac{1}{2 \times 3.14 \times 4.5 \times 10^{-12} \times 80}$$

$$= \frac{1}{2260.8 \times 10^{-12}}$$

$$= 4.4 \times 10^{10} \text{ Hz}$$

$$\boxed{B_m = 0.44 \text{ GHz}} \quad \textbf{Ans.}$$

Example 10.17. A photodiode has quantum efficiency of 50% at 0.90 µm. Calculate responsivity and received optical power if mean photocurrent is 10^{-6} A.

Solution. Given that,

$$\eta = 0.50$$
$$\lambda = 0.90 \text{ µm}$$
$$I_P = 10^{-6} \text{ Amp.}$$

$$\text{Responsivity, } R = \frac{\eta e \lambda}{hc}$$

$$= \frac{0.5 \times 1.6 \times 10^{-19} \times 0.90 \times 10^{-6}}{6.626 \times 10^{-34} \times 3 \times 10^{8}}$$

$$= \frac{0.7200 \times 10^{-25}}{19.878 \times 10^{-26}}$$

$$\boxed{R = 0.36} \quad \textbf{Ans.}$$

$$\text{Received optical power, } P_0 = \frac{I_P}{R} = \frac{10^{-6}}{0.36}$$

$$\boxed{P_0 = 2.77 \times 10^{-6} \text{ W}} \quad \textbf{Ans.}$$

Example 10.18. Calculate the efficiency of a PIN silicon photodiode if the responsivity is 0.374 A/W at 1300 nm wavelength.

Solution. Given that,

$$R = 0.374 \text{ A/W}$$
$$\lambda = 1300 \text{ nm}$$

The responsivity of a photodiode is given by

$$R = \frac{\eta e \lambda}{hc}$$

$$\Rightarrow \quad \eta = \frac{Rhc}{e\lambda}$$

$$= \frac{0.374 \times 6.626 \times 10^{-34} \times 3 \times 10^{8}}{1.6 \times 10^{-19} \times 1300 \times 10^{-9}}$$

$$= \frac{7.434 \times 10^{-26}}{2080 \times 10^{-28}}$$

$$= 0.357$$

$$\boxed{\eta = 35.7\%} \quad \textbf{Ans.}$$

Example 10.19. A p-n photodiode has a quantum efficiency of 50% at a wavelength of 0.9 μm. Calculate :

(*a*) Its responsivity at 0.9 μm

(*b*) The received optical power if the mean photocurrent is 10^{-6} A.

Solution. Given that, $n = 0.5$

$\lambda = 0.9$ μm

(*a*) Responsivity is given by

$$R = \frac{\eta e \lambda}{hc}$$

$$= \frac{0.5 \times 1.602 \times 10^{-19} \times 0.9 \times 10^{-6}}{6.626 \times 10^{-23} \times 2.998 \times 10^{8}}$$

$$\boxed{R = 0.36 \text{ AW}^{-1}} \quad \textbf{Ans.}$$

(*b*) Received optical power is given by

$$P_0 = \frac{I_P}{R}$$

$$= \frac{10^{-6}}{0.36}$$

$$= 2.78 \times 10^{-6} \text{ W}$$

$$\boxed{P_0 = 2.78 \text{ μW}} \quad \textbf{Ans.}$$

Example 10.20. A silicon p-i-n photodiode with an area of 1.5 mm² is to be used in conjunction with a load resistor of 100 Ω. If the requirement for the device is a fast response time, estimate the thickness of the intrinsic region that should be provided. It may be assumed that the permittivity for silicon is 1.04 × 10^{-10} Fm^{-1} and the electron saturation velocity is 10^{7} ms^{-1}.

Solution. Given that, $A = 1.5 \text{ mm}^2$

$R = 100\ \Omega$

$$\varepsilon_0 = 1.04 \times 10^{-10}\ \text{Fm}^{-1}$$

$$v_d = 10^7\ \text{ms}^{-1}$$

B is given by, $$B = \frac{1}{2\pi RC}$$

$$\therefore \quad \frac{1}{2\pi RC} = \frac{v_d}{2\pi w} \qquad \left[\text{but } c = \frac{\varepsilon_0 A}{w}\right]$$

Therefore, $$\frac{w}{R\varepsilon_0 A} = \frac{v_d}{w}$$

$$w = (R\varepsilon_0 A v_d)^{1/2}$$

$$= (100 \times 1.04 \times 10^{-10} \times 1.5 \times 10^{-3} \times 10^7)^{1/2}$$

$$= 3.95 \times 10^{-4}$$

$$\boxed{w = 395\ \mu\text{m}}\ \textbf{Ans.}$$

Example 10.21. A silicon p-i-n photodiode with active dimensions 10 μm has specific detectivity of 7 × 10^{10} MHz$^{1/2}$ W^{-1} when operating at a wavelength of 0.85 μm. The device quantum efficiency is 64%. Assuming that it is dominant noise source, calculate the dark current over 1 Hz bandwidth in the device.

Solution. Given that, $\lambda = 0.85\ \mu\text{m}$

$$\eta = 0.64$$

$$B = 1\ \text{Hz}$$

$$D^* = 7 \times 10^{10}\ \text{m Hz}^{1/2}\ \text{W}^{-1}$$

$$A = 10\ \mu\text{m}$$

The specific detectivity of p-i-n photodiode is given by

$$D^* = \frac{\eta e\lambda}{hc\left(\frac{2eI_d}{A}\right)^{1/2}}$$

$$= \frac{\eta.\sqrt{eA}.\lambda}{hc(2I_d)^{1/2}}$$

$$\Rightarrow \quad (I_d)^{1/2} = \frac{\eta\sqrt{eA}\,\lambda}{hc.\sqrt{2}.D^*}$$

$$(I_d)^{1/2} = \frac{0.64 \times \sqrt{1.602 \times 10^{-19} \times 10 \times 10^{-6}} \times 0.85 \times 10^{-6}}{6.626 \times 10^{-34} \times 2.998 \times 10^8 \times \sqrt{2} \times 7 \times 10^{10}}$$

$$(I_d)^{1/2} = 1.1 \times 10^{-7}$$

$$\boxed{I_d = 1.2 \times 10^{-14}\ \text{A}} \quad \textbf{Ans.}$$

SUMMARY

1. An OPTICAL DETECTOR is a transducer that converts an optical signal into an electrical signal. It does this by generating an electrical current proportional to the intensity of incident optical radiation.
2. The semiconductor **Positive-intrinsic-negative (PIN) Photodiode** and **Avalanche Photodiode (APD)** are the principal optical detectors used in fiber optic systems.
3. A **Fiber Optic Receiver** is an electro-optic device that accepts optical signals from an optical fiber and converts them into electrical signals. A typical fiber optic receiver consists of an optical detector, a low-noise amplifier, and other circuitry used to produce the output electrical signal.
4. **Receiver Spectral Response, Sensitivity, Frequency Response** and **Dynamic Range** are key receiver performance parameters that can affect overall system operation.
5. **Receiver Sensitivity** is the minimum amount of optical power required to achieve a specific receiver performance. For digital transmission at a given data rate and coding, this performance is described by a maximum bit-error rate (BER). In analog systems, for a given modulation and bandwidth, it is described by a minimum signal-to-noise ratio (SNR).
6. **Dynamic Range** refers to the range of optical power levels over which the receiver operates within the specified values. It usually is described by the ratio of the maximum input power to the sensitivity.
7. A **Transducer** is a device that converts input energy of one form into output energy of another.
8. A **Photocurrent** is generated when photons are absorbed by a photodiode.
9. **Responsivity** is the ratio of the optical detector's output photocurrent in amperes to the incident optical power in watts.
10. **Dark Current**, or reverse-leakage current, is the current that continues to flow in the Photodetector when there is no incident light.
11. The **Response Time** of a photodiode and its output circuitry depends primarily on the thickness of the detector active area and the detector RC time constant.
12. The **Transit Time** is the time it takes electrons to travel out of the detector active area.
13. The RC **Time Constant** is defined by the capacitance (C) of the photodiode and the resistance (R) of the load. The RC time constant is given by $t_{RC} = RC$.
14. A **High-speed Response** requires short transit times and low capacitance. However any change in photodiode parameters to optimize the transit time and capacitance can also affect quantum efficiency, dark current, and coupling efficiency.
15. Detector **Linearity** means that the output electrical current (photocurrent) of the photodiode is linearly proportional to the input optical power.
16. An **Avalanche Photodiode (APD)** is a photodiode that internally amplifies the photocurrent by an avalanche process.
17. In APDs, a large **Reverse-bias Voltage**, typically over 100 volts, is applied across the active region.

18. **Avalanche Multiplication** occurs when accelerated electrons collide with other electrons in the semiconductor material, causing a fraction of them to become part of the photocurrent.
19. **Trade-offs** are made in APD design to optimize responsivity and gain, dark current, response time, and linearity.
20. The **Response Time** of APDs accounts for the avalanche build-up time in addition to transit time and RC time constant.
21. The **Preamplifier** is defined as the first stage of amplification following the optical detector.
22. The **Postamplifier** is defined as the remaining stages of amplification required to raise the detectors electrical signal to a level suitable for further signal processing.
23. **Receiver Sensitivity, Bandwidth** and **Dynamic Range** are key operational parameters used to define receiver performance.
24. **Noise** is the main factor that determines receiver sensitivity.
25. **Receiver Noise** includes thermal noise, dark current noise, and quantum noise.
26. **Thermal Noise** is the noise resulting from the random motion of electrons in a conducting medium.
27. **Shot Noise** is noise caused by current fluctuations due to the discrete nature of charge carriers.
28. **Dark Current Noise** results from dark current that continues to flow in the photodiode when there is no incident light.
29. **Quantum Noise** results from the random generation of electrons by the incident optical radiation.
30. **The High-impedance Amplifier** and the **Transimpedance Amplifier** are the two basic types of amplifiers used in fiber optic receivers.
31. The **High-impedance Preamplifier** provides a high sensitivity, but limits receiver bandwidth and dynamic range.
32. The **Transimpedance Preamplifier** provides improvements in bandwidth and dynamic range with some degradation in sensitivity from an increase in noise.
33. A **PIN photodiode** is a semiconductor positive-negative (*p-n*) structure with an intrinsic region sandwiched between the other two regions. It is normally operated by applying a reverse-bias voltage.
34. **Pin Photodiodes** are used as the detector in most applications.
35. **Avalanche Photodiodes** are only used in high-speed applications and applications where detectors with extremely low sensitivities are required.
36. Detective quantum efficiency is given by :

$$DQE = \left[\frac{\dfrac{S}{N_{out}}}{\dfrac{S}{N_{in}}}\right]$$

So $$\frac{S}{N_{out}} = \frac{S}{N_{in}} * DQE^{1/2}$$

Poisson statistics $$\frac{S}{N_{in}} = N^{1/2}$$

37. Photoconductors are the simplest conceivable optical detector. The device consists of a piece of (undoped) semiconductor material with electrical contacts attached. A voltage is applied across the contacts.

38. P-N photodiodes are used in similar applications to other photodetectors, such as photoconductors, charge-coupled devices and photomultiplier tubes.

39. PIN diodes are much faster and more sensitive than ordinary *p-n* junction diodes, and hence are often used for optical communications and in lighting regulation.

40. The main structural difference between an APD and a *p-i-n* diode are that the *i* the (which is lightly *n*-doped in a *p-i-n* structure) is lightly *p*-doped and renamed the layer.

41. Devices for the 1550 nm wavelength band are quite difficult to build as the InP/ lnGaAsP system must be used and this gives very low RI contrast steps with which to build the mirrors.

42. Phototransistors are very similar to ordinary BJTs except they are designed for use as detectors. Transistors are amplifiers and in a photo transistor the amplifier gain is controlled by the amount of light striking the device.

43. Photo-emissive detectors are either evacuated or gas-filled tubes containing a cathode and one or more anodes.

44. An alternative to travelling wave devices for getting higher quantum efficiency at very high speeds is the Resonant-Cavity Photodetector.

45. The idea of a hetero-junction APD is to replace the *p*-layer silicon material with some material that will absorb light in the long wavelength bands and GaAs is one such material.

REVIEW QUESTIONS

1. What do you understand by optical detector ? Discuss its various types also.
2. What are the important parameters of photodetectors ?
3. List the properties of optical detectors.
4. Briefly explain the principle of optical detection.
5. What do you meant by quantum efficiency and responsivity of a photodiode ? Calculate the wavelength at which quantum efficiency and responsivity are equal.
6. Discuss the concept of negative resistance in APD photodetector.
7. Explain the working principle of a P-N photodiode.
8. Discuss the concept of p-i-n photodiode. What is the significance of the intrinsic layer inserted in between the P and N layer ?
9. What do you understand by quantum efficiency ? Derive an expression for responsivity and prove that responsivity is directly proportional to the quantum efficiency at a particular wavelength.
10. Explain the various measures of efficiency in PIN photodiode.
11. Briefly explain the working principle of Schottky barrier photodiodes.
12. Discuss the impact ionization in avalanche photodiode. Explain the multiplication factor and photomultiplication factors also.

13. Name the three layers of a pin photodiode. What is the purpose of the middle layer ? **[PTU 2005]**

14. Derive the relationship between noise equivalent power and noise current in a photodetector.

15. Differentiate between photovoltaic and photoconductive modes of operation of a photodiode.

16. List some important semiconductor materials and indicate the wavelength regions of each for photodetector applications. **[PTU 2003]**

17. What is the position of the Fermi level with respect to energy bands for unbiased, forward and reverse biased semiconductor ? Explain.

18. Why is modal noise not a problem for an LED transmitter or for a single mode fiber ? Can a laser diode be used with a multimode fiber ? **[PTU 2007]**

19. What is the major advantage of an APD over a p-i-n photodiode ?

20. What is the cutoff wavelength of a photodiode ? What is the physical reason for the existence of a cutoff wavelength ?

21. Explain the principle of photodetection in an optical system.

22. Explain the principle of availanche photodiode.

23. Explain the basic detection process in n-p-n phototransistor.

24. Define quantum efficiency and responsivit of a p-n diode. How are the two related to each other ?

25. What do you understand by a silicon reach through avalanche photodiode ? Explain.

26. What do you understand by a photo-transistor ? Explain its working.

27. Draw the layer diagram of a p-i-n photodiode layer diagram for a pin photodiode. **[PTU Dec. 2007]**

27. The responsivity of a photodiode is 0.9 A/W and its saturation power is 2 mW. What is the photocurrent if the received power is

(*i*) 1 mW (*ii*) 2 mW (*iii*) 3 mW

28. Calculate the risetime of a detector circuit having a capacitance of 3 PF and a load resistance of 160 Ω.

29. Determine the thickness of the depletion layer of an Si photodiode be to make its quantum efficiency 85%.

30. Calculate the external quantum efficiency of a laser diode if its slope efficiency is equal to 0.08 mW/mA and the operating wavelength is 1550 nm. **[PTU 2006]**

31. Determine the laser noise power detected by a receiver for the given data :

$$R_{IN} = -150 \text{ dB/Hz}$$
$$P_{received} = 80 \text{ mW}$$
$$BW = 140 \text{ MHz}$$

32. Calculate the bandwidth of germanium photodiode with w = 20 mm and $v_{sat} \simeq 10^5$ m/sec.

33. When 3×10^{11} photons each with a wavelength of 0.85 μm are incident on a photodiode, on average 1.2×10^{11} electrons are collected at the terminals of device. Determine the quantum efficiency and the responsivity of the photodiode at 0.85 μm. **[PTU 2007]**

34. A photodetectors dark current is 5 nA and its responsivity is 0.5 A/W. At what value of optic power is the signal generated shot noise equal to the dark current generated shot noise ?

35. Calculate and plot the responsivity of an ideal photodetector for wavelengths between 0.5 and 1.7 μm. Why does responsivity increase with wavelength ?

36. Determine the responsivity of an InGaAs APD operating at 1.55 μm and having a quantum efficiency of 0.7 if its gain is 10. How much optical power is required by this detector to produce 20 nA ?

37. Explain the working principle of a photo-transistor and hence prove that collector current is given by

$$i_C = i_{OP}(1 + h_{FE})$$

38. What is a phototransistor ? Derive an expression for base transport factor in terms of base width and diffusion length.

❑❑❑

Light Wave Systems

INSIDE THIS CHAPTER

The three main components of a fiber optical communication system are :

- Optical fiber
- Optical transmitter
- Optical receiver.

In this chapter, we consider the issues related to system design and performance when the three components are put together to form a practical light wave system.

11.1. System Architecture

According to the architectural point of view, fiber optical communication system can be classified into **three broad categories :**

- Point to point links
- Distributed networks
- Local area networks

11.1.1. Point-to-Point Links

It is the simplest kind of lightwave systems. Its function is to transport information available in the form of digital bit stream, from one place to another as accurate as possible.

The **link length** can vary from less than a kilometer (short haul) to thousand of kilometer (long haul), depending on the specific application. Example, Optical data links are used to connect computers and terminals within the same building or between two buildings with a relatively short transmission distance (< 10 km).

When the **link length exceeds a certain value, in the range 20–100 km** depending upon the operating wavelength then it is the necessary to compensate the fiber losses because the signal may become too weak to be detected reliably.

Commonly **two techniques used for loss compensation,** which are :

- Loss compensation through regenerators.
- Loss compensation through optical amplifiers.

Figure 11.1 shows these two schemes.

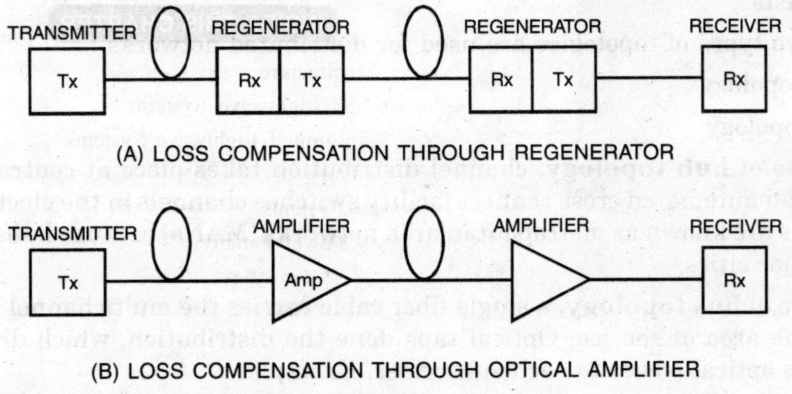

Fig. 11.1

- As seen in Fig. 11.1(*a*), a **regenerator** is a receiver-transmitter pair that detects the incoming optical signal, recovers the electrical bit stream, and then converts it back into optical form by modulating an optical source.
- Optical amplifier can also compensate fiber losses. They directly amplify the optical bit stream without requiring the conversion of the signal to the electrical domain.
- For wavelength division multiplex (WDM) light wave systems, amplifiers are especially valuable because they can amplify many channels simultaneously.
- Optical amplifiers solve the problem of loss but they add noise.
- **Optoelectronic repeater** do not suffer from this problem because they generate the original bit stream and thus effectively compensate for all sources of signal degradation automatically.
- To replace an optoelectronics repeater, an optical regenerator should perform the same three functions : reamplification, reshaping and retiming.

According to the developing research, most terrestrial systems use a combination of the two techniques as shown in Fig. 11.1 and place an optoelectronic regenerator after a certain number of optical amplifiers. The spacing L between regenerators or optical amplifiers is known as **repeater spacing.** It is a major design parameter because the system cost reduces as L increases.

11.1.2. Distributed Network

Many optical fiber communication applications require that information is not only transmitted but is also distributed to a group of subscribers.

Examples :

- Local loop distribution o: :elephone services.
- Broadcast of multiple video channels over cable television.

Some efforts have been made towards the integration of audio and video services through a broadband **integrated services digital network (ISDN).** This network has the ability to distribute a wide range of services including telephone, facsimile, computer data and video broadcasts.

Mainly two types of topologies are used for distributed networks :

- Hub topology
- Bus topology

In the case of **hub topology,** channel distribution takes place at central location (or hubs), where an automated cross connect facility switches channels in the electrical domain. Such networks are known as metropolitan area networks (MANs) because hubs are typically located in major cities.

In the case of **bus topology,** a single fiber cable carries the multichannel optical signal throughout the area of service. Optical taps done the distribution, which diverts a small fraction of the optical power to each subscriber. Examples :

- **CATV application :** It consists of distributing multiple video channels within a city.
- **HDTV :** High definition television (HDTV) also requires light wave transmission because of a large bandwidth of each video channel unless a compression technique is used.

There is a problem with bus topology i.e. signal loss increases exponentially with the number of taps and limits the numbers of subscribers served by a single optical bus. When the fiber losses are neglected, the power avialable at the Nth tap is given by

$$P_N = P_T C[(1-\delta)(1-C)]^{N-1} \quad ...(11.1)$$

where, P_T = Transmitted power

C = Fraction of power coupled out of each tap

δ = Insertion loss

11.1.3. Local-Area Networks

Many application of fiber optical communication technology require networks where a large number of users within a local area are interconnected in such a way that any user can access the network randomly to transmit data to any other user. Such networks are known as local area networks (LANs). **Three commonly used topologies** are :

- Bus topology
- Ring topology
- Star topology

An example of **bus topology** is provided by the **Ethernet,** a network protocol used to connect multiple computers and used by the internet. The Ethernet operates at speeds upto 1 Gb/s by using a protocol based on carrier-sense multiple access (CSMA) with collision detection. When optical fibers are used a number of difficulties arise. A major limitation is related to the losses occurring at each tap, which limits the number of users.

In the **ring topology,** consecutive nodes are connected by point to point links to form a closed ring. Each node can transmit and receive data by using a transmitter-receives pair, which also acts as a repeater. The use of ring topology for fiber-optical LANs have been commercialized with the standardized interface known as the **fiber distributed data interface (FDDI).** The FDDI operates at 100 Mb/s by using multimode fibers and 1.3 μm transmitters based on light emitting diodes (LEDs).

In the **star toplogy**, all nodes are connected through point to point links to a central node called a **hub** or a **star.** Further such LANs are classified as :

(1) Active star networks

(2) Passive star networks

Active Star Configuration : Here all incoming optical signals are converted to the electrical domain through optical receivers. Then electrical signal is distributed to drive individual node transmitters. Switching operations can also be performed at the central node since distribution takes place in the electrical domain.

Passive Star Configuration : Distribution takes place in the optical domain through devices such as directional couplers. The input from one node is distributed to many output nodes, therefore, the power transmitted to each node depends on the number of users. Similar to the bus topology, the number of users supported by passive star LANs is limited by the distribution losses.

For a passive star (composed by directional couplers), the power is further reduced because of insertion losses and can be written as :

$$P_N = \left(\frac{P_T}{N}\right)(1-\delta)^{\log_2 N} \quad \text{...(11.2)}$$

where, δ = Insertion loss of each directional coupler.

11.2. Loss-Limited Lightwave Systems

Fiber losses play an important role in the system design except for some short haul fiber links. Let us consider an optical transmitter which is capable of launching an average power $\overline{P}_{tr}$. The receiver detects a signal with a minimum average power $\overline{P}_{rec}$ at the bit rate B, the **maximum transmission distance** is limited by

$$L = \frac{10}{\alpha_f}\log_{10}\left(\frac{\overline{P}_{tr}}{\overline{P}_{rec}}\right) \quad \text{...(11.3)}$$

where, α_f = Net loss of the fiber cable (in dB/km)

and $$\overline{P}_{rec} = N_P h\upsilon B \quad \text{...(11.4)}$$

where, N_P = Average number of photons per bit required by receiver

$h\upsilon$ = Photon energy

The distance L decreases logarithmically as B increases at a given operating wavelength. In Fig. 11.2, the solid lines show the dependence of L on B for three common operating wavelength of 0.85, 1.3 and 1.55 μm by using $\alpha_f = 2.5$, 0.4 and 0.25 dB/km respectively. The dotted line shows the bit rate dependence of L for coaxial cables by assuming that the loss increases as $\sqrt{B}$.

Typically the system requirements are :

- Bit rate, *B*
- Transmission distance, *L*

The **performance criterion** is specified through the Bit Error Rate (BER), a typical requirement being BER < 10^{-9}. Many other issues are there, which are to be considered while designing a realistic fiber-optic communication system. Among them are :

- Choice of operating wavelength
- Selection of appropriate transmitter, receivers and fibers.
- Compatibility of various components
- Issue of cost versus performance.
- System realibility and upgradability concern.

11.3. Dispersion Limited Lightwave Systems

Fiber dispersion limits the bit rate distance produce *BL* because of pulse broadening. When

Dispersion limited transmission distance < Loss limited distance	...(11.5)

The system is known as **dispersion limited.**

In Fig. 11.2, the dashed lines show the dispersion limited transmission distance as a function of the bit rate. Dispersion limitations can be different for different operating wavelength.

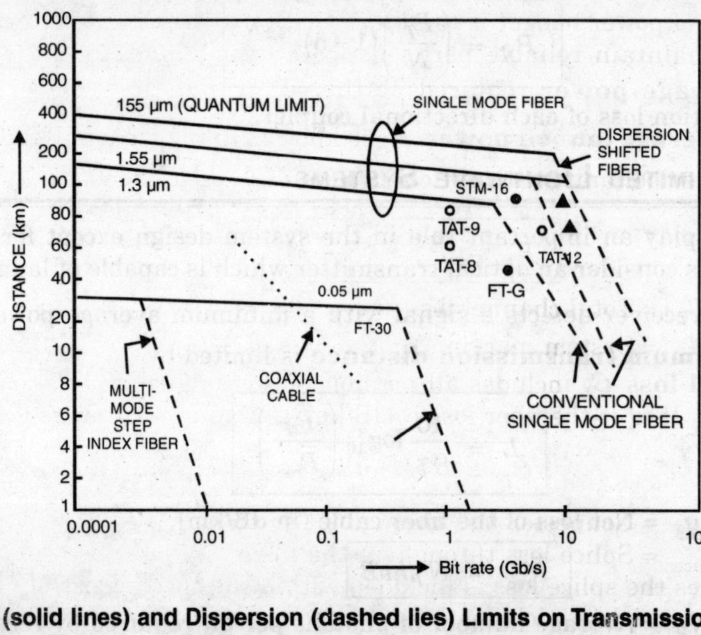

Fig. 11.2. Loss (solid lines) and Dispersion (dashed lies) Limits on Transmission Distance L as a Function of Bit Rate B for the Three Wavelength Windows.

Let us consider the **case of 0.85 μm lightwave systems,** which often use multimode fibers to minimize the system cost. The most limiting factor for multimode fiber is intermodal dispersion. In the case of step index multimode fibers, a slightly more restrictive condition.

$BL = \left(\frac{C}{2n_1\Delta}\right)$ is plotted in Fig. 11.2 by using typical values $n_1 = 1.46$, $\Delta = 0.01$. At a low bit rate of 1 Mb/s, multimode systems are dispersion limited and their transmission distance is limited to below 10 km. Therefore multimode step index fibers are rarely used in the design of fiber optical communication system.

Some improvement can be realized by using graded index fibers for which intermodal dispersion limits the *BL* product given by

$$BL = \frac{2C}{(n_1\Delta)^2} \quad \text{...(11.6)}$$

It shows that **0.85 μm lightwave systems are loss limited rather than dispersion limited** for bit rates upto 100 Mb/s when graded index fibers are used.

- The **first generation** of terrestrial telecommunication systems took advantage of such improvement and used graded index fibers.
- The **second generation** of light wave systems used single mode fibers near the minimum dispersion wavelength occring at about 1.31 μm. For this system, the most limiting factor is dispersion induced pulse broadening.
- **Third and Fourth generation** light wave systems operate near 1.55 μm to take advantage of smallest fiber losses occurring in this wavelength region. Fiber dispersion becomes the major problem for such systems.

11.4. Power Budget

The purpose of power budget is to ensure that enough power will reach at the receiver so that it can maintain reliable performance during the entire system life time. The **minimum average power** required by the receiver is the receiver sensitivity $\overline{P}_{rec}$. Generally the **average launch power** $\overline{P}_{tr}$ is known for any transmitter. The power budget is expressed in decibel units with optical powers expressed in dBm units.

$$\overline{P}_{tr} = \overline{P}_{rec} + C_L + M_s \quad \text{...(11.7)}$$

where, C_L = Total channel loss

M_S = System margin

The **channel loss C_L** includes all possible sources of power loss, including connector and splice losses. If αf is the fiber loss in dB/km. C_L can be written as

$$C_L = \alpha_f L + \alpha_{con} + \alpha_{splice} \quad \text{...(11.8)}$$

where, α_{con} = Connector loss throughout the fiber

α_{splice} = Splice loss throughout the fiber

- Sometimes the splice loss is included within the specific loss of fiber cable.
- The connector loss includes connectors at the transmitter and receiver ends but should include other connectors if used within the fiber link.

TABLE 11.1. Power Budget of 0.85 μm Lightwave Systems

S. No.	Quantity	Laser	LED
1.	Transmitter power, $\overline{P}_{tr}$	0 dBm	–13 dBm
2.	Receiver Sensitivity, $\overline{P}_{rec}$	–42 dBm	–42 dBm
3.	System margin, M_s	6 dB	6 dB
4.	Available channel loss, C_L	36 dB	23 dB
5.	Connector loss, α_{con}	2 dB	2 dB
6.	Fiber cable loss, α_f	3.5 dB/km	3.5 dB/km
7.	Maximum fiber length, L	9.7 km	6 km

System margin allocates a certain amount of power to additinal sources of power penalty that may develop during system lifetime because of component degradation or other unforeseen events.

Equation (11.7) and (11.8) can also be used to estimate the maximum transmission distance for a given choice of the components. Table 11.1 gives the power budget for the two transmitters by assuming that the splice loss is included within the cable loss.

11.5. Long Haul Systems

Fiber losses can be compensated by inserting optical amplifiers periodically along a long haul fiber link. Similarly the effect of fiber dispersion (GVD) can be reduced by using dispersion management. This topic is based on the factors that limit the performance of amplified fiber links and provide a few design guidelines.

11.5.1. Performance Limiting Factors

The most important parameter in designing a periodically amplified fiber link is because of the **nonlinear effects** occurring inside all optical fibers.

For single channel lightwave systems, the dominant nonlinear phenomenon which limits the system performance is **self phase modulation (SPM).** When optoelectronic regenerators are used, the SPM effects accumulate only over one repeater spacing (*i.e.* < 100 km) and is of little concern if P_{in} << 22 mW when NA = 1.

In contrast, SPM effects accumulate over long length (~ 1000 km) when in line amplifiers are used periodically for loss compensation.

The effect of SPM on pulses propagating inside an optical fiber can be included by using the **nonlinear schrodinger (NLS) equation,** given as

$$\frac{\partial A}{\partial z}+\frac{i\beta_2}{2}\cdot\frac{\partial^2 A}{\partial t^2} = -\frac{\alpha}{2}A+\sigma|A|^2 A \qquad ...(11.9)$$

where fiber losses are included through the $ term. This term can also include periodic amplification of the signal by treating α as a function of *z*.

The **NLS equation is mainly used for designing modern lightwave systems.**

- The combine effect of GVD and SPM also depend on the sign of the dispersion parameter β_2. In the case of anomalous dispersion ($\beta_2 < 0$), the nonlinear phenomenon of **modulation instability** can affect the system performance drastically. This problem can be overcome by using a combination of fibers with normal and anomalous GVD so that the average dispersion over the entire fiber link is normal. While another kind of modulation instability, known as **sideband instability** can occur in both the normal and anomalous GVD region.
- Another important factor is noise added by optical amplifiers. Noise has a much larger bandwidth than the signal, its impact can reduced bu using optical filters.

11.5.2. Terrestrial Lightwave Systems

The important application of fiber optical communication links is for enhancing the capacity of telecommunication networks world wide Terrestrial lightwave systems became available commercially in 1980 after a successful chicago field trial in 1980. Table 11.2 gives the operating characteristics of several terrestrial systems developed since then.

TABLE 11.2. Terrestrial Lightwave Systems

System	Year	λ (μm)	B(Mb/s)	L (km)	Voice channels
FT-3	1980	0.85	45	< 10	672
FE-3C	1983	0.85	90	< 15	1,344
FT-3X	1984	1.30	180	< 25	2,688
FT-G	1985	1.30	417	< 40	6,048
FT - G-1.7	1987	1.30	1,668	< 46	24,192
STM-16	1991	1.55	2,488	< 85	32,256
STM-64	1996	1.55	9,953	< 90	129,024
STM-256	2002	1.55	39,813	< 90	516,096

- The **first generation** systems operated near 0.85 μm and used multimode graded index fibers as the transmission medium. The BL product of such systems is limited to 2 (Gb/s) km.

A commercial lightwave system (FT-3C) operating at 90 Mb/s with a repeater spacing of about 12 km realized a BL product of nearly 1 (Gb/s) km.

- In **second generation** lightwave systems, the operating wavelength moved to 1.3 μm to take advantage of low fiber losses and low dispersion near this wavelength. The BL product of 1.3 μm lightwave systems is limited to about 100 (Gb/s) km when a multimode semiconductor laser is used inside the transmitter.

In 1987, a commercial 1.3 μm lightwave system provided data transmission at 1.7 Gb/s with a repeater spacing of about 45 km.

- The **third generation** of light wave systems became commercially available in 1991. They operate near 1.55 μm at bit rates in excess of 2 Gb/s, typically at 2.488 Gb/s, corresponding to the OC-48 level of the synchronized optical network (SONET) specifications.

This wavelength helps to increase the loss limited transmission distance to more than 100 km because of fiber losses of less than 0.25 dB/km in this wavelength region.

- The **fourth generation** light-wave systems appeared around 1996. They operate in the 1.55 μm region at a bit rate as high as 40 Gb/s by using dispersion shifted fibers in combination with optical amplifiers. However, more than 50 million kilometers of the standard telecommunication fiber is already installed in the world wide telephone network.

Two approaches are used to solve the dispersion problem :

(1) Several dispersion management schemes make it possible to extend the bit rate to 10 Gb/s while maintaining an amplifier spacing of upto 100 km.

(2) Several 10 Gb/s signals can be transmitted simultaneously by using the WDM techniques.

- The **fifth generation of** light wave systems was just beginning to emerge in 2001. In this generation the bit rate of each channel WDM systems is 40 Gb/s. Several new techniques developed in recent year make it possible to transmit a 40 Gb/s optical signal over long distances.

New fibers known as **reverse dispersion fibers** have been developed with a negative GVD slope. There use in combination with tunable dispersion compensating techniques can compensate the GVD for all channels simultaneously.

11.5.3. Long Haul Undersea Networks

There is a little difference between the economics of long haul under sea links and long haul terrestrial links. There are **several types of undersea links** commonly used.

- One type spans several thousands of kilometers across the atlantic or pacific oceans to interconnect North America with Europe or Asia.
- Another one is relatively shorter haul (a few hundred kilometers), interconnection countries either by **festoon type** of arrnagement or by **direct links** across short stretches of water.

The term festoon means a string suspended in a loop between two points. In this context, it is an under sea cable used to connect two locations that are not separated by a body of water, usually neighbouring countries.

A **trunk and branch configuration** is also popular, where an under sea trunk cable serves several countries. Each country is connected to the trunk cable by a branching cable, with passive optical components, used to perform the branching at the branching units. If a branch cable is cut, access to a particular country is lost, but other countries continue to communicate via the trunk cable. In these types of links, WDM is widely used.

- The **long haul undersea systems** operates at the leading edge of technology and have to overcome significant impairments to attain the distances involved. The links use the dispersion management techniques.
- **The main objective of shorter distance undersea link** is to eliminate any undersea amplifiers or repeater stations, because of their higher cost of installation and maintenance. Therefore these systems use relatively high power transmitters.
- Undersea systems are designed due to the high cost of servicing or replacing failed parts of the network and they provide :

- Very high levels of reliability.
- Availability

■ Optical amplifiers with redundant pumping arrangements are highly reliable devices and their failure rates are much lower than those of electronic regenerators. Likewise, optical add/drop multiplexers using passive WDM devices have been qualified for use in undersea branching configurations.

■ Undersed networks are very expensive to built, and the capacity of these networks is shared among a number of users.

■ There is a **one key difference between undersea links and terrestrial links :** In most cases undersea links are deployed from scratch with new fibers rather than over existing fiber plant. It is rare to upgrade an exciting long haulf amplified undersea link because the cost of lying a new link is not significantly higher than the cost of upgrading an existing link. This gives more flexibility in design choices.

11.6. Sources of Power Penalty

In a realistic lightwave system, the receiver sensitivity is affected by various factors which is in combination with fiber dispersion, degrade the SNR at the decision circuit. Some factors are there which degrade the receiver sensitivity :

■ Modal noise
■ Dispersion broadening and intersymbol interference.
■ Mode partition noise
■ Frequency chirp
■ Reflection feedback.

11.6.1. Modal Noise

The major problem occurs when laser is used with multimode fiber, this phenomenon is known as **Modal Noise.** Modal noise is a distortion of the signal caused by the channel (the fiber and connectors) itself interacting with the signal in such a way as to produce random fluctuations in signal power at the receiver. In many situations it can be a serious effect for the signal-to-noise performance of the communication link. The problem is confined to the use of high quality laser sources on multimode fiber. Modal Noise comes from the signal itself so it only happens when there is a signal and it is more pronounced at low bit rates. . There is no modal noise in the "0" state.

Modal noise is produced in two stages :

1. Mode coupling
2. Mode selective loss

Mode Coupling

When we couple a laser to a multimode fiber the light will travel in many modes along the fiber. If the laser involved is narrow in line width and stable, the light launched into the fiber will be coherent and in phase in all of the excited fiber modes. After travelling a very short distance along the fiber the light in each mode will be still at exactly the same wavelength but the modes will be out of phase with each other.

Mode coupling effect can cause light from multiple modes to merge into the same mode. When this happens the phase differences will cause interference in the new mode. This is a complex process. If two newly mixed signals happen to be about the same amplitude and 180 degrees out of phase then the signals will destructively interfere with one another in the new mode. In this case no power can travel in the new mode.

Mode coupling behaviour at an irregularity will change when the wavelength or the temperature changes. Even the slightest change in laser wavelength will cause very large changes in the phases of the mixed signals. All of this can change the amplitude of the mixed signal significantly and abruptly. Noise is created when mode selective loss is present.

Mode Selective Loss

A large irregularity in the fiber causes not only the coupling of modes but also the loss of modes completely from the fiber. If we consider the loss characteristic some modes will be completely lost and some will couple their power to other modes, but the power will stay in the fiber and other modes may be minimally affected. Thus when the signal is changing randomly and abruptly from mode to mode then it meets a bad connector, we get a loss of signal power which varies randomly and abruptly. Even though the signal power loss from each mode at a particular connector is constant, the power arriving moves rapidly from mode to mode. For example: if one particular mode is lost at a connector then no signal is lost when no power is being carried in that mode and a lot of power is lost when there is a lot of power in the mode.

Because the signal power is shifting into and out of that mode quickly and randomly, there is a loss of total signal power which occurs at random and this is noise. The transmitter signal becomes more and more mixed and the power present in each mode tends to become the same as the amount of power in each other mode. When such a signal meets a source of mode selective loss although power loss occurs there is no noise effect because the power loss does not vary.

The process is shown in Fig. 11.3. The speckle patterns in the top row of the figure represent the changing power distribution of a signal as it passes a particular point on the fiber. Although the speckles change the total amount of power in the fiber remains constant.

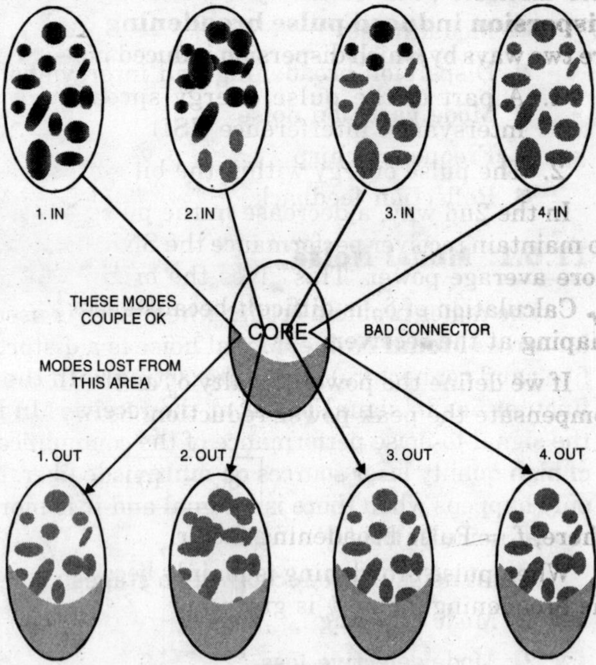

Fig. 11.3. Origin of Modal Noise.

There are many ways to **minimize modal noise effect :**

1. Keep connector loss to a minimum. Connectors with a loss of less than 1 dB create very little modal noise.
2. Use a laser that has a relatively large line width and hops between lasing modes very frequently.

3. Use an ASE (amplified spontaneous emission) laser source with an external modulator. This source is not really a laser, it is an amplified LED. These are known as SLDs (Super Luminescent Diodes), sometimes it requires the use of an external modulator.

4. Destroy coherence of the input signal of the laser. One technique is to impose a high frequency microwave signal on to the bit stream at the input of the laser. This means that the signal level at the laser input is made to vary sinusoidally over a few cycles in each bit time.

5. Short wavelength should be used so that the fiber has a large number of possible modes.

6. Use a large core fiber (to increase the number of modes). Example: use 62.5 micron fiber to have fewer problems with modal noise than 50 micron.

7. Self-pulsating laser diode can be used.

11.6.2. Dispersive Pulse Broadening

Single mode fibers used for lightwave systems nearly avoides the problem of intermodal dispersion and associated modal noise. The group velocity dispersion stills limits the bit rate distance product BL by broadening optical pulses beyond their allocated bit slot. **Dispersion induced pulse broadening** can also decreases the receiver sensitivity. There are two ways by which dispersion induced pulse broadening affects the receiver performance :

1. A part of the pulse energy spreads beyond the allocated bit slot and leads to intersymbol interference (ISI).

2. The pulse energy within the bit slot is reduced when the optical pulse broadens.

In the 2nd way, a decrease in the pulse energy reduces the SNR at the decision circuit. To maintain receiver performance the SNR should remain constant thus the receiver requires more average power. This gives the origination of **dispersion induced power penalty δ_d.** Calculation of δ_d is difficult because it depends of many factors like the extent of pulse shaping at the receiver.

If we define the power penalty δ_d as the increase (in dB) in the received power that will compensate the peak power reduction, δ_d is given by

$$\boxed{\delta_d = 10 \log_{10} (f_b)} \qquad \text{...(11.10)}$$

where, f_b = Pulse broadening factor

When pulse broadening is mainly because of a wide source spectrum at the transmitter, the broadening factor f_b is given by

$$f_b = \frac{\sigma}{\sigma_0}$$

or

$$\boxed{f_b = \left[1 + \left(\frac{DL\sigma_\lambda}{\sigma_0}\right)^2\right]^{1/2}} \qquad \text{...(11.11)}$$

where, σ_0 = Rms width of optical pulse at the fiber input

σ_1 = Rms width of source spectrum assumed to be gaussian

Equation (11.10) and (11.11) can be used to estimate the dispersion penalty for lightwave systems that use single mode fiber together with a multimode laser on an LED.

SUMMARY

1. Channel multiplexing can be done in the time or the frequency domain through:
   - Time-division multiplexing (TDM).
   - Frequency-division multiplexing (FDM).
2. WDM corresponds to the scheme in which multiple optical carriers at different wavelengths are modulated by using independent electrical bit streams and are then transmitted over the same fiber.
3. The concept of WDM has been pursued since the first commercial lightwave system became available in 1980. In its simplest form, WDM was used to transmit two channels in different transmission windows of an optical fiber.
4. The ultimate capacity of WDM fiber links depends on how closely channels can be packed in the wavelength domain. The minimum channel spacing is limited by inter channel crosstalk.
5. Optical networks are used to connect a large group of users spread over a geographical area. They can be classified as :
   - Local-area network (LAN)
   - Metropolitan-area network (MAN)
   - Wide-area network (WAN)
6. The test-bed implementation of an optically switched metro network is known as the **multi-wavelength optical network (MONET).**
7. MONET incorporated diverse switching technologies [synchronous digital hierarchy (SDH), asynchronous transfer mode (ATM), etc.] into an all-optical ring network using cross-connect switches based on the LiNbO3 technology.
8. Broadly speaking, WDMA networks can be classified into two categories:
   - Single-hop optical networks
   - Multi-hop optical networks
9. WDM networks making use of a passive star coupler are often called **passive optical networks** (PONs) because they avoid active switching.
10. A **waveguide-grating router** (WGR) can be used for wavelength routing. Spectral slicing and WGR devices are discussed in the next section devoted to WDM components.
11. The role of a tunable optical filter in a WDM system is to select a desired channel at the receiver.
12. All optical filters require a wavelength-selective mechanism and can be classified into two broad categories depending on the underlying physical mechanism:
    - Optical interference
    - Diffraction.
13. A Fabry-Perot (FP) interferometer—a cavity formed by using twomirrors—can act as a tunable optical filter if its length is controlled electronically by using a piezoelectric transducer.

14. Tunable FP filters can also be made using several other materials such as liquid crystals and semiconductor waveguides.

15. A chain of Mach–Zehnder (MZ) interferometers can also be used for making a tunable optical filter. A MZ interferometer can be constructed simply by connecting the two output ports of a 3 dB coupler to the two input ports of another 3 dB coupler.

16. The reflective nature of fiber gratings is often a limitation in practice and requires the use of an **optical circulator**.

17. Acousto-optic tunable filters can be made by using bulk components as well as waveguides, and both kinds are available commercially.

18. Another category of tunable optical filters operates on the principle of amplification of a selected channel. Any amplifier with a gain bandwidth smaller than the channel spacing can be used as an optical filter.

19. Multiplexers and demultiplexers are the essential components of a WDM system. Similar to the case of optical filters, demultiplexers require a wavelength-selective mechanism.

20. Demultiplexers can be classified into two broad categories:

- Diffraction-based demultiplexers
- Interference-based demultiplexers

21. Grating-based demultiplexers use the phenomenon of Bragg diffraction from an optical grating.

22. A problem with grating demultiplexers is that their bandpass characteristics depend on the dimensions of the input and output fibers.

23. Add–drop multiplexers are needed for wide-area and metro-area networks in which one or more channels need to be dropped or added while preserving the integrity of other channels.

24. The wavelength selectivity of Bragg gratings can also be used to make add–drop filters.

25. The role of a star coupler is to combine the optical signals entering from its multiple input ports and divide it equally among its output ports.

26. An important WDM component is an $N \times N$ wavelength router, a device that combines the functionality of a star coupler with multiplexing and demultiplexing operations.

27. A wavelength converter changes the input wavelength to a new wavelength without modifying the data content of the signal.

28. The most important issue in the design of WDM lightwave systems is the **interchannel crosstalk**. The system performance degrades whenever crosstalk leads to transfer of power from one channel to another.

29. Several nonlinear effects in optical fibers can lead to interchannel and intrachannel crosstalk that affects the system performance considerably.

30. Raman crosstalk can be avoided if channel powers are made so small that SRS induced amplification is negligible over the entire fiber length.

31. Stimulated Brillouin scattering (SBS) can also transfer energy from a high-frequency channel to a low-frequency one when the channel spacing equals the Brillouin shift.

32. The electrical TDM becomes difficult to implement at bit rates above 10 Gb/s because of the limitations imposed by high-speed electronics. A solution is offered by the *optical* TDM (OTDM), a scheme that can increase the bit rate of a single optical carrier to values above 1 Tb/s.
33. In OTDM lightwave systems, several optical signals at a bit rate B share the same carrier frequency and are multiplexed optically to form a composite bit stream at the bit rate NB, where N is the number of channels.
34. Demultiplexing of individual channels from an OTDM signal requires electro-optic or all-optical techniques.
35. The main limitation of a fiber-based demultiplexer stems from the weak fiber nonlinearity. Typically, fiber length should be 5 km or more for the device to function at practical power levels of the clock signal.
36. The basic concept behind **subcarrier multiplexing (SCM)** is borrowed from microwave technology, which employs multiple microwave carriers for transmission of multiple channels (electrical FDM) over coaxial cables or free space.
37. In analog SCM lightwave systems, each microwave subcarrier is modulated using an analog format, and the output of all subcarriers is summed using a microwave power combiner.
38. A multiplexing scheme well known in the domain of wireless communications makes use of the **spread-spectrum technique**. It is referred to as **code-division multiplexing (CDM)** because each channel is coded in such a way that its spectrum spreads over a much wider region than occupied by the original signal.

REVIEW QUESTIONS

1. What do you understand by lightwave systems? List the various issues related to system design and performance to form a practical lightwave system.
2. How fiber-optic communication systems can be classified? Briefly explain each of them.
3. What is distributed network and how it is useful in the designing of lightwave system?
4. What do you meant by local area network? Differentiate between LANs and MANs.
5. Briefly explain the various design guidelines of fiber optic communication systems. What is link budget and bandwidth budget?
6. In a power budget analysis determine the expression for average launch power for any transmitter and channel loss.
7. Define the term modal noise and briefly discuss the attempts have been made to estimate the extent of sensitivity degradation induced by modal noise.
8. Which of the dominant nonlinear phenomenon limits the system performance in a long haul system? Also analyze the combined effects of GVD and SPM.

8\. Prove that the rise time Tr and the 3 dB bandwidth Δf of a RC circuit are related by $Tr\Delta f = 0.35$.

10\. obtain an expression for the total receiver noise by including thermal noise, shot noise, intensity noise, mode-partition noise, chirp noise, and reflection noise.

❑❑❑

Multichannel Systems

Chapter 12

INSIDE THIS CHAPTER

12.1. Introduction

Practically the bit rate of optical communication systems was limited to 10 Gb/s or less until 1995 because of the limitations imposed by the :

- Dispersive and nonlinear effects.
- By the speed of electronic components.

Principally the capacity of optical communication systems can exceed 10 Tb/s because of large frequency associated with optical carrier. But the transmission of multiple optical channels over the same fiber has provided a simple way for extending the system capacity to beyond 1 Tb/s.

Channel multiplexing can be done :

- In time domain through Time Division Multiplexing (TDM).
- In frequency domain through Frequency Domain Multiplexing (FDM).

The TDM and FDM techniques can also be used in the electrical domain. Commonly with reference to two **optical domain techniques** it is known as :

- Optical TDM (OTDM)
- Wavelength division multiplexing (WDM).

These multichannel systems were gained attention during the 1990s. In fact WDM lightwave systems were available commercially by 1996. This chapter is devoted to WDM lightwave systems by considering in different sections the architectural aspects of such systems, the optical components required for their implementations and the performance issues.

12.2. WAVELENGTH DIVISION MULTIPLEXING (WDM)

Wavelength Division Multiplexing (WDM) is the basic technology of optical networking. This technique is used for fiber (or optical device) to carry many separate and independent optical channels.

The principle is identical to that used when we tune our television receiver to one of many TV channels. Each channel is transmitted at a different radio frequency and we select between them using a tuner which is just a resonant circuit within the TV set.

Another way of investigating WDM is to consider that each channel consists of light of a different colours. Thus a WDM system transmits a **rainbow**. Actually at the wavelengths involved, the light is invisible but it's a good way of describing the principle.

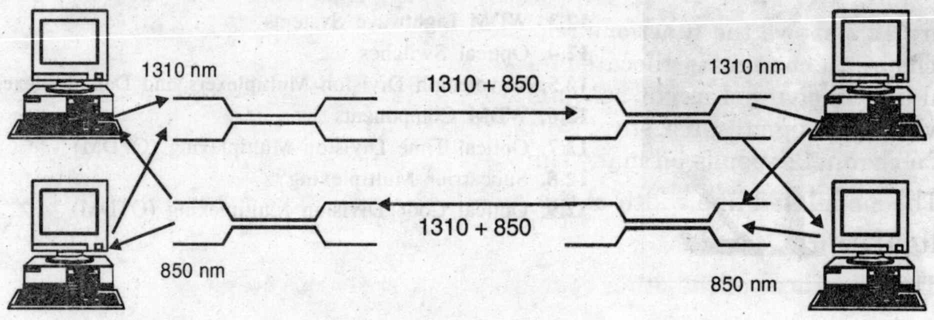

Fig. 12.1. Simple WDM.

There are many **types of WDM:**

(1) A **simple WDM** can be constructed using 1310 nm as one wavelength and 1550 as the other or 850 and 1310. This type of WDM can be built using relatively simple and inexpensive components. Some applications have been in operation for a number of years using this principle.

(2) **Dense WDM** refers to the close spacing of channels. Dense is a qualitative measure and just what dense means is largely in the mind of a particular individual. To some, a series of WDM channels spaced at 3.6 nm apart qualifies for the description. Others use the term to distinguish systems where the wavelength spacing is 1 nm per channel or less.

12.2.1. Simple WDM

Figure 12.1 shows an example of a very simple sparse WDM system using multimode fibers. Wavelength selective couplers are used both to mix (multiplex) and to separate (demultiplex) the signals. Here the distinguishing characteristic is the very wide separation of wavelengths used (different bands rather than different wavelengths in the same band).

There are many variations around this. Some systems use a single fiber bidirectionally while others use separate fibers for each direction. Other systems use different wavelength bands from those shown in the figure (1310 and 1550 for example). The most common systems run at very low data rates.

12.2.2. Dense WDM

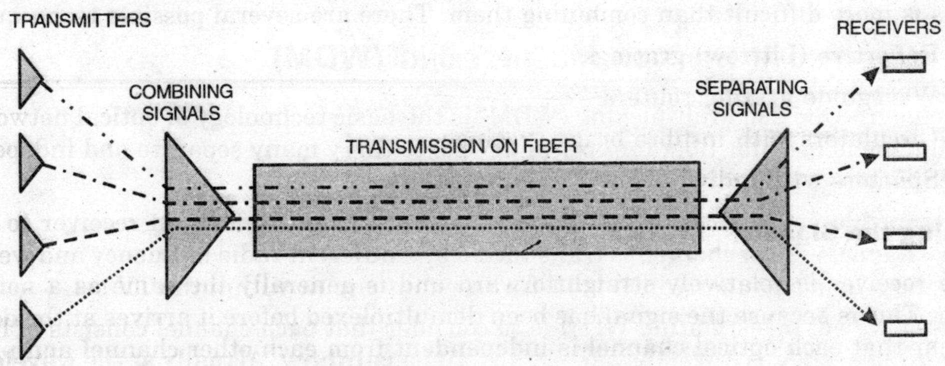

Fig. 12.2. WDM Functions Schematic.

Figure 12.2 shows the functions required to build a simple dense WDM shared optical link. Each optical channel is allocated its own wavelength - or rather range of wavelengths. A typical optical channel might be 1 nm wide. This channel is really a wavelength range within which the signal must stay. Normally it is much wider than the signal itself. The width of a channel depends on many things such as:

- The modulated line width of the transmitter
- Its stability
- The tolerances of the other components in the system.

Transmitters

Practically the transmitter is a laser. It must have a line width which (after modulation) fits easily within its allocated band. It must not go outside the allocated band so it should have chirp and drift characteristics that ensure this.

Combining the Signals (Channels)

There are several ways of combining the signals. The most obvious is to use a number of 3 dB splitters or Y-junctions connected in cascade. With a large number (say 32) signals each one will be reduced to 1/32nd of its initial strength. This is fine with a small number of channels but for a large number we have to amplify the combined signal immediately after it is mixed.

Gratings and planar waveguide gratings have much lower loss and their loss is not dependent on the number of channels so these are most often used in systems with more than four channels.

Transmission and Amplification

In transmission on a fibre the main issue is controlling crosstalk effects. Channel spacings, widths and power levels are system variables that can be used to minimize crosstalk.

Amplification is a major issue. The ability to amplify the mixed signal is one of the things that makes WDM possible. However, when multiple amplifiers are used in a long link their non-linearities add up and cause significant difficulty.

Separating the Channels at the Receiver

This is more difficult than combining them. There are several possible techniques :

- Reflective (Littrow) gratings
- Waveguide grating routers
- Circulators with in-fibre bragg gratings
- Splitters with individual Fabry-Perot filters

Receiving the Signals

The receiver is relatively straightforward and is generally the same as a non-WDM receiver. This is because the signal has been demultiplexed before it arrives at the detector. It is clear that each optical channel is independent from each other channel and provided the signals stay within their allocated bands, there is no necessary relationship between them. For example: one channel might run at 2 Gbps, another at 622 Mbps and a number of others at 200 Mbps.

12.3. WDM Lightwave Systems

The concept of WDM has been pursued since the first commercial lightwave system became available in 1980. In its simplest form, WDM was used to transmit two channels in different transmission windows of an optical fiber. For example, an existing 1.3-μm lightwave system can be upgraded in capacity by adding another channel near 1.55 μm, resulting in a channel spacing of 250 nm.

Considerable attention was directed during the 1980s toward reducing the channel spacing, and multichannel systems with a channel spacing of less than 0.1 nm had been demonstrated by 1990. However during the decade of the 1990s that WDM systems were developed most aggressively. Commercial WDM systems first appeared around 1995, and their total capacity exceeded 1.6 Tb/s by the year 2000. The advent of WDM has led to a virtual revolution in designing lightwave systems. The **WDM systems can be classified into three categories :**

12.3.1. High-Capacity Point-to-Point Links

For long-haul fiber links forming the backbone or the core of a telecommunication network, the role of WDM is simply to increase the total bit rate. Figure 12.3 shows a point-to-point, high-capacity, WDM link. The output of several transmitters, each operating at its own carrier frequency (or wavelength), is multiplexed together. The multiplexed signal is launched into the op-

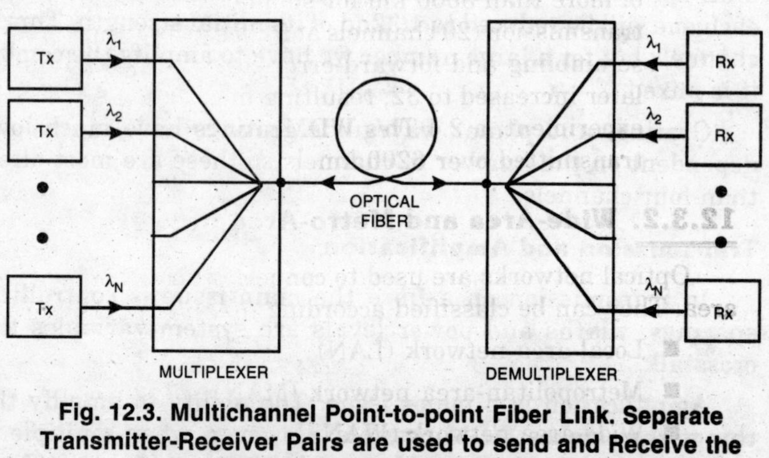

Fig. 12.3. Multichannel Point-to-point Fiber Link. Separate Transmitter-Receiver Pairs are used to send and Receive the Signal at Different Wavelengths.

tical fiber for transmission to the other end, where a demultiplexer sends each channel to its own receiver.

When N channels at bit rates B_1, B_2, ..., and B_N are transmitted simultaneously over a fiber of length L, the total bit rate–distance product, B_L becomes

$$BL = (B_1 + B_2 + ... + B_N)L \quad ...(12.1)$$

For equal bit rates, the system capacity is enhanced by a factor of N. An experiment in 1985 demonstrated the BL product of 1.37 (Tb/s) km by transmitting 10 channels at 2 Gb/s over 68.3 km of standard fiber with a channel spacing of 1.35 nm.

The ultimate capacity of WDM fiber links depends on the closeness of the channels packed in the wavelength domain. The minimum channel spacing is limited by **inter channel crosstalk.** Channel spacing ΔV_{ch} should exceed $2B$ at the bit rate B but this requirement wastes considerable bandwidth. It is common to introduce a measure of the spectral efficiency of a WDM system as

$$\eta_s = B/\Delta V_{ch} \quad ...(12.2)$$

(Attempts are made to make η_s as large as possible.)

The channel frequencies (or wavelengths) of WDM systems have been standardized by the International Telecommunication Union (ITU) on a 100-GHz grid in the frequency range 186–196 THz (covering the C and L bands in the wavelength range 1530–1612 nm). Therefore channel spacing for most commercial WDM systems is 100 GHz (0.8 nm at 1552 nm). This value leads to only 10% spectral efficiency at the bit rate of 10 Gb/s.

Experimental results on **WDM systems can be divided into two groups:**

(1) In **first group** the transmission distance is ~100 km or exceeds 1000 km. Since the 1985 experiment in which ten 2 Gb/s channels were transmitted over 68 km, both the number of channels and the bit rate of individual channels have increased considerably. A capacity of 340 Gb/s was demonstrated in 1995 by transmitting 17 channels, each operating at 20 Gb/s over 150 km. By 2001, the capacity of WDM systems exceeded 10 Tb/s in several laboratory experiments.

(2) The **second group** of WDM experiments is concerned with transmission distance of more than 5000 km for submarine applications. In a 1996 experiment, 100 Gb/s transmission (20 channels at 5 Gb/s) over 9100 km was realized using the polarization scrambling and forward-error-correction techniques. The number of channels was later increased to 32, resulting in a 160-Gb/s transmission over 9300 km. In a 2001 experiment, a 2.4 Tb/s WDM signal (120 channels, each operating at 20 Gb/s) was transmitted over 6200 km.

12.3.2. Wide-Area and Metro-Area Networks

Optical networks are used to connect a large group of users spread over a geographical area. They can be classified according to the area they cover:

- Local-area network (LAN)
- Metropolitan-area network (MAN)
- Wide-area network (WAN)

All three types of networks can benefit from the WDM technology. They can be designed using the hub, ring, or star topology. A **ring topology** is most practical for MANs and WANs, while the **star topology** is commonly used for LANs.

- At the LAN level, a broadcast star is used to combine multiple channels. At the next level, several LANs are connected to a MAN by using passive wavelength routing.
- At the highest level, several MANs connect to a WAN whose nodes are interconnected in a mesh topology.
- At the WAN level, the network makes extensive use of switches and wavelength-shifting devices so that it is dynamically configurable.

Opaque Network

Hubs or nodes located in large metropolitan areas house electronic switches, which connect any two nodes either by creating a **virtual circuit** between them or by using **packet switching** through protocols such as **TCP/IP (transmission control protocol/Internet protocol)** and **asynchronous transfer mode (ATM)**. With the advent of WDM during the 1990s, the nodes were connected through point-to-point WDM links, but the switching was being done electronically even in 2001. Such transport networks are termed **opaque networks** because they require optical-to-electronic conversion. As a result, neither the bit rate nor the modulation format can be changed without changing the switching equipment.

Optically Transparent WDM Network

An optical network in which a WDM signal can pass through multiple nodes (possibly modified by adding or dropping certain channels) is called **optically transparent**. Transparent WDM networks are desirable because they do not require demultiplexing and optical-to-electronic conversion of all WDM channels.

An alternative topology implements a regional WDM network in the form of several interconnected rings. Figure 12.4 shows such a scheme.

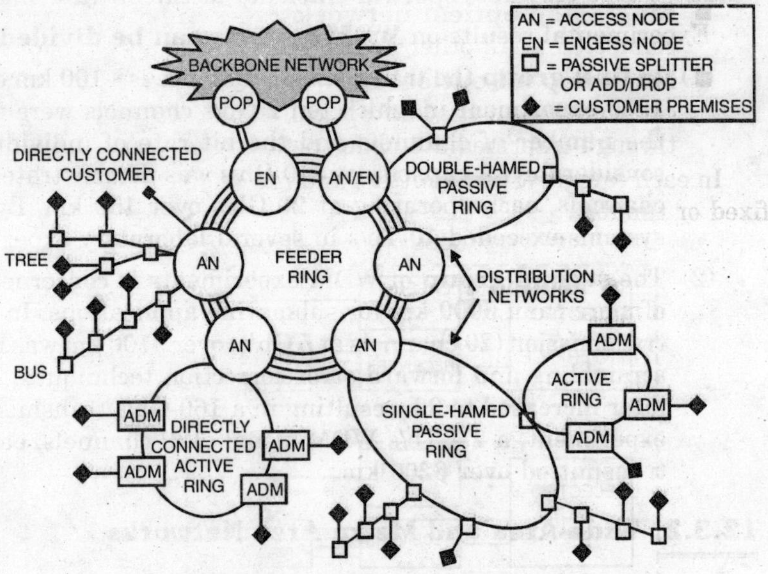

Fig. 12.4. A WDM Network with a Feeder Ring Connected to Several Local Distribution Networks.

- The feeder ring connects to the backbone of the network through an egress node. This ring employs four fibers to ensure robustness.
- Two of the fibers are used to route the data in the clockwise and counterclockwise directions. The other two fibers are called **protection fibers** and are used in case a point-to-point link fails (self-healing).

- The feeder ring supplies data to several other rings through access nodes.
- An add-drop multiplexer can be used at all nodes to drop and to add individual WDM channels.
- Dropped channels can be distributed to users using bus, tree, or ring networks. Nodes are not always directly connected and require data transfer at multiple hubs. Such networks are called **multihop networks.**

12.3.3. Multiple-Access WDM Networks

Multiple-access networks offer a random bidirectional access to each subscriber. Each user can receive and transmit information to any other user of the network at all times.

Telephone networks provide one example: they are known as subscriber loop, local loop or access networks. Another example is provided by the Internet used for connecting multiple computers. In 2001, both the local-loop and computer networks were using electrical techniques to provide bidirectional access through circuit or packet switching. The main **limitation of such techniques** is that each node on the network must be capable of processing the entire network traffic. Since it is difficult to achieve electronic processing speeds in excess of 10 Gb/s, such networks are inherently limited by the electronics.

The **use of WDM** permits a novel approach in which the channel wavelength itself can be used for switching, routing, or distributing each channel to its destination, resulting in an all-optical network. Since wavelength is used for multiple access, such a WDM approach is referred to as **wavelength-division multiple access (WDMA)**. WDMA networks can be classified into two categories:

- **Single-hop optical network :** In a **single-hop network** every node is directly connected to all other nodes, resulting in a fully connected network.
- **Multi-hop all-optical network :** While **multi-hop networks** are only partially connected such that an optical signal sent by one node may require several hops through intermediate nodes before reaching its destination.

In each category, transmitters and receivers can have their operating frequencies either fixed or tunable.

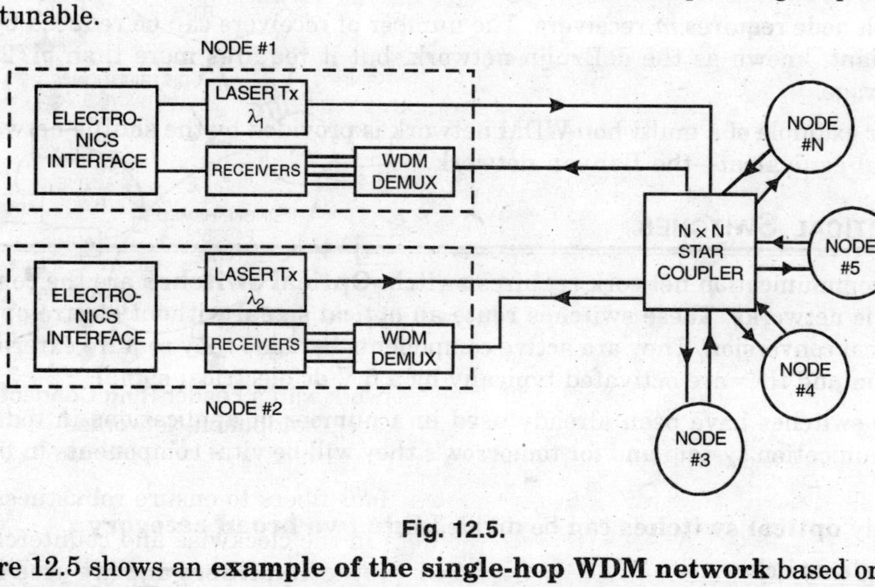

Fig. 12.5.

Figure 12.5 shows an **example of the single-hop WDM network** based on the use of a **broadcast star**. This network is known as **Lambdanet**, an example of the **broadcast-and-select** network.

- Schematic of the Lambdanet with N nodes. Each node consists of one transmitter and N receivers.
- The new feature of the Lambdanet is that each node is equipped with one transmitter emitting at a unique wavelength and N receivers operating at the N wavelengths, where N is the number of nodes.
- The output of all transmitters is combined in a passive star and distributed to all receivers equally. Each node receives the entire traffic flowing across the network.
- A tunable optical filter can be used to select the desired channel.
- In the case of the Lambdanet, each node uses a bank of receivers in place of a tunable filter. This feature creates a nonblocking network whose capacity and connectivity can be reconfigured electronically depending on the application.
- This network is also transparent to the bit rate or the modulation format.
- Different users can transmit data at different bit rates with different modulation formats.
- The flexibility of the Lambdanet makes it suitable for many applications.
- The main drawback of the Lambdanet is that the number of users is limited by the number of available wavelengths.

Several architectures can be used for all-optical **multihop networks**. **Hypercube architecture** provides one example :

- It has been used for interconnecting multiple processors in a supercomputer. The hypercube configuration can be easily visualized in three dimensions such that eight nodes are located at eight corners of a simple cube.
- In general the number of nodes N must be of the form 2^m where, m — the dimensionality of the hypercube.
- Each node is connected to m different nodes. The maximum number of hops is limited to m, while the average number of hops is about $m/2$ for large N.
- Each node requires m receivers. The number of receivers can be reduced by using a variant, known as the deBruijn network, but it requires more than $m/2$ hops on average.

Another example of a multi-hop WDM network is provided by the shuffle network or its bidirectional equivalent—the Banyan network.

12.4. Optical Switches

Every communication network requires switch. **Optical switches** are the components of fiber optic networks. These switches route an optical signal without electro-optical and opti-electrical conversion. They are active components because they require external power for operation and they are activated typically by a 5 V dc electrical signal.

Optical switches have been already used in a number of applications in today's fiber optic communication system and for tomorrow's they will be vital components in fiber optic networks.

Basically **optical switches** can be divided **into two broad category :**

- Single switches
- Multistage (large) switches.

12.4.1. Single Switches

There are two main **functional categories** of single switches :

(1) On/off switch

(2) Passing switch

- An of/off (1×1) switch, as shown in Fig. 12.6(*a*), either allows or does not allow a light signal to pass.
- A (1×2) passing switch directs the light signal from fiber 1 to fiber 2 or to fiber 3, as shown in Fig. 12.6(*b*). This switch is of course no more than a simple example. Single (1×N) passing switches are commercially avialable.
- A (2×2) passing switch can connect two fibers to two other fibers. A (2×2) passing switch can be either in the **bypass (bar) state** or in the **cross (inserted) state,** as shown in Fig. 12.6(*c*).
- The term nonblocking means that a switch can connect any of its inpus to any of its outputs. Example, a (2×2) blocking switch, as shown in Fig. 12.5(*d*) where the switch connect only fiber 1 to fiber 4.

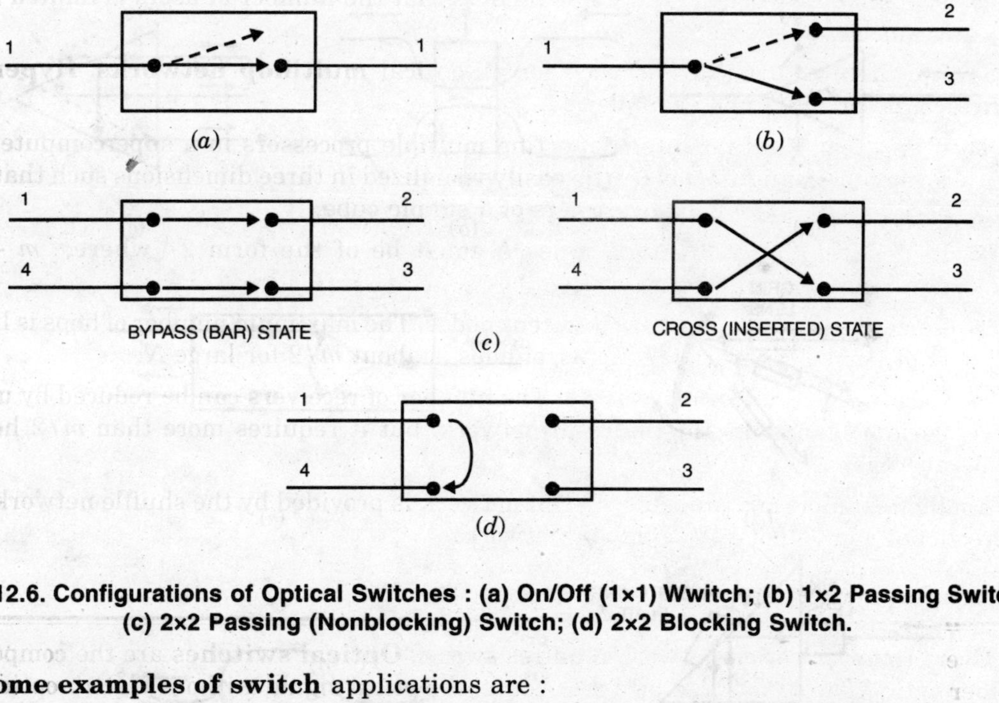

Fig. 12.6. Configurations of Optical Switches : (a) On/Off (1×1) Wwitch; (b) 1×2 Passing Switch; (c) 2×2 Passing (Nonblocking) Switch; (d) 2×2 Blocking Switch.

Some examples of switch applications are :

- On/off switches are used to isolate sources and receivers in test equipment A (1×2) switches permit channel selection and are used to for protection switching.
- A (1×N) switches are used for testing and measurement of fiber optic components, the remote testing of fiber opticl communication syste, and fiber optic netwυrk restoration.
- A (2×2) switches are used for node by passing in fiber optic network. A typical application of a 2×2 switch is in FDDI networks.

Operation of Optical Switches

Single switches are manufactured in latching and nonlatching styles. The latching operation means that the switch remains in its selected position if power is lost. The principle of operation of commercially available switches can be understand if we will be aware that they come in the forms as :

- Electro-optical
- Optomechanical
- Thermo-optical
- Acousto-optical

Some of the principles of switches operation are shown in Fig. 12.7.

If we switch the bias voltage of a semiconductor optical amplifier (SOA) on and off, we can create an electro-optical (EO) switch, as shown in Fig. 12.7(*a*). Another example of an EO switch is shown in Fig. 12.7(*b*), where the coupling ratio of a waveguide coupler is dependent on the applied voltage. The change in coupling ratio is achieved by fabricating the coupler on the top of $LiNbO_3$.

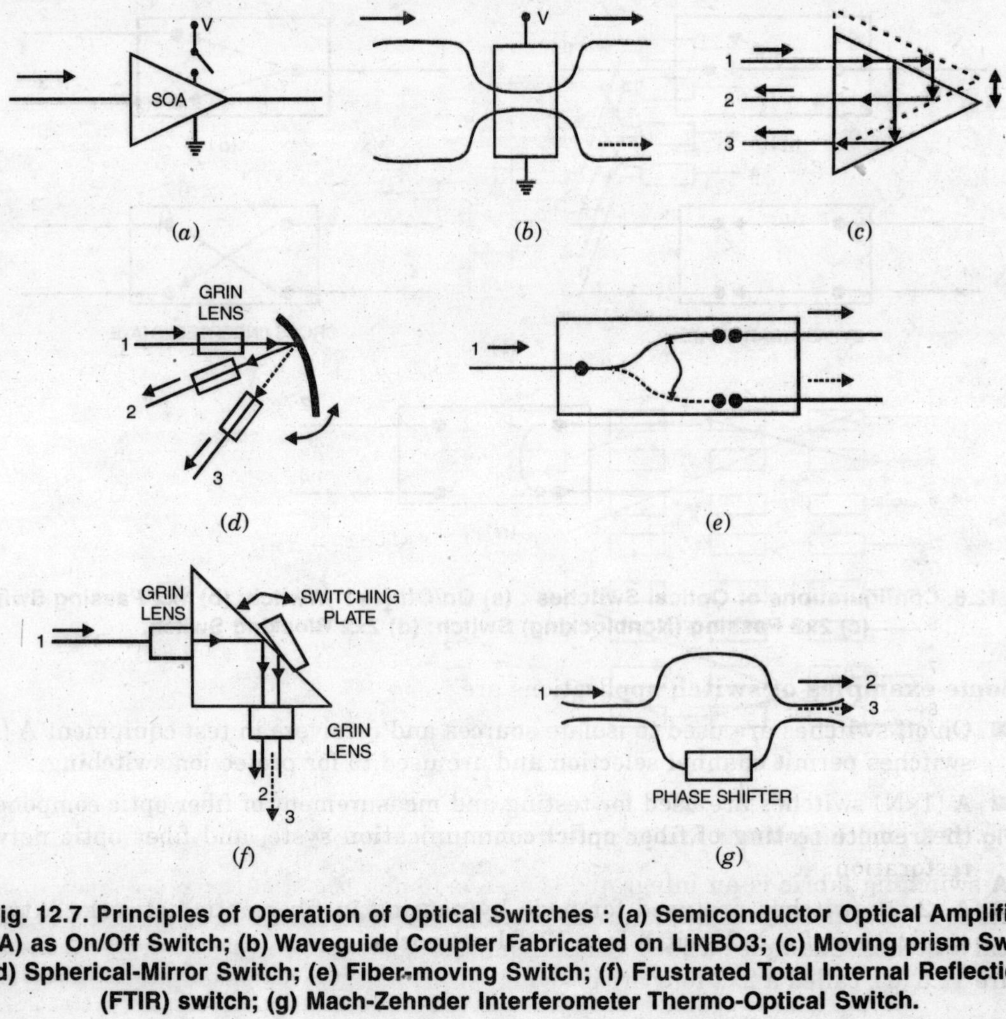

Fig. 12.7. Principles of Operation of Optical Switches : (a) Semiconductor Optical Amplifier (SOA) as On/Off Switch; (b) Waveguide Coupler Fabricated on LiNBO3; (c) Moving prism Switch; (d) Spherical-Mirror Switch; (e) Fiber-moving Switch; (f) Frustrated Total Internal Reflection (FTIR) switch; (g) Mach-Zehnder Interferometer Thermo-Optical Switch.

Opto mechanical switches are based on the mechanical motion of the optical components. Example, In Fig. 12.7(*c*), vertically moving the prism enables the switching of an optical signal from fiber 2 to fiber 3. The same result can be achieved by slightly pivoting a spherical mirror, as shown in Fig. 12.7(*d*). Graded-index (GRIN) lenses facilitate passage of the coupling light from and into optical fibers. One such example of optical switch is shown in Fig. 12.7(*e*). The principle of frustation of total internal reflection (FTIR) is used to manufacture the switch shown in Fig. 12.7(*f*). Figure 12.7(*g*) shows an example of thermo optical switch. A mach-zehnder inferometer incorporates a phase shifter in one of its legs. By applying heat, we can control the amount of phase shift, which means we can direct onoptical signal either into fiber 2 or into fiber 3. Thermo optical switches have faster switching speeds than the optomechanical counterparts.

12.4.2. Multistage (Large) Switches

For multichannel fiber optic networks. There is a requirements of switch capable of switching many individual channels simultaneously. Therefore the need for **integrated switches or switching fabrics arises.**

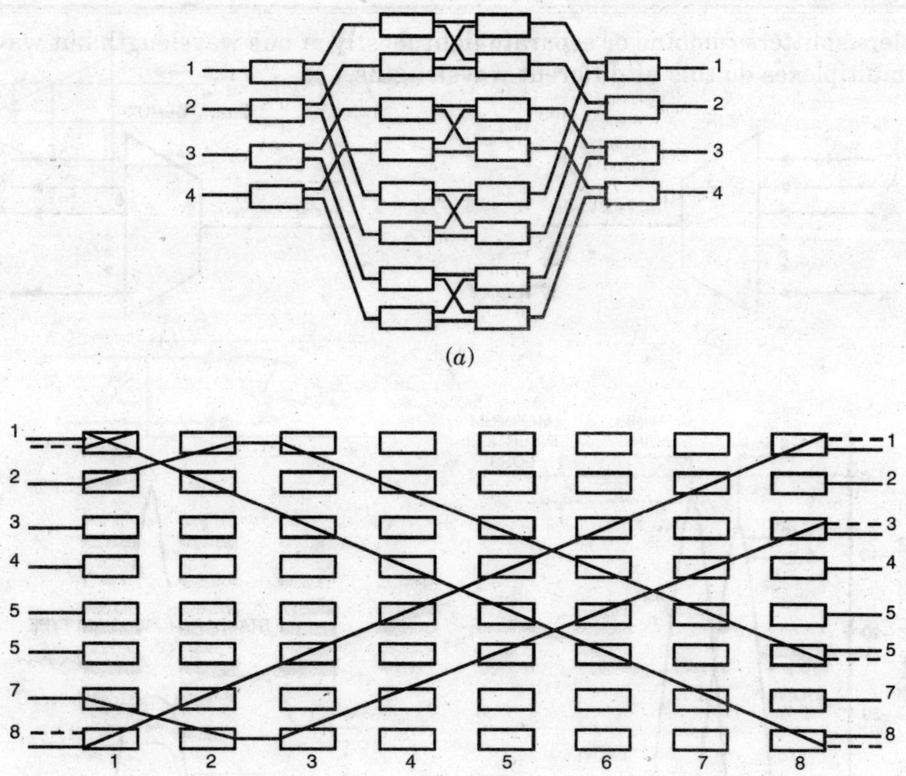

Fig. 12.8. Switching-Matrix Architectures : (a) 2 × 2 4-Array; (b) 8 × 8 Optical Matrix Switch.

A switching fabric is an integrated switch enabling the simultaneous connection of N inputs to M outputs. The architecture of switching fabric is based on combinations of single optical switches. Example : a 4 × 4 switching fabric can be built using 2 × 2 optical switches. Figure 12.8 (*a*), called a 2×2 four array switch, it is implemented as a solid state integrated unit. The same architecture is used to manufacture a 2×2 eight array switching matrix on a single chip.

Another example of an 8×8 optical matrix switch is shown in Fig. 12.8(*b*), where only several characteristic connections are present. The switch has the following characteristics :

Dimension : 32 × 32

Insertion Loss : 3.5 dB for directed light and 15 dB in a 512 × 512 matrix.

Switching Speed : Less than 10 ms.

PDL : Less than 0.25 dB

Cross talk : Less than –50 dB

In general, $N \times M$ switches can be built for two functions :

(1) to switch each input fiber to each output fiber (**a directional switch**).

(2) To switch each input fiber to one or more output fibers simultaneously in a broadcast configuration. (Fig. 12.8 **a distributional switch).**

12.5. Wavelength Division Multiplexers and Demultiplexers

Couplers/splitters combine or separate light mostly at one wavelength but wavelength-division multiplexes do this at different wavelengths.

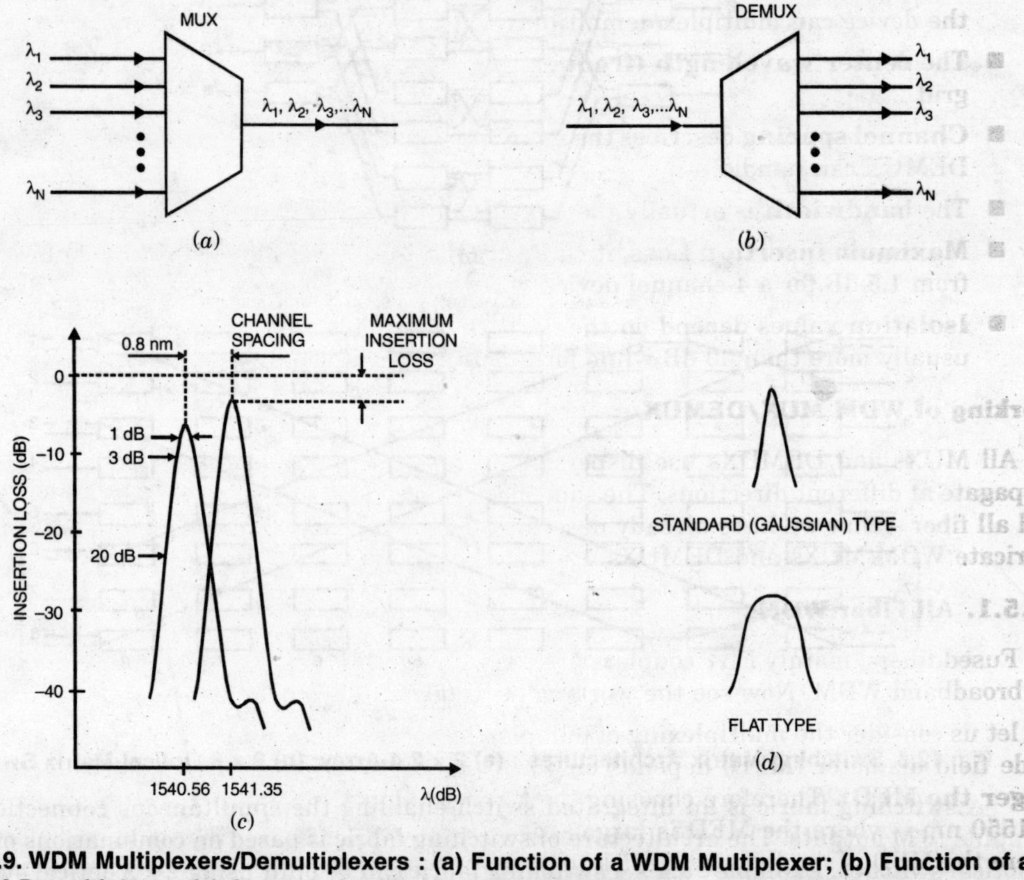

Fig. 12.9. WDM Multiplexers/Demultiplexers : (a) Function of a WDM Multiplexer; (b) Function of a WDM Demultiplexer; (c) Characterization of a WDM MUX/DEMUX; (d) Channel-Profile Types.

Wavelength division multiplexers and demultiplexers (**also called WDM MUXs and WDM DEMUXs)** are devices that combine (couple) and separate (split) signals at different wavelengths a WDM MUX combines several wavelength channels into one fiber; a WDM DEMUX does just the opposite. WDM multiplexers and demultiplexers are crucial elements of WDM fiber optic networks. Figure 12.9 (*a*) and (*b*) shows their function.

Sometimes one device can perform both multiplexing and demultiplexing i.e. we have a bidirectional WDM MUX/DEMUX, at other times a device is designed specifically for multiplexing and demultiplexing.

Basically there are **three types of WDM MUXs and DEMUXs** in terms of wavelength spacing with which they are dealing.

(1) **Broadband WDMs (BWDMs)** combine and separate 1310 nm and 1550 channels or even 850 nm and 1310 channels.

(2) **Narrowband WDMs (NWDMs)** combine and separate wavelength channels with center-to-center spacing greater than 200 GHz.

(3) **Dense WDMs (DWDMs)** are combined MUX and DEMUX devices operating with wavelength channel spacing not more than 200 GHz.

A typical WDM MUX/DEMUX include many characteristics but some of new parameters used to characterize these devices are :

- The **number of channels** usually describes the number of wavelength channels the device can multiplex/demultiplex.
- The **center wavelength (frequencies)** of these channels are given by the ITU-T grid.
- **Channel spacing** describes the minimum distance between channels a WDM MUX/ DEMUX can handle.
- The **bandwidth** is actually the linewidth of a specific wavelength channel.
- **Maximum insertion Loss,** it varies from channel to channel. Typical values range from 1.5 dB for a 4-channel device to 6 dB for a 32 channel MUX/DEMUX.
- **Isolation** values depend on the device's function. For a demultiplexer isolation is usually more than 30 dB while for a multiplexer it could be as low as 18 dB.

Working of WDM MUX/DEMUX

All MUXs and DEMUXs use dispersive elements, which make different wavelengths propagate at different directions. The same basic techniques – micro optic, integrated optic, and all fiber – that we have already described for the manufacture of couplers are used to fabricate WDM MUXs and DEMUXs.

12.5.1. All Fiber WDMs

Fused fibers, mainly FBT couplers are the examples of these devices. They can be used for broadband WDM. Now see the working of EDFA WDM couplers work.

let us consider the multiplexing of pumping light at 980 nm with a 1550 nm signal. The mode field diameter (MFD) depends on wavelength. **(The longer the wavelength, the larger the MFD).** Therefore choosing the distance between two cores, we can make light at 1550 nm – where the MFD is large – penetrate another core, while light at 980 nm – where the MFD is smaller – will stick to its core. This is shown in Fig. 12.10 (*a*) and (*b*).

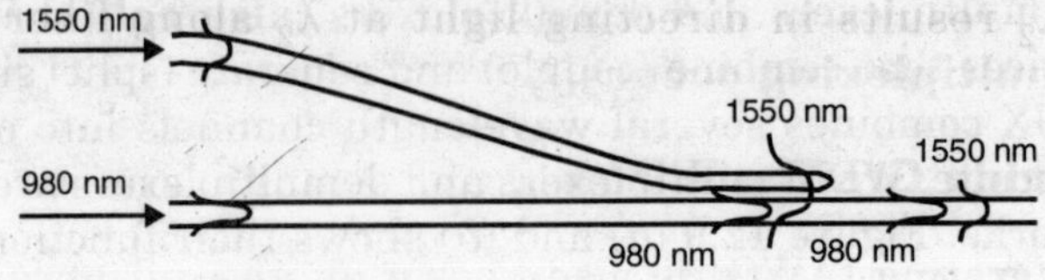

(a)

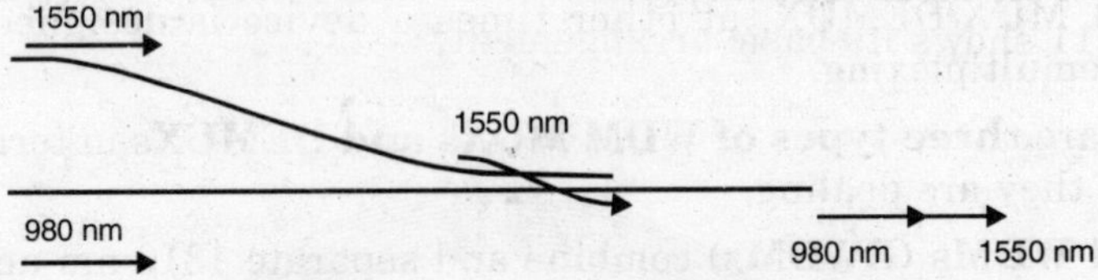

(b)

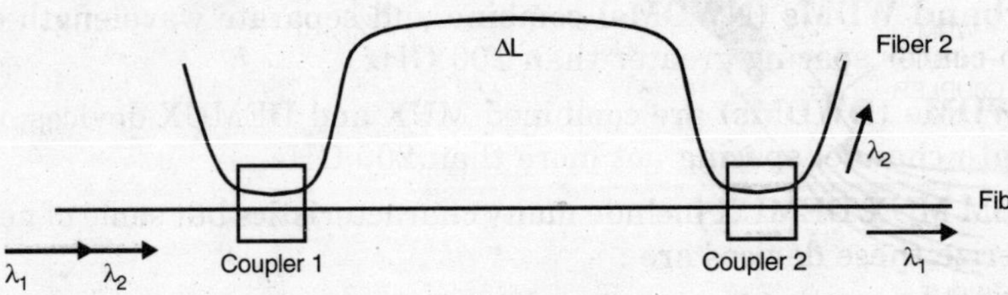

(c)

Fig. 12.10. All-Fiber Wavelength Division Multiplexer/Demultiplexer : (a) Coupling Process in EDFA WDM Coupler; (b) EDFA WDM Coupler Diagram; (c) WDM DEMUX Using Mach-Zehnder Interferometer.

The isolation of pump coupler typically ranges from 14 to 24 dB. The multipumping technique is used to increase pump power. Narrowband (NWDM) couplers can be used to accomplish this.

Another example of an all-fiber interferometer WDM coupler is a WDM MUX/DEMUX **based on the unbalanced Mach-Zehnder interferometer.** This structure is shown in Fig. 12.10 (*c*).

In this case not only the couplers but the entire structure provides excellent wavelength division multiplexing/demultiplexing. **Example :** The demultiplexing of the two wavelengths shown in Fig. 12.10 (*c*). The first coupler splits the input signal equally and directs it along two paths having different lengths. The longer arm of interferometer, having the additional length (ΔL), introduces an additional phase shift for both wavelengths. This phase shift can be calculated by :

$$\Delta Q_i = \frac{[2\pi n_{eff} . \Delta L]}{\lambda_i}$$

$$\boxed{\Delta Q_i = \beta \Delta L} \qquad \text{...(12.3)}$$

Here a key point is that a light wave acquires an additional phase shift and waves at different wavelength travelling the same extra distance (ΔL) experience different phase shift.

At second coupler, two beams at λ_1 that have travelled different distances constructively interfere with each other. This interference gives the result that the maximum intensity of light at λ_1 is directed along fiber 1. At the same time, at coupler 2, the interference of two beams at wavelength λ_2 results in directing light at λ_2 along fiber 2. **This is how wavelength division multiplexing occurs.**

12.5.2. Array Waveguide Grating WDMs

An array waveguide grating (AWG) is also known as **phased array waveguide** (or phaser). It is a type of off shoot of a Mach Zehnder interferometer but works like a diffraction grating. Figure 12.11 shows its basic arrangement.

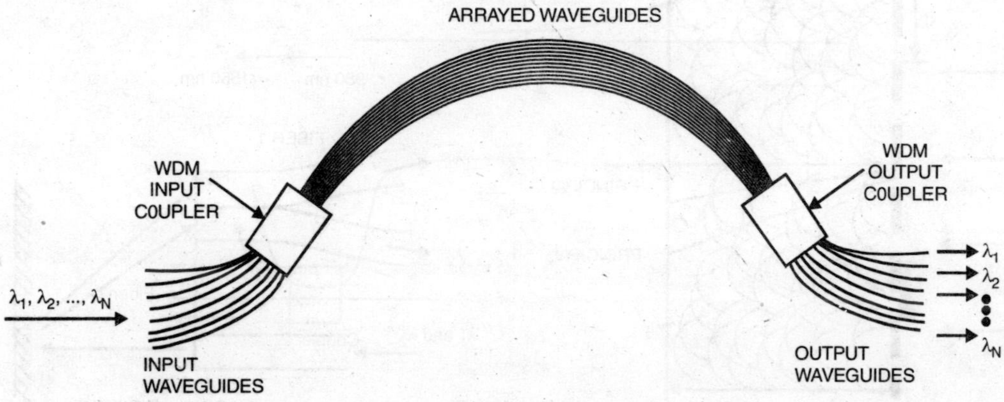

Fig. 12.11. Arrayed-Waveguide Grating (AWG).

Usually an AWG is fabricated as a planar structure. It consists of input and output waveguides, input and output WDM couplers and arrayed waveguides. The length of any array waveguide is distinguished from its adjacent waveguide by a constant (ΔL).

Wavelength channels enter the AWG, where an input WDM coupler splits them equally, among the arrayed waveguides. Each portion of the input light travelling through an arrayed waveguide includes all the wavelengths that have entered the device. Thus each wavelength acquires an individual phase shift. In addition each wavelength receives phase shifts at the input and output couplers. Therefore, every portion of light at a given wavelength acquires different phase shifts and all these portions interfere at the output coupler. The net result is a series or set of maximum light intensities. The direction of each maximum intensity depends on the wavelength. Thus each wavelength is directed into an individual fiber at the output of the device.

AWG MUXs and DEMUXs can combine and separate 48 channels (and more) and they provide multiplexing/demultiplexing of wavelength channels with spacing as low as 0.4 nm (50 GHz).

12.5.3. Diffraction Grating WDMs

It is a set of closely spaced slits. It can provide trnasmission or reflection of incident light. The distance between the slits (d) is known as **grating pitch (period).**

Let us see **how a diffraction grating works.** Figure 12.12(*a*) describes the principle of operation of a transmission diffraction grating.

Light from a light source (LED and LD) falls on the transmission diffraction grating as a plane wave. After passing through the individual grating slits, the light spreads in all directions. In Fig. 12.12, it is shown as a dotted semisphere. The interference of light with the same wavelength at the imaging plane results in a pattern of maximum and minimum intensity. The direction of the principal maximum intensity is given by

$$d \sin \Theta = m\lambda \qquad ...(12.4)$$

where $m = 0, \pm 1, \pm 2, \pm 3, ...$

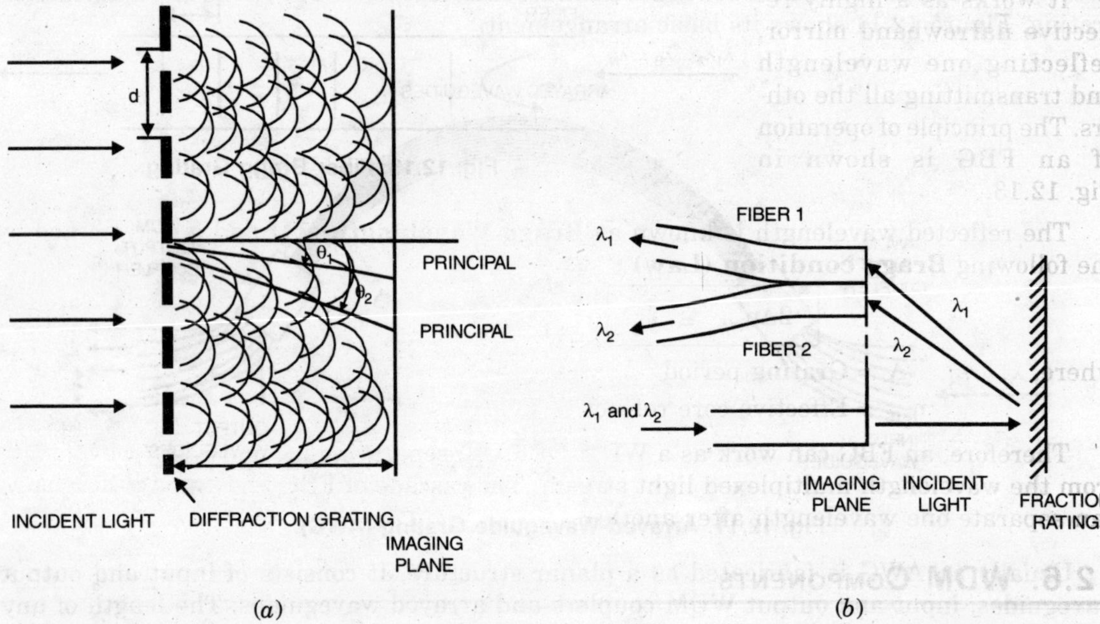

Fig. 12.12. Operation of Diffraction Grating : (a) Transmission Grating; (b) Reflection Grating.

Let us consider the first order principal maxima *i.e.* $m = 1$. Thus for wavelength λ_i, we get the following expression from above

$$\sin \Theta = \frac{\lambda_i}{d} \qquad ...(12.5)$$

It means that an individual wavelength has its principal maximum at a certain angle. In other words, the principal maxima of the different wavelengths are separated from one another by some angle. This is how a diffraction grating directs different wavelengths in different directions.

Angular separation is also known as angular dispersion and can be obtained from Eq. (12.5)

$$\frac{\Delta\Theta}{\Delta\lambda} = \frac{m}{d} \cos \Theta \qquad ...(12.6)$$

where, $\Delta\Theta$ = Angular sepration of two principal maximia for two wavelengths that differ by $\Delta\lambda$

Figure 12.12 (*b*) shows the operation of a reflective diffraction grating with concentration on the net result : light from an input fiber containing two wavelengths falls on the reflective diffraction grating and is separated at the output fiber ends. This separation is caused by the dispersive properties of the diffraction grating as described above.

12.5.4. Fiber Bragg Grating (FBG) WDMs

It is a further development of a diffraction grating concept. Basically the fiber Bragg grating is the diffraction grating implemented as a variation of the refractive index of the fiber core.

It works as a highly reflective narrowband mirror, reflecting one wavelength and transmitting all the others. The principle of operation of an FBG is shown in Fig. 12.13.

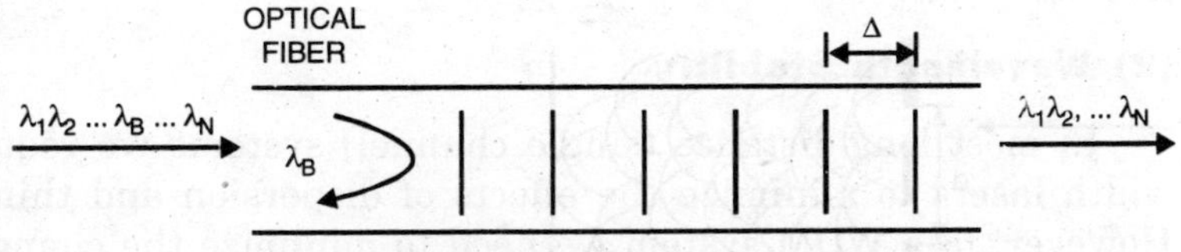

Fig. 12.13. Fiber Bragg Grating.

The reflected wavelength is known as **Bragg Wavelength (λ_B)** and is determined by the following **Bragg condition (Law) :**

$$2\Delta\eta_{eff} = \lambda_B \qquad \text{...(12.7)}$$

where, Δ = Grating period

η_{eff} = Effective core refractive index

Therefore, an FBG can work as a WDM DEMUX, separating an individual wavelength from the wavelength multiplexed light stream. The cascade of FBGs having a different Λ_B can separate one wavelength after another.

12.6. WDM Components

The implementation of WDM technology for fiber-optic communication systems requires several new optical components. The components are :

- **Multiplexers :** Which combine the output of several transmitters and launch it into an optical fiber.
- **Demultiplexers :** Which split the received multi-channel signal into individual channels destined to different receivers.
- **Star couplers :** Which mix the output of several transmitters and broadcast the mixed signal to multiple receivers.
- **Tunable optical filters :** Which filter out one channel at a specific wavelength that can be changed by tuning the pass band of the optical filter.
- **Multi wavelength optical transmitters :** Whose wavelength can be tuned over a few nanometers.
- **Add-drop multiplexers and WGRs :** Which can distribute the WDM signal to different ports.
- **Wavelength shifters :** Which switch the channel wavelength.

This section focuses on all such WDM components.

12.6.1. Light Sources for WDM

Lasers required for WDM systems are almost the same as lasers for ordinary long distance communication. However, some requirements are more critical with WDM and a number of new requirements become apparent.

(1) Spectral Width and Line Width

Generally in a dense WDM system laser is required with only one line in its spectrum. This will mean either a DFB or a DBR laser. The required line width will depend on the number of channels in the proposed system and the tolerance of other components (such as grating demultiplexers).

(2) Wavelength Stability

In most long-distance (single channel) systems we required very stable, narrow line width lasers to minimize the effects of dispersion and things like mode partition noise. However, in a WDM system we need to minimize the change in wavelength over time. A shift of a nm or two taking place over a few seconds might not bother a regular WAN single channel system but it would disrupt a WDM one.

(3) Tunable Lasers

Tunable lasers may be very important in future lightwave networks:

- In **Broadcast and Select LANs and MANs** fast tuning of either the transmitter or the receiver is essential to the overall performance of the system.
- In **wavelength-routed networks** (predominantly WANs) the sender will be told which channel to use (which wavelength) before a connection is established. It then needs to tune to that wavelength. In this case tuning need not be very fast but it should be very accurate.
- In most **WDM systems** tunability either at the transmitter or the receiver is not required. However, it is very difficult to manufacture lasers to an exact desired wavelength.

(4) Multi-Wavelength Lasers

One approach that allows very fast tuning is to build a number of lasers of different wavelengths together on the same substrate. Tuning can be done very quickly by selecting which laser is to transmit. Alternatively, multiple signals may be sent simultaneously.

12.6.2. Tunable Optical Filters

Optical filters are often the building blocks of more complex WDM components. The role of a tunable optical filter in a WDM system is to select a desired channel at the receiver.

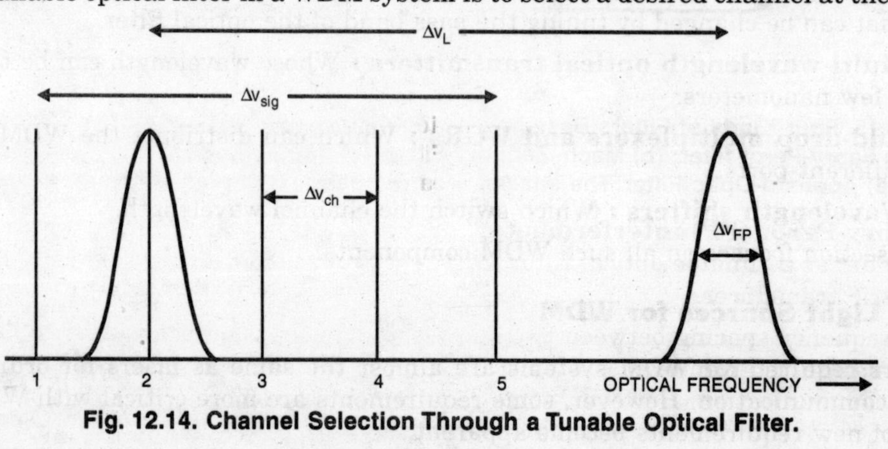

Fig. 12.14. Channel Selection Through a Tunable Optical Filter.

Figure 12.14 shows the selection mechanism. The filter bandwidth must be large enough to transmit the desired channel but, at the same time small enough to block the neighbouring channels.

All optical filters require a wavelength-selective mechanism and can be classified into two broad categories depending on the physical mechanism :

- Optical interference
- Diffraction.

Each category can be further subdivided according to the scheme adopted. Here we consider four kinds of optical filters; Fig. 12.15 shows an example of each kind.

The desirable properties of a tunable optical filter include:

(1) Wide tuning range to maximize the number of channels that can be selected.

(2) Negligible crosstalk to avoid interference from adjacent channels.

(3) Fast tuning speed to minimize the access time.

(4) Small insertion loss.

(5) Polarization insensitivity.

(6) Stability against environmental changes (humidity, temperature, vibrations, etc.).

(7) Low cost.

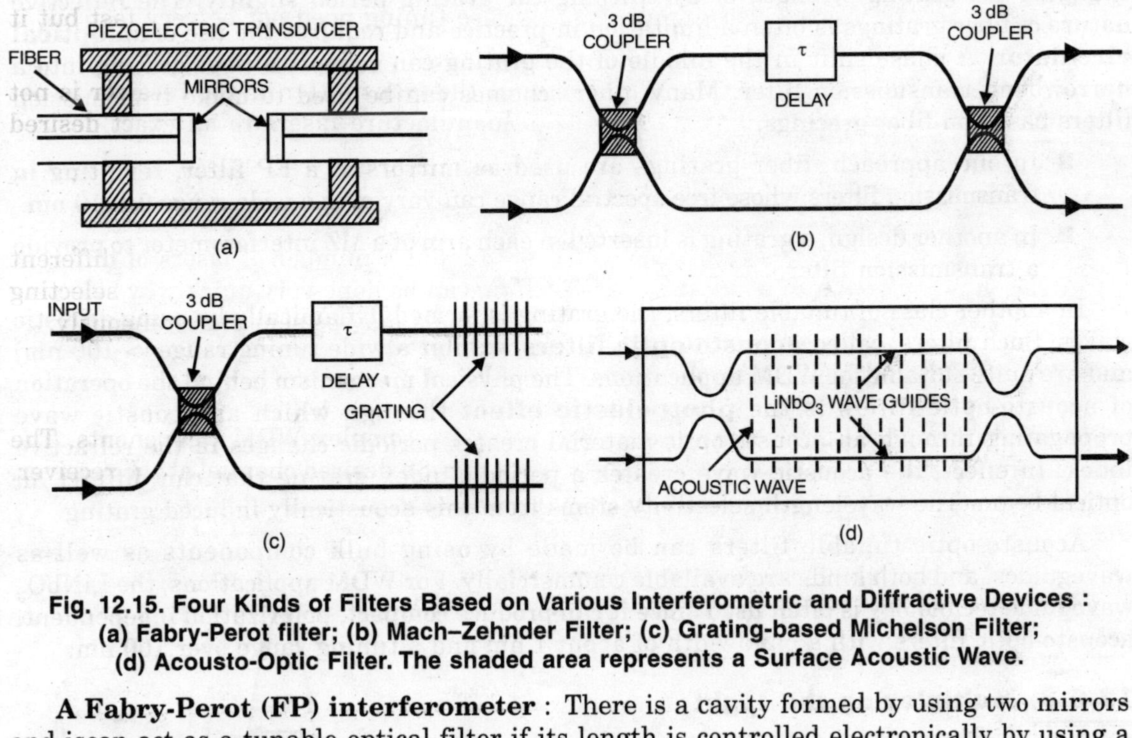

Fig. 12.15. Four Kinds of Filters Based on Various Interferometric and Diffractive Devices : (a) Fabry-Perot filter; (b) Mach–Zehnder Filter; (c) Grating-based Michelson Filter; (d) Acousto-Optic Filter. The shaded area represents a Surface Acoustic Wave.

A Fabry-Perot (FP) interferometer : There is a cavity formed by using two mirrors and iscan act as a tunable optical filter if its length is controlled electronically by using a piezoelectric transducer.

The frequency spacing between two successive transmission peaks, known as the **free spectral range**, is given by :

$$\Delta \varsigma L = \frac{c}{(2n_g L)}$$

where, n_g is the group index of the intra cavity material for a FP filter of length L.

Mach-Zehnder (MZ) Interferometers : A chain of M2 in ferometer can also be used for making a tunable optical filter. It can be constructed simply by connecting the two output ports of a 3 dB coupler to the two input ports of another 3 dB coupler. The first coupler splits the input signal equally into two parts, which acquire different phase shifts (if the arm lengths are made different) before they interfere at the second coupler. Since the relative phase shift is wavelength dependent, the **transmittivity** (T) is also wavelength dependent.

Other Kinds of Interferometers : Such as the **Sagnac** and **Michelson interferometers**, can also be used to realize transmission filters. Figure 12.7(c) shows an example of the Michelson interferometer made by using a 3 dB fiber coupler and two fiber gratings acting as mirrors for the two arms of the Michelson interferometer.

Another class of tunable optical filters makes use of the wavelength selectivity provided by a **Bragg grating**. Fiber Bragg gratings provide a simple example of grating-based optical filters. In its simplest form a fiber grating acts as a reflection filter whose central wavelength can be controlled by changing the grating period and whose bandwidth can be tailored by changing the grating strength or by chirping the grating period slightly. The reflective nature of fiber gratings is often a limitation in practice and requires the use of an **optical circulator**. A phase shift in the middle of the grating can convert a fiber grating into a narrowband transmission filter. Many other schemes can be used to make transmission filters based on fiber gratings.

- In one approach, fiber gratings are used as mirrors of a FP filter, resulting in transmission filters whose free spectral range can vary over a wide range 0.1–10 nm.
- In another design, a grating is inserted in each arm of a MZ interferometer to provide a transmission filter.

In another class of tunable filters, the grating is formed dynamically by using acoustic waves. Such filters, called **acousto-optic filters**, exhibit a wide tuning range (> 100 nm) and are quite suitable for WDM applications. The physical mechanism behind the operation of acousto-optic filters is the **photoelastic effect** through which an acoustic wave propagating through an acousto-optic material creates periodic changes in the refractive index. In effect, the acoustic wave creates a periodic index grating that can diffract an optical beam. The wavelength selectivity stems from this acoustically induced grating.

Acousto-optic tunable filters can be made by using bulk components as well as waveguides, and both kinds are available commercially. For WDM applications, the $LiNbO_3$ waveguide technology is often used since it can produce compact, polarization independent, acousto-optic filters with a bandwidth of about 1 nm and a tuning range over 100 nm.

12.6.3. Multiplexing the Light

Superficially, combining a number of different light sources into a single stream should be easy. Electronically we could join many different signal sources together very simply and almost without loss of any signal. Splitters and Couplers combine the light in the same way as required but every time we have to join two fibers in a coupler and we lose half the light.

If we combine eight signals on different fibers onto a single fiber using simple couplers then each individual channel in the combined output will be one eighth of the strength it started out at (a loss of 9 dB). If we combine 64 signals this way then each channel ends up at one-sixty-fourth of its original strength (18 dB of loss).

We can use **Littrow gratings** by assuming that these signals are all on different wavelengths, Array Waveguide Gratings (AWGs) and similar devices to integrate the signal with a much lower loss. A typical commercial Littrow grating combining 32 channels has a loss (per channel) of around 6 dB (3/4 of each signal is lost). AWGs are quoted with total loss levels of around 5 dB for devices with up to 64 channels.

In systems with only a few channels (say four), we can use simple couplers. If larger numbers of channels are involved then there is a requirement of grating device or an amplifier after the multiplexer.

12.6.4. Demultiplexing the Light

There are three basic approaches to demultiplexing:

1. Split the mixed light up into many mixed outputs (one per required output port) and then filter each port individually.
2. Split off a single channel at a time.
3. Demultiplex the whole bundle of optical channels in one operation.

The following are examples of each approach.

12.6.4.1. 3 dB Splitter Array with Fabry-Perot Filters

In this configuration cascaded 3 dB splitters are used to divide the mixed signal up into as many equal outputs as necessary. Figure 12.8 shows an 8-port configuration. It is then necessary to separate each individual signal from the others. In this case the separation is achieved with separate Fabry-Perot filters.

This configuration has been used in research systems but it has a number of disadvantages:

- Loss of signal power in the system accumulates very quickly indeed. In the figure the total loss (if the splitters are perfect) is 9 dB per port.

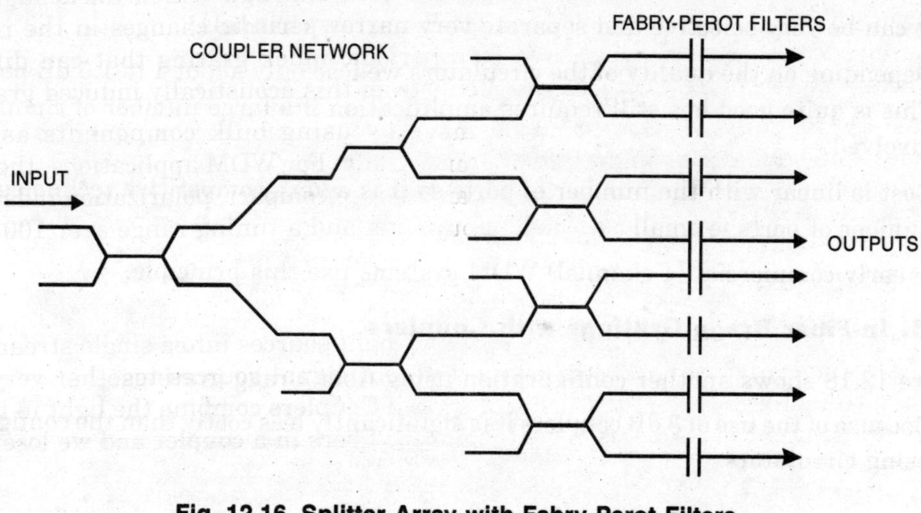

Fig. 12.16. Splitter Array with Fabry-Perot Filters.

- Built with fiber couplers this becomes messy and hard to manage. However, if built in planar technology this disadvantage is overcome.
- Cost tends to be linear with the number of ports. With a small number of ports this is an attractive approach. With a large number other approaches tend to be more cost-effective.

12.6.4.2. Circulators with FBGs

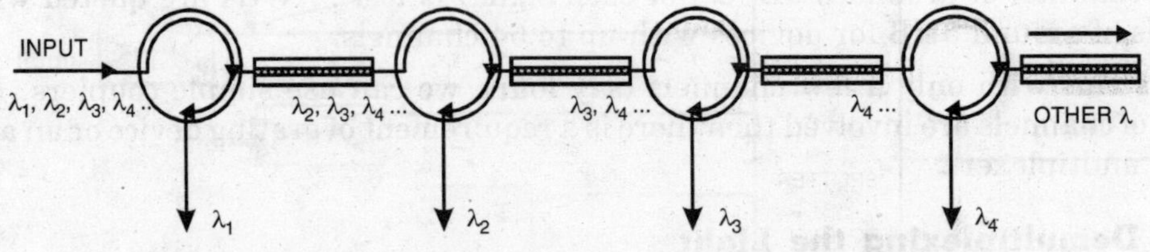

Fig. 12.17. Demultiplexor Using in-Fibre Bragg Gratings.

In this configuration each wavelength is separated from the multiplexed stream individually.

- Input consisting of many mixed wavelengths arrives.
- It enters the first circulator and is output into the first FBG.
- The FBG reflects the selected wavelength back to the circulator but allows all other wavelengths to pass. With many types of circulators this operation can involve an attenuation of 1 dB or less.
- The selected wavelength travels around the circulator to Port 3 where it is output.

All other wavelengths continue through the FBG to the next circulator where the process is repeated for l2. Individual wavelengths are demultiplexed as we proceed from one stage to another.

There are several considerations for this type of device :

- We can separate the circulators and FBGs so that each is located with its end device. The fiber is then connected in a ring configuration from device to device.
- It can be very selective and separate very narrow channels.
- Depending on the quality of the circulators we lose only about 1 to 1.5 dB per stage. This is quite good but still requires amplification if a large number of channels are involved.
- Cost is linear with the number of ports so it is a very competitive technology if the number of ports is small.

Some early commercial (4-channel) WDM systems use this principle.

12.6.4.3. In-Fiber Bragg Gratings with Couplers

Figure 12.18 shows another configuration using fiber Bragg gratings.

- Because of the use of 3 dB couplers it is significantly less costly than the configuration using circulators.

- In fact this configuration offers a narrowband filter with a "flat top" shape (ideal for WDM).
- The **major problem** is the extremely high insertion loss which is 6 + 10 log (n) dB per channel (where n = the number of channels).
- In **practical applications** the input would need to be amplified to a high level for the device to be considered practical.

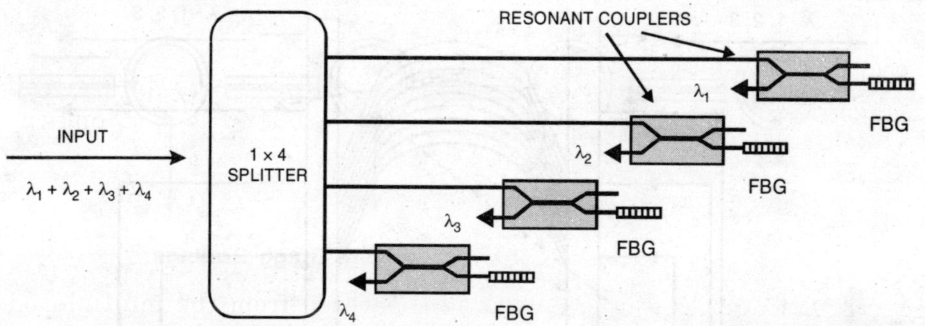

Fig. 12.18. Demultiplexer Using In-Fiber Bragg Gratings with 3 dB Couplers.

12.6.4.4. Littrow Gratings

Littrow gratings may be used for both multiplexing and demultiplexing. Commercially available in versions with up to 132 ports their main problem is cost. The mechanical problem of positioning many fibers very accurately means that these devices have a high cost. However if the number of ports required is large (20 or more) then Littrow gratings can be cost-competitive with other approaches.

The IBM 9729 WDM concentrator uses this type of grating because it can be bidirectional with input and output on the same fiber.

12.6.5. Add-Drop Multiplexers

An add-drop multiplexer adds and/or removes a single channel from a combined WDM signal without interfering with the other channels on the fiber. This function is shown in Fig. 12.19.

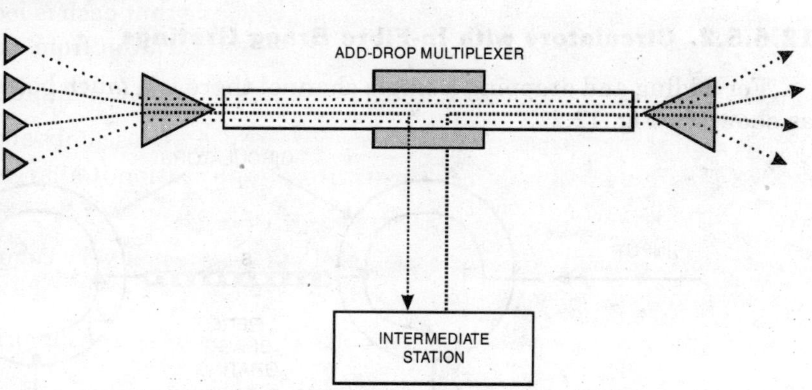

Fig. 12.19. Add-Drop Multiplexer Function.

There are several devices which may perform this function such as :

1. Array waveguide gratings
2. Circulators with FBGs
3. A Cascade of MZIs

Operation of these devices as add/drop multiplexers is described in the following sections.

12.6.5.1. Array Waveguide Gratings

In Fig. 12.20, a WGR configured as an add-drop multiplexer is shown.

- Wavelength λ_1 is added to the multiplexed stream on the left of the figure and dropped (demultiplexed) from the input stream on the right of the figure.

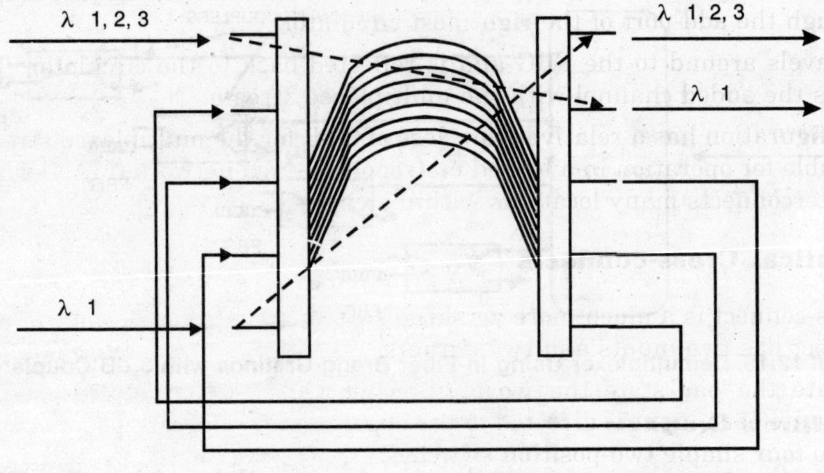

Fig. 12.20. Array Waveguide Add-Drop Multiplexer.

- This is a very versatile function.
- For example multiple channels can be added or dropped in the same operation.
- However, the signal loss is about 5 dB per pass through the device. So channels that are not added or dropped experience an insertion loss of around 10 dB.
- In a system we require to equalize the signal power between channels (so that the newly added channels are about the same strength as the dropped ones) and probably to amplify the whole stream as well.

12.6.5.2. Circulators with In-Fibre Bragg Gratings

For adding and dropping a single channel there is a much better configuration available as shown in Fig. 12.21.

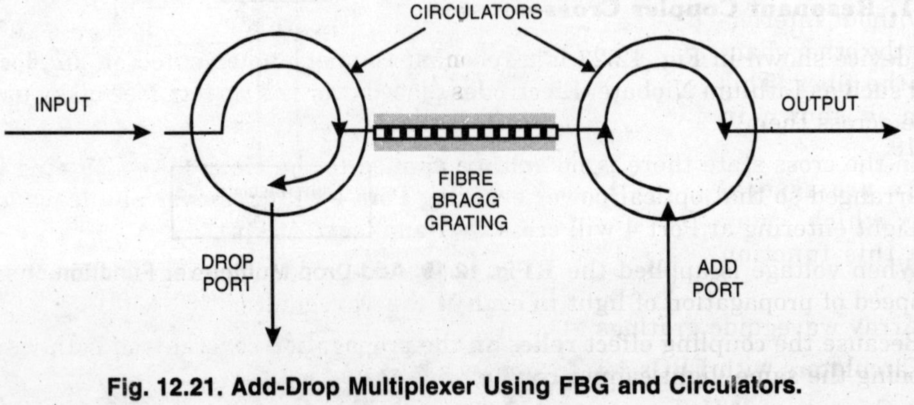

Fig. 12.21. Add-Drop Multiplexer Using FBG and Circulators.

Here we are using an FBG with a pair of circulators to add and drop a single channel. Operation is as follows:

- The signal enters at the left of the figure and is routed through the circulator to the FBG.
- The non-selected wavelengths pass through the FBG to the next circulator.
- The selected wavelength is reflected by the FBG and then directed out of the next circulator port.
- The wavelength to be added (which must be the same as the one just dropped) enters through the add-port of the rightmost circulator.
- It travels around to the FBG and is reflected back to the circulator. This process mixes the added channel with the multiplexed stream.

This configuration has a relatively low loss of 3 dB for the multiplexed stream. It could be very suitable for operation in a looped metropolitan area network (MAN) where a single fiber loop interconnects many locations within a city area.

12.6.6. Optical Cross-connects

The cross-connect is a much more versatile switching element. As shown in Fig. 12.22, a cross-connect has two inputs and two outputs.

In one state (the "bar" state) the two inputs connect directly to two corresponding outputs. In the cross state each input is directed to the opposite output. A cross-connect can take the place of up to four simple two-position switches.

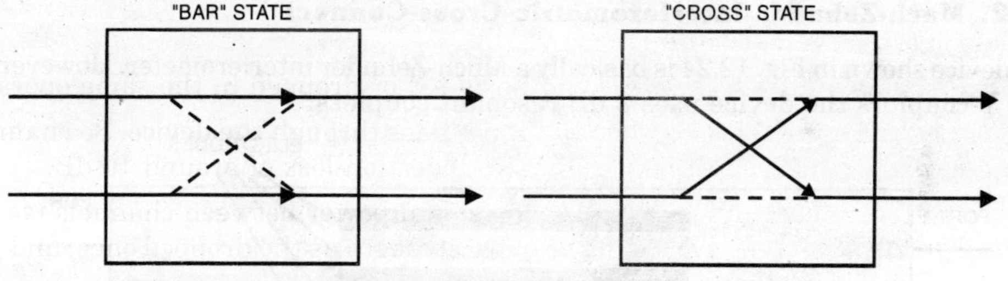

Fig. 12.22. Cross-connect Switch Element - Function.

Cross-connects can be realized in optical technology in a number of ways, two of which are described here:

12.6.6.1. Resonant Coupler Cross-connect

The device shown in Fig. 12.23 is a resonant coupler implemented on an electro optic material such as Lithium Niobate. Electrodes placed over the waveguides allow us to apply a voltage across them.

- In the cross state there is no voltage applied to the electrodes. Coupling length is arranged so that optical power entering Port 1 will crossover and leave at Port 3. Light entering at Port 4 will cross over and leave at Port 2.
- When voltage is applied the RI in both waveguides is changed. This changes the speed of propagation of light in each of the waveguides.
- Because the coupling effect relies on the propagation constants of both waveguides being the same, light cannot couple.

- Thus when voltage is applied light entering Port 1 leaves at Port 2 and light entering at Port 4 leaves at Port 3. This is the bar state.
- The **major difficulty** with this device is that it requires extreme accuracy in device fabrication.
- Coupling length is critical and has very little tolerance. The RI of the waveguide is likewise critical.
- In the right hand part of Fig. 12.23 a modification of the cross-connect is shown which makes the fabrication tolerances significantly less critical. In this case one of the electrodes is built in sections. Each section has an equal but opposite voltage on it.

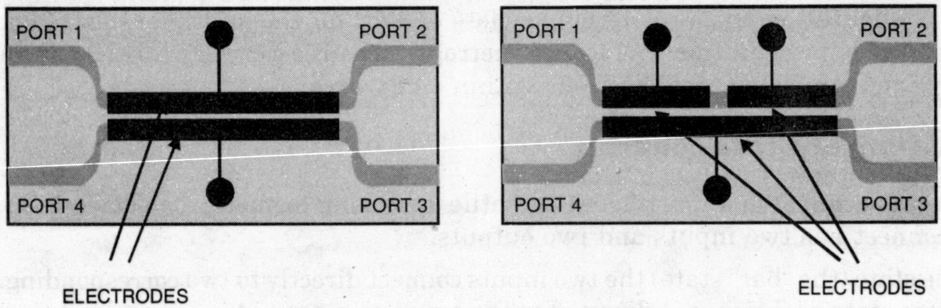

Fig. 12.23. Cross-Connect Using a Resonant Coupler.

12.6.6.2. Mach-Zehnder Interferometric Cross-Connect

The device shown in Fig. 12.24 is basically a Mach-Zehnder interferometer. However, instead of using Y-couplers the device uses 3 dB resonant couplers.

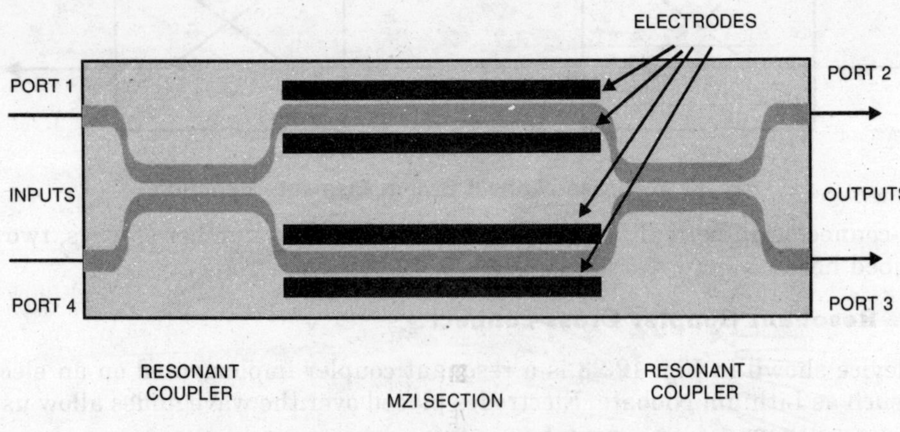

Fig. 12.24. Cross-Connect Using an MZI Configuration.

- When light is input at Port 1, it is split evenly between both arms of the interferometer.
- If there is no voltage applied the light arrives at the second 3 dB coupler in phase (since the arms of the interferometer are equal in length).
- The second coupler then acts as a continuation of the first and all of the light is output at Port 3.

- Likewise light entering at Port 4 will be output at Port 2. This is the cross **state.**
- When a voltage is applied across the electrodes, the RI in each waveguide is changed.
- The voltage is arranged to be such that the change in RI results in the light on the two arms arriving at the second coupler 180° out of phase with each other.
- There is now destructive interference and light which entered at Port 1 will now exit at Port 2. Light entering at Port 4 will now leave at Port 3.

12.6.7. Wavelength Routers

An important WDM component is an $N \times N$ wavelength router. It is a device that combines the functionality of a star coupler with multiplexing and demultiplexing operations. Figure 12.25(a) shows the operation of such a wavelength router schematically for $N = 5$.

- The WDM signals entering from N input ports are demultiplexed into individual channels and directed toward the N output ports of the router in such a way that the WDM signal at each port is composed of channels entering at different input ports.
- This operation results in a cyclic form of demultiplexing. Such a device is an example of a passive router since its use does not involve any active element requiring electrical power. It is also called a **static router**.

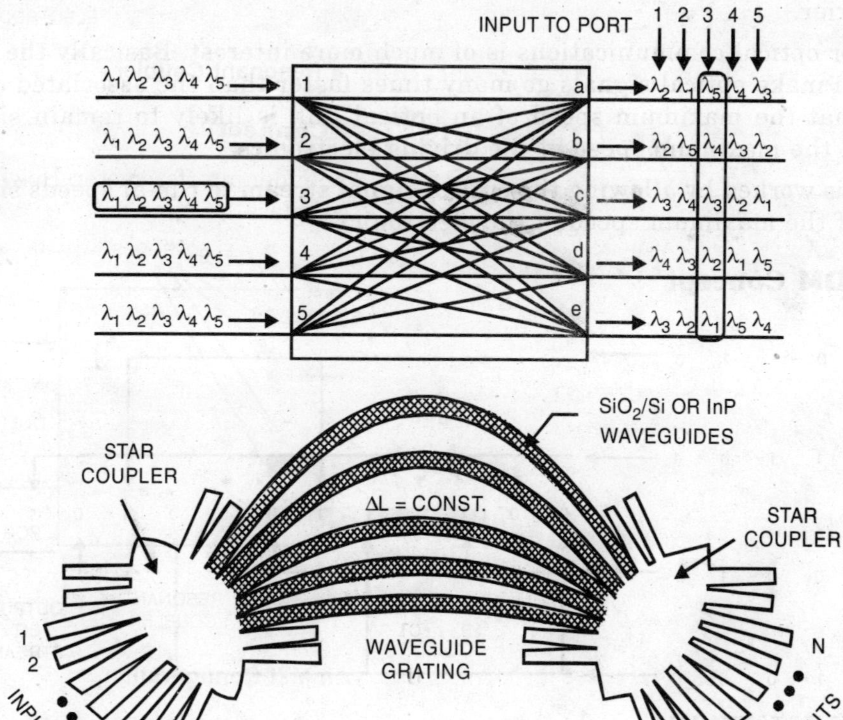

Fig. 12.25. (a) Schematic Illustration of a Wavelength Router (b) Its Implementation Using an AWG.

The most common design of a wavelength router uses a AWG demultiplexer modified to provide multiple input ports. Such a device, called the **waveguide-grating router (WGR)**, is shown schematically in Fig. 12.25(b).

- It consists of two $N \times M$ star couplers such that M output ports of one star coupler are connected with M input ports of another star coupler through an array of M waveguides that acts as an AWG.
- Such a device is a generalization of the MZ interferometer in the sense that a single input is divided coherently into M parts (rather than two), which acquire different phase shifts and interfere in the second free-propagation region to come out of N different ports depending on their wavelengths.
- The symmetric nature of the WGR permits to launch N WDM signals containing N different wavelengths simultaneously, and each WDM signal is demultiplexed to N output ports in a periodic fashion.

12.7. Optical Time Division Multiplexing (OTDM)

Time Division Multiplexing (TDM) is a time-honoured technique in electronic communications. It provides a very simple and effective way of sub-dividing a high-speed digital data stream into many slower speed data streams.

Indeed most optical communications links are really TDM data streams but the TDM is done electronically rather than optically and the optical transmission system processes a single high-speed stream. SDH and SONET are standards for electronic TDM over an optical carrier.

TDM for optical communications is of much more interest. Basically the problem is that we can make optical signals go many times faster than the associated electronics. It seems that the maximum speed of an optical link is likely to remain significantly faster than the maximum speed of the driving electronics.

OTDM is worked by allowing the optical signal stream to run at speeds significantly in excess of the maximum speed of the electronics.

12.7.1. TDM Concept

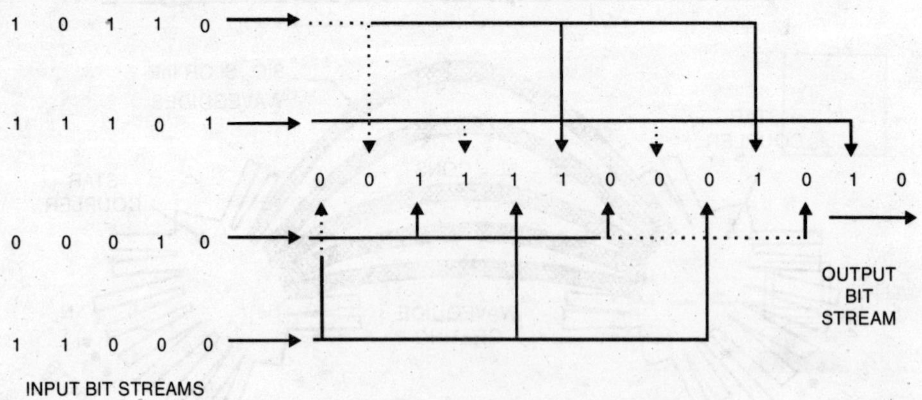

Fig. 12.26. Time Division Multiplexing (TDM)—Concept.

Figure 12.26 shows the principle of time division multiplexing. In the description there are four slow-speed bit streams merged into a single high speed stream at four times the speed of any one of the component signals. Each input stream is assigned one bit in every four in turn. There are a number of important points:

- In the illustration we are allocating places (time slots) in the high-speed data stream at the individual bit level. However, optical TDM proposed systems use the bit-interleaving technique exclusively.
- The data stream is arranged in repeating patterns of time slots usually called frames. In the example a frame would be just four bits. Thus input channel x might be allocated bit number 3 in every frame.
- It is not necessary for each of the slow-speed streams to be the same. For example: we could TDM three signals at different rates by allocating a different number of bits in each time frame to each stream. Thus input stream 1 might be allocated bit numbers 1, 3, 5, 7..., stream 2 might be allocated bits 2, 6, 10... and stream 3 bits 4, 8, 12... In this example stream 1 would be twice the rate of either stream 2 or stream 3.
- There is very little delay experienced by the slow speed streams due to their travel over the higher speed trunk.
- Once the time slots are allocated each subordinate signal stream has a fixed and invariant data rate. This can be changed by reallocating the bits but this is difficult to do dynamically, takes time and wastes resources.
- Each signal stream must be synchronized to the higher speed stream. This is the most significant problem in TDM.

12.7.2. TDM Network Principles

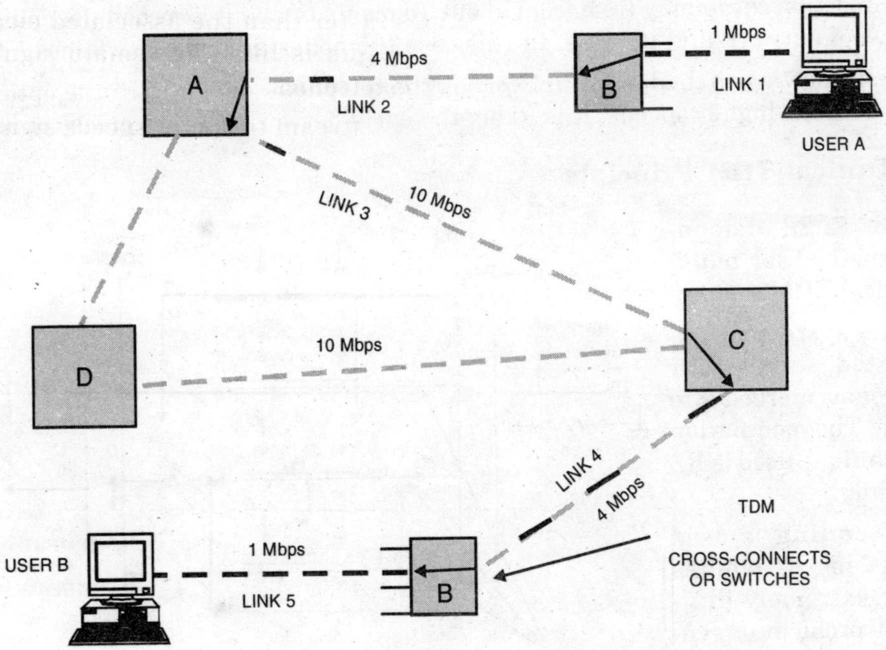

Fig. 12.27. Principle of a TDM Network.

Figure 12.27 shows general principle of a TDM network. Let us assume that data is multiplexed in units of a single byte. In the figure we have illustrated a 1 Mbps synchronous connection between the two end users (User A and User B). The network is configured as follows :

- User A is connected on a dedicated link to Node B at a speed of 1 Mbps (125,000 bytes/sec). The timing for the link is provided by Node B not by the end user. This means that Node B sends a clock signal to User A each time that a bit is to be sent.
- Node B has been set up with a rule that says whenever a byte is received on Link 1 place it into time-slot x of Link 2.
- Node B has a connection with Node A at a speed of 4 Mbps. Our connection between end users is allocated to this link and so gets every fourth byte on Link 2.
- Node A has a rule that says whenever time-slot x of Link 2 arrives take the data in it and place it into time-slot y of Link 3.
- Node A to Node C is a 10 Mbps link which will also carry our end-user connection. Thus only 1 byte in 10 on the link will belong to this particular end-user connection.
- Node C to Node E is at 4 Mbps (same as B to A) and again we get every fourth byte.
- Node E is connected directly to User B at a dedicated speed of 1 Mbps. Clocking for this link is again provided by the network not the end user.
- Each connection is bi-directional.

Connections can be set up in two ways :

1. **By Network Management :** In this case the node is called a cross-connect. Connection setup or tear-down may take from a few minutes to a few days.
2. **By Signalling :** Signalling means by request in real-time by the end user. Connection setup may take from about 10 ms up to about a second depending on the type of network and it's size. In this case the nodes are called switches.

The best known switched TDM network is the telephone network. In the optical world, operating at much higher speeds, it is likely the early networks will be cross-connects only.

12.7.3. Optical TDM Principle

Figure 12.28 demonstrates a method for building an optical TDM system.

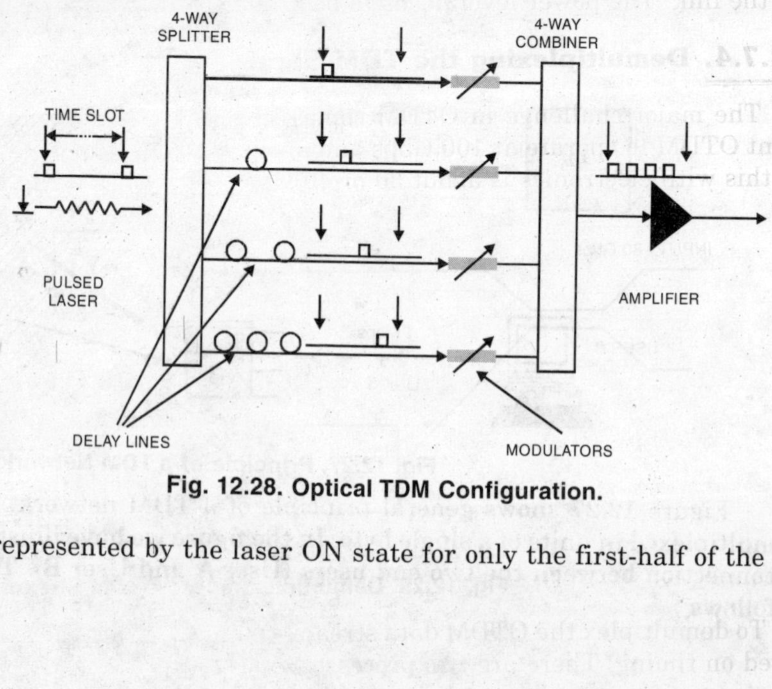

Fig. 12.28. Optical TDM Configuration.

- The system illustrated shows four streams merged into one. The modulation technique used is RZ coding.
- **RZ coding** is used because it avoides the extremely difficult problem of synchronising different bit streams into adjacent time slots. (In RZ coding a 1-bit is represented by the laser ON state for only the first-half of the bit-time.)

- There will always be some jitter in the slow stream bit stream as it is mixed into the faster stream.
- In addition any optical pulse will be bell shaped rather than square. Thus no matter what we will do there for gaps between the bits.

The system operates in the following way :

1. Each time slot is sub-divided into 4 bit times.
2. Each bit-time is further divided into two halves. For a "1" bit the first half of the bit time will be occupied by an optical pulse (and the second half will be dark). For a "0" bit the whole bit time will be dark.
3. A laser produces a short pulse (for half a bit time) at the beginning of each time slot. In this example the laser is ON for one-eighth of the time slot.
4. The laser signal is split 4-ways. A planar free-space coupler will also do this and would be used if the whole TDM device were built on a single planar substrate.
5. Each signal (except one) is then delayed by a fixed amount. This delay is easily and conveniently provided by using a loop of standard fibre. Of course each signal is delayed by a different amount.
6. Then each signal is separately modulated to carry it's own unique information stream. The trick here is to synchronise the modulators accurately given that their response will be much slower than a single bit time.
7. The signals are then re-combined to form a single data stream.
8. During all this the original signal has lost a very large amount of power. Each pulse will lose a minimum of 6 dB in each of the 4-way splitter and the combiner. In addition there will be loss in the modulator.

The whole stream then must be amplified to reach a strength suitable for transmission on the link. The power level needs to be around 3 mw or above for a soliton to form.

12.7.4. Demultiplexing the TDM Signal

The major challenge in OTDM comes in receiving and demultiplexing the signal. We want OTDM to operate at 100 Gbps and above, and currently the maximum speed we could do this with electronics is about 30 or 40 Gbps. Therefore it can be done optically.

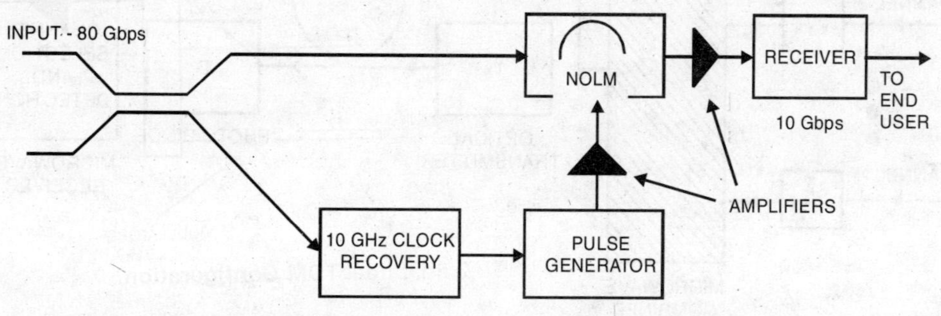

Fig. 12.29. Demultiplexing the OTDM Stream.

To demultiplex the OTDM data stream we must extract each slow(er) speed data stream based on timing. There are two processes :

Extracting the Data

Data for a single channel may be extracted as shown in Fig. 12.29. A Non-linear Optical Loop Mirror (NOLM) is used as an AND gate to select every n^{th} pulse (in the figure n = 8) from the TDM stream.

Add/Drop Multiplexing

A device capable of being either an add/drop multiplexor or a simple demultiplexor is discussed in section 12.4.5.

Extracting the Timing

Each bit in the higher speed signal is identified by its arrival time. Thus the first problem is to derive an electronic clock signal (at the slower rate of one of the component streams) and ensure that it is accurately synchronised to the incoming bit stream. This is the most difficult aspect of OTDM.

12.8. SUBCARRIER MULTIPLEXING

In some LAN and MAN applications the bit rate of each channel should be relatively low but the number of channels can become quite large. Example: common-antenna (cable) television (CATV) networks that have used historically electrical communication techniques.

The basic concept behind **subcarrier multiplexing (SCM)** is borrowed from microwave technology, which employs multiple microwave carriers for transmission of multiple channels (electrical FDM) over coaxial cables or free space. The total bandwidth is limited to below 1 GHz when coaxial cables are used to transmit a multichannel microwave signal. However, if the microwave signal is transmitted optically by using optical fibers, the signal bandwidth can easily exceed 10 GHz for a single optical carrier. Such a scheme is referred to as SCM, since multiplexing is done by using microwave subcarriers rather than the optical carrier.

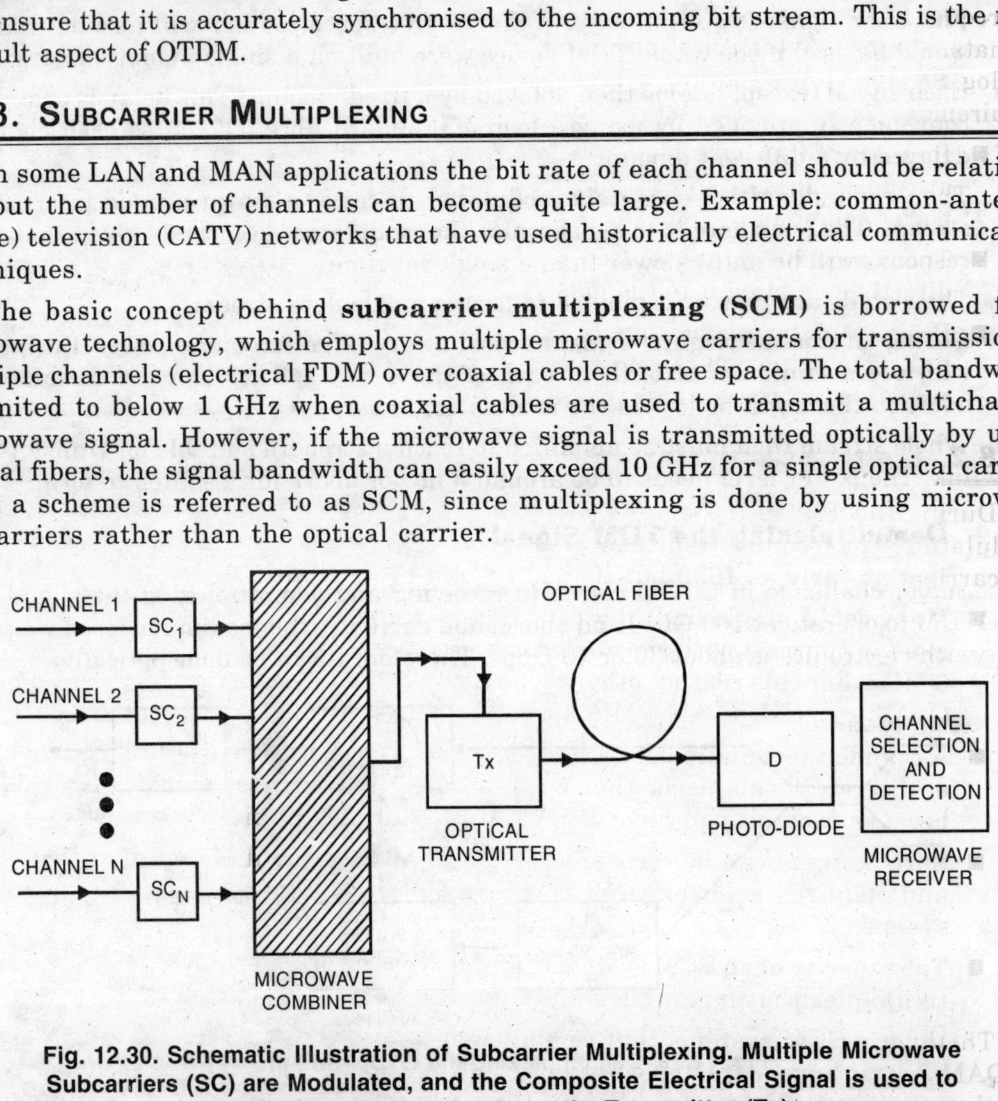

Fig. 12.30. Schematic Illustration of Subcarrier Multiplexing. Multiple Microwave Subcarriers (SC) are Modulated, and the Composite Electrical Signal is used to Modulate an Optical Carrier at the Transmitter (Tx).

Figure 12.30 shows schematically a SCM lightwave system designed with a single optical carrier. The main advantage of SCM in the design of broadband networks is:

- Flexibility
- Upgradability

One can use analog or digital modulation, or a combination of the two, to transmit multiple voice, data, and video signals to a large number of users. Each user can be served by a single subcarrier, or the multichannel signal can be distributed to all users as done commonly by the CATV industry. In this section we describe both the analog and digital SCM systems with emphasis on their design and performance.

12.8.1. Analog SCM Systems

Most CATV networks distribute television channels by using analog techniques based on frequency modulation (FM) or amplitude modulation with vestigial sideband (AM-VSB) formats. As the wave form of an analog signal must be preserved during transmission, analog SCM systems require a high SNR at the receiver and impose strict linearity requirements on the optical source and the communication channel.

- In analog SCM lightwave systems, each microwave subcarrier is modulated using an analog format, and the output of all subcarriers is summed using a microwave power combiner.
- The composite signal is used to modulate the intensity of a semiconductor laser directly by adding it to the bias current.
- Practically the analog signal is distorted during its transmission through the fiber link. The distortion is referred to as **intermodulation distortion (IMD)** and is similar in nature to the FWM distortion.

12.8.2. Digital SCM Systems

During the 1990s, the emphasis of SCM systems shifted from analog to digital modulation. The frequency-shift keying (FSK) format was used for modulating microwave subcarriers as early as 1990 but its use requires coherent detection techniques.

- Moreover, a single digital video channel requires a bit rate of more 100 Mb/s or more in contrast with the analog channel that occupies a bandwidth of only about 6 MHz. For this reason, other modulation formats such as quadrature AM (called QAM), carrierless AM/PM, and quadrature PSK have been explored.
- A common technique uses a multilevel QAM format. If M represents the number of discrete levels used, the resulting nonbinary digital signal is called **M-ary** because each bit can have M possible amplitudes (typically $M = 64$).
- Such a signal can be recovered at the receiver without using coherent detection and requires a lower CNR compared with that needed for analog AM-VSB systems.
- The capacity of an SCM system can be increased considerably by employing hybrid techniques that mix analog and digital formats.

The hybrid SCM systems that combine the analog AM-VSB format with the digital M-QAM format have attracted considerable attention because they can transmit a large number of video channels over the same fiber simultaneously. The performance of such systems is affected by many reasons like :

- Clipping noise
- Multiple optical reflections
- The nonlinear mechanisms such as self-phase modulation (SPM) and SBS.

All of which limit the total power and the number of channels that can be multiplexed. Nevertheless, hybrid SCM systems can transport up to 80 analog and 30 digital channels using a single optical transmitter. If only QAM format is employed, the number of digital channels is limited to about 80.

12.8.3. Multiwavelength SCM Systems

The combination of WDM and SCM provides the potential of designing broadband passive optical networks capable of providing integrated services (audio, video, data, etc.) to a large number of subscribers. This scheme is shown in Fig. 12.31 where multiple optical carriers are launched into the same optical fiber through theWDM technique.

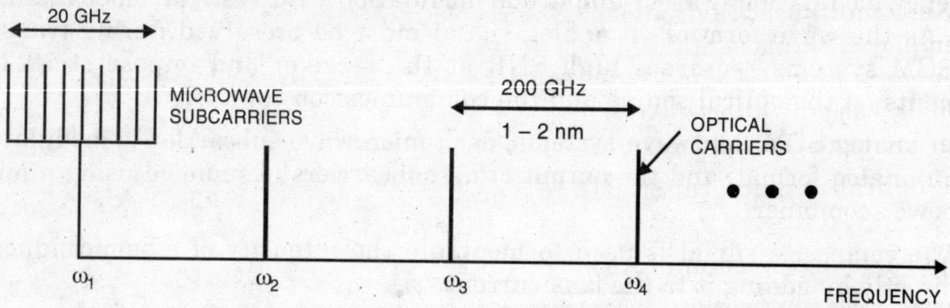

Fig. 12.31. Frequency Allocation in a Multiwavelength SCM Network.

- Each optical carrier carries multiple SCM channels using several microwave subcarriers.
- One can mix analog and digital signals using different subcarriers or different optical carriers. Such networks are extremely flexible and easy to upgrade as the demand grows.
- As early as 1990, 16 DFB lasers with a wavelength spacing of 2 nm in the 1.55-mm region were modulated with 100 analog video channels and six 622-Mb/s digital channels.
- Video channels were multiplexed using the SCM technique such that one DFB laser carried 10 SCM channels over the bandwidth 300–700 MHz.
- The limiting factor for multiwavelength SCM networks is **interchannel crosstalk** resulting from both the linear and nonlinear processes. The nonlinear effects that produce interchannel crosstalk are SRS and XPM, both of which have been analyzed.
- The **linear crosstalk** results from the phenomenon of optical beat interference. It occurs when two or more users transmit simultaneously on the same optical channel using different subcarrier frequencies.
- Multiwavelength SCM systems are quite useful for LAN and MAN applications. They can provide multiple services (telephone, analog and digital TV channels, computer data, etc.) with only one optical transmitter and one optical receiver per user because different services can use different microwave subcarriers. This approach lowers the cost of terminal equipment in access networks.

- The main **advantage** of multiwavelength SCM is that the network can serve *NM* users, where *N* is the number of optical wavelengths and *M* is the number of microwave carriers by using only *N* distinct transmitter wavelengths.

12.9. Optical Code Division Multiplexing (OTDM)

A multiplexing scheme well known in the domain of wireless communications makes use of the **spread-spectrum technique**. It is referred to as **code-division multiplexing (CDM)** because each channel is coded in such a way that its spectrum spreads over a much wider region than occupied by the original signal.

Although spectrum spreading may appear counter-intuitive from a spectral point of view but this is not the case because all users share the same spectrum. In fact, CDM is used extensively in microwave communications (e.g., cell phones) as it provides the most flexibility in a multi user environment. The term **code-division multiple access (CDMA)** is often employed in place of CDM to emphasize the asynchronous and random nature of multi user connections. The use of CDMA for fiber-optic communications has been studied since 1986, it was only after 1995 that the technique of optical CDM was pursued seriously as an alternative to OTDM. It can be easily combined with the WDM technique.

In the optical communication CDMA has been suggested as a possible technology in two roles :

1. Optical shared medium LANs
2. Local access networks

In most communications systems we try to maximize the amount of useful signal that can be fit into a minimum bandwidth. In spread spectrum we try to artificially spread a signal over a bandwidth much wider than required. In CDMA we transmit multiple signals over the same frequency band, using the same modulation techniques at the same time.

In a spread spectrum system we use some artificial technique to broaden the amount of bandwidth used. This has the following effects :

(1) Capacity Gain

Using the Shannon-Hartley law for the capacity of a band limited channel it is easy to see that for a given signal power the wider the bandwidth used, the greater the channel capacity. So if we broaden the spectrum of a given signal we get an increase in channel capacity and/or an improvement in the signal-to-noise ratio.

(2) Interference Rejection

Spread spectrum signals can be received even in the presence of very strong narrowband interfering signals (up to perhaps 30 dB above the wanted signal).

(3) Multiplexing Technique (CDMA)

Some techniques of frequency spreading enable the transmission of many completely separate and unrelated channels on the same frequency and at the same time as other, similar signals.

(4) Security

Spread spectrum was invented by military communications people for the purpose of battlefield communications. Spread spectrum signals have an excellent rejection of intentional jamming.

(5) Immunity to Multipath Distortion

Some spectrum spreading techniques have a significantly better performance in the presence of multipath spreading than any available narrowband technique.

12.9.1. Direct Sequence Spread Spectrum (DSSS)

Also known as **Pseudo Noise (PN),** DSSS is a popular technique for spreading the spectrum. Figure 12.32 shows how the signal is generated.

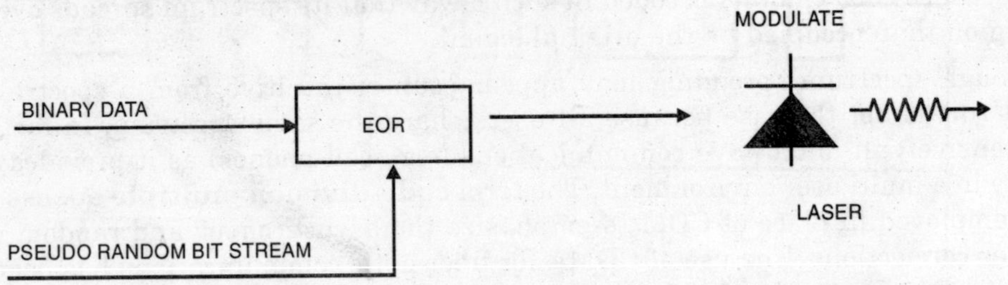

Fig. 12.32. Direct Sequence Spread Spectrum Modulation – Transmitter.

- The binary data stream (user data) is used to modulate a pseudo-random bit stream. The rate of this pseudo-random bit stream is much faster (from 9 to 100 times) as shown in Fig. 12.33 than the user data rate. The bits of the pseudo-random stream are called **chips.** The ratio between the speed of the chip stream and the data stream is called the **spread ratio.**
- The form of modulation used is typically just an EOR operation performed between the two bit streams.
- The output of the faster bit stream is used to modulate a radio frequency (RF) or optical carrier.
- Any suitable modulation technique could be used but :
  - In practical radio systems a very simple bipolar phase shift keying (BPSK) approach is usually adopted.
  - In optical systems NRZ coding is typically used.
- Whenever a carrier is modulated the result is a spread signal with two sidebands above and below the carrier frequency. These sidebands are spread over a range plus or minus the modulating frequency. The sidebands carry the information and it is common to suppress the transmission of the carrier (and sometimes one of the sidebands). It can be easily seen that **the width (spread) of each sideband has been multiplied by the spread ratio.**
- The **advantage of DSSS** is in the way the signal is received. The receiver knows the pseudo-random bit stream (because it has the same random number generator). Incoming signals in the analogue processing stage are correlated with the known pseudo-random stream. Thus the chip stream performs the function of a known waveform against which we correlate the input.
- **Correlational receivers** can be constructed in several ways but they almost always take the form of a tapped delay line. There are many ways to do this. **Two examples are as follows :**

1. In an optical device we might detect the optical signal first, and then use it to drive an electronic Surface Acoustic Wave (SAW) Filter. This works but requires a receiver with a very high frequency response since it must respond to the chip rate rather than the bit rate.
2. All-optical correlational filters are possible and have the great advantage that the electronics associated with them don't need to be very fast.

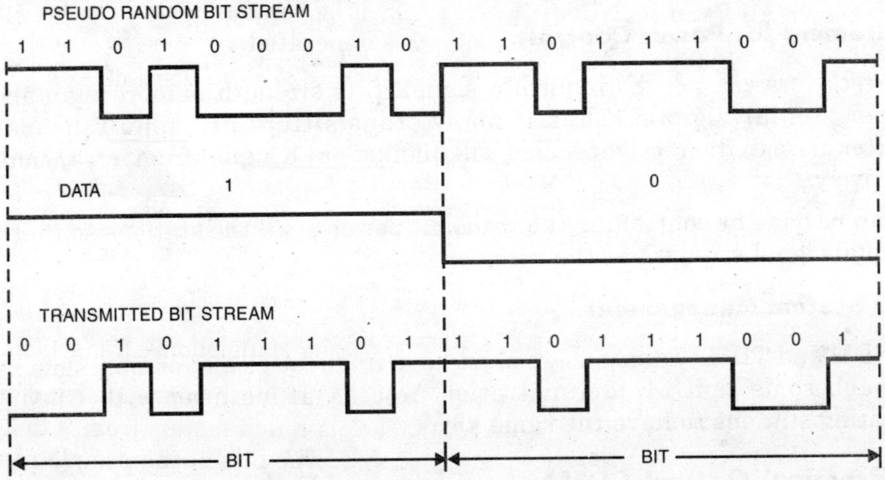

Fig. 12.33. Direct Sequence Spread Spectrum Modulation. A pseudo-Random Bit Stream Much Faster (here 9 times the speed) than the Data Rate is EORed with the Data. The Resulting bit Stream is then used to Modulate a Carrier Signal. This Results in a Much Broader Signal.

12.9.2. Code Division Multiple Access (CDMA)

The DSSS technique gives rise to a way of sharing the bandwidth. Multiple transmitters and receivers are able to use the same frequencies at the same time without interfering with each other. This is a by-product of the DSSS technique. The receiver correlates its received signal with a known random sequence - all other signals are filtered out.

If we consider a DSSS signal as a modulated waveform, when there are many overlapping DSSS signals then the filtering process required to select one of them from many except that we have waveforms that are not sinusoidal in shape. However, the DSSS **chipping sequences** (pseudo-random number sequences) must be orthogonal. Fortunately there are several ways of generating orthogonal pseudo-random sequences.

For this a receiving filter is required which can select a single DSSS signal from among all the intermixed ones. Practically we require a filter that can correlate the complex signal with a known **chipping sequence** (and reject all others). There are several available filtering techniques which will do this. The usual device used for this filtering process is known as **Surface Acoustic Wave (SAW) filter.**

CDMA has a number of important characteristics:

(1) Allocation of Capacity

Any particular DSSS receiver experiences other DSSS signals as noise. This means that we can continue adding channels until the signal-to-noise ratio gets too great and we start getting bit errors.

(2) No Guard Time or Guard Bands

In a TDM system when multiple users share the same channel certain time must be allowed between the end of one user's transmission and the beginning of the next. This is called **guard time**. At slow data rates it is not too important but as speed gets higher it comes to dominate the system throughput.

CDMA does not require a guard time - stations simply transmit whenever they are ready.

(3) Requirement for Power Control

DSSS receivers can not distinguish a signal if its strength is more than about 20 dB below other similar signals. Thus if many transmitters are simultaneously active, a transmitter close to the receiver (near) will blanket out a signal from a transmitter which is farther away.

This can be done by controlling the transmit power of all the stations so that they have roughly equal signal strength at the receiver.

(4) Easier System Management

With FDM and TDM systems users must have frequencies and/or time slots assigned to them through some central administration process. If we need with CDMA it is for communicating stations to have the same key.

12.9.3. Practical Optical CDMA

Optical CDMA is still very much a research technology. In the early 1990's it was proposed as a technology for shared medium LANs but since then the shared medium LAN itself has proven more costly than switch based star networks. Thus there is not a lot of interest in shared medium LANs in either the optical or electronic world.

Today it is very difficult and important challenge to find out a low cost technology for the upstream transport in a passive optical network. CDMA might be a good choice here.

Practical optical CDMA systems have some differences from RF ones :

1. Instead of using a random number generator to generate the chipping sequence a fixed sequence only as long as one data bit is likely to be used. Example: we may have 31 chips per bit and the chipping sequence would be the same for every bit transmitted.
2. Zero data bits are not transmitted at all. This nets out to saying that a 1 bit (for a particular end-user) is transmitted as an invariant 31-chip sequence.
3. The coding used in an optical system need to be different from those used in an RF system because we do not have a negative signal state in optical communications. We only have positive (or zero) states.

12.9.4. Spectral Encoding

Spectrum spreading can also be done by using the technique of frequency hopping in which the carrier frequency is shifted periodically according to a preassigned code. The situation differs from WDM in the sense that a fixed frequency is not assigned to a given channel. Rather, all channels share the entire bandwidth by using different carrier frequencies at different times according to a code.

Spectrally encoding of CDM lightwave systems requires a rapid change in the carrier frequency. It is difficult to make tunable semiconductor lasers whose wavelength can be changed over a wide range in a sub-nanosecond time scale. This approach has the advantage that coding and decoding is done in the electrical domain, where the existing commercial microwave components can be used.

Depending on the type of optical source used for the CDMA system. **Spectral encoding can be classified as :**

- Incoherent
- Coherent

Incoherent CDM : In the case of **incoherent CDM**, a broadband optical source such as an LED is used in combination with a multi peak optical filter (such as an AWG) to create multi wavelength output.

Optical switches are then used to select different wavelengths for different chip slots. This technique can also be used to make CDM add-drop multiplexers.

Coherent CDM : In **coherent CDM**, a broadband optical source is used in combination with an unbalanced MZ interferometer that introduces a delay longer than the **coherence time** in one of its branches. Such CDM systems rely on coherence to discriminate among channels and are affected severely by the optical beat noise.

Hybrid WED-CDMA System : The coherent CDMA systems designed with spectral encoding make use of ultra short optical pulses with a relatively broad spectrum. The encoder splits the broad spectrum into many distinct wavelengths using a periodic optical filter (such as an AWG) and then assembles different frequency signals according to the code used for that channel. A matched-filter decoder at the receiver end performs the reverse operation so that a single ultra short pulse is regenerated in a coherent fashion.

This technique has a distinct advantage that the CDMA signal can be overlaid over a WDM signal such that both signals occupy the same wavelength range. Figure 12.26 shows schematically how such a hybrid scheme works.

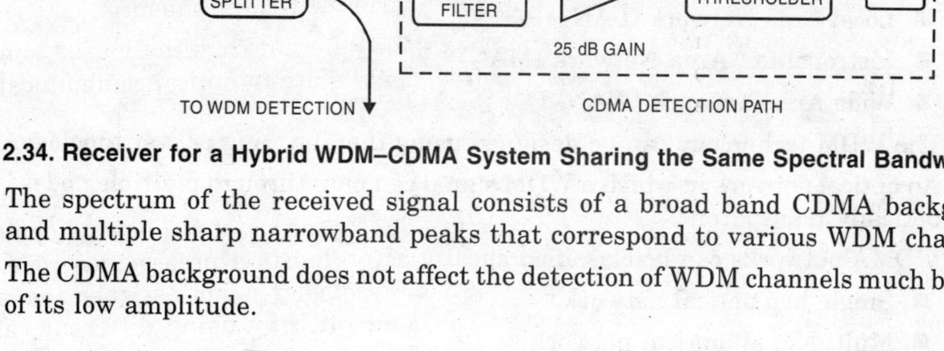

Fig. 12.34. Receiver for a Hybrid WDM–CDMA System Sharing the Same Spectral Bandwidth.

- The spectrum of the received signal consists of a broad band CDMA background and multiple sharp narrowband peaks that correspond to various WDM channels.
- The CDMA background does not affect the detection of WDM channels much because of its low amplitude.

- The CDMA receiver employs a notch filter to remove the WDM signal before decoding it.
- The hybrid WDM–CDMA scheme is spectrally efficient as it makes use of the unused extra bandwidth around each WDM channel.

SUMMARY

1. Until 1995 the bit rate of optical communication systems was limited to 10 Gb/s because of limitations imposed by the dispersive and nonlinear effects.
2. Principally the capacity of optical communication systems can exceed 10 Tb/s because of large frequency associated with optical carrier.
3. Channel multiplexing can be done :
   - In time domain
   - In frequency domain
4. With reference to TDM and FDM techniques are :
   - Optical TDM (OTDM)
   - Wavelength Division Multiplexing (WDM)
5. Wavelength division multiplexing (WDM) is the basic technology of optical networking, and is used for fiber to carry many separate and independent optical channels.
6. The types of WDM are :
   - Simple WDM
   - Dense WDM
7. In the simplest form, WDM was used to transmit two channels in different trnsmission windows of an optical fiber.
8. The spectral efficiency of a WDM system is given as

$$\eta_s = \frac{B}{\Delta V_{ch}}$$

9. The channel frequencies of WDM systems have been standardized by the International Telecommunication Union (ITU) on a 100 GHz grid in the frequency range 186-196 THz.
10. Optical network can be classified according to the area they cover :
    - Local Area Network (LAN)
    - Metropolitan Area Network (MAN)
    - Wide Area Network (WAN)
11. The WDM technology can be designed using the hub, ring or star topology.
12. An optical network in which a WDM signal can pass through multiple nodes is called optically transparent.
13. WDMA networks can be classified as :
    - Single hop optical network
    - Multi-hop all optical network

14. Feeder ring connects to the backbone of the network through an egress node. This ring employs four fibers to ensure robustness.
15. An add-drop multiplexer can be used at all nodes to drop and to add individual WDM channels.
16. In a single hop network every node is directly connected to all other nodes, resulting in a fully connected network while multi-hop networks are only partially connected.
17. The main drawback of the Lambdnet is that the number of users is limited by the number of available wavelengths.
18. The architectures used for all-optical multi-hop networks :
    - Hypercube architecture
    - De Bruijn Network
    - Shuffle network
    - Banyan network
19. The WDM components are :
    - Multiplexer/demultiplexer
    - Star coupler
    - Tunable optical filter
    - Multiwavelength optical transmitter
    - Add-drop multiplexer
    - Wavelength shifter
20. Optical filters are the basic building blocks of WDM components and its role in a WDM system is to select a desired channel at the receiver.
21. Optical filters can be classified into two broad categories depending upon
    - Optical interference
    - Diffraction
22. There are several devices which may perform add-drop multiplexer function such as :
    - Array waveguide gratings
    - Circulators with FBGs
    - A cascade of MZIs
23. Littlow gratings may be used for both multiplexing and demultiplexing. Commercially it is available with 132 ports and their main problem is cost.
24. Wavelength router is a device that combines the functionality of a star coupler with multiplexing and demultiplexing operations.
25. The most common design of a wavelength router uses a AWG demultiplexer modified to provide multiple input ports. Such a device is known as waveguide grating router (WGR).
26. OTDM is worked by allowing the optical signal stream to run at speeds significantly in excess of the maximum speed of the electronics.
27. Connections in TDM network can be setup in two ways :
    - By network management
    - By signalling

28. The modulation technique used in optical TDM principle is RZ coding.

29. The subcarrier multiplexing (SCM) employs multiple microwave carrier for transmission of multiple channels over coaxial cables or free space.

30. The main advantage of SCM in the design of broadband network is :

- Flexibility
- Upgradibility

31. The analog signal is distorted during its transmission through the fiber link and is known as intermodulation distortion (IMD).

32. The performance of hybrid SCM systems is affected by many reasons like :

- Clipping noise
- Multiple optical reflections
- The nonlinear mechanism.

33. Optical fiber is used purely as a transmission medium, serving as a replacement for copper cable, and all the switching and processing of the bits is handled by electronics. These networks are known as first generation optical networks.

34. Examples of first generation optical networks are SONET (synchronous optical network) and SDH (synchronous digital hierarchy) networks.

35. To push TDM technology beyond 10 Gb/s and 40 Gb/s, the approach is known as optical time division multiplexing (OTDM).

36. A directional coupler is used to combined and split signals in an optical network.

37. The components used in modern optical communication systems include :

- Couplers
- Lasers
- Photodetectors
- Optical amplifiers
- Optical switches
- Filters
- Multiplexure

38. Couplers are also used to tap off a small portion of the power from a light stream for monitoring purposes or other reasons. Such couplers are also known as taps.

39. Couplers are the principle components used to construct Mach-Zehnder interferometers. They can be used as optical filters and multiplexers/demultiplexers.

40. An isolator is a passive non-reciprocal device. Its main function is to allow transmission in one direction through it but block all transmission in other direction.

41. Isolators are used in systems in front of optical amplifiers and lasers primarily to prevent reflections from entering these devices.

42. The two key parameters of an isolator are its :

- Insertion loss
- Isolation

43. A wavelength router is the key component for a class of optical networks called wavelength routing networks.

44. An add/drop multiplexer is essentially a simple form of a wavelength router with one input port and one output port with an additional local port wherein wavelengths are added to/dropped from the incoming light stream.
45. The term grating is used describe almost any device whose operation involves interference among multiple optical signals originating from the same source but with different relative phase shifts.
46. Bragg gratings are widely used in fiber optic communication systems.
47. The acousto-optic tunable fiber is also known as tunable filter, which is capable of selecting several wavelengths simultaneously.
48. A regerator converts the optical signal to an electrical signal, cleans it up, and converts it back into an optical signal for onward transmission.
49. Three different types of amplifiers are :
    - Erbium-doped fiber amplifiers (EDFAs).
    - Praseodymium-doped fiber amplifiers (PDFAs).
    - Semiconductor optical amplifiers (SOAs).
50. The photodetector generates an electrical current proportional to the incident optical power.
51. In digital communication systems, the front end amplifier is followed by a decision circuit that estimates the data from the output of the front end amplifier.
52. A wavelength converter is a device that converts data from one incoming wavelength to another outgoing wavelength.
53. Wavelength converters may be classified based on the :
    - The range of wavelengths
    - Their inputs and outputs
54. SONET (synchronous optical network) is the current transmission and multiplexing standard for high speed signals within the carrier infrastructure in North America.
55. SDH (Synchronous digital hierarchy) has been adopted in Europe and Japan.
56. SONET is deployed in three types of network configurations :
    - Ring
    - Linear configurations
    - Point to point links
57. Terminal multiplexers are also sometimes known as line terminating equipment (LTE).
58. SONET equipment can be configured to work in any of these three configurations :
    - Ring ADM
    - Linear ADM
    - Terminal multiplexer
59. Two types of ring architecture are used :
    - Unidirectional path switched rings (UPRs)
    - Bidirectional line switched rings (BLSRs)

60. A major component in the SONET infrastructure is a digital cross connect (DCS).

61. FDDI is widely deployed methopolitan area network.

62. FDDI is commonly deployed in the form of two fiber rings, similar to SONET BLSR/2S, and incorporates a similar protection mechanism as well.

REVIEW QUESTIONS

1. What do you understand by multichannel systems ? List the different ways in which multiplexing can be done.

2. Discuss the principle of wavelength division multiplexing. Explain the types of WDM also.

3. Draw and discuss the functions required to build a dense WDM shared optical link.

4. What do you meant by WDM lightwave systems ? Briefly explain the various categories of WDM systems.

5. Discuss the various topologies used in the implementation of a regional WDM network.

6. Differentiate between single hop and multihop optical network.

7. Draw and discuss the various architectures used for all optical multihop networks.

8. Briefly explain the various components used for the implementation of WDM technology.

9. What are the basic requirements of WDM system ?

10. What is the function of optical filter ? List the various properties of a tunable optical filter.

11. Discuss the various kinds of optical filters.

12. What do you undersatnd by the multiplexing of light and demultiplexing of light ?

13. What is luttrow gratings ? List their advantages and disadvantages also.

14. What is the requirement of optical cross connect ? By how many ways it can be realized in optical technology ?

15. Which is the most common design of a wavelength router uses a AWG demultiplexer ?

16. Explain the various principles that a fully optical wavelength converters may use.

17. Discuss the concept of optical time division multiplexing. Explain the way a network is configured.

18. What do you understand by demultiplexing the TDM signal ?

19. What is subcarrier multiplexing ? What are its advantage in the design of broadband network ? Discuss the various types of subcarrier multiplexing.

20. What do you meant by optical code division multiplexing ? List the possible technologies used in the optical communication CDMA.

21. Briefly discuss the way of signal generation in Direct Sequence Spread Spectrum (DSSS).

22. Explain how an array of planar waveguides can be used for demultiplexing WDM channels. Draw the necessary diagrams.

23. Use a single fiber coupler and two fiber grating to design an add-drop filter. Briefly explain how such a device works.

24. Use a waveguide grating router to design an integrated WDM transmitter. How would the design change for a WDM receiver ?

25. What do you meant by the in-band linear crosstalk ?

26. Explain how stimulated roman scattering can cause crosstalk in multichannel lightwave systems.

27. Design a 4 × 4 demultiplexer with as few couplers as possible if we are having ten 3 dB fiber couplers.

28. Which fiber optic component (splice, connector, or coupler) makes a permanent connection in a distributed system ?

29. Fiber-to-fiber coupling loss is affected by intrinsic and extrinsic coupling losses. Can intrinsic coupling losses be limited by limiting fiber mismatches ? **[PTU 2003]**
30. In fiber-to-fiber connections, Fresnel reflection is one source of coupling losses. Light is reflected back into the source fiber and is lost. What causes Fresnel reflection ?
31. Reduction of Fresnel reflection is possible by reducing the step change in the refractive index at the fiber interface. What material reduces the step change in refractive index at a fiber interface ?
32. What is the difference between passive and active fiber optic couplers ?
33. Which type of optical splitter (Y-coupler or T-coupler) splits only a small amount of power from the input fiber to one of the output fibers ? **[PTU 2005]**
34. Describe a directional coupler.
35. What connection properties result in fiber optic connector coupling loss ?
36. Which is the more critical parameter in maintaining total insertion loss below the required level, fiber alignment or fiber mismatch ?
37. Fiber optic connectors can reduce system performance by increasing what two types of noise ?
38. Which type of fiber optic connector (butt-jointed or expanded beam) brings the prepared ends of two optical fibers into close contact ?
39. Is coupling loss from fiber separation and lateral misalignment more critical in expanded-beam or butt-jointed connectors ?
40. Is coupling loss from angular misalignment more critical in expanded beam or butt-jointed connectors ?
41. What is WDM and list the reasons for its development as a major communication technology. **[PTU 2008]**

❑❑❑

APPENDICES

APPENDICES

APPENDIX I : The Role of Optical Networks

APPENDIX II : The Visible Spectrum

APPENDIX III : Some Basic Constants

APPENDIX IV : Acronyms

Appendix **I**

The Role of Optical Networks

The reason for the above discussion was to show that different forms of networking can (and should) be built "on top" of one another. In the case of practical optical networks today we have four different possible structures:

1. Point-to-Point WDM
2. Add-Drop Optical WDM Networks
3. Optical TDM Networks (with an add/drop function).
4. Wavelength Routed Networks

At the present time we can define three optical networking layers :

Optical Link Layer

This is the lowest layer in an optical network and includes the protocols necessary to manage a point-to-point optical link. However, it does not contain the transmission protocols or the coding structures as this is done at the next layer.

WDM Networking Layer

This layer provides end-to-end transparent optical transport through the network. It applies to both add-drop and nodal networks.

OTOM Layer

An OTDM networking system can be (optionally) built above a WDM structure.

Appendix II

The Visible Spectrum

In optical communications we almost never use visible light. However, it is useful to know the wavelengths of the colours in the visible spectrum.

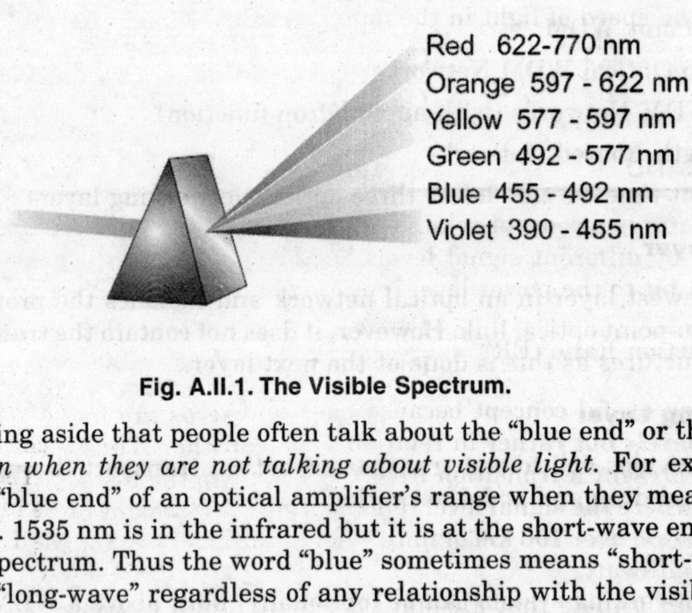

Fig. A.II.1. The Visible Spectrum.

It's an interesting aside that people often talk about the "blue end" or the "red end" of a spectral range *even when they are not talking about visible light.* For example someone might refer to the "blue end" of an optical amplifier's range when they mean a wavelength of around 1535 nm. 1535 nm is in the infrared but it is at the short-wave end (the blue end) of the amplifier's spectrum. Thus the word "blue" sometimes means "short-wave" and "red" sometimes means "long-wave" regardless of any relationship with the visible spectrum.

Appendix III
Some Basic Concepts

Refractive Index

The speed of light in a vacuum (in free space) is 2.998 × 108 metres per second. Whenever a material is present the speed of light is reduced by an amount depending on the material characteristics. The "Refractive Index" (RI) of a material is the ratio of the speed of light in a vacuum over the speed of light in the material in question.

$$R_f = \frac{C_{frequency}}{C_{freespace}}$$

Extinction Ratio

Extinction ratio is a concept related to digital signals where a digital 1 and a digital 0 are represented by different signal levels. The ER is simply the ratio of the power level representing a 1 bit to the power level representing a 0 bit.

$$\text{Extinction Ratio (ER)} = \frac{\text{Power}_{1\ \text{bit}}}{\text{Power}_{0\ \text{bit}}}$$

This is a very useful concept because ones and zeros are usually not represented by absolute power levels but rather in relation to one another. Thus a change in line state by –15 dB might represent a transition from the 1-state to the 0-state. This would be true at the transmitter where the signal level representing a 1-bit might be as high as 10 milliwatts or after transmission over 100 km of fibre where the signal level of the 1-bit may be perhaps only 0.001 of a milliwatt.

It is important to note that a 0-state is usually not a complete absence of power. The extinction ratio is *never* infinite.In many systems the laser is run (slightly) above threshold in the 0-state and thus some power is always transmitted. At the receiver there is always some current in the zero state caused by thermal noise in the detector and in the receiver circuit.

Optical Decibels

Throughout almost every activity in the planning and design of communications systems and devices we are constantly faced with calculating losses and gains of signal power. This is fundamental to everything we do. The decibel is a very convenient way of expressing an amount of loss or gain within a system or the amount of loss or gain caused by some component of a system.

When power is lost such as during the transit of a signal along a transmission line *you never* lose a fixed amount of power - you always lose a proportion. For example if an electrical signal is attenuated from 1 milliwatt to 112 of a milliwatt in its transport on a transmission line then the transmission line has an attenuation of .5. If we sent a signal of 114 of a milliwatt on the same transmission line then it will be attenuated to 118 of a milliwatt. In each case a fixed proportion (112) of the power is lost but the absolute amount of the loss is quite different. The same applies for example in amplifiers where power is gained.

When you have to calculate the total losses and gains in a system you have to *multiply* the loss and gain proportions for each element of the system. This is hard to do mentally - especially when many stages are involved.

Decibels are logarithms. This allows us to easily and conveniently calculate the total loss/gain characteristics of a system by just adding them up? This is very convenient and once you get used to it, natural.

The Decibel originated as the "Bell" and was first defined by Alexander Graham Bell. A Bell is the logarithm of the ratio of a signal's *power* with some reference power level. A deci-bel is just one tenth of a Bell and was invented because the Bell unit was too large for convenient manipulation.

Usually a dB is the amount of loss or gain of signal *power*. In the case of a component that attenuates a signal, the attenuation in dB is 10 times the log (base 10) of the ratio of output power to input power.

$$\text{Attenuation} = 10 \log_{\text{base } 10} \left(\frac{\text{Power}_{\text{output}}}{\text{Power}_{\text{input}}} \right)$$

A very misleading situation can arise when optical power ratios (quoted in dB) are quoted in conjunction with electrical power ratios (also quoted in dB). This can be illustrated very easily. If you take an optical signal of a particular power level and feed it to an electronic detector you get a certain amount of photocurrent. If you double the amount of optical power and feed it to the same detector then you get (or we hope we get) twice the photocurrent output. So the optical signal level has doubled (increased by approx. 3 dB).

But now if we look at electrical power we get something different. Electrical power is proportional to *the square of the current not current itself.* Thus:

$$\text{Attenuation}_{\text{electrical}} = 10 \log \left(\frac{\text{Output power}_{\text{electrical}}}{\text{Input power}_{\text{electrical}}} \right)$$

Expressed as current this is :

$$\text{Attenuation}_{\text{electrical}} = 20 \log \left(\frac{\text{Output power}_{\text{electrical}}}{\text{Input power}_{\text{electrical}}} \right)$$

Thus if you measure (say) the loss on a transmission link by taking the ratio of the optical powers involved you will get *half* of the figure you would have obtained if you measured the current received at a detector. The ratio dBopticai is half as large as the ratio dB electrical. This can be confusing in a number of situations. For example a loss of 20 dB on an electrical cable is equivalent to the loss of 10 dB in an optical one.

When losses/gains are quoted in dB you should always be sure of the context in which the measurement has been made. Never mix optical and electrical dB.

Appendix IV

Acronyms

ABF	Air Blown Fibre
ACK	Acknowledgement
ADM	Add-Drop Multiplexor
ADSL	Asymmetric Digital Subscriber Line
AEL	Accessible Emission Limit
AON	All-Optical Network
AOTF	Acousto-Optical Tunable Filter
APD	Avalanche Photodiode
ASE	Amplified Spontaneous Emission
ATM	Asynchronous Transfer Mode
AWG	Array Waveguide Grating
BER	Bit Error Rate
BEDSFA	Broadband Erbium Doped Silicon Fibre Amplifier
BISDN	Broadband Integrated Services Digital Network
CATV	Community Antenna Television
CBS	Central Base Station
CCITT	Comite Consultatif International Telegraphique et Telephonique (a.k.a. ITU-TS)
CD	Compact Disk
CDMA	Code Division Multiple Access
CFR	Cross-coupling Fibre Ring resonator
CIP	Carrier-Induced Phase Modulation
CNR	Carrier to Noise Ratio
CMOS	Complementary Metal Oxide Semiconductor

CW	Continuous Wave
dB	Decibel
dBm	Decibels above/below one milliwatt
DBR	Distributed Bragg Reflector (laser)
DCF	Dispersion Compensating Fibre
DCM	4-(dicyanomethylene)-2-methyl-6 (4-dimethylaminostyryl)-4H-pyran
DFB	Distributed FeedBack (laser)
DFG	Difference Frequency Generation
DFR	Direct-coupling Fibre Ring resonator
DH	Double Heterostructure
DMD	Differential Mode Delay
DSF	Dispersion Shifted Fibre
DSHI	Delayed Self-Heterodyne Interferometer
DUT	Device Under Test
DWDM	Dense Wavelength Division Multiplexing EA Electro-Absorption
ECL	Emitter Coupled Logic
EDFA	Erbium Doped Fibre Amplifier
EDFFA	Erbium Doped Fluoride Fibre Amplifier
EDSFA	Erbium Doped Silicon Fibre Amplifier
EH	Electric Magnetic (H = magnetic)
EIA	Electronic Industries Association
EMC	ElectroMagnetic Compatibility
EMD	Equilibrium Mode Distribution
EMI	ElectroMagnetic Interference
EMV	Effective Mode Volume
FBT	Fused Biconical Taper
FC	Fibre Channel
FDDI	Fibre Distributed Data Interface
FDHM	Full Duration at Half Maximum
FDM	Frequency Division Multiplexing
FEC	Forward Error Correction
FEL	Free Electron Laser
FET	Field Effect Transistor
FFP	Fibre Fabry-Perot filter
FITL	Fibre In The Loop
FOTP	Fibre Optic Test Procedure FP Fabry-Perot
FPI	Fabry-Perot Interferometer
FPM	Four Photon Mixing

FSC	Free Space Coupler
FSK	Frequency Shift Keying
FSR	Free Spectral Range
FTTC	Fibre To The Curb
FTTH	Fibre To The Home
FTTN	Fibre To The Neighbourhood
FWHM	Full Width Half Maximum
FWM	Four Wave Mixing
GbE	Gigabit Ethernet
GI	Graded-Index
GRIN	GRaded-INdex
GVD	Group Velocity Dispersion
HDSL	High-Speed Digital Subscriber Line
HE	Magnetic Electric (H = magnetic)
HPCF	Hard Plastic Coated silica Fibre or Hard Polymer Clad silica Fibre
HPCS	Hard Plastic Coated Silica or Hard Polymer Clad Silica
HWHM	Half-Width Half-Maximum
IC	Integrated Circuit
IEEE	Institute of Electrical and Electronic Engineers
IP	Internet Protocol
IR	Infra-Red
ISDN	Integrated Services Digital Network
IVPO	Inside Vapour Phase Oxidation
LAN	Local Area Network
LASER	Light Amplification by the Stimulated Emission of Radiation
LEOS	(IEEE) Lasers and Electro-Optics Society
LD	Laser Diode
LED	Light Emitting diode
LOL	Loss Of Light
LP	Linearly Polarised
LW	Long Wavelength
MAN	Metropolitan Area Network
Mb	Mega bit
MB	Mega Byte
MBd	Mega Baud
MBE	Molecular Beam Epitaxy
MCVD	Modified Chemical Vapour Deposition
MESFET	Metal Schottky Field Effect Transistor

MFD	Mode Field Diameter
MISFET	Metal Integrated Semiconductor Field
MM	Multimode
MOS	Metal Oxide Semiconductor
MOSFET	Metal Oxide Semiconductor Field Effect Transistor
MQW	Multi-Quantum-Well
MTBF	Mean Time Between Failures (or Before Failure)
MZ	Mach-Zehnder
MZI	Mach-Zehnder Interferometer
n-ISDN	Narrowband Integrated Services Digital Network
NA	Numerical Aperture
NOLM	Non-linear Optical Loop Mirror ns nanosecond
NTSC	National Television Standards Committee
OADM	Optical Add-Drop Multiplexor
OEIC	Opto-Electronic Integrated Circuit
OFC	Open Fibre Control
OFDM	Optical Frequency Division Multiplexing
OFL	Over-Filled Launch
OILED	Organic Light-Emitting Diodes
OOK	On-Off Keying
OSL	Organic Semiconductor Laser
OTDM	Optical Time Division Multiplexing
OTDR	Optical Time-Domain Reflectometer
OVD	Outside Vapour Deposition
OVPO	Outside Vapour-Phase Oxidation
OXC	Optical Cross-Connect
PAL	Phase Alternating Line
PANDA	Polarisation maintaining AND Absorption reducing
PCN	Personal Communication Network
PCOF	Primary Coated Optical Fibre
PCs	Personal Communication System
PCVD	Plasma-activated Chemical Vapour Deposition
PDG	Polarisation Dependent Gain
PDL	Polarisation Dependent Loss
PIC	Photonic Integrated Circuit
PIN	P-doped silicon, Intrinsic silicon, N-doped silicon junction
PMD	Polarisation Mode Dispersion
PMF	Polarisation Maintaining Fibre

PMMA	Poly-Methyl Methyl Acrylate
POF	Plastic Optical Fibre
PON	Passive Optical Network
POTS	Plain Old Telephone System
PRBS	Passive Radio Base Station
PrDFA	Praseodymium Doped Fibre Amplifier
QCSE	Quantum Confined Stark Effect
RAPID	Reach-through Avalanche Photodiode
RECAP	REsonant CAvity Photodetector
REDFA	Rare-Earth Doped Fibre Amplifier
RFI	Radio Frequency Interference
RI	Refractive Index
RIN	Relative Intensity Noise
RMS	Root Mean Square
RO	Relaxation Oscillation
ROFL	Radial Over-Filled Launch Rx Receiver
SAW	Surface Acoustic Wave
SBS	Stimulated Brillouin Scattering
SCM	Sub-Carrier Multiplexing
SCMOL	Sub-Carrier Multiplexed Optical Link
SCOF	Secondary Coated Optical Fibre
SDH	Synchronous Digital Hierarchy
SHG	Second Harmonic Generation
SHIP	Silicon Hetero-Interface Photodetector SI Step-Index
SLA	Semiconductor Laser Amplifier
SLID	Super Luminescent Diode
SM	Single Mode
SNR	Signal-to-Noise Ratio
SOA	Semiconductor Optical Amplifier
SOI	Silicon On Insulator or Silica On Insulator
SONET	Synchronous Optical Network
SQW	Single Quantum Well
SRS	Stimulated Raman Scattering
STP	Shielded Twisted Pair
SW	Short Wavelength
TDM	Time Division Multiplexing
TE	Transverse Electric
TEM	Transverse Electro-magnetic

TM	Transverse Magnetic
TP	Twisted Pair TR Taken-Ring
TWA	Travelling Wave Amplifier
TWSLA	Travelling Wave Semiconductor Laser Amplifier
Tx	Transmitter
UV	Ultra-Violet
VAD	Vapour-phase Axial Deposition
VCSEL	Vertical-Cavity Surface-Emitting Laser
VDSL	Very high speed Digital Subscriber Line
WDM	Wavelength Division Multiplexing
WGR	Waveguide Grating Router
XGM	Cross Gain Modulation
XPM	Cross Phase Modulation
YLF	Yttrium Lithium Fluoride
ZBGA	Zirconium Barium Gadolinium Aluminium (Fluoride)
ZBLA	Zirconium Barium Lanthium Aluminium (Fluoride)
ZBLAN	Zirconium Barium Lanthium Aluminium Sodium (Fluoride)
ZMD	Zero Material Dispersion.

❑❑❑

GLOSSARY

A

Absorption : That portion of optical attenuation in optical fiber resulting from the conversion of optical power to heat. Caused by impurities in the fiber such as hydroxyl ions.

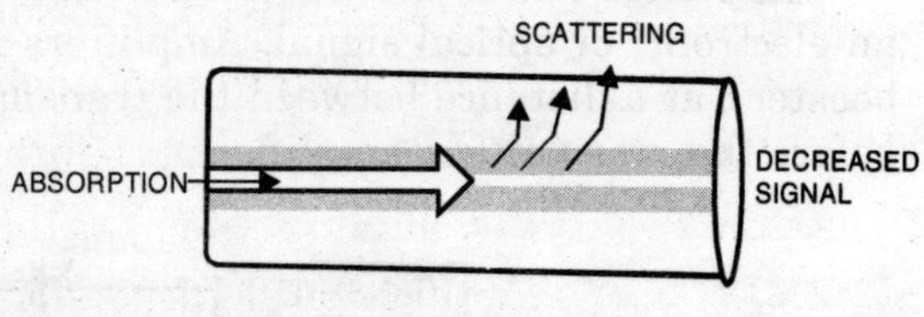

A/B Switch : A device that accepts inputs (optical or electrical) from a primary path and a secondary path to provide automatic or manual switching in the event that the primary path signal is broken or otherwise disrupted. In optical A/B switches, optical signal power thresholds dictate whether the primary path is functioning and signals a switch to the secondary path until optical power is restored to the primary path.

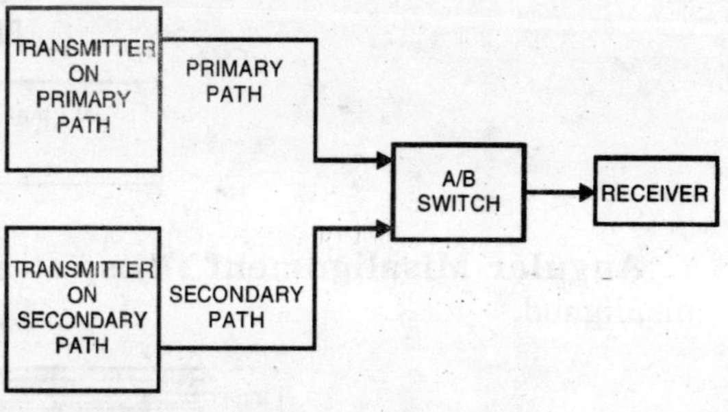

Acceptance Angle : The half-angle of the cone (α) within which incident light is totally internally reflected by the fiber core. It is equal to $\sin^{-1}$(NA).

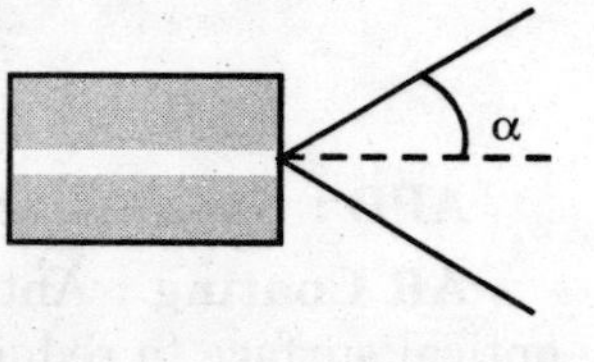

Active Device : A device that requires a source of energy for its operation and has an output that is a function of present and past input signals. Examples include controlled power supplies, transistors, LEDs, amplifiers, and transmitters.

A/D or ADC : Abbreviation for analog-to-digital converter. A device used to convert analog signals to digital signals.

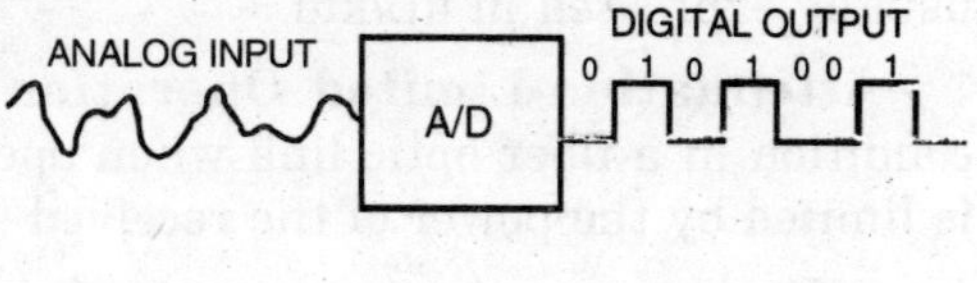

Add/Drop Multiplexing : a multiplexing function offered in connection with SONET that allows lower level signals to be added or dropped from a high-speed optical carrier in a wire center. The connection to the add/drop multiplexer is via a channel to a central office port at a specific digital speed (DS3, DS1, etc.)

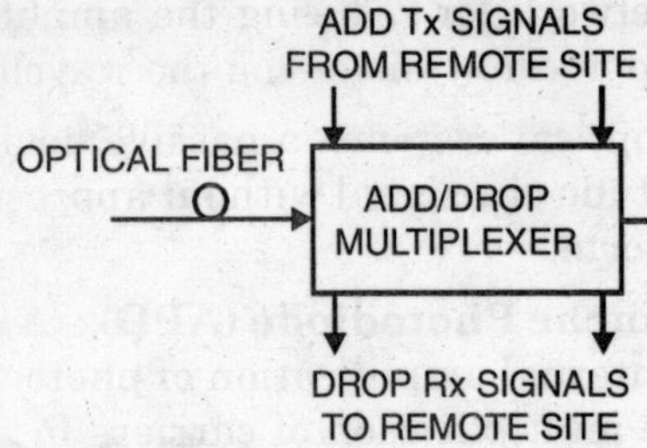

AGC : Abbreviation for automatic gain control. A process or means by which gain is automatically adjusted in a specified manner as a function of input level or another specified parameter.

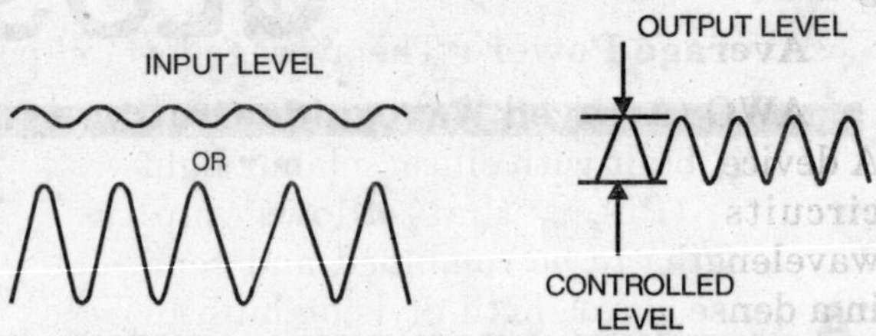

Amplifier : A device, inserted within a transmission path, that boosts the strength of an electronic or optical signal. Amplifiers may be placed just after the transmitter (power booster), at a distance between the transmitter and the receiver (in-line amplifier), or just before the receiver (preamplifier).

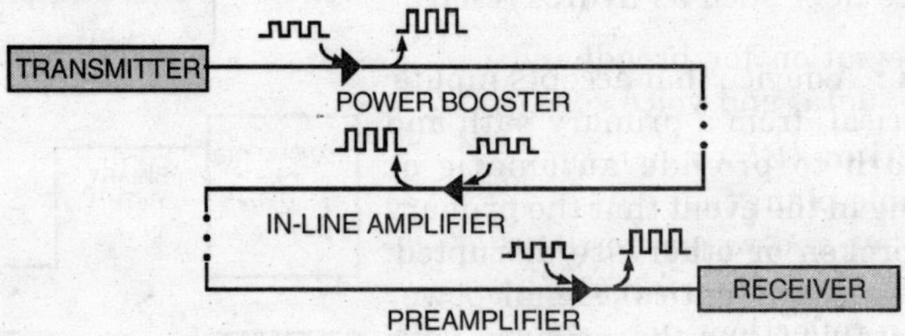

Angular Misalignment : Loss at a connector due to fiber end face angles being misaligned.

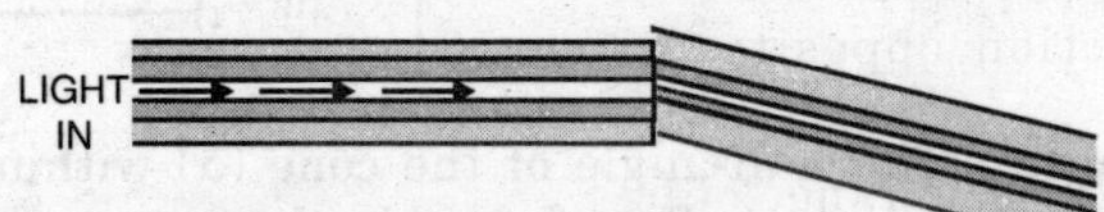

APD : See avalanche photodiode.

AR Coating : Antireflection coating. A thin, dielectric or metallic film applied to an optical surface to reduce its reflectance and thereby increase its transmittance.

Attenuation : The decrease in signal strength along a fiber optic waveguide caused by absorption and scattering. Attenuation is usually expressed in dB/km.

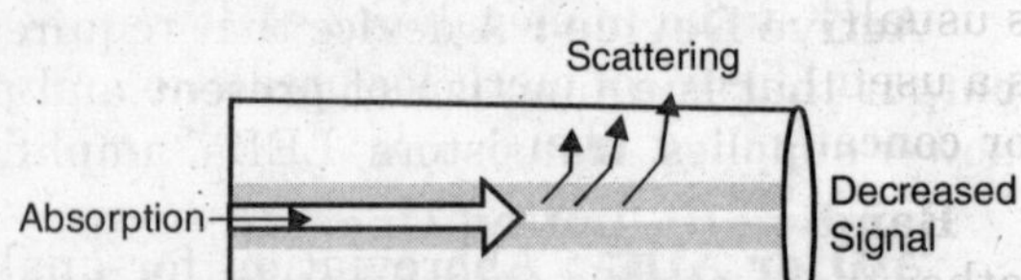

Attenuation-Limited Operation : The condition in a fiber optic link when operation is limited by the power of the received signal (rather than by bandwidth or distortion).

Attenuator : 1. In electrical systems, a usually passive network for reducing the amplitude of a signal without appreciably distorting the waveform.

2. In optical systems, a passive device for reducing the amplitude of a signal without appreciably distorting the waveform.

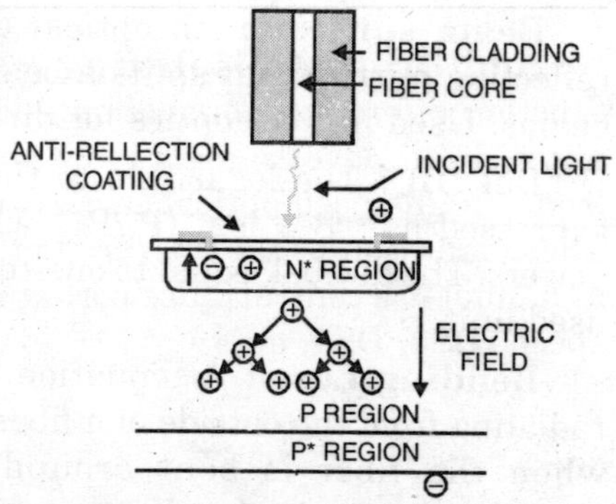

Avalanche Photodiode (APD) : A photodiode that exhibits internal amplification of photocurrent through avalanche multiplication of carriers in the junction region.

Average Power : The average level of power in a signal that varies with time.

AWG (Arrayed Waveguide Grating) : A device, built with silicon planar lightwave circuits (PLC), that allows multiple wavelengths to be combined and separated in a dense wavelength-division multiplexing (DWDM) system.

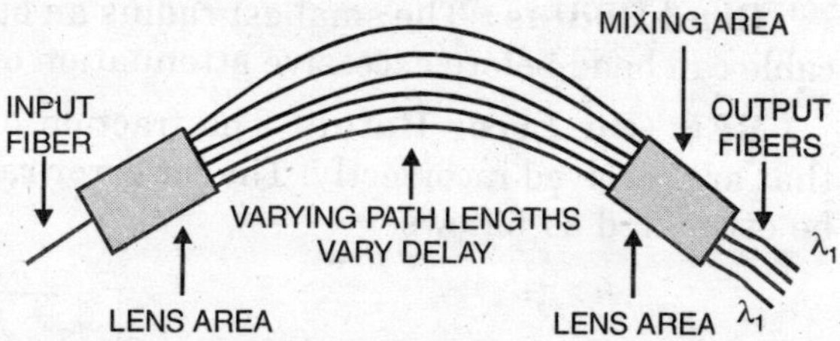

Axis : The center of an optical fiber.

B

BB-I : Abbreviation for broadband interactive services. The delivery of all types of interactive video, data, and voice services over a broadband communications network.

Back reflection (BR) : A term applied to any process in the cable plant that causes light to change directions in a fiber and return to the source. Occurs most often at connector interfaces where a glass-air interface causes a reflection.

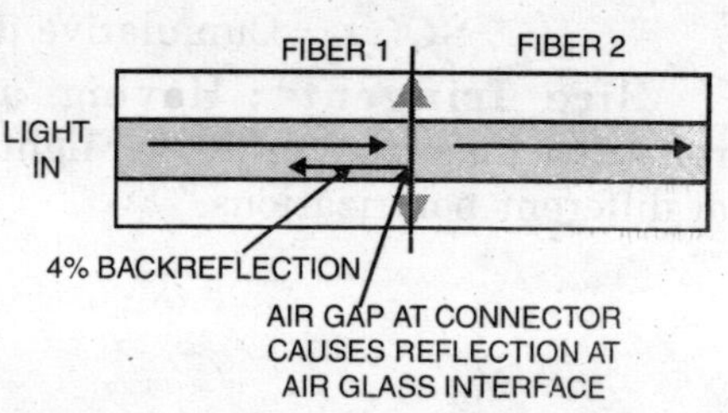

Back scattering : The return of a portion of scattered light to the input end of a fiber; the scattering of light in the direction opposite to its original propagation.

Band width (BW) : The range of frequencies within which a fiber optic waveguide or terminal device can transmit data or information.

Bandwidth•Distance Product : Of an optical fiber, under specified launching and cabling conditions, at a specified wavelength, a figure of merit equal to the product of the fiber's length and the 3 dB bandwidth of the optical signal. The bandwidth distance product is usually stated in megahertz •kilometer (MHz•km) or gigahertz•kilometer (GHz•km). It is a useful figure of merit for predicting the effective fiber bandwidth for other lengths, and for concatenated fibers.

Bandwidth-limited Operation : The condition in a fiber optic link when bandwidth, rather than received optical power, limits performance. This condition is reached when the signal becomes distorted, principally by dispersion, beyond specified limits.

Baud : A unit of signaling speed equal to the number of signal symbols per second, which may or may not be equal to the data rate in bits per second.

Beam splitter : An optical device, such as a partially reflecting mirror, that splits a beam of light into two or more beams. Used in fiber optics for directional couplers.

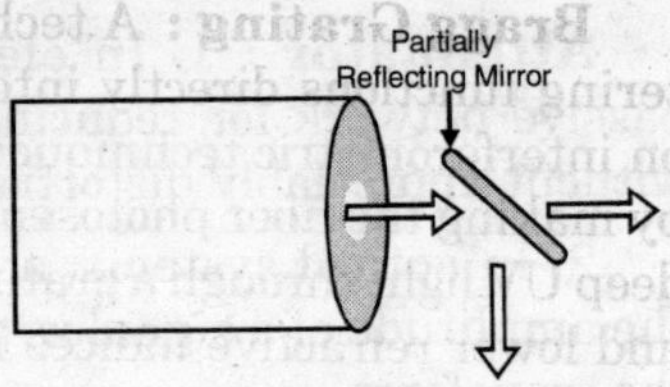

Bel (B) : The logarithm to the base 10 of a power ratio, expressed as : $B = \log_{10}(P_1/P_2)$, where P_1 and P_2 are distinct powers. The decibel, equal to one-tenth bel, is a more commonly used unit.

Bending Loss : Attenuation caused by high-order modes radiating from the outside of a fiber optic waveguide which occur when the fiber is bent around a small radius. See also macrobending, microbending.

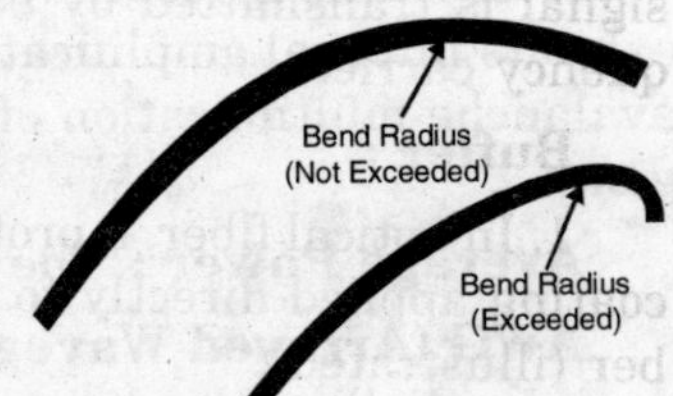

Bend Radius : The smallest radius an optical fiber or fiber cable can bend before excessive attenuation or breakage occurs.

BER (Bit Error Rate) : The fraction of bits transmitted that are received incorrectly. The bit error rate of a system can be estimated as follows:

$$\text{BER} = Q\left[\sqrt{\frac{I_{\min}^2}{4 \times N_0 \times B}}\right]$$

where N_0 = Noise power spectral density (A^2/Hz).

I_{MIN} = Minimum effective signal amplitude (Amps).

B = Bandwidth (Hz).

$Q(x)$ = Cumulative distribution function (Gaussian distribution).

Bire fringent : Having a refractive index that differs for light of different polarizations.

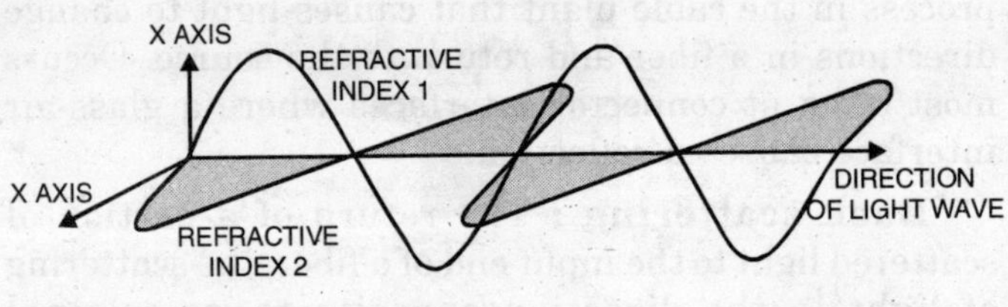

Bit : The smallest unit of information upon which digital communications are based; also an electrical or optical pulse that carries this information.

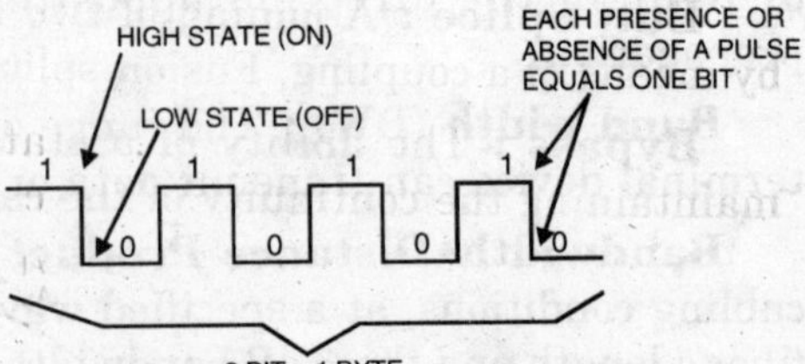

BITE : Abbreviation for built-in test equipment. Features designed into a piece of equipment that allow on-line diagnosis of failures and operating status. Status LEDs are one example.

Bit Period (T) : The amount of time required to transmit a logical one or a logical zero.

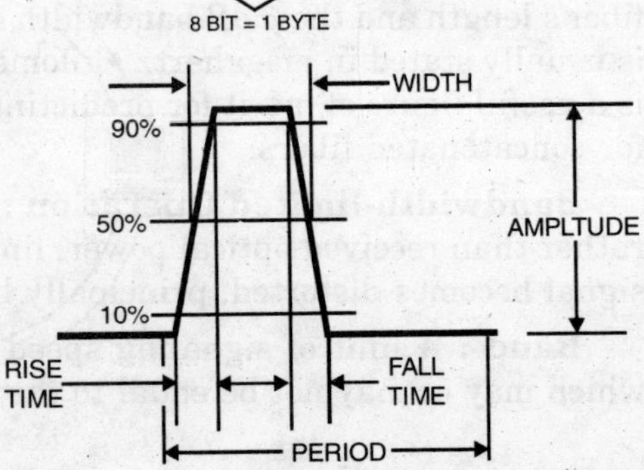

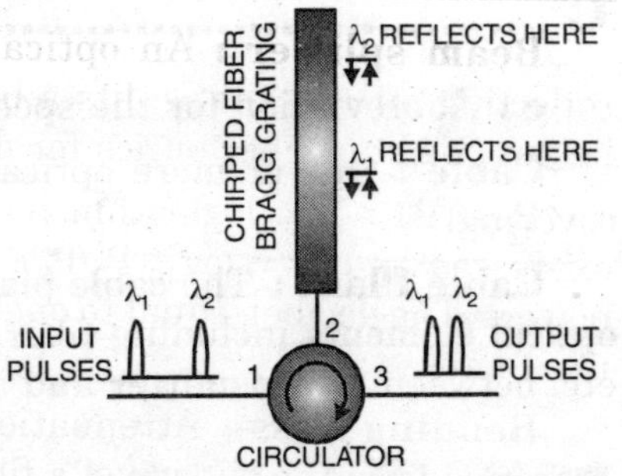

Bragg Grating : A technique for building optical filtering functions directly into a piece of optical fiber based on interferometric techniques. Usually this is accomplished by making the fiber photosensitive and exposing the fiber to deep UV light through a grating. This forms regions of higher and lower refractive indices in the fiber core.

Broadband : A method of communication where the signal is transmitted by being impressed on a high-frequency carrier.

Buffer :

1. In optical fiber, a protective coating applied directly to the fiber (illustrated).

2. A routine or storage used to compensate for a difference in rate of flow of data, or time of occurrence of events, when transferring data from one device to another.

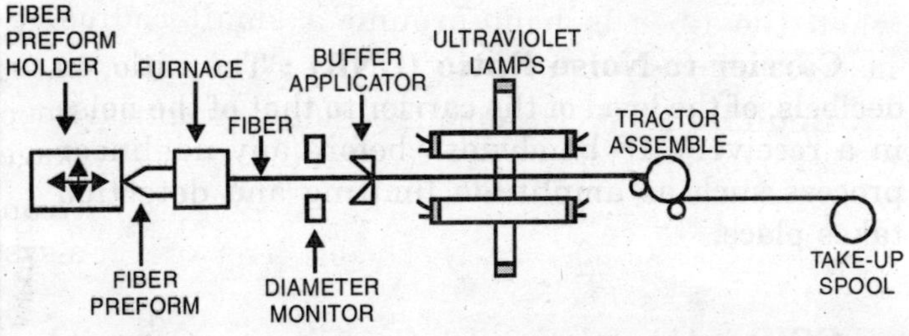

Bus Network : A network topology in which all terminals are attached to a transmission medium serving as a bus. Also called a daisy-chain configuration.

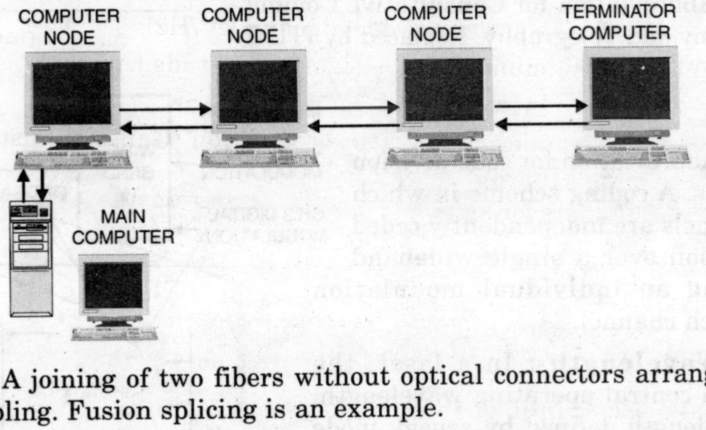

Butt Splice : A joining of two fibers without optical connectors arranged end-to-end by means of a coupling. Fusion splicing is an example.

Bypass : The ability of a station to isolate itself optically from a network while maintaining the continuity of the cable plant.

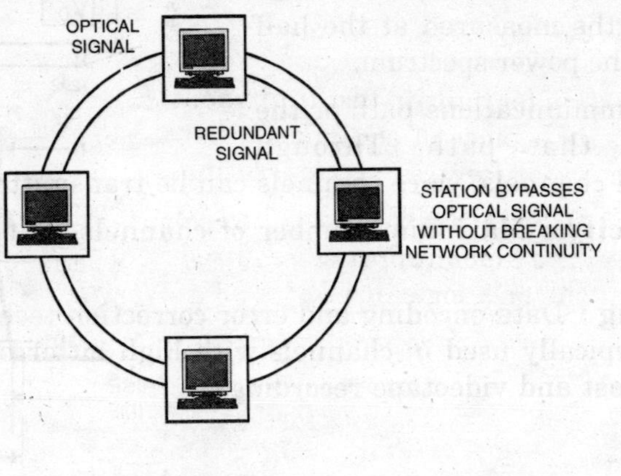

C

c : Abbreviation for the speed of light. 299,792.5 km per second in a vacuum.

Cable : One or more optical fibers enclosed, with strength members, in a protective covering.

Cable Plant : The cable plant consists of all the optical elements including fiber, connectors, splices, etc. between a transmitter and a receiver.

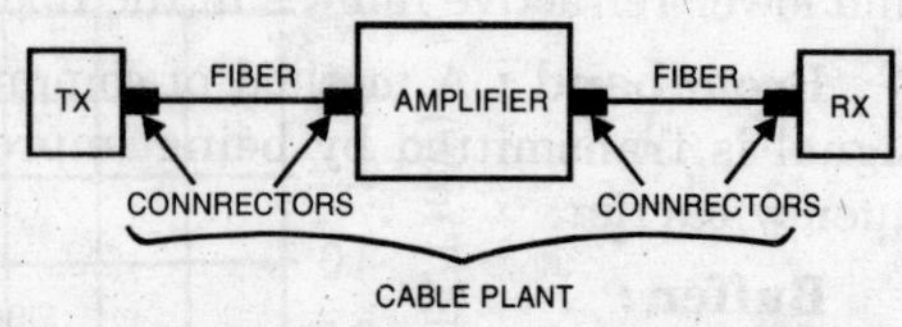

Carrier-to-Noise Ratio (CNR) : The ratio, in decibels, of the level of the carrier to that of the noise in a receiver's IF bandwidth before any nonlinear process such as amplitude limiting and detection takes place.

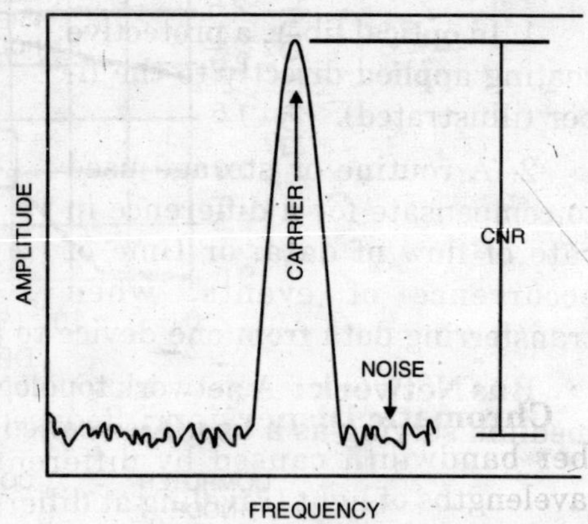

CCIR : Abbreviation for Consultative Committee on Radio. Replaced by ITU-R.

CCITT : Abbreviation for Consultative Committee on Telephony and Telegraphy. Replaced by ITU-T.

CDMA : Abbreviation for code-division multiple access. A coding scheme in which multiple channels are independently coded for transmission over a single wideband channel using an individual modulation scheme for each channel.

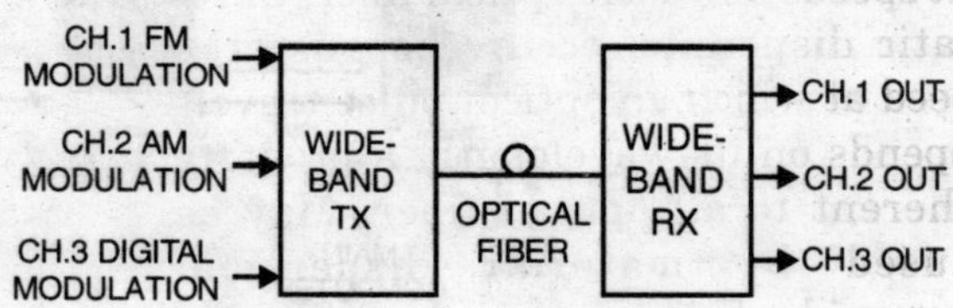

Center Wavelength : In a laser, the nominal value central operating wavelength. It is the wavelength defined by a peak mode measurement where the effective optical power resides (see illustration). In an LED, the average of the two wavelengths measured at the half amplitude points of the power spectrum.

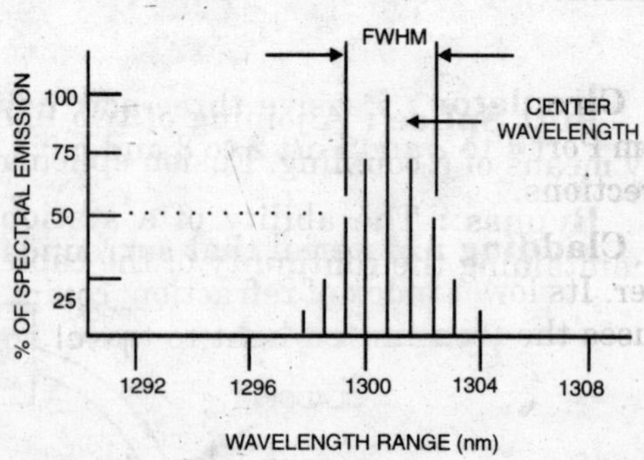

Channel : A communications path or the signal sent over that path. Through multiplexing several channels, voice channels can be transmitted over an optical channel.

Channel Capacity : Maximum number of channels that a cable system can carry simultaneously.

Channel Coding : Data encoding and error correction techniques used to protect the integrity of data. Typically used in channels with high bit error rates such as terrestrial and satellite broadcast and videotape recording.

Chirp : In laser diodes, the shift of the laser's center wavelength during single pulse durations.

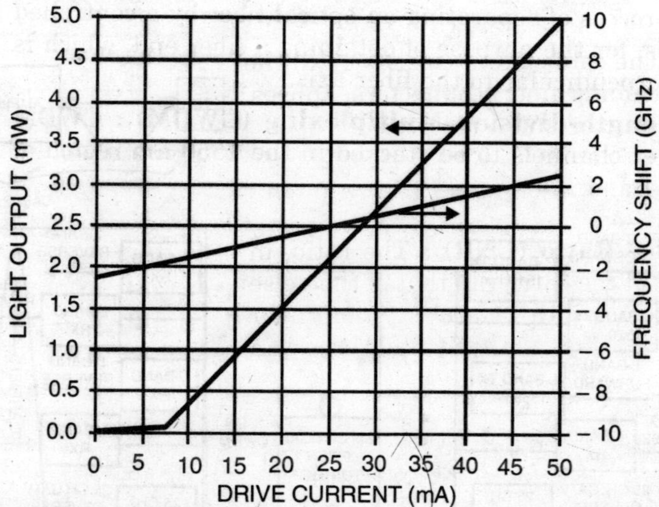

Chromatic Dispersion : Reduced fiber bandwidth caused by different wavelengths of light traveling at different speeds down the optical fiber. Chromatic dispersion occurs because the speed at which an optical pulse travels depends on its wavelength, a property inherent to all optical fiber. May be caused by material dispersion, waveguide dispersion, and profile dispersion.

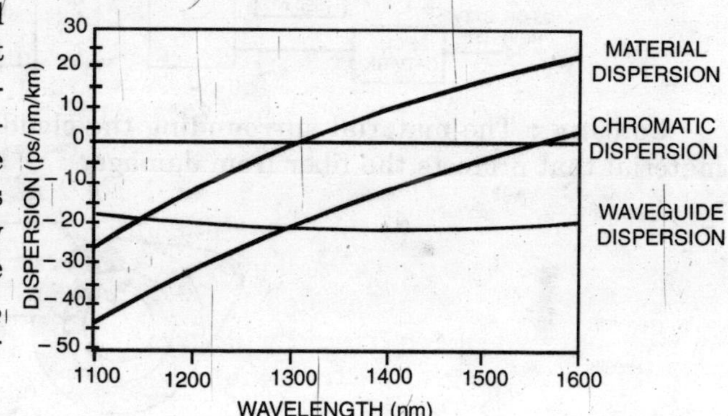

Circulator : Passive three-port devices that couple light from Port 1 to 2 and Port 2 to 3 and have high isolation in other directions.

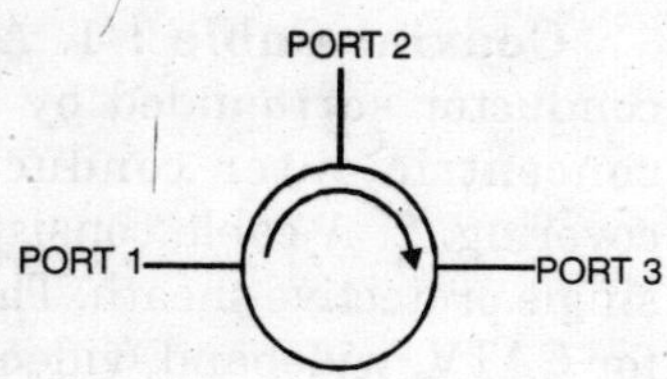

Cladding : Material that surrounds the core of an optical fiber. Its lower index of refraction, compared to that of the core, causes the transmitted light to travel down the core.

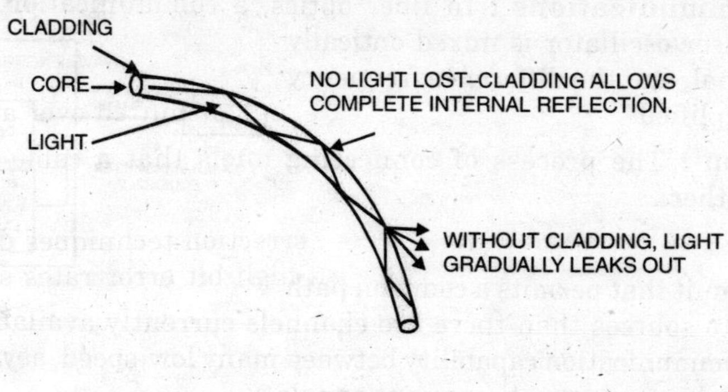

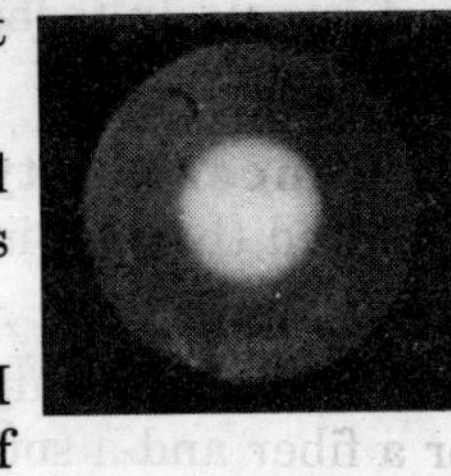

Cladding Mode : A mode confined to the cladding; a light ray that propagates in the cladding.

Cleave : The process of separating an optical fiber by a controlled fracture of the glass, for the purpose of obtaining a fiber end, which is flat, smooth, and perpendicular to the fiber axis.

Coarse Wavelength-division Multiplexing (CWDM) : CWDM allows eight or fewer channels to be stacked in the 1550 nm region of optical fiber, the C-Band.

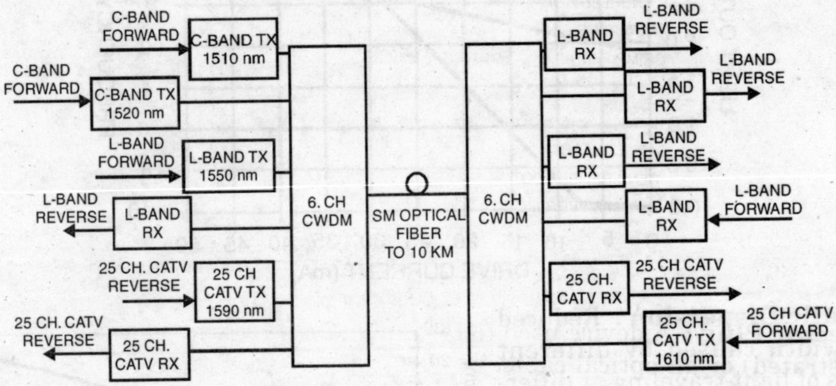

Coating : The material surrounding the cladding of a fiber. Generally a soft plastic material that protects the fiber from damage.

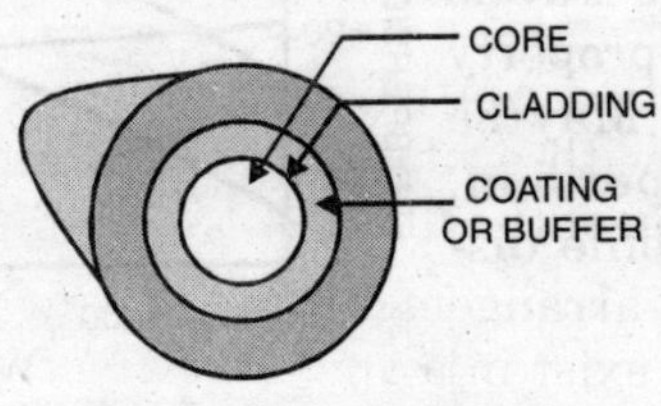

Coaxial Cable : 1. A cable consisting of a center conductor surrounded by an insulating material and a concentric outer conductor and optional protective covering. 2. A cable consisting of multiple tubes under a single protective sheath. This type of cable is typically used for CATV, wideband, video, or RF applications.

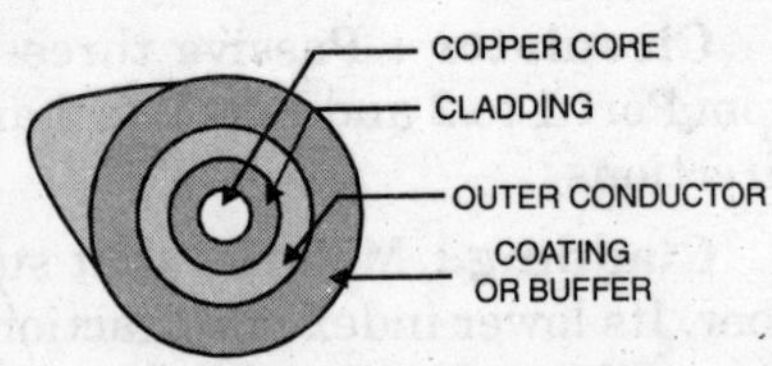

Coherent Communications : In fiber optics, a communication system where the output of a local laser oscillator is mixed optically with a received signal, and the difference frequency is detected and amplified.

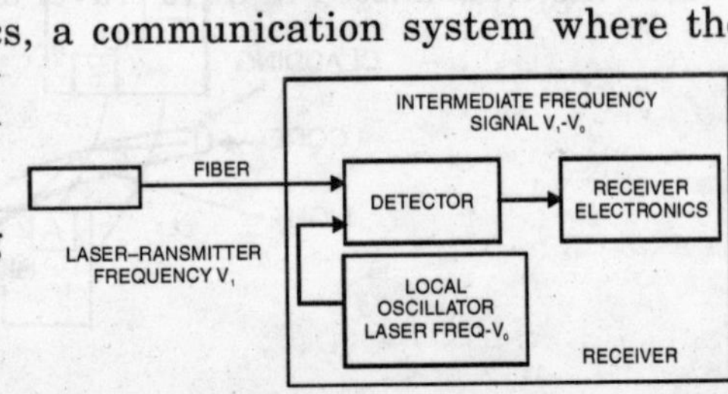

Concatenation : The process of connecting pieces of fiber together.

Concentrator :

1. A functional unit that permits a common path to handle more data sources than there are channels currently available within the path. Usually provides communication capability between many low-speed, asynchronous channels and one or more high-speed, synchronous channels.

2. A device that connects a number of circuits, which are not all used at once, to a smaller group of circuits for economy.

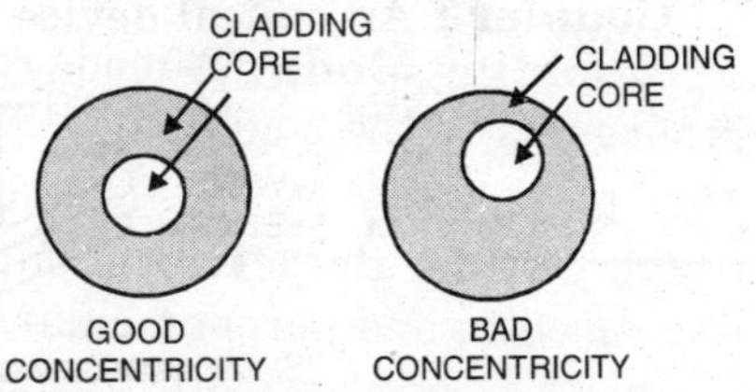

Concentricity : The measurement of how well-centered the core is within the cladding.

Connector : A mechanical or optical device that provides a demountable connection between two fibers or a fiber and a source or detector.

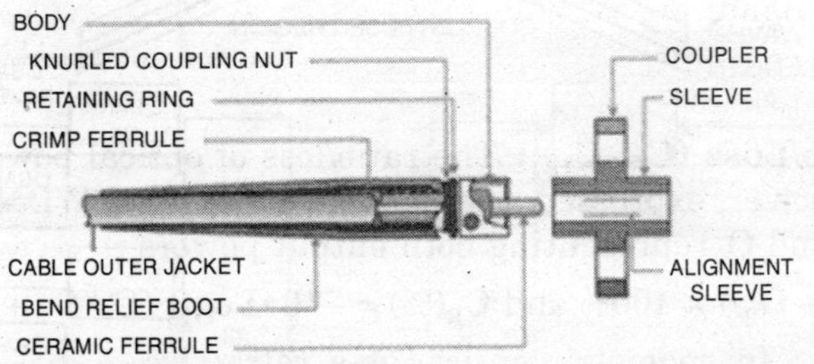

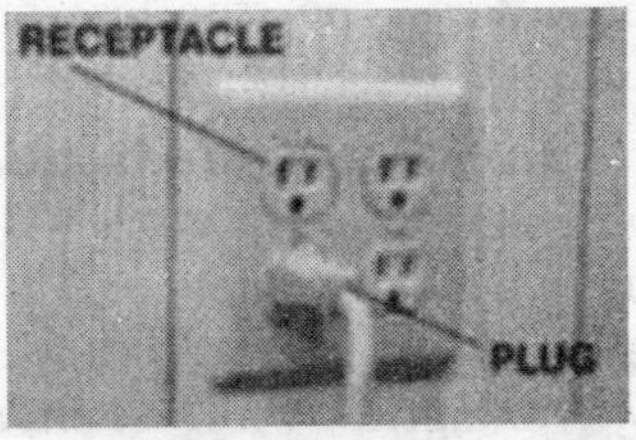

Connector Plug : A device used to terminate an electrical cable (illustrated) or an optical cable.

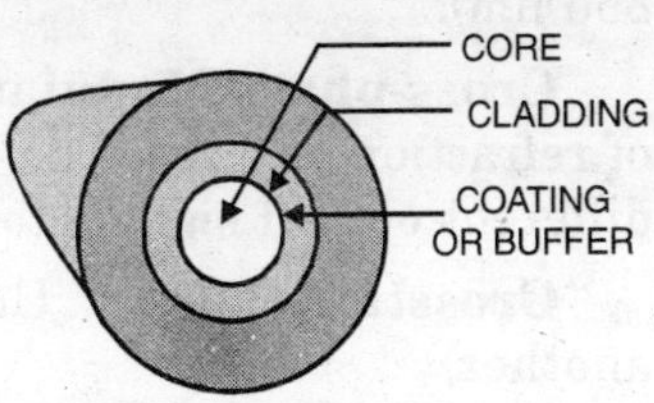

Core : The light-conducting central portion of an optical fiber, composed of material with a higher index of refraction than the cladding. The portion of the fiber that transmits light.

Counter-Rotating : An arrangement whereby two signal paths, one in each direction, exist in a ring topology.

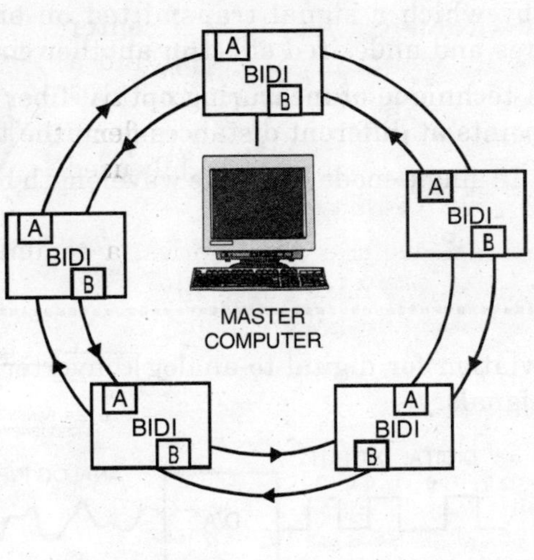

Coupler : An optical device that combines or splits power from optical fibers.

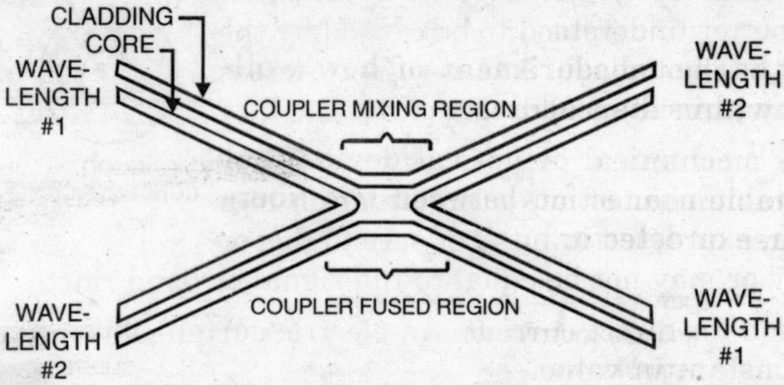

Coupling Ratio/Loss (C_R, C_L) : The ratio/loss of optical power from one output port to the total output power, expressed as a percent. For a 1 x 2 WDM or coupler with output powers O_1 and O_2, and O_i representing both output powers :

$C_R(\%) = (O_i/(O_1 + O_2)) \times 100\%$ and $C_R(\%) = -10 \bullet \text{Log}_{10} (O_i/(O_1 + O_2))$.

Critical Angle : In geometric optics, at a refractive boundary, the smallest angle of incidence at which total internal reflection occurs.

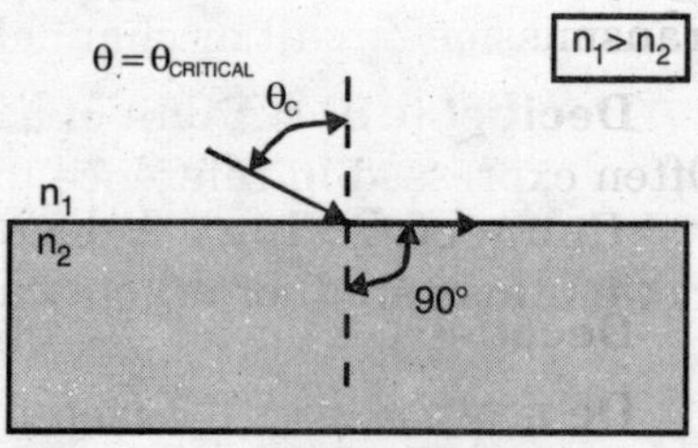

Cross-gain Modulation (XGM) : A technique used in wavelength converters where gain saturation effects in an active optical device, such as a semiconductor optical amplifier (SOA), allow the conversion of the optical wavelength. Better at shorter wavelengths (e.g. 780 nm or 850 nm).

Cross-phase Modulation (XPM) : A fiber nonlinearity caused by the nonlinear index of refraction of glass. The index of refraction varies with optical power level which causes different optical signals to interact.

Crosstalk (XT) : 1. Undesired coupling from one circuit, part of a circuit, or channel to another.

2. Any phenomenon by which a signal transmitted on one circuit or channel of a transmission system creates and undesired effect in another circuit or channel.

Cutback Method : A technique of measuring optical fiber attenuation by measuring the optical power at two points at different distances from the test source.

Cutoff Wavelength : In single-mode fiber, the wavelength below which the fiber ceases to be single-mode.

D

D/A or DAC : Abbreviation for digital-to-analog converter. A device used to convert digital signals to analog signals.

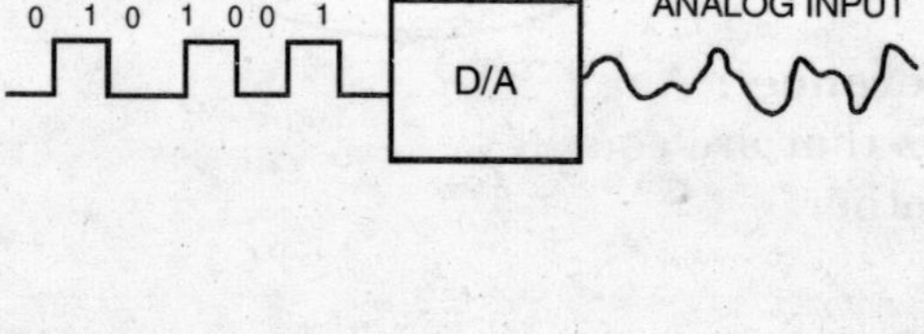

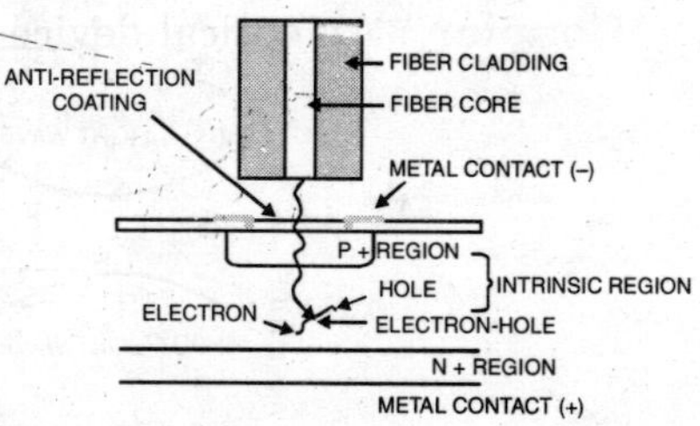

Dark Current : The induced current that exists in a reversed biased photodiode in the absence of incident optical power. It is better understood to be caused by the shunt resistance of the photodiode. A bias voltage across the diode (and the shunt resistance) causes current to flow in the absence of light.

Data Rate : The number of bits of information in a transmission system, expressed in bits per second (b/s or bps), and which may or may not be equal to the signal or baud rate.

DC : Abbreviation for direct current. An electric current flowing in one direction only and substantially constant in value.

DCE : Abbreviation for **data circuit-terminating equipment.** (1) In a data station, the equipment that performs functions such as signal conversion and coding, at the network end of the line between the data terminal equipment (DTE) and the line, and may be a separate or an integral part of the DTE or of intermediate equipment. (2) The interfacing equipment that may be required to couple the data terminal equipment (DTE) into a transmission circuit or channel and from a transmission circuit of a channel into the DTE.

Decibel (dB) : A unit of measurement indicating relative power on a logarithmic scale. Often expressed in reference to a fixed value, such as dBm or dBμ .

$$dB = 10 \, Log_{10} \, (P_1/P_2)$$

Decoder : A device used to convert data by reversing the effect of previous coding.

De multiplexer : A module that separates two or more signals previously combined by compatible multiplexing equipment.

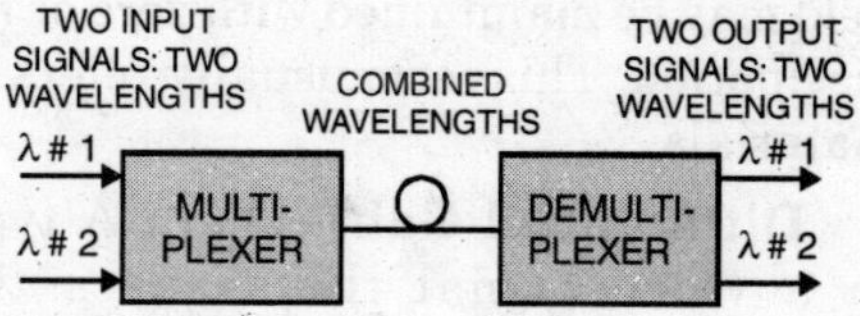

Dense Wavelength-division Multiplexing (DWDM) : The transmission of many of closely spaced wavelengths in the 1550 nm region over a single optical fiber. Wavelength spacings are usually 100 GHz or 200 GHz which corresponds to 0.8 nm or 1.6 nm. DWDM bands include the C-Band, the S-Band, and the L-Band.

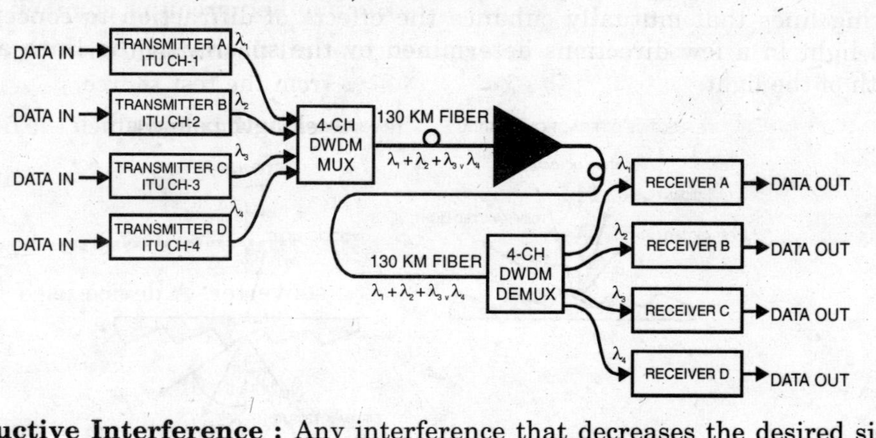

Destructive Interference : Any interference that decreases the desired signal. For example, two light waves that are equal in amplitude and frequency, and out of phase by 180°, will negate one another.

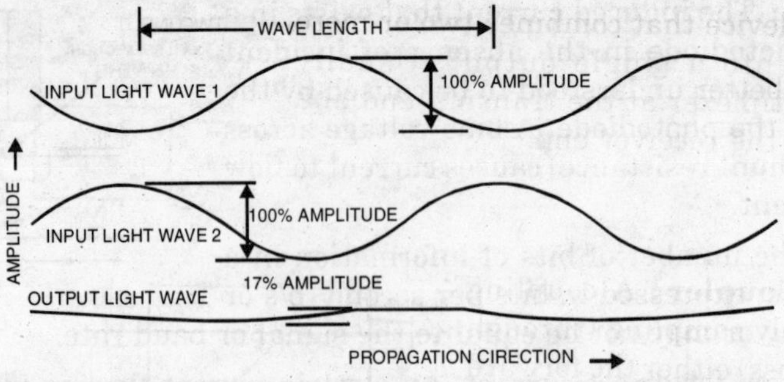

Diameter-mismatch Loss : The loss of power at a joint that occurs when the transmitting fiber has a diameter greater than the diameter of the receiving fiber. The loss occurs when coupling light from a source to fiber, from fiber to fiber, or from fiber to detector.

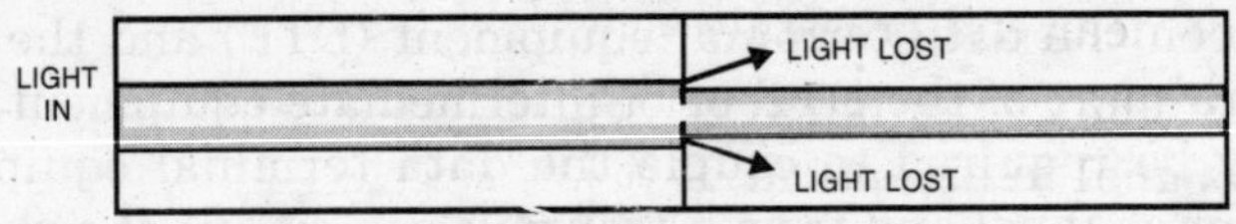

Dichroic Filter : An optical filter that transmits light according to wavelength. Dichroic filters reflect light that they do not transmit. Used in bulk optics wavelength-division multiplexers.

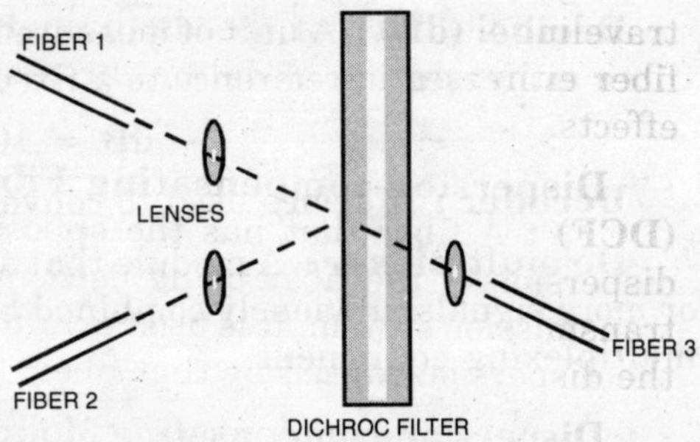

Dielectric : Any substance in which an electric field may be maintained with zero or near-zero power dissipation. This term usually refers to non-metallic materials.

Differential Gain (DG) : A type of distortion in a video signal that causes the brightness information to be incorrectly interpreted.

Differential Phase (DP) : A type of distortion in a video signal that causes the color information to be incorrectly interpreted.

Diffraction Grating : An array of fine, parallel, equally spaced reflecting or transmitting lines that mutually enhance the effects of diffraction to concentrate the diffracted light in a few directions determined by the spacing of the lines and by the wavelength of the light.

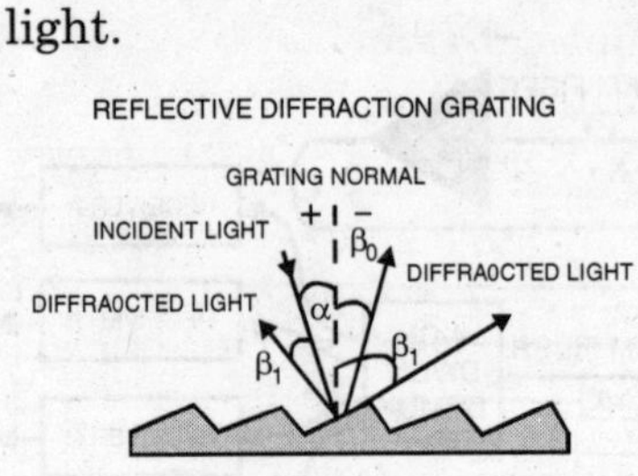

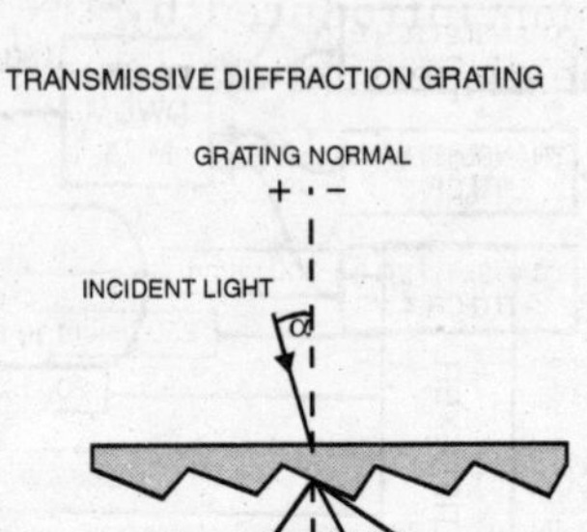

Diode Laser : Synonymous with injection laser diode.

Diplexer : A device that combines two or more types of signals into a single output. Usually incorporates a multiplexer at the transmit end and a demultiplexer at the receiver end.

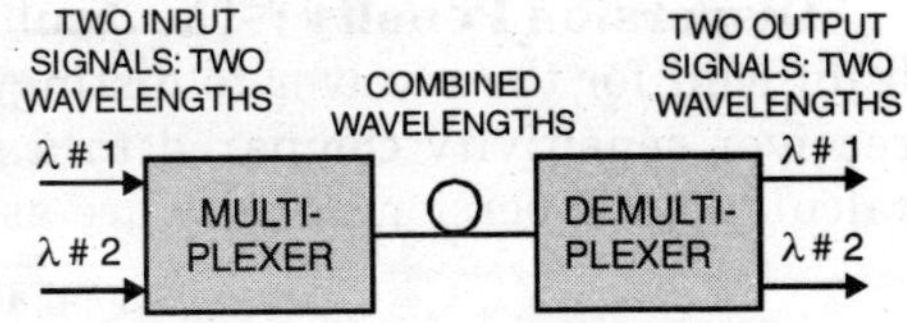

Directional Coupler : A coupling device for separately sampling (through a known coupling loss) either the forward (incident) or the backward (reflected) wave in a transmission line.

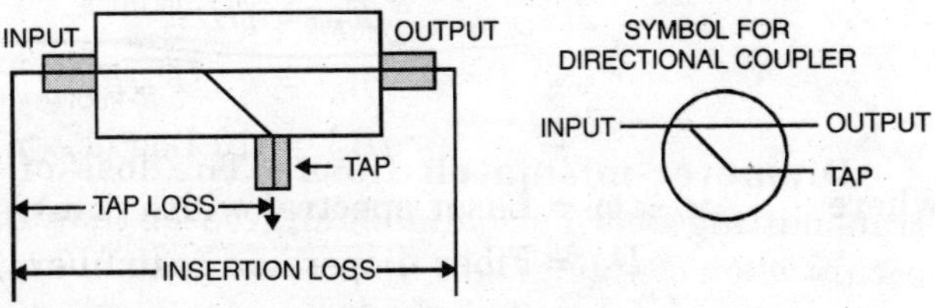

Directivity : See near-end crosstalk.

Dispersion : The temporal spreading of a light signal in an optical waveguide caused by light signals traveling at different speeds through a fiber either due to modal or chromatic effects.

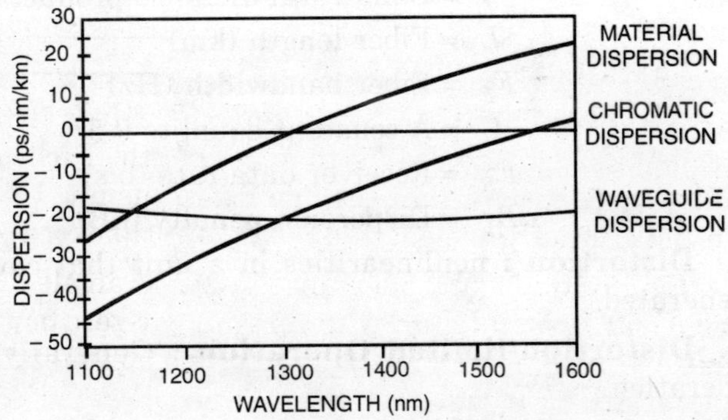

Dispersion-compensating Fiber (DCF) : A fiber that has the opposite dispersion of the fiber being used in a transmission system. It is used to nullify the dispersion caused by that fiber.

Dispersion-compensating Module (DCM) : This module has the opposite dispersion of the fiber being used in a transmission system. It is used to nullify the dispersion caused by that fiber. It can be either a spool of a special fiber or a grating based module.

Dispersion-shifted Fiber (DSF) : A type of single-mode fiber designed to have zero dispersion near 1550 nm. This fiber type works very poorly for DWDM applications because of high fiber nonlinearity at the zero-dispersion wavelength.

Dispersion Management : A technique used in a fiber optic system design to cope with the dispersion introduced by the optical fiber. A dispersion slope compensator (illustrated) is one dispersion management technique.

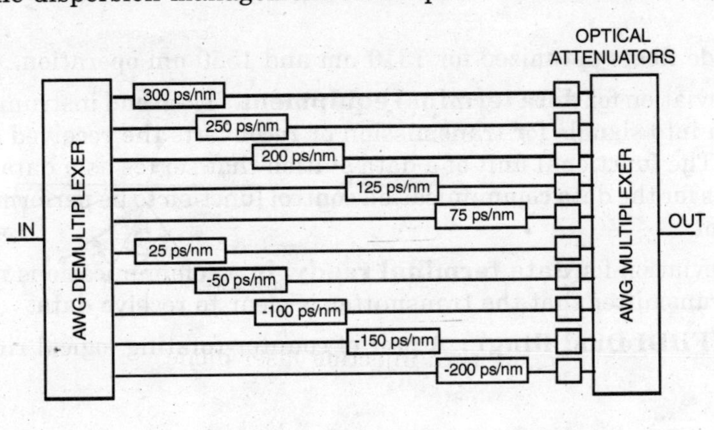

Dispersion Penalty : The result of dispersion in which pulses and edges smear making it difficult for the receiver to distinguish between ones and zeros. This results in a loss of receiver sensitivity compared to a short fiber and measured in dB. The equations for calculating dispersion penalty are as follows:

$$\tau = \omega . D_\lambda \text{ ps/km}$$
$$f = 1\eta(4)/(\tau \times \pi) \text{ Hz . km}$$
$$F_F = f/L$$
$$\eta_L = c \cdot (F_R/F_F)^2$$
$$dB_L = 10 \text{ Log } (1 + \eta)_L \text{ dB}$$

where :
- ω = Laser spectral width (nm)
- D_λ = Fiber dispersion (ps/nm/km)
- τ = System dispersion (ps/km)
- f = Bandwidth-distance product of the fiber (Hz • km)
- L = Fiber length (km)
- F_F = Fiber bandwidth (Hz)
- C = A constant equal to 0.5
- F_R = Receiver data rate (b/s)
- dB_L = Dispersion penalty (dB)

Distortion : nonlinearities in a unit that cause harmonics and beat products to be generated.

Distortion-limited Operation : Generally synonymous with bandwidth-limited operation.

Distributed Feedback Laser (DFB) : An injection laser diode which has a Bragg reflection grating in the active region in order to suppress multiple longitudinal modes and enhance a single longitudinal mode.

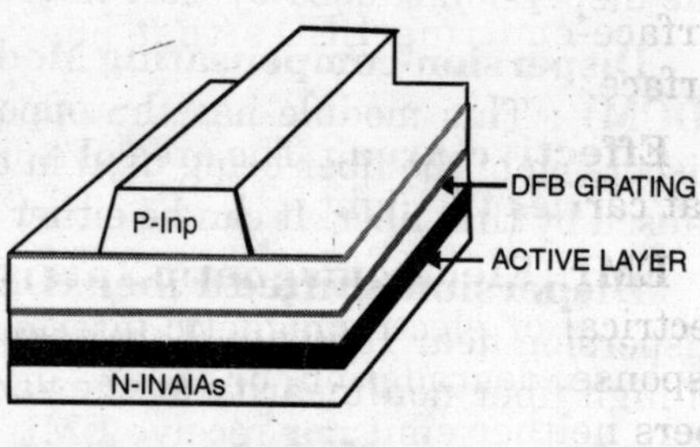

Dominant Mode : The mode in an optical device spectrum with the most power.

Dopant : An impurity added to an optical medium to change its optical properties. EDFAs use erbium as a dopant for optical fiber.

Double-window Fiber : 1. Multimode fibers optimized for 850 nm and 1310 nm operation.

2. Single-mode fibers optimized for 1310 nm and 1550 nm operation.

DTE : Abbreviation for **data terminal equipment**. (1) An end instrument that converts user information into signals for transmission or reconverts the received signals into user information. (2) The functional unit of a data station that serves as a data source or a data sink and provides for the data communication control function to be performed in accordance with link protocol.

DTR : Abbreviation for **data terminal ready.** In a communications network, a signal from a remote transmitter that the transmitter is clear to receive data.

Dual Ring (FDDI Dual Ring) : A pair of counter-rotating logical rings.

Duplex Cable : A two-fiber cable suitable for duplex transmission.

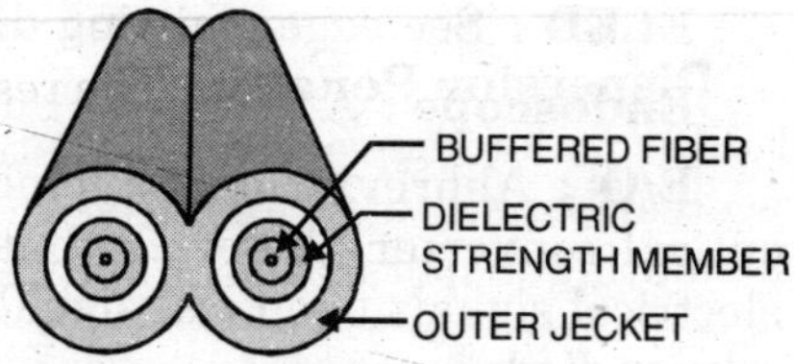

Duplex Transmission : Transmission in both directions, either one direction at a time (half-duplex) or both directions simultaneously (full-duplex).

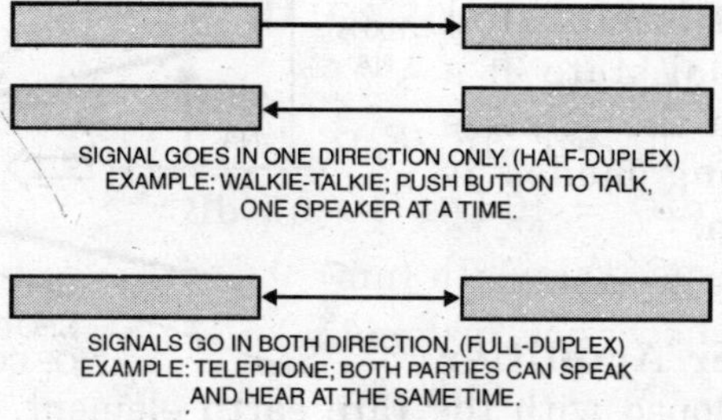

Duty Cycle : In a digital transmission, the fraction of time a signal is at the high level.

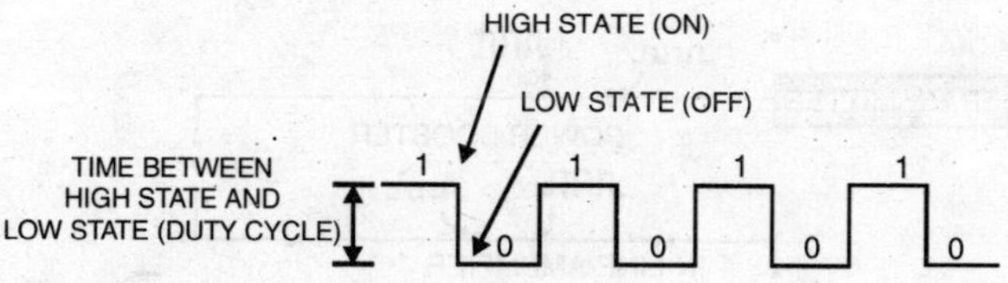

E

Edge-emitting Diode : An LED that emits light from its edge, producing more directional output than surface-emitting LED's that emit from their top surface.

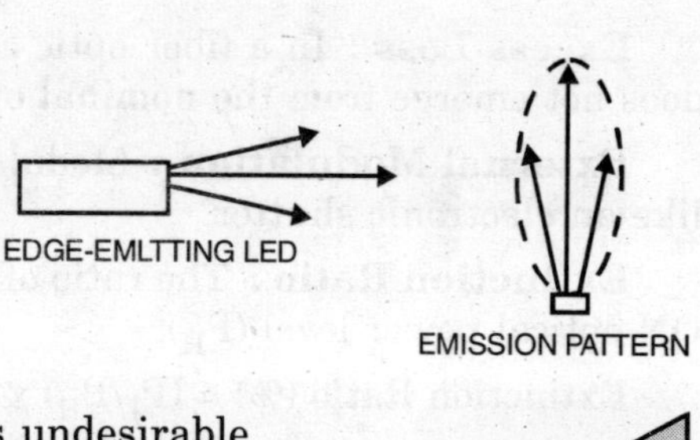

Effective Area : The area of a single-mode fiber that carries the light.

EMI (Electromagnetic Interference) : Any electrical or electromagnetic interference that causes undesirable response, degradation, or failure in electronic equipment. Optical fibers neither emit nor receive EMI.

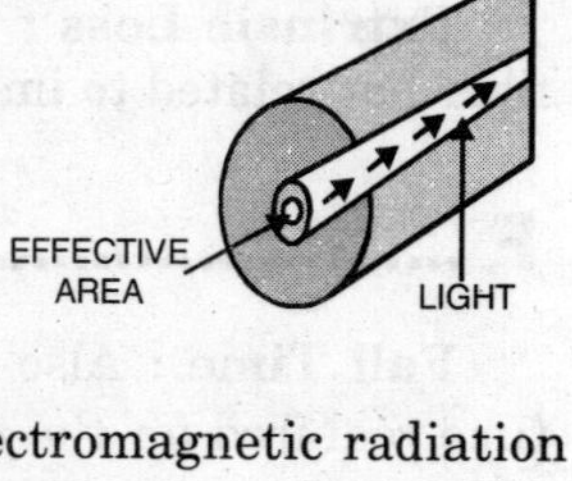

EMR (Electromagnetic Radiation) : Radiation made up of oscillating electric and magnetic fields and propagated with the speed of light. Includes gamma radiation, X-rays, ultraviolet, visible, and infrared radiation, and radar and radio waves.

Electromagnetic Spectrum : The range of frequencies of electromagnetic radiation from zero to infinity.

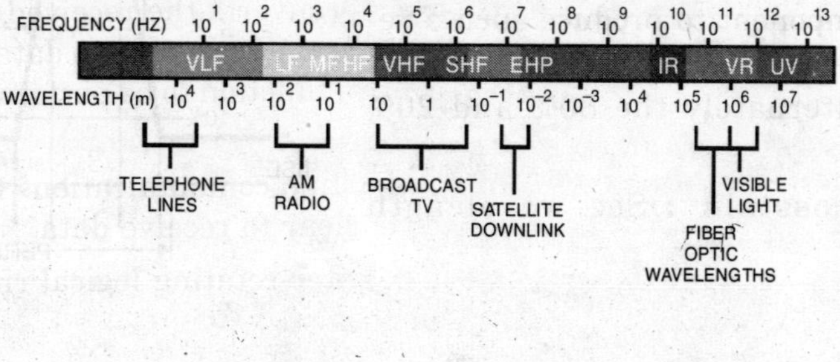

ELED : See edge-emitting diode.

Endoscope : A fiber optic bundle used for imaging and viewing inside the human body.

E/O : Abbreviation for electrical-to-optical converter. A device that converts electrical signals to optical signals, such as a laser diode.

Equilibrium Mode Distribution (EMD) : The steady modal state of a multimode fiber in which the relative power distribution among modes is independent of fiber length.

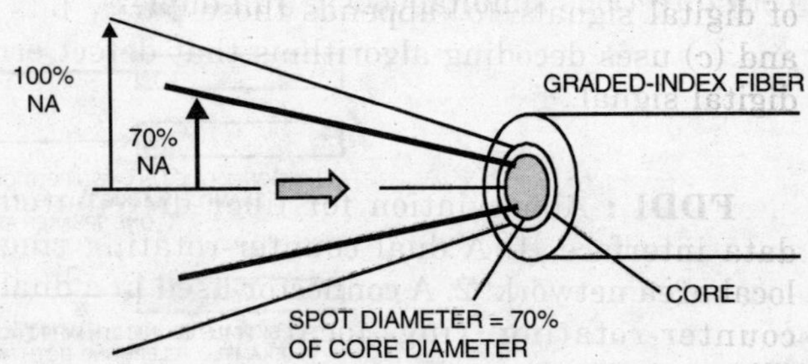

Erbium-doped Fiber Amplifier (EDFA) : Optical fibers doped with the rare earth element, erbium, which can amplify light in the 1550 nm region when pumped by an external light source.

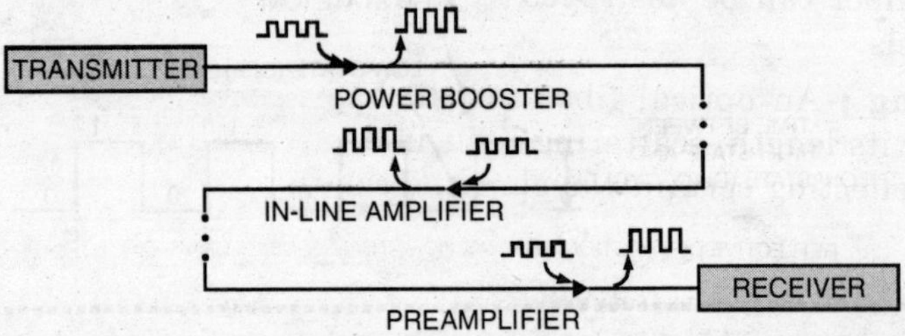

Excess Loss : In a fiber optic coupler, the optical loss from that portion of light that does not emerge from the nominal operation ports of the device.

External Modulation : Modulation of a light source by an external device that acts like an electronic shutter.

Extinction Ratio : The ratio of the low, or OFF optical power level (P_L) to the high, or ON optical power level (P_H):

Extinction Ratio (%) = $(P_L/P_H) \times 100$

Extrinsic Loss : In a fiber interconnection, that portion of loss not intrinsic to the fiber but related to imperfect joining of a connector or splice.

F

Fall Time : Also called turn-off time. The time required for the trailing edge of a pulse to fall from 90% to 10% of its amplitude; the time required for a component to produce such a result. Typically measured between the 90% and 10% points or alternately the 80% and 20% points.

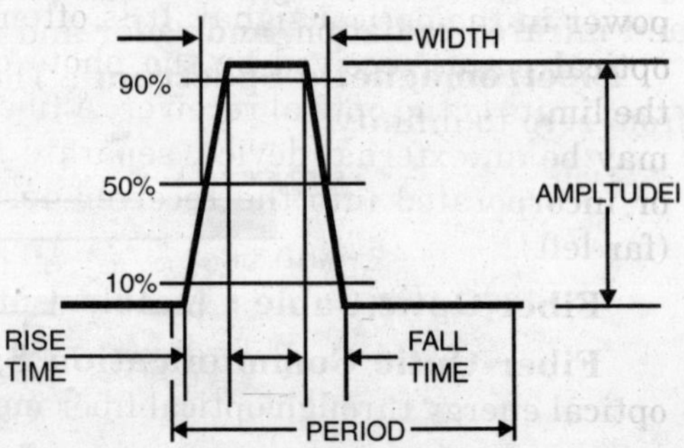

Far-End Crosstalk : See wavelength isolation.

FC/PC : See FC. A threaded optical connector that uses a special curved polish on the connector for very low back reflection. Good for single-mode or multimode fiber.

FCS : Abbreviation for frame check sequence. An error-detection scheme that (*a*) uses parity bits generated by polynomial encoding of digital signals, (*b*) appends those parity bits to a digital signal, and (*c*) uses decoding algorithms that detect errors in the received digital signal.

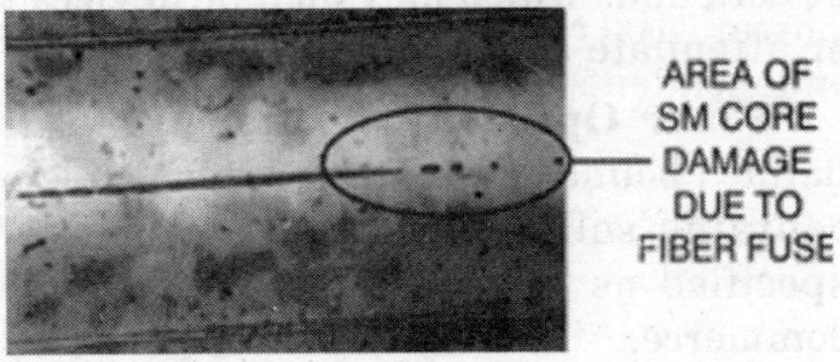

FDDI : Abbreviation for fiber distributed data interface. 1. A dual counter-rotating ring local area network. 2. A connector used in a dual counter-rotating ring local area network (illustrated).

Fiber Fuse : A mechanism whereby the core of a single-mode fiber can be destroyed at high optical power levels.

Fiber Grating : An optical fiber in which the refractive index of the core varies periodically along its length, scattering light in a way similar to a diffraction grating, and transmitting or reflecting certain wavelengths selectively.

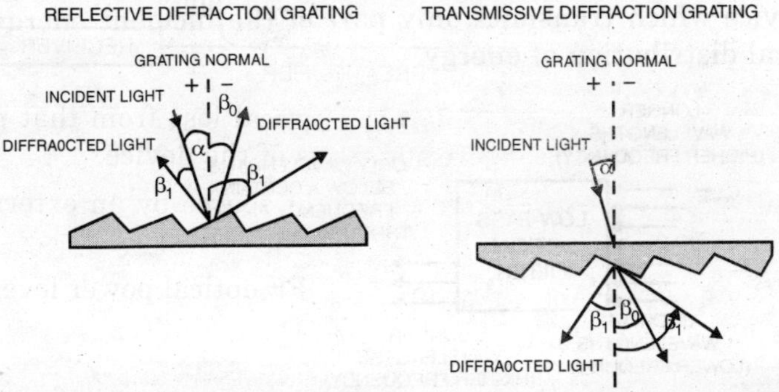

Fiber-in-the-loop (FITL) : Fiber optic service to a node that is located in a neighborhood.

Fiber Optic Attenuator : A component installed in a fiber optic transmission system that reduces the power in the optical signal. It is often used to limit the optical power received by the photo detector to within the limits of the optical receiver. A fiber optic attenuator may be an external device, separate from the receiver, or incorporated into the receiver design as illustrated (far left).

Fiber Optic Cable : A cable containing one or more optical fibers.

Fiber Optic Communication System : The transfer of modulated or unmodulated optical energy through optical fiber media which terminates in the same or different media.

Fiber Optic Link : A transmitter, receiver, and cable assembly that can transmit information between two points.

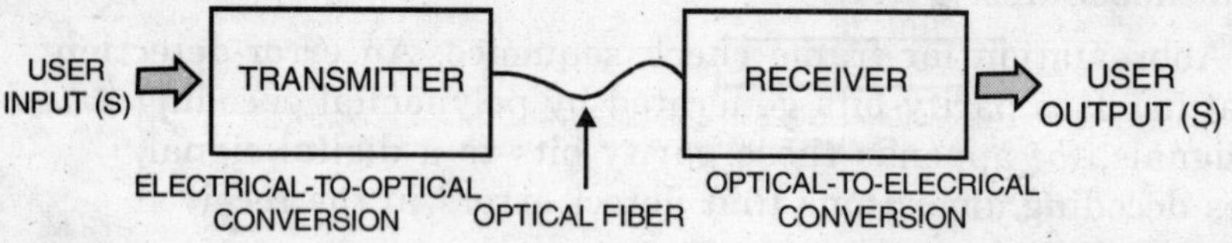

Fiber Optic Span : An optical fiber/cable terminated at both ends which may include devices that add, subtract, or attenuate optical signals.

Fiber Optic Subsystem : A functional entity with defined bounds and interfaces which is part of a system. It contains solid state and/or other components and is specified as a subsystem for the purpose of trade and commerce.

Fiber Channel : An industry-standard specification that originated in Great Britain which details computer channel communications over fiber optics at transmission speeds from 132 Mb/s to 1062.5 Mb/s at distances of up to 10 kilometers.

Filter : A device which transmits only part of the incident energy and may thereby change the spectral distribution of energy.

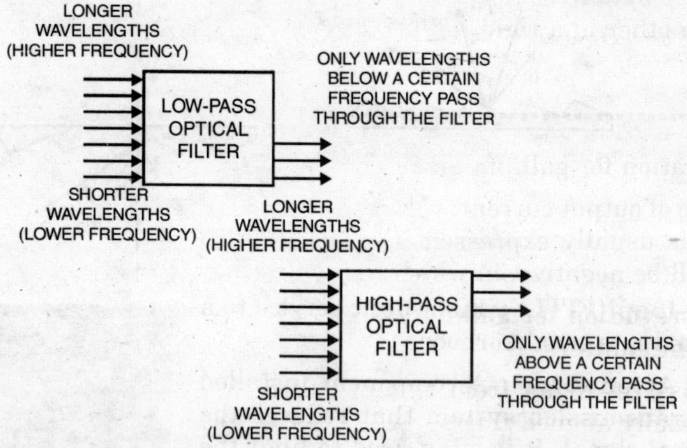

Frequency-division Multiplexing (FDM) : A method of deriving two or more simultaneous, continuous channels from a transmission medium by assigning separate portions of the available frequency spectrum to each of the individual channels.

Frequency-shift Keying (FSK) : Frequency modulation in which the modulating signal shifts the output frequency between predetermined values. Also called frequency-shift modulation, frequency-shift signaling.

Fresnel Reflection Loss : Reflection losses at the ends of fibers caused by differences in the refractive index between glass and air. The maximum reflection caused by a perpendicular air-glass interface is about 4% or about –14 dB.

Full-duplex Transmission : Simultaneous bidirectional transfer of data.

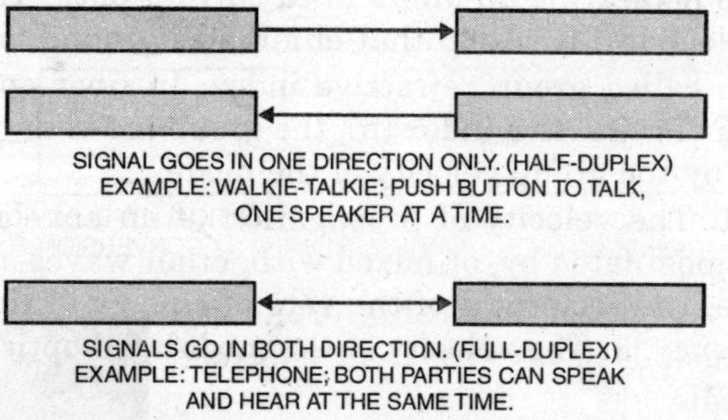

Fused Coupler : A method of making a multimode or single-mode coupler by wrapping fibers together, heating them, and pulling them to form a central unified mass so that light on any input fiber is coupled to all output fibers.

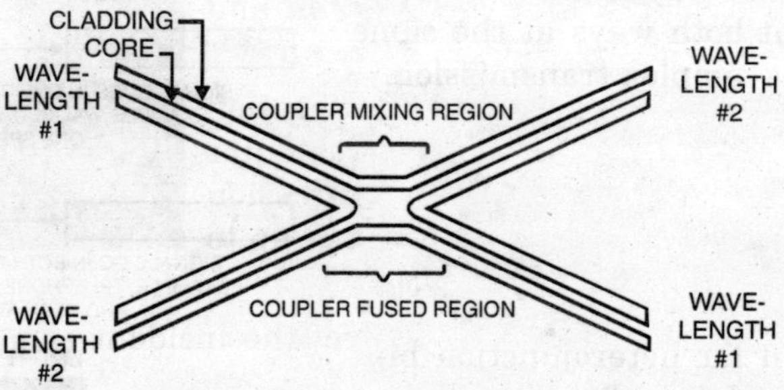

Fused Fiber : A bundle of fibers fused together so they maintain a fixed alignment with respect to each other in a rigid rod.

G ....

GaAs : Abbreviation for gallium arsenide. Used in light emitters.

Gain : The ratio of output current, voltage, or power to input current, voltage, or power, respectively. Gain is usually expressed in dB. If the ratio is less than unity, the gain, expressed in dB, will be negative, in which case there is a loss between input and output.

GaInAsP : Abbreviation for gallium indium arsenide phosphide. Generally used for long wavelength light emitters.

Gap Loss : Loss resulting from the end separation of two axially aligned fibers.

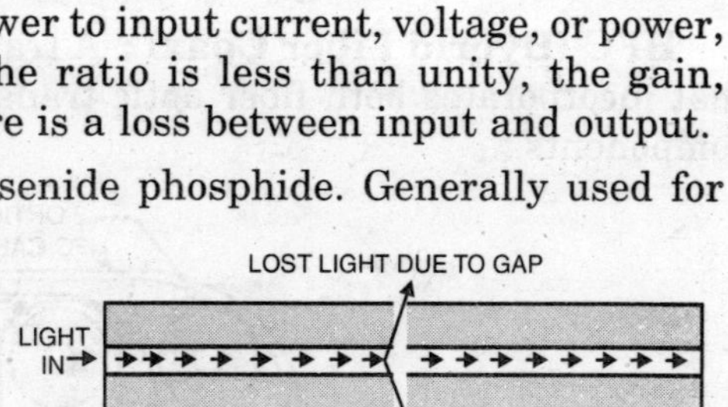

Graded-index Fiber : Optical fiber in which the refractive index of the core is in the form of a parabolic curve, decreasing toward the cladding.

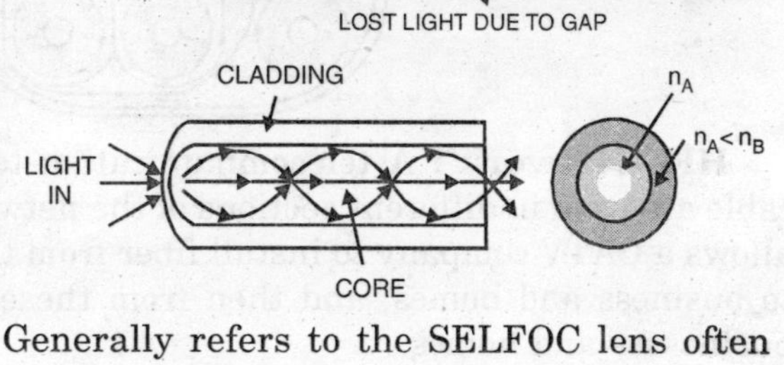

GRIN : Abbreviation for gradient index. Generally refers to the SELFOC lens often used in fiber optics.

Ground Loop Noise : Noise that results when equipment is grounded at points having different potentials thereby creating an unintended current path. The dielectric properties of optical fiber provide electrical isolation that eliminates ground loops.

Group Index : Also called group refractive index. In fiber optics, for a given mode propagating in a medium of refractive index (n), the group index (N), is the velocity of light in a vacuum (c), divided by the group velocity of the mode.

Group Velocity : 1. The velocity of propagation of an envelope produced when an electromagnetic wave is modulated by, or mixed with, other waves of different frequencies. 2. For a particular mode, the reciprocal of the rate of change of the phase constant with respect to angular frequency. 3. The velocity of the modulated optical power.

H

Half-Duplex Transmission : A bidirectional link that is limited to one-way transfer of data, *i.e.*, data cannot be sent both ways at the same time. Also referred to as simplex transmission.

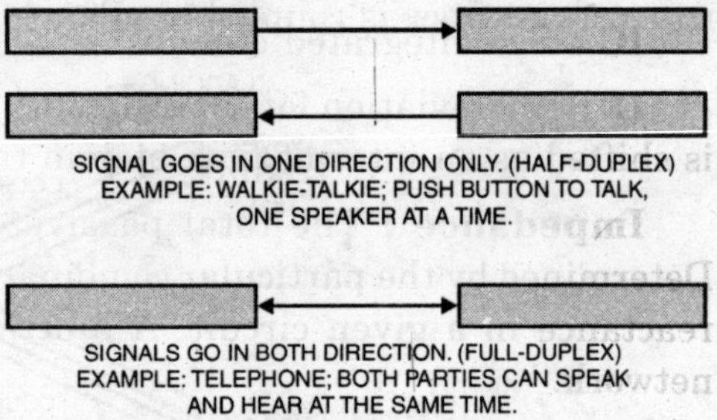

HBT : Abbreviation for heterojunction bipolar transistors. A very high performance transistor structure built using more than one semiconductor material. Used in high performance wireless telecommunications circuits such as those used in digital cell phone handsets and high-bandwidth fiber optic systems.

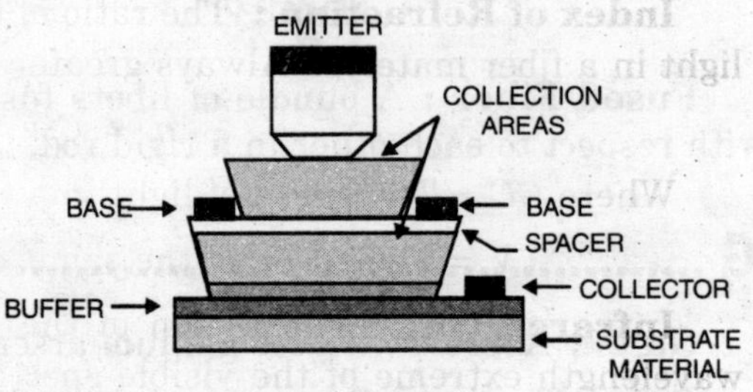

Hertz (Hz) : One cycle per second.

HFC (Hybrid Fiber Coax) : A transmission system or cable construction (illustrated) that incorporates both fiber optic transmission components and copper coax transmission components.

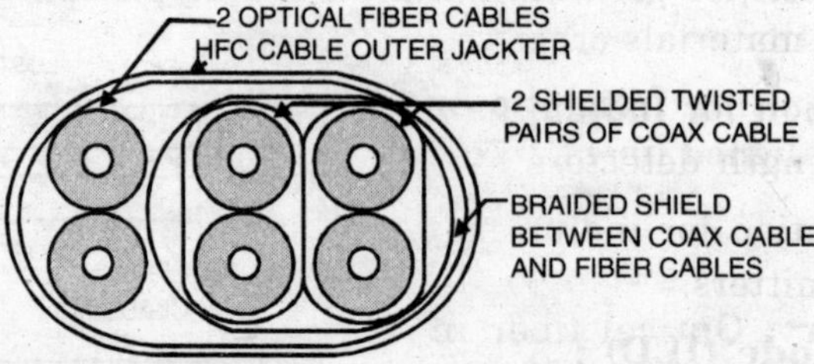

HFC Network : A telecommunication technology in which optical fiber and coaxial cable are used in different sections of the network to carry broadband content. The network allows a CATV company to install fiber from the cable headend to serve nodes located close to business and homes, and then from these fiber nodes, use coaxial cable to individual businesses and homes.

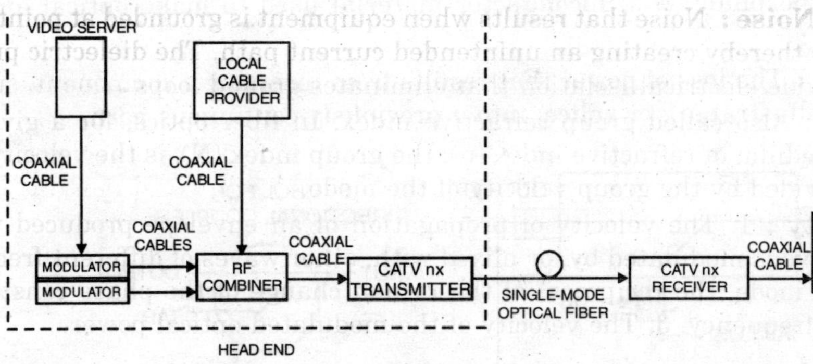

I

IC : See integrated circuit.

IF : Abbreviation for Intermediate Frequency. A frequency to which a carrier frequency is shifted as an intermediate step in transmission or reception.

Impedance : The total passive opposition offered to the flow of electric current. Determined by the particular combination of resistance, inductive reactance, and capacitive reactance in a given circuit. A function of frequency, except when in a purely resistive network.

Index of Refraction : The ratio of the velocity of light in free space to the velocity of light in a fiber material. Always greater than or equal to one. Also called refractive index.

$$n = C/V$$

Where : C = The speed of light in a vacuum.

V = The speed of the same wavelength in the fiber material.

Infrared (IR) : The region of the electromagnetic spectrum bounded by the long-wavelength extreme of the visible spectrum (about 0.7 μm) and the shortest microwaves (about 0.1 μm).

Infrared Emitting Diodes : LEDs that emit infrared energy (830 nm or longer).

Infrared Fiber : Colloquially, optical fibers with best transmission at wavelengths of 2 mm or longer, made of materials other than silica glass.

InGaAs : Abbreviation for indium gallium arsenide. Generally used to make high-performance long-wavelength detectors.

InGaAsP : Abbreviation for indium gallium arsenide phosphide. Generally used for long-wavelength light emitters.

Injection Laser Diode (ILD) : A laser employing a forward-biased semiconductor junction as the active medium. Stimulated emission of coherent light occurs at a PIN junction where electrons and holes are driven into the junction.

In-line Amplifier : An EDFA or other type of amplifier placed in a transmission line to strengthen the attenuated signal for transmission onto the next, distant site. In-line amplifiers are all-optical devices.

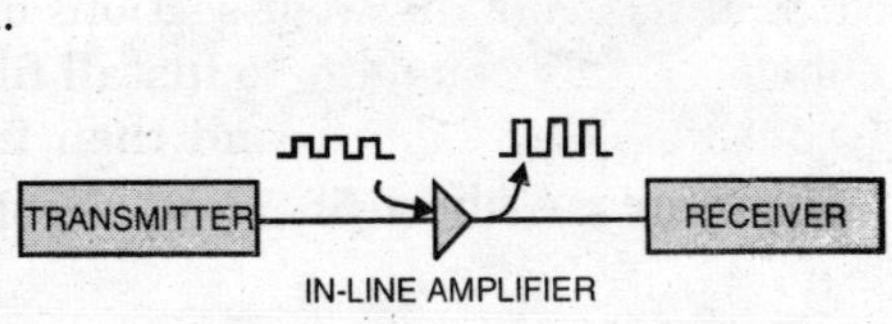

InP : Indium Phosphide. A semiconductor material used to make optical amplifiers and HBTs.

Insertion Loss : The loss of power that results from inserting a component, such as a connector, coupler (illustrated), or splice, into a previously continuous path.

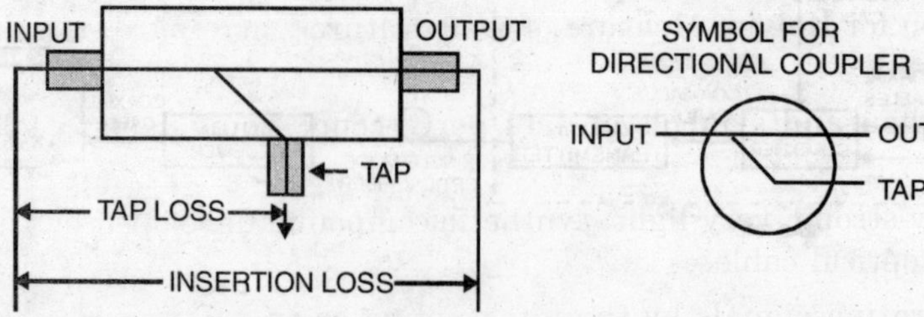

Integrated Detector/Preamplifier (IDP) : A detector package containing a PIN photodiode and transimpedance amplifier.

Interchannel Isolation : The ability to prevent undesired optical energy from appearing in one signal path as a result of coupling from another signal path. Also called crosstalk.

Interference : Any extraneous energy, from natural or manmade sources, that impedes the reception of desired signals. The interference may be constructive or destructive, resulting in increased or decreased amplitude, respectively.

Interferometric Intensity Noise (IIN) : Noise generated in optical fiber caused by the distributed backreflection that all fiber generates mainly due to Rayleigh scattering. OTDRs make use of this scattering power to deduce the fiber loss over distance.

Interferometric Sensors : Fiber optic sensors that rely on interferometric detection.

Intrinsic Losses : Splice losses arising from differences in the fibers being spliced.

Irradiance : Power per unit area.

J

Jacket : The outer, protective covering of the cable. Also called the cable sheath.

Jitter : Small and rapid variations in the timing of a waveform due to noise, changes in component characteristics, supply voltages, imperfect synchronizing circuits, etc. See also DDJ, DCD, and RJ.

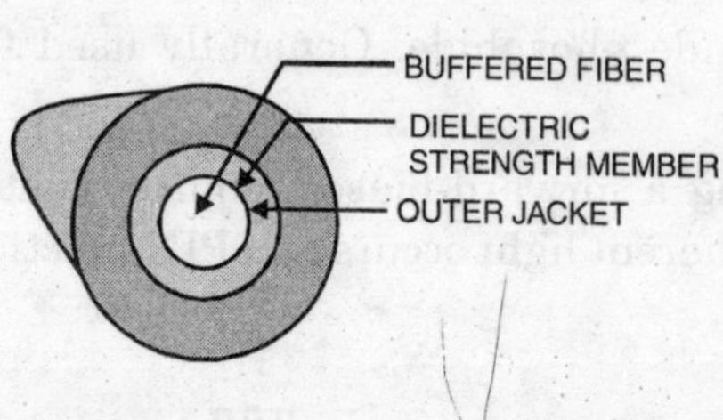

Jumper : A short fiber optic cable with connectors on both ends.

K

k : Abbreviation for kilo. One thousand or 10^3.

K : Abbreviation for Kelvin. Measure of temperature where pure water freezes at 273° and boils at 373°.

kBaud : One thousand symbols of data per second. Equivalent to 1 kb/s for binary signaling.

Kevlar : A very strong, very light, synthetic compound developed by DuPont which is used to strengthen optical cables.

Keying : Generating signals by the interruption or modulation of a steady signal or carrier.

kg : Abbreviation for kilogram. Approximately 2.2 pounds.

kHz : One thousand cycles per second.

km : Abbreviation for kilometer. 1 km = 3,280 feet or 0.62 miles.

L

LAN (Local Area Network) : A communication link between two or more points within a small geographic area, such as between buildings. Smaller than a metropolitan area network (MAN) or a wide area network (WAN).

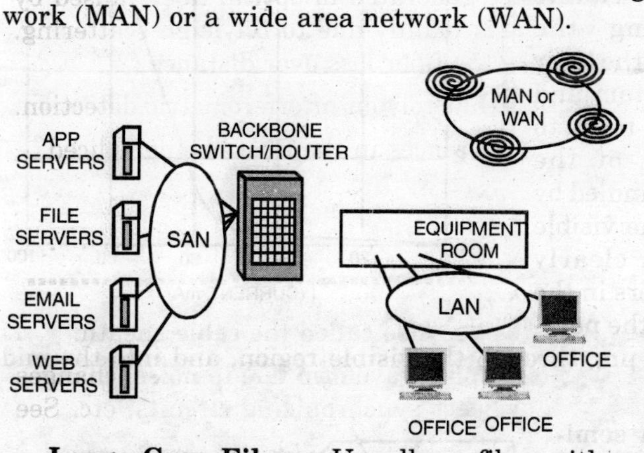

Large Core Fiber : Usually, a fiber with a core of 200 μm or more.

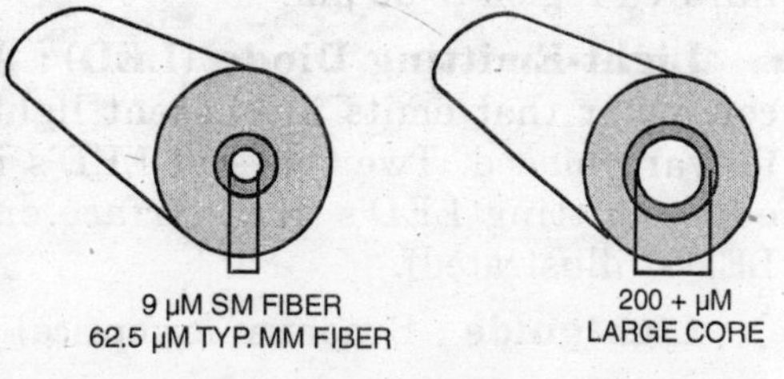

Large Effective Area Fiber (LEAF) : An optical fiber, developed by Corning, designed to have a large area in the core, which carries the light.

Laser : Acronym for light amplification by stimulated emission of radiation. A light source that produces, through stimulated emission, coherent, near monochromatic light.

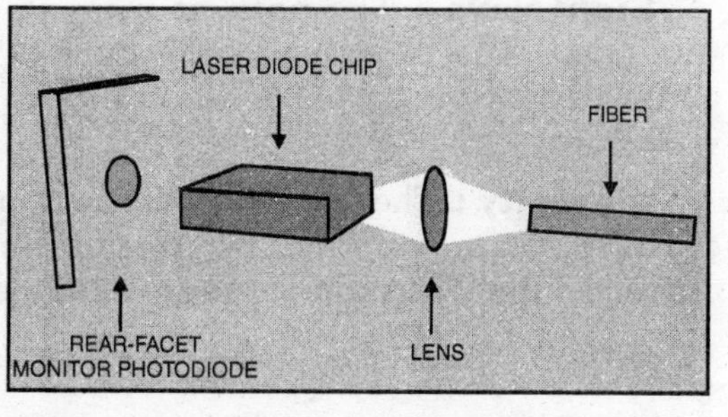

Laser Diode (LD) : A semiconductor that emits coherent light when forward biased.

Lateral Displacement Loss : The loss of power that results from lateral displacement of optimum alignment between two fibers or between a fiber and an active device.

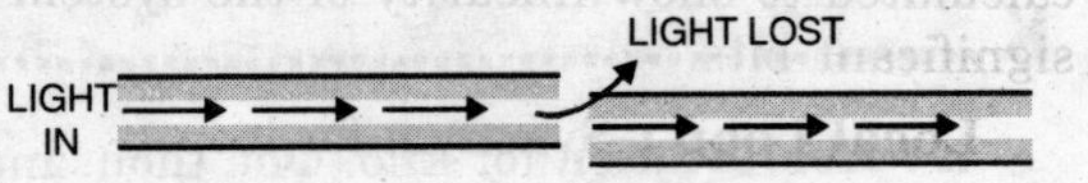

Launch Fiber : An optical fiber used to couple and condition light from an optical source into an optical fiber. Often the launch fiber is used to create an equilibrium mode distribution in multimode fiber. Also called launching fiber.

L-Band : The wavelength range between 1570 nm and 1610 nm used in some CWDM and DWDM applications.

LD : See laser diode.

LEAF : See large effective area fiber.

LED : See light-emitting diode.

L-I Curve : The plot of optical output (L) as a function of current (I) which characterizes an electrical-to-optical converter. A typical L-I curve is shown at right.

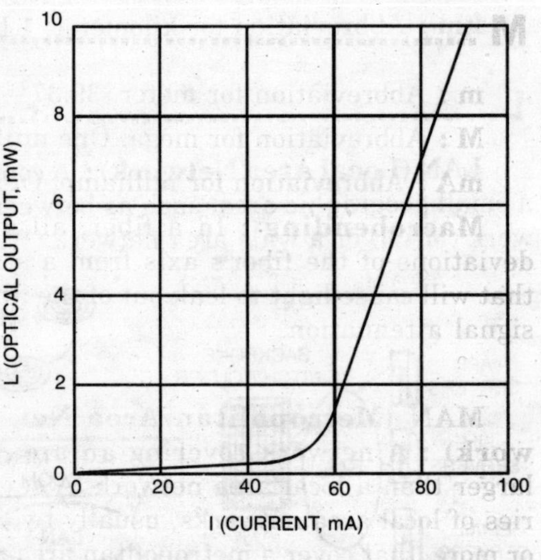

Light : In a strict sense, the region of the electromagnetic spectrum that can be perceived by human vision, designated the visible spectrum, and nominally covering the wavelength range of 0.4 μm to 0.7 μm. In the laser and optical communication fields, custom and practice have extended usage of the term to include the much broader portion of the electromagnetic spectrum that can be handled by the basic optical techniques used for the visible spectrum. This region has not been clearly defined, but, as employed by most workers in the field, may be considered to extend from the near-ultraviolet region of approximately 0.3 μm, through the visible region, and into the mid-infrared region to 30 μm.

Light-Emitting Diode (LED) : A semiconductor that emits incoherent light when forward biased. Two types of LED's include edge-emitting LED's and surface-emitting LED's (illustrated).

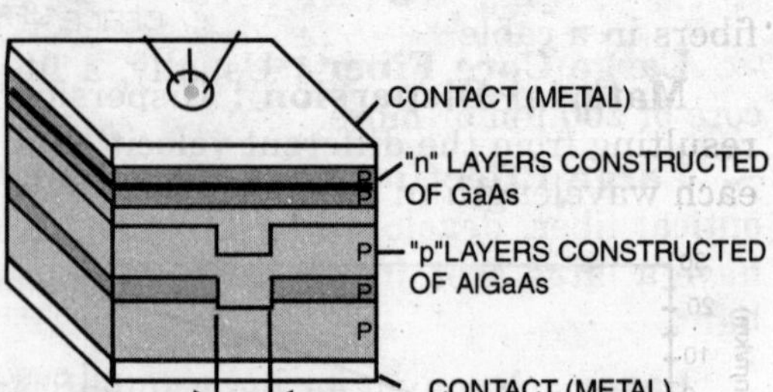

Lightguide : Synonym for optical fiber.

Lightwave : The path of a point on a wavefront. The direction of the lightwave is generally normal (perpendicular) to the wavefront.

Linearity : The basic measurement of how well analog-to-digital and digital-to-analog conversions are performed. To test for linearity, a mathematically perfect diagonal line is converted and then compared to a copy of itself. The difference between the two lines is calculated to show linearity of the system and is given as a percentage or range of least significant bits.

Local Loop : Synonym for loop.

Longitudinal Mode : An optical waveguide mode with boundary condition determined along the length of the optical cavity.

Loose-Tube : A type of fiber optic cable construction where the fiber is contained within a loose tube in the cable jacket.

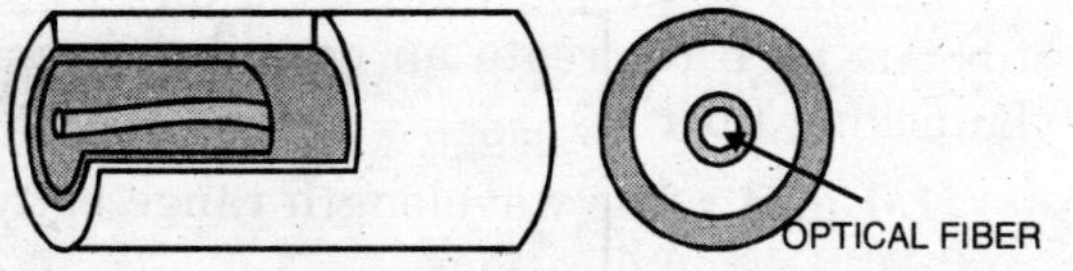

Loss : The amount of a signal's power, expressed in dB, that is lost in connectors, splices, or fiber defects.

Loss Budget : An accounting of overall attenuation in a system. See optical link loss budget.

M

m : Abbreviation for meter. 39.37".

M : Abbreviation for mega. One million or 10^6.

mA : Abbreviation for milliamp. One thousandth of an Amp or 10^{-3} Amps.

Macrobending : In a fiber, all macroscopic deviations of the fiber's axis from a straight line, that will cause light to leak out of the fiber, causing signal attenuation.

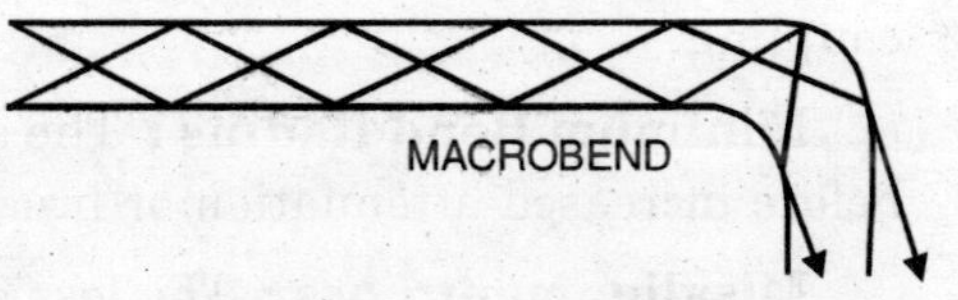

MAN (Metropolitan Area Network) : A network covering an area larger than a local area network. A series of local area networks, usually two or more, that cover a metropolitan area.

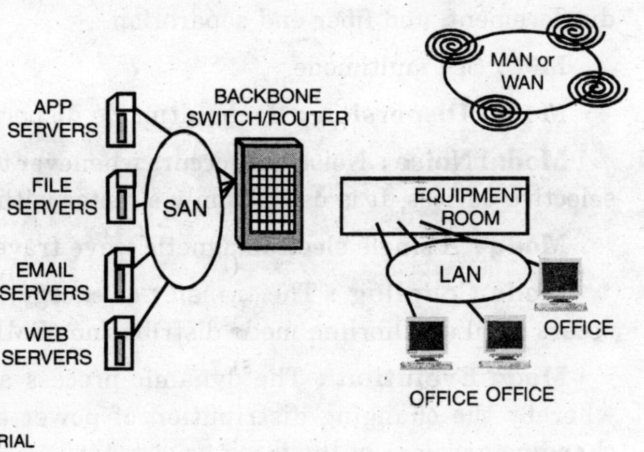

Mass Splicing : splicing of many fibers in a cable.

Material Dispersion : Dispersion resulting from the different velocities of each wavelength in a material.

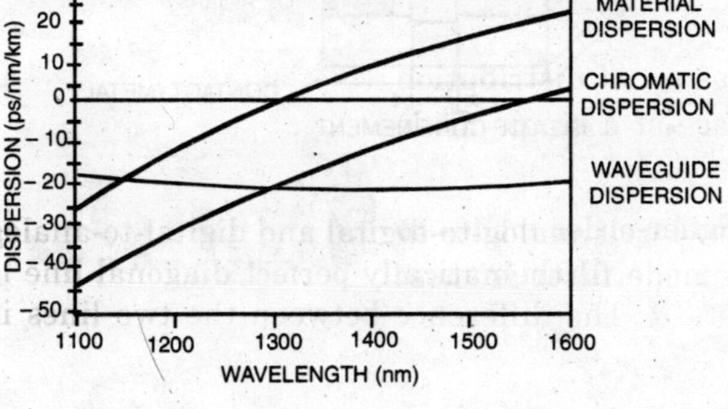

Mean Launched Power : The average power for a continuous valid symbol sequence coupled into a fiber.

Mechanical Splice : An optical fiber splice accomplished by fixtures or materials, rather than by thermal fusion. The capillary splice, illustrated, is one example of a mechanical splice.

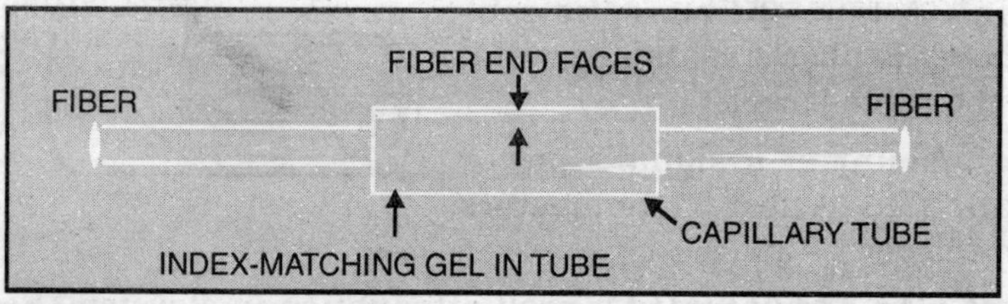

MH : Abbreviation for medium-haul. A classification of video performance under RS-250C. Higher performance than long-haul and lower performance than short-haul.

MHz : Abbreviation for megahertz. One million Hertz (cycles per second).

Microbending : Mechanical stress on a fiber that introduces local discontinuities, which results in light leaking from the core to the cladding by a process called mode coupling.

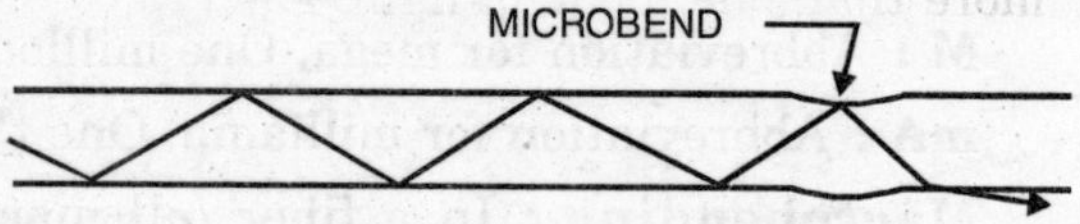

Minimum Bend Radius : The smallest radius an optical fiber or fiber cable can bend before increased attenuation or breakage occurs.

Misalignment Loss : The loss of power resulting from angular misalignment, lateral displacement, and fiber end separation.

MM : See multimode.

Modal Dispersion : See multimode dispersion.

Modal Noise : Noise that occurs whenever the optical power propagates through mode-selective devices. It is usually only a factor with laser light sources.

Mode : A single electromagnetic wave traveling in a fiber.

Mode Coupling : The transfer of energy between modes. In a fiber, mode coupling occurs until equilibrium mode distribution (EMD) is reached.

Mode Evolution : The dynamic process a multilongitudinal mode laser undergoes whereby the changing distribution of power among the modes creates a continuously changing envelope of the laser's spectrum.

Mode Field Diameter (MFD) : A measure of distribution of optical power intensity across the end face of a single-mode fiber.

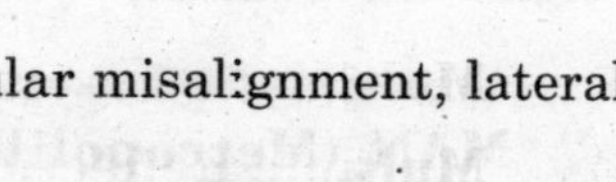

Mode Filter : A device that removes higher-order modes to simulate equilibrium mode distribution. A mode filter is most easily constructed

Mode Stripper : A device that removes cladding modes.

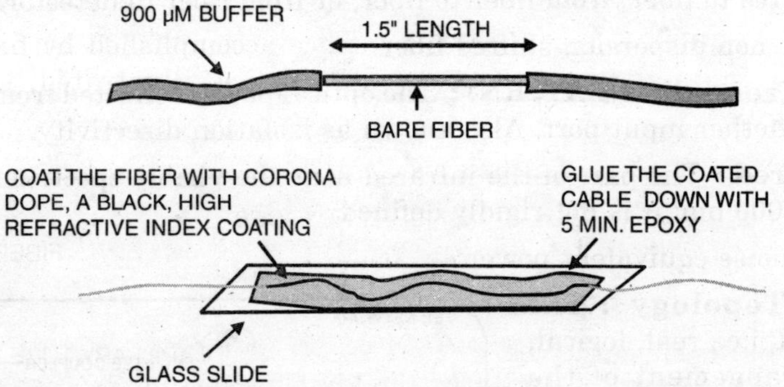

Multilongitudinal Mode (MLM) Laser : An injection laser diode which has a number of longitudinal modes.

Multimode Dispersion : Dispersion resulting from the different transit lengths of different propagating modes in a multimode optical fiber. Also called modal dispersion.

Multimode (MM) Fiber : An optical fiber that has a core large enough to propagate more than one mode of light The typical diameter is 62.5 micrometers.

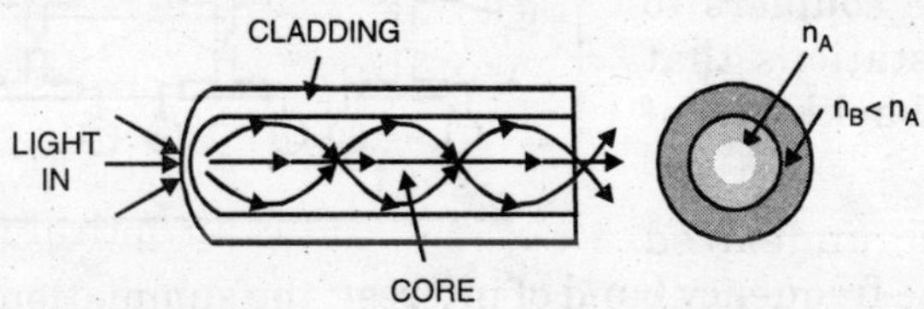

Multimode Laser Diode (MMLD) : Synonym for multilongitudinal mode laser.

Multiple Reflection Noise (MRN) : The fiber optic receiver noise resulting from the interference of delayed signals from two or more reflection points in a fiber optic span. Also known as multipath interference.

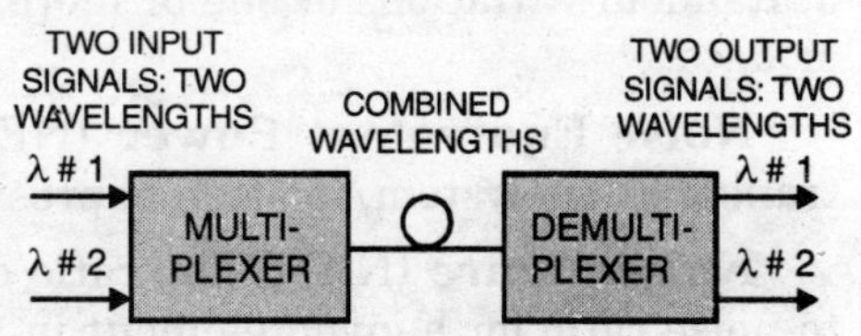

Multiplexer : A device that combines two or more signals into one output.

Multiplexing : The process by which two or more signals are transmitted over a single communications channel. Examples include time-division multiplexing (TDM) and wavelength-division multiplexing (WDM).

N

n : Abbreviation for nano. One billionth or 10^{-9}.

nA : Abbreviation for nanoamp. One billionth of an Amp or 10^{-9} Amps.

NA : See numerical aperture.

NA Mismatch Loss : The loss of power at a joint that occurs when the transmitting half has a

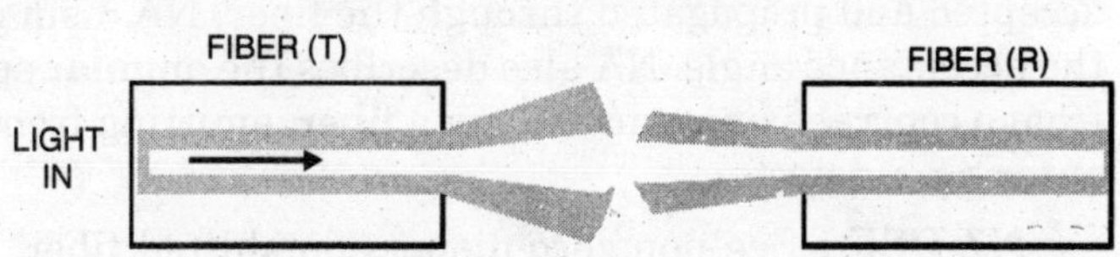

numerical aperture greater than the NA of the receiving half. The loss occurs when coupling light from a source to fiber, from fiber to fiber, or from fiber to detector.

NDSF : See non dispersion-shifted fiber.

Near-end Crosstalk (NEXT, RN) : The optical power reflected from one or more input ports, back to another input port. Also known as isolation directivity.

Near Infrared : The part of the infrared near the visible spectrum, typically 700 nm to 1500 nm or 2000 nm; it is not rigidly defined.

NEP : See noise equivalent power.

Network Topology : The specific physical, i.e., real, logical, or virtual, arrangement of the elements of a network. Common network topologies include a bus (or linear) topology, a ring topology, and a hybrid topology, which can be a combination of any two or more network topologies. Illustrated to the right is a bus topology utilizing tee couplers to connect a series of stations that listen to a single backbone of cable.

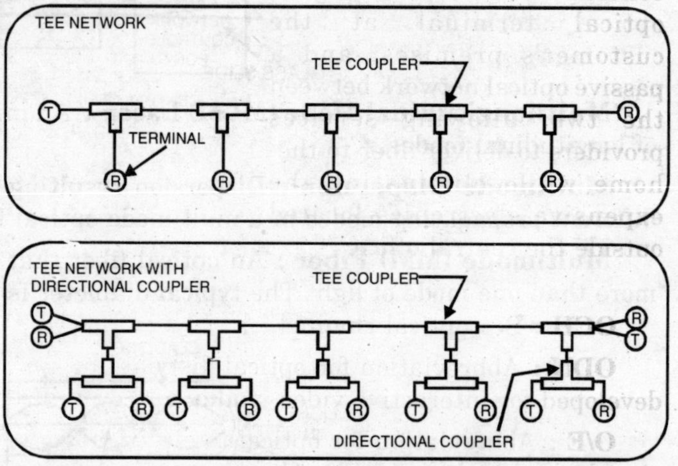

Noise : 1. An undesired disturbance within the frequency band of interest; the summation of unwanted or disturbing energy introduced into a communications system from man-made and natural sources. 2. A disturbance that affects a signal and that may distort the information carried by the signal. 3. Random variations of one or more characteristics of any entity such as voltage, current, or data.

Noise Equivalent Power (NEP) : The noise of optical receivers, or of an entire transmission system, is often expressed in terms of noise equivalent optical power.

Noise Figure (NF) : The ratio of the output signal-to-noise ratio to the input signal-to-noise ratio for a given element in a transmission system. Used for optical and electrical components.

Non Dispersion-shifted Fiber (NDSF) : The most popular type of single-mode fiber deployed. It is designed to have a zero-dispersion wavelength near 1310 nm.

Non Zero-dispersion-shifted Fiber (NZ-DSF) : A dispersion-shifted single-mode fiber that has the zero-dispersion wavelength near the 1550 nm window, but outside the window actually used to transmit signals. This strategy maximizes bandwidth while minimizing fiber nonlinearities.

Numerical Aperture (NA) : The light-gathering ability of a fiber; the maximum angle to the fiber axis at which light will be accepted and propagated through the fiber. NA = sin α, where α is the acceptance angle. NA also describes the angular spread of light from a central axis, as in exiting a fiber, emitting from a source, or entering a detector.

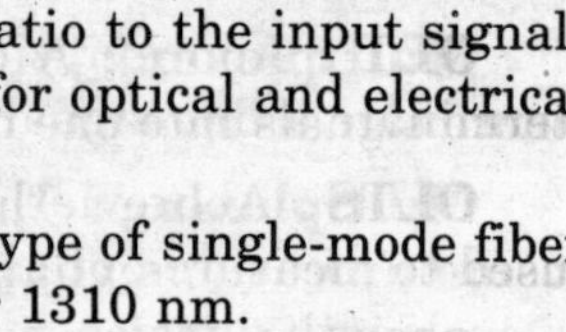

$$NA = \sin\alpha = \sqrt{n_1^2 - n_2^2}$$

NZ-DSF : See non zero-dispersion-shifted fiber.

O

OADM : See optical add/drop multiplexer.

OAM : Abbreviation for operation, administration, and maintenance. Refers to telecommunications networks.

OAN : Abbreviation for optical access network. A network technology, based on passive optical networks (PONs), that includes an optical switch at the central office, an intelligent optical terminal at the customer's premises, and a passive optical network between the two, allowing services providers to deliver fiber-to-the-home while eliminating the expensive electronics located outside the central office.

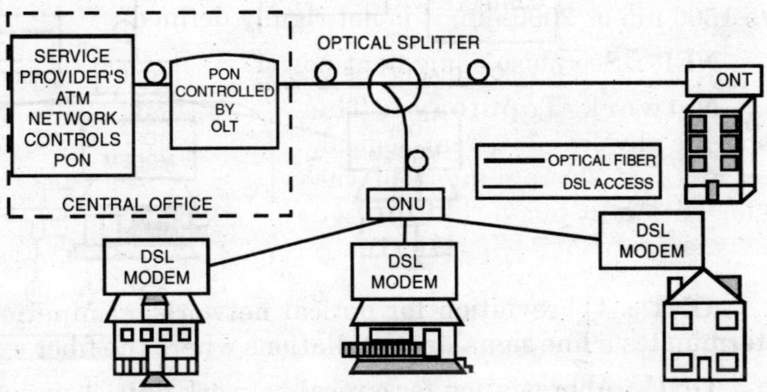

OCH : See optical channel.

ODN : Abbreviation for optical distribution network. Term for optical networks being developed for interactive video, audio, and data distribution.

O/E : Abbreviation for optical-to-electrical converter. A device used to convert optical signals to electrical signals. Also known as OEC.

OEIC : Abbreviation for opto-electronic integrated circuit. An integrated circuit that includes both optical and electrical elements.

OEM : Abbreviation for original equipment manufacturer. The manufacturer of any device that is designed and built to be distributed under the label of another company.

OLT : Abbreviation for optical line termination. Optical network elements that terminate a line signal.

OLTS : Abbreviation for optical loss test set. A source and optical power meter combined used to measure optical loss.

OMS : Abbreviation for optical multiplex section. A section of a DWDM system that incorporates an optical add/drop multiplexer.

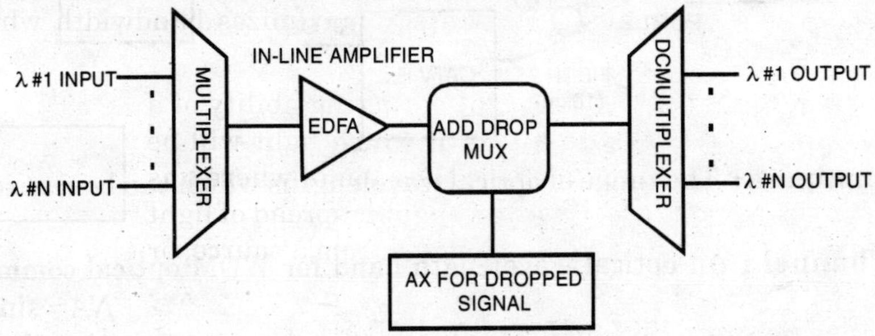

ONI : Abbreviation for optical network interface. A device used in an optical distribution network to connect two parts of that network.

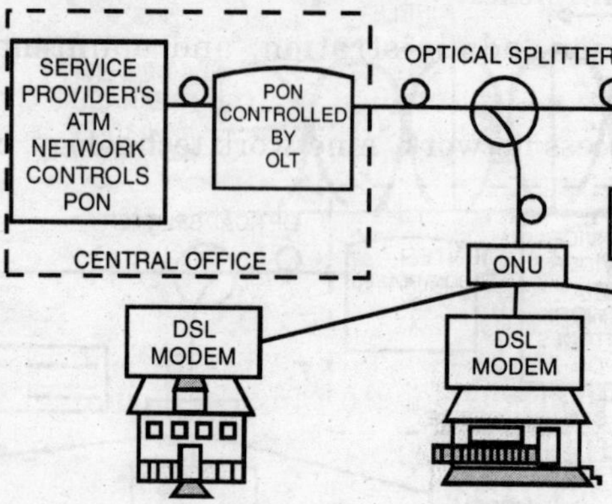

ONT : Abbreviation for optical network termination. Optical network element that terminates a line signal in installations where the fiber extends into the customer premises.

ONU : Abbreviation for optical network unit. A network element that is part of a fiber-in-the-loop system.

Open Systems Interconnection (OSI) : Pertaining to the logical structure for communications networks standardized by the International Organization for Standardization (ISO).

Optical Add/Drop Multiplexer (OADM) : A device which adds or drops individual wavelengths from a DWDM system.

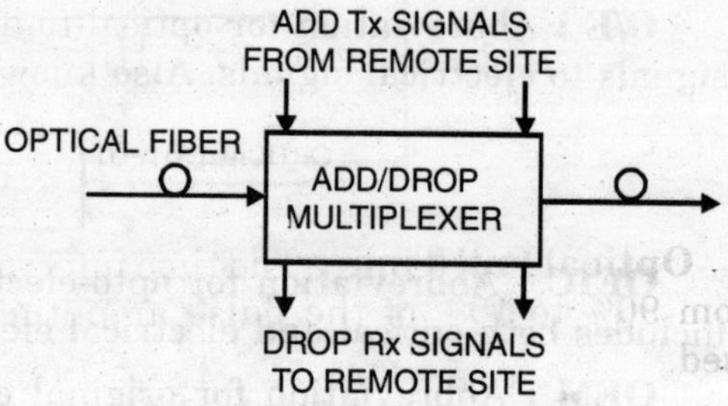

Optical Amplifier : A device that amplifies an input optical signal without converting it into electrical form. The best developed are optical fibers doped with the rare earth element, erbium. See also EDFA.

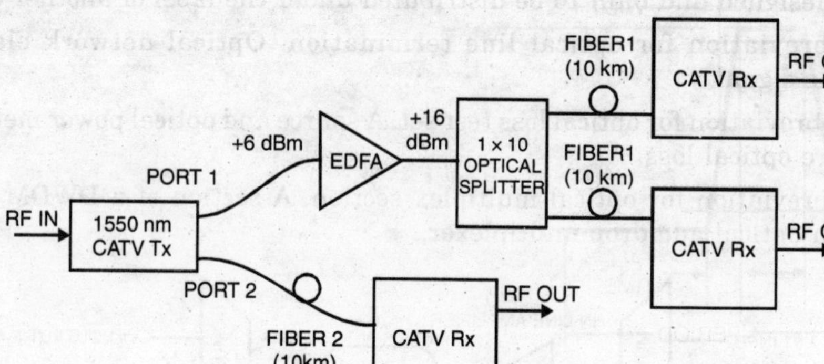

Optical Bandpass : The range of optical wavelengths which can be transmitted through a component.

Optical Channel : An optical wavelength band for WDM optical communications.

Optical Channel Spacing : The wavelength separation between adjacent WDM channels.

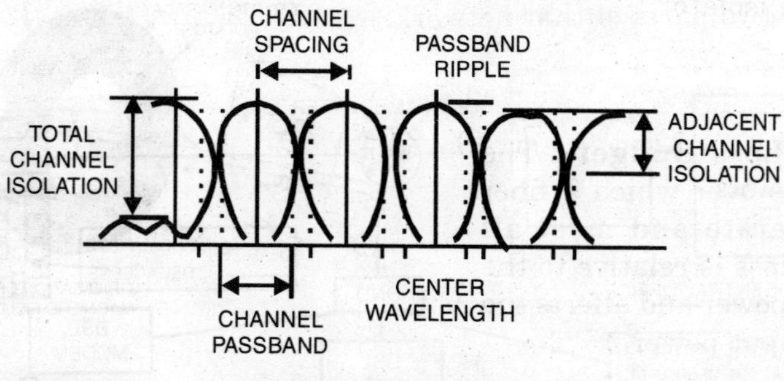

Optical Channel Width : The optical wavelength range of a channel.

Optical Continuous Wave Reflectometer (OCWR) : An instrument used to characterize a fiber optic link wherein an unmodulated signal is transmitted through the link, and the resulting light scattered and reflected back to the input is measured. Useful in estimating component reflectance and link optical return loss.

Optical Directional Coupler (ODC) : A component used to combine and separate optical power.

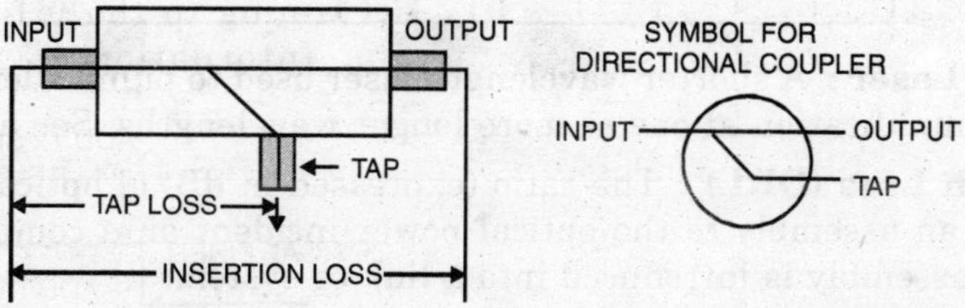

Optical Fall Time : The time interval for the falling edge of an optical pulse to transition from 90% to 10% of the pulse amplitude. Alternatively, values of 80% and 20% may be used.

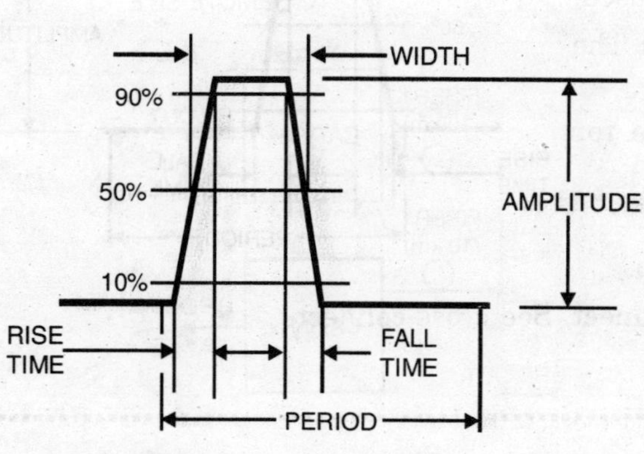

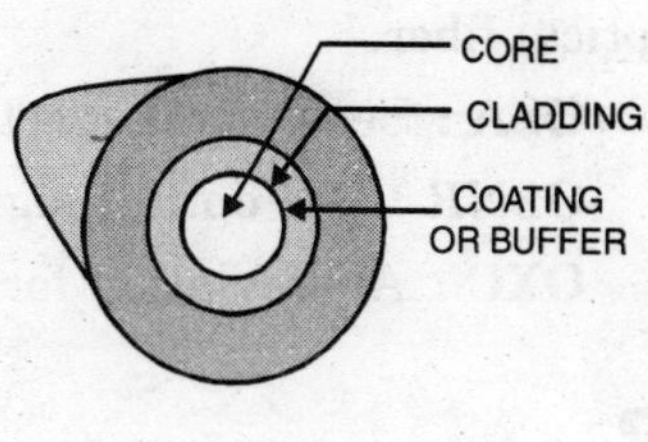

Optical Fiber : A glass or plastic fiber that has the ability to guide light along its axis. The three parts of an optical fiber are the core, the cladding, and the coating or buffer.

Optical Isolator : A component used to block out reflected and unwanted light. Also called an isolator.

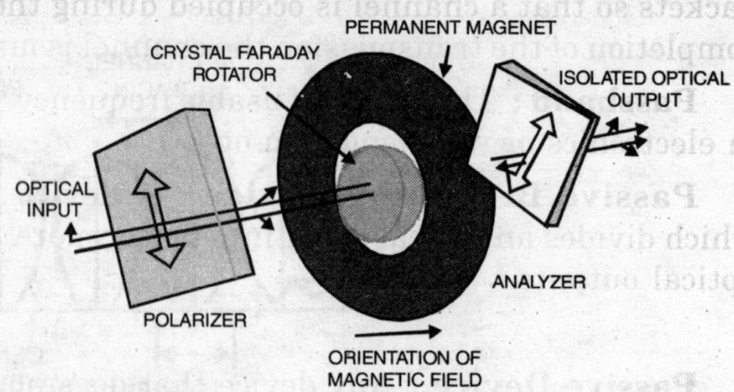

Optical Link Loss Budget : The range of optical loss over which a fiber optic link will operate and meet all specifications. The loss is relative to the transmitter output power and affects the required receiver input power.

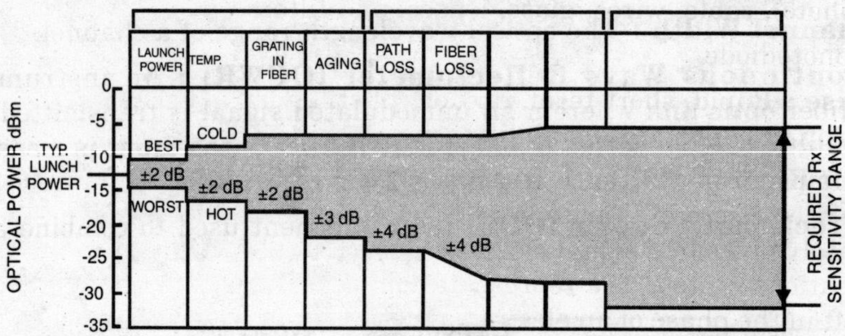

Optical Pump Laser : A shorter wavelength laser used to pump a length of fiber with energy to provide amplification at one or more longer wavelengths. See also EDFA.

Optical Return Loss (ORL) : The ratio (expressed in dB) of optical power reflected by a component or an assembly to the optical power incident on a component port when that component or assembly is introduced into a link or system.

Optical Rise Time : The time interval for the rising edge of an optical pulse to transition from 10% to 90% of the pulse amplitude. Alternatively, values of 20% and 80% may be used.

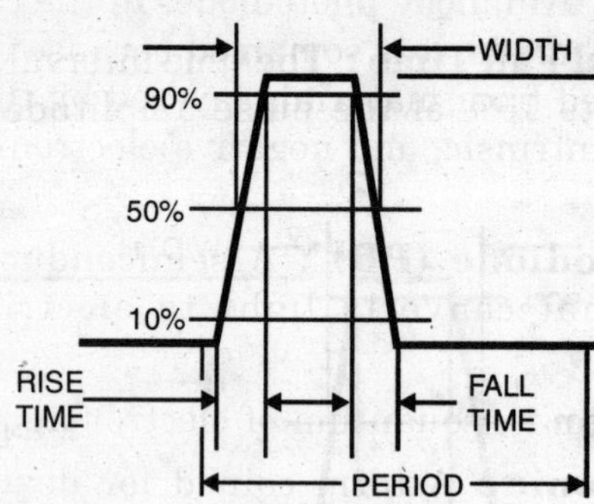

Optical Signal-to-Noise-Ratio (OSNR) : The optical equivalent of SNR.

Optical Waveguide : Another name for optical fiber.

OSA : See optical spectrum analyzer.

OSNR : See optical signal-to-noise ratio.

OXC : Abbreviation for optical cross-connect. See cross-connect.

P

p : Abbreviation for pico. One trillionth or 10^{-12}.

Packet : In data communications, a sequence of binary digits, including data and control signals, that is transmitted and switched as a composite whole. The packet contains data, control signals, and possibly error control information, arranged in a specific format.

Packet Switching : The process of routing and transferring data by means of addressed packets so that a channel is occupied during the transmission of the packet only, and upon completion of the transmission the channel is made available for the transfer of other traffic.

Passband : The region of usable frequency in electronics or wavelength in optics.

Passive Branching Device : A device which divides an optical input into two or more optical outputs.

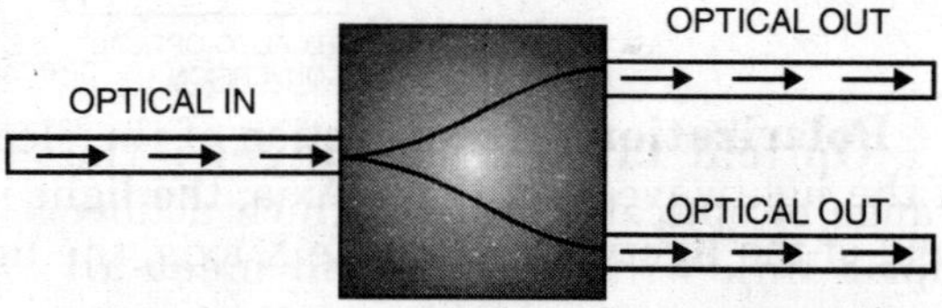

Passive Device : Any device that does not require a source of energy for its operation. Examples include electrical resistors or capacitors, diodes, optical fiber (photo), cable, wires, glass, lenses, and filters.

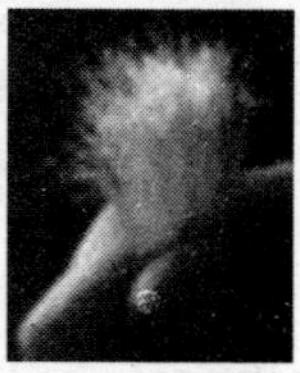

PD : See photodiode.

Phase Noise : Rapid, short-term, random fluctuations in the phase of a wave caused by time-domain instabilities in an oscillator.

Phase-shift Keying (PSK) : 1. In digital transmission, angle modulation in which the phase of the carrier discretely varies in relation, either to a reference phase or to the phase of the immediately preceding signal element, in accordance with data being transmitted. 2. In a communications system, the representation of characters, such as bits or quaternary digits, by a shift in the phase of an electromagnetic carrier wave with respect to a reference, by an amount corresponding to the symbol being encoded. Also called biphase modulation, phase-shift signaling.

Photoconductive : Losing an electrical charge on exposure to light.

Photodetector : An optoelectronic transducer such as a PIN photodiode (illustrated) or avalanche photodiode. In the case of the PIN diode, it is so named because it is constructed from materials layered by their positive, intrinsic, and negative electron regions.

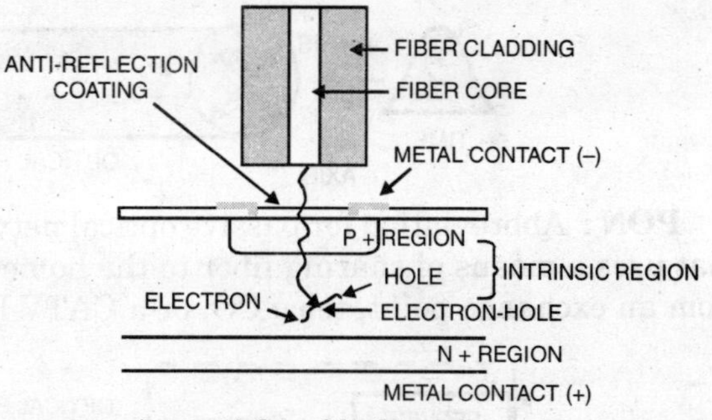

Photodiode (PD) : A semiconductor device that converts light to electrical current.

Photon : A quantum of electromagnetic energy. A particle of light.

Photonic : A term coined for devices that work using photons, analogous to the electronic for devices working with electrons.

Pigtail : A short optical fiber permanently attached to a source, detector, or other fiber optic device at one end and an optical connector at the other.

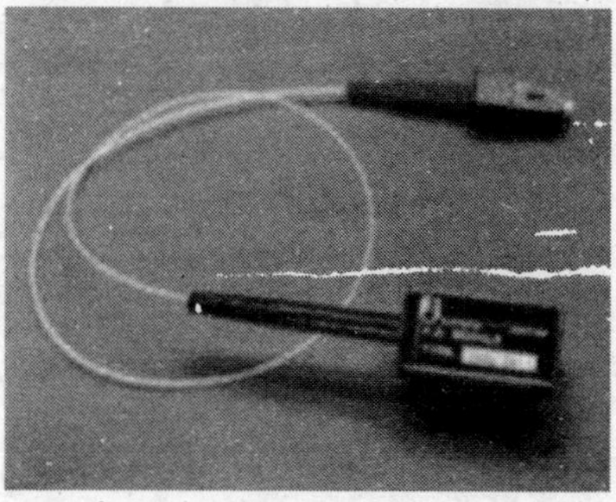

PIN Photodiode : See photodiode.

Planar Waveguide : A waveguide fabricated in a flat material such as thin film.

Plastic Fiber : An optical fiber having a plastic core and plastic cladding.

Point-to-Point Transmission : Transmission between two designated stations.

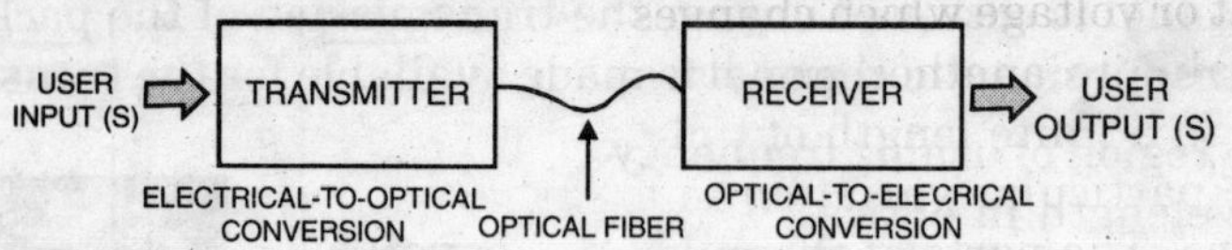

Polarization : The direction of the electric field in the lightwave. If the electric field of the lightwave is in the Y Axis, the light is said to be vertically polarized. If the electric field of the lightwave is in the X axis, the light is said to be horizontally polarized.

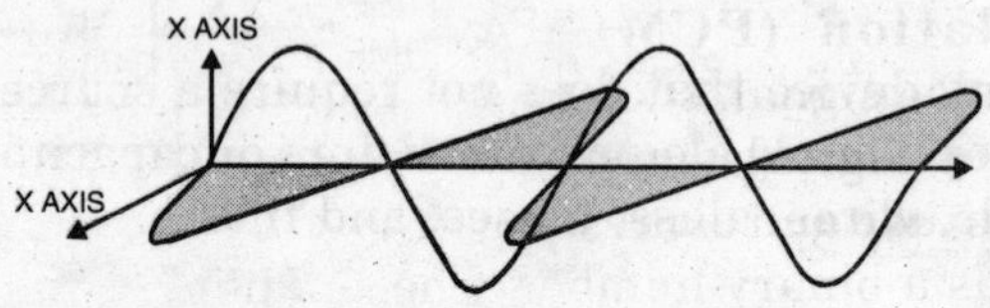

Polarization Maintaining Fiber : Fiber designed to propagate only one polarization of light that enters it.

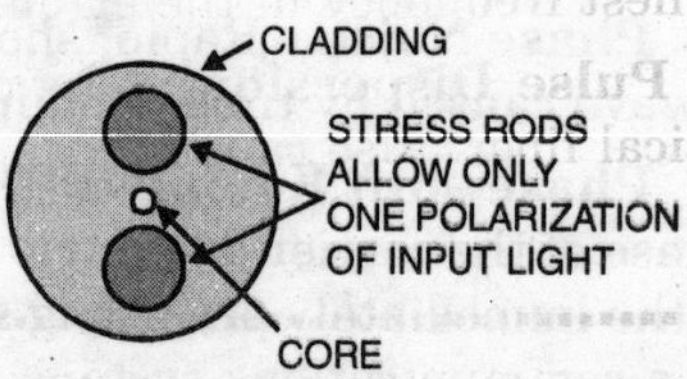

Polarization Mode Dispersion (PMD) : Polarization mode dispersion is an inherent property of all optical media. It is caused by the difference in the propagation velocities of light in the orthogonal principal polarization states of the transmission medium. The net effect is that if an optical pulse contains both polarization components, then the different polarization components will travel at different speeds and arrive at different times, smearing the received optical signal.

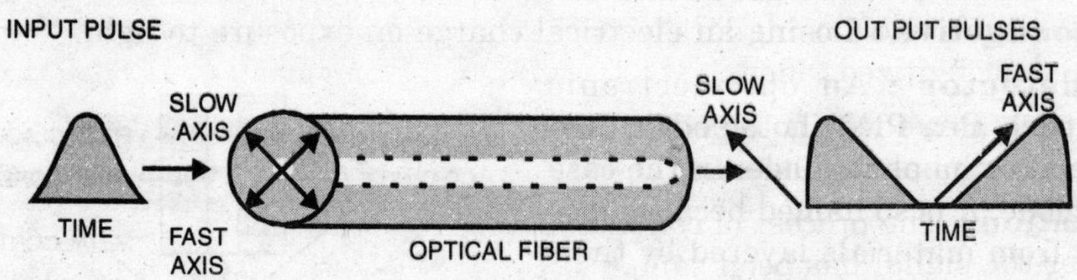

PON : Abbreviation for passive optical network. A broadband fiber optic access network that uses a means of sharing fiber to the home without running individual fiber optic lines from an exchange point, telco CO, or a CATV headend and the subscriber's home.

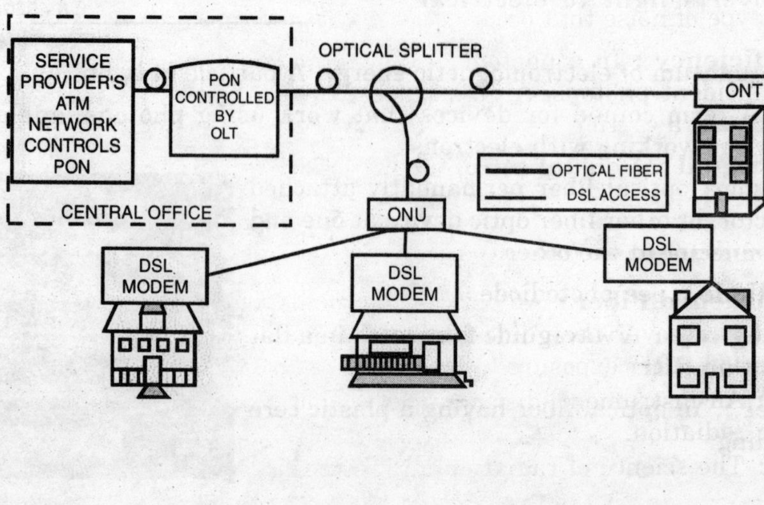

Pulse : A current or voltage which changes abruptly from one value to another and back to the original value in a finite length of time. Used to describe one particular variation in a series of wave motions. The parts of the pulse include the rise time, fall time, and pulse width, pulse amplitude. The period of a pulse refers to the amount of time between pulses.

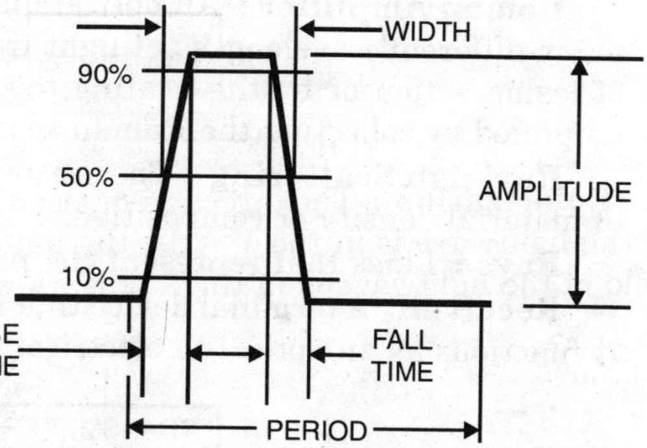

Pulse-code Modulation (PCM) : A technique in which an analog signal, such as a voice, is converted into a digital signal by sampling the signal's amplitude and expressing the different amplitudes as a binary number. The sampling rate must be at least twice the highest frequency in the signal.

Pulse Dispersion : The dispersion of an optical signal as it propagates through an optical fiber. Also called pulse spreading.

Q

QAM : See quadrature amplitude modulation.

QPSK : See quadrature phase-shift keying.

Quadrature Amplitude Modulation (QAM) : A coding technique that uses many discrete digital levels to transmit data with minimum bandwidth. QAM256 uses 256 discrete levels to transmit digitized video.

Quadrature Phase-shift Keying (QPSK) : Phase-shift keying uses four different phase angles out of phase by 90°. Also called quadriphase or quaternary phase-shift keying.

Quantization : The process of converting the voltage level of a signal into digital data before or after the signal has been sampled.

Quantizing Error : Inaccuracies in the digital representation of an analog signal. These errors occur because of limitations in the resolution of the digitizing process.

Quantizing Noise : Noise which results from the quantization process. In serial digital video, a granular type of noise that occurs only in the presence of a signal.

Quantum Efficiency : In a photodiode, the ratio of primary carriers (electron-hole pairs) created to incident photons. A quantum efficiency of 70% means seven out of ten incident photons create a carrier.

Quaternary Signal : A digital signal having four significant conditions.

R

Radiation-hardened Fiber : An optical fiber made with core and cladding materials that are designed to recover their intrinsic value of attenuation coefficient, within an acceptable time period, after exposure to a radiation pulse.

Radiometer : An instrument, distinct from a photometer, to measure power (Watts) of electromagnetic radiation.

Radiometry : The science of radiation measurement.

Raman Amplifier : An optical amplifier based on Raman scattering which generates many different wavelengths of light from a nominally single-wavelength source by means of lasing action or by the beating together of two frequencies. The optical signal can be amplified by collecting the Raman scattered light.

Rayleigh Scattering : The scattering of light that results from small inhomogeneities of material density or composition.

Rays : Lines that represent the path taken by light.

Receiver : A terminal device that includes a detector and signal processing electronics. It functions as an optical-to-electrical converter.

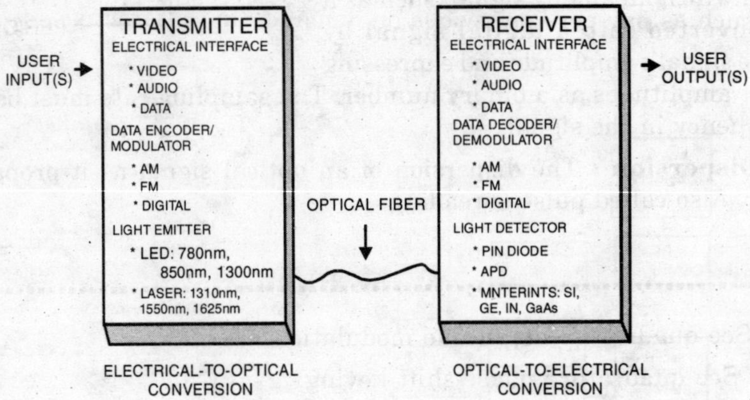

Receiver Overload : The maximum acceptable value of average received power for an acceptable BER or performance.

Receiver Sensitivity : The minimum acceptable value of received power needed to achieve an acceptable BER or performance. It takes into account power penalties caused by use of a transmitter with worst-case values of extinction ratio, jitter, pulse rise times and fall times, optical return loss, receiver connector degradations, and measurement tolerances. The receiver sensitivity does not include power penalties associated with dispersion, or back reflections from the optical path; these effects are specified separately in the allocation of maximum optical path penalty. Sensitivity usually takes into account worst-case operating and end-of-life (EOL) conditions.

Refraction : The changing of direction of a light wave in passing through a boundary between two dissimilar media, or in a graded-index medium where refractive index is a continuous function of position.

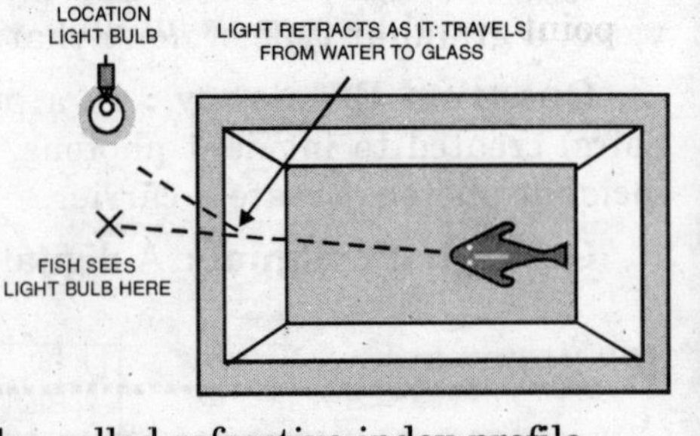

Refractive Index : A property of optical materials that relates to the speed of light in the material versus the speed of light in a vacuum.

Refractive Index Gradient : The description of the value of the refractive index as a function of distance from the optical axis along an optical fiber diameter. Also called refractive index profile.

Regenerative Repeater : A repeater, designed for digital transmission, in which digital signals are amplified, reshaped, retimed, and retransmitted.

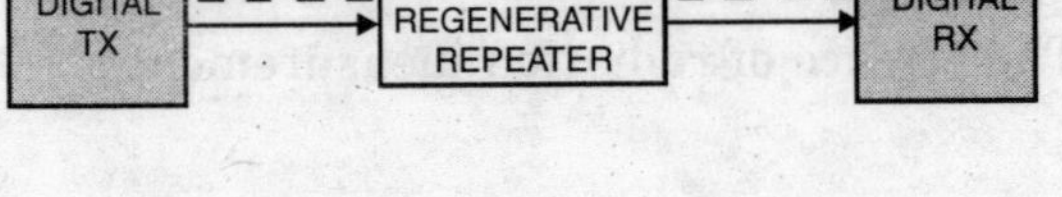

Regenerator : Synonym for regenerative repeater.

Repeater : A receiver and transmitter set designed to regenerate attenuated signals. Used to extend operating range.

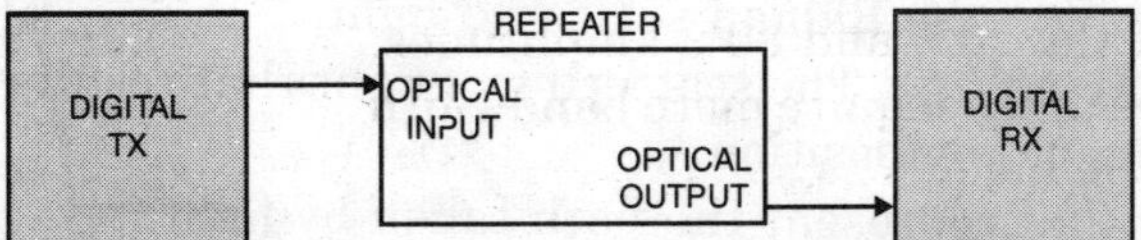

Return Path : A communications connection that carries signals from the subscriber back to the operator. The return path allows for interactive television and on-demand services, such as pay-per-view, video on demand, and interactive games.

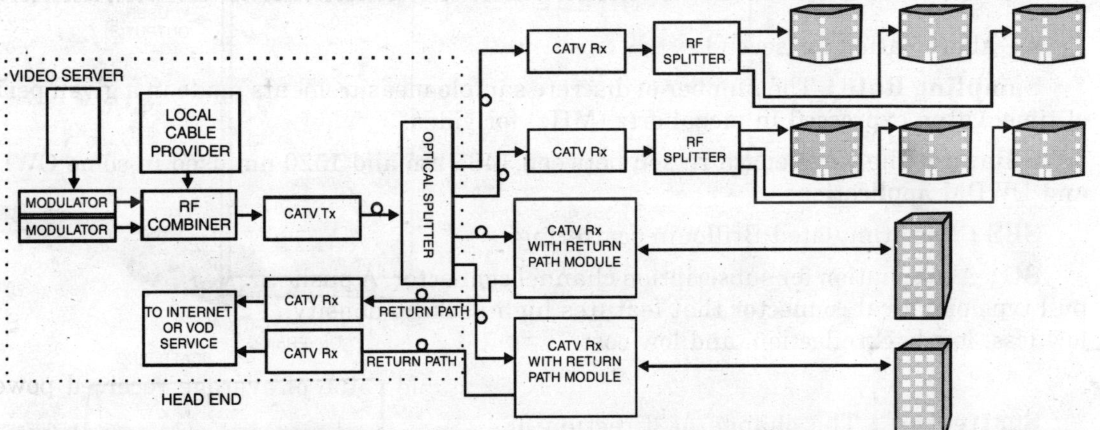

Responsivity : The ratio of a photodetector's electrical output to its optical input in Amperes/Watt (A/W).

Return Loss : See optical return loss.

Ring : A set of stations wherein information is passed sequentially between stations, each station in turn examining or copying the information, and finally returning it to the originating station.

Ring Network : A network topology in which terminals are connected in a point-to-point serial fashion in an unbroken circular configuration.

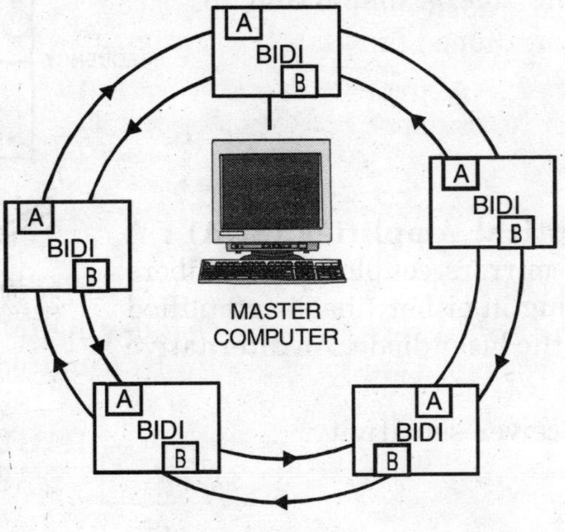

Rise Time : The time taken to make a transition from one state to another, usually measured between the 10% and 90% completion points of the transition. Alternatively the rise time may be specified at the 20% and 80% amplitudes. Shorter or faster rise times require more bandwidth in a transmission channel.

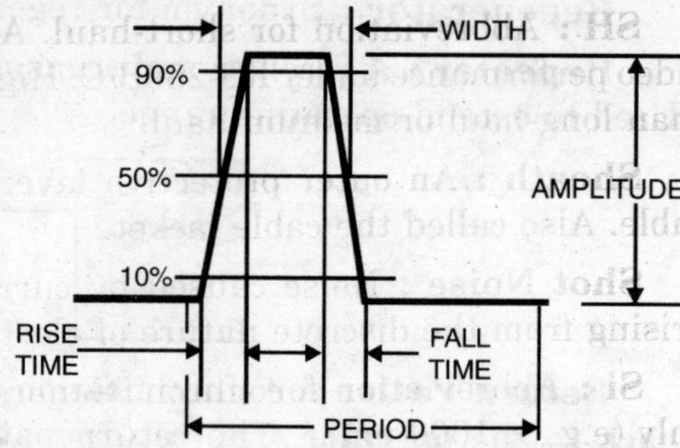

S

s : Abbreviation for second.

Sampling Rate : The number of discrete sample measurements made in a given period of time. Often expressed in megahertz (MHz) for video.

S-Band : The wavelength region between 1485 nm and 1520 nm used in some CWDM and DWDM applications.

SBS : See stimulated Brillouin scattering.

SC : Abbreviation for subscription channel connector. A push-pull type of optical connector that features high packing density, low loss, low backreflection, and low cost.

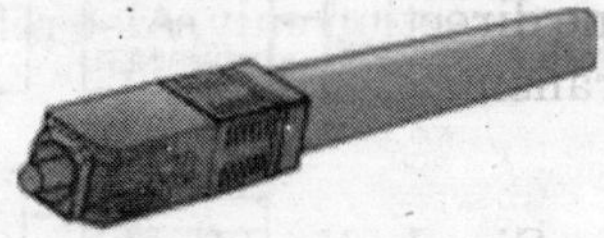

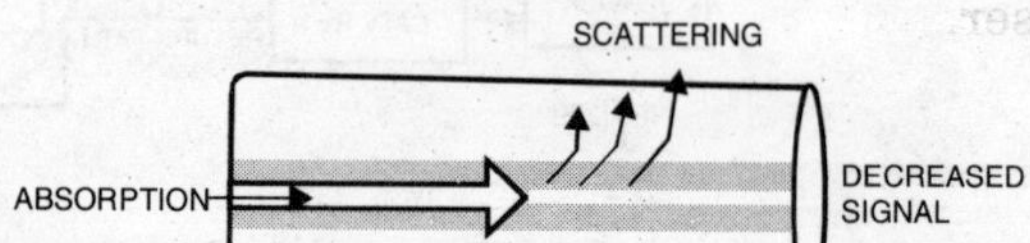

Scattering : The change of direction of light rays or photons after striking small particles. It may also be regarded as the diffusion of a light beam caused by the inhomogeneity of the transmitting material.

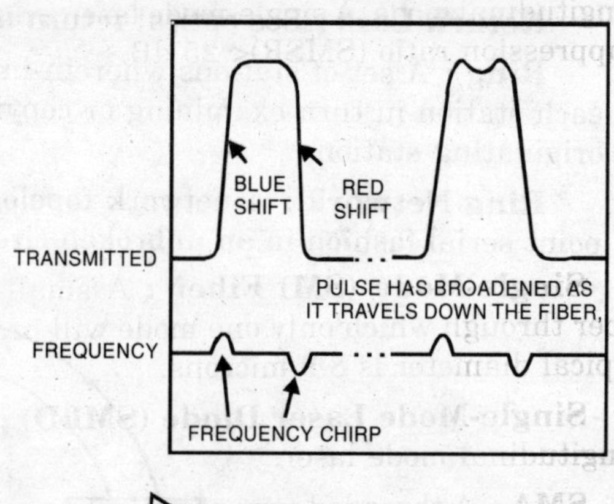

Self-phase modulation (SPM) : A fiber nonlinearity caused by the nonlinear index of refraction of glass. The index of refraction varies with optical power level causing a frequency chirp which interacts with the fiber's dispersion to broaden the pulse.

Semiconductor Optical Amplifier (SOA) : A laser diode without end mirrors coupled to the fibers on both ends. Light coming in either fiber is amplified by a single pass through the laser diode. An alternative to EDFAs.

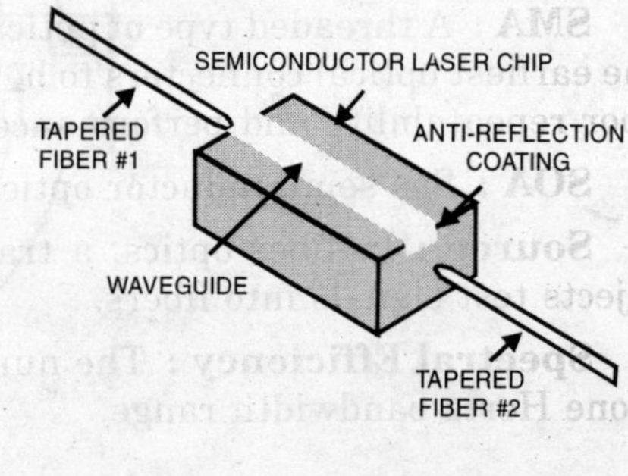

Sensitivity : See receiver sensitivity.

SH : Abbreviation for short-haul. A classification of video performance under RS-250B/C. Higher performance than long-haul or medium-haul.

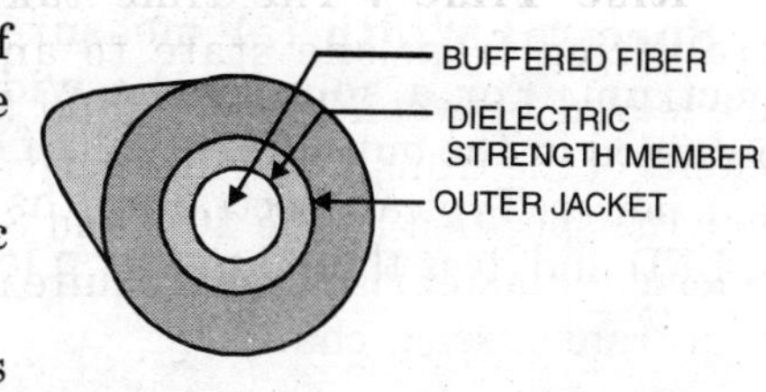

Sheath : An outer protective layer of a fiber optic cable. Also called the cable jacket.

Shot Noise : Noise caused by current fluctuations arising from the discrete nature of electrons.

Si : Abbreviation for silicon. Generally used in detectors. Good for short wavelengths only (e.g., < 1000 nm).

Signal-to-Noise Ratio (SNR) : The ratio of the total signal to the total noise which shows how much higher the signal level is than the level of the noise. A measure of signal quality.

Simplex : Single element (e.g., a simplex connector is a single-fiber connector).

Simplex Cable : A term sometimes used for a single-fiber cable.

Simplex Transmission : Transmission in one direction only. Also referred to as half-duplex transmission.

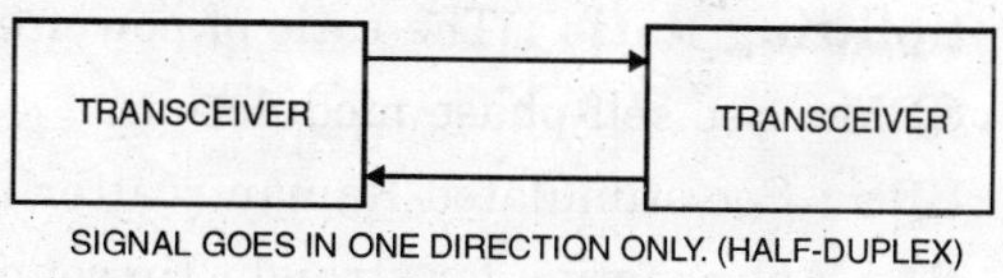

Single-line Laser : See single-longitudinal mode laser.

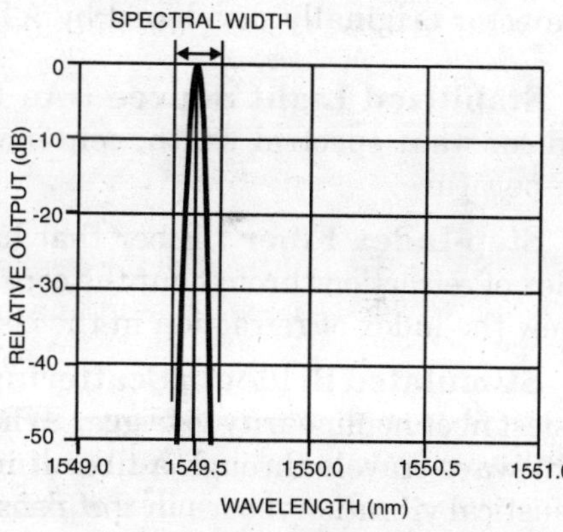

Single-longitudinal Mode Laser (SLM) : An injection laser diode which has a single dominant longitudinal mode. A single-mode laser with a side mode suppression ratio (SMSR)< 25 dB.

Single-Mode (SM) Fiber : A small-core optical fiber through which only one mode will propagate. The typical diameter is 8-9 microns.

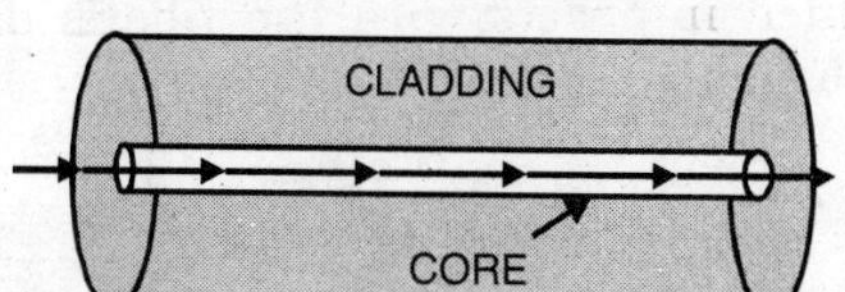

Single-Mode Laser Diode (SMLD) : See single-longitudinal mode laser.

SMA : A threaded type of optical connector. One of the earliest optical connectors to be widely used. Offers poor repeatability and performance.

SOA : See semiconductor optical amplifier.

Source : In fiber optics, a transmitting LED or laser diode, or an instrument that injects test signals into fibers.

Spectral Efficiency : The number of data bits per second that can be transmitted in a one Hertz bandwidth range.

Spectral Width : A measure of the extent of a spectrum. For a source, the width of wavelengths contained in the output at one half of the wavelength of peak power. Typical spectral widths are 50 to 160 nm for an LED and less than 5 nm for a laser diode.

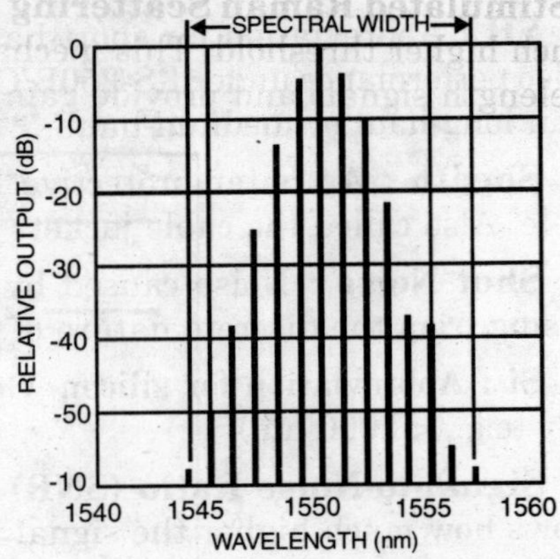

Splice : A permanent connection of two optical fibers through fusion or mechanical means.

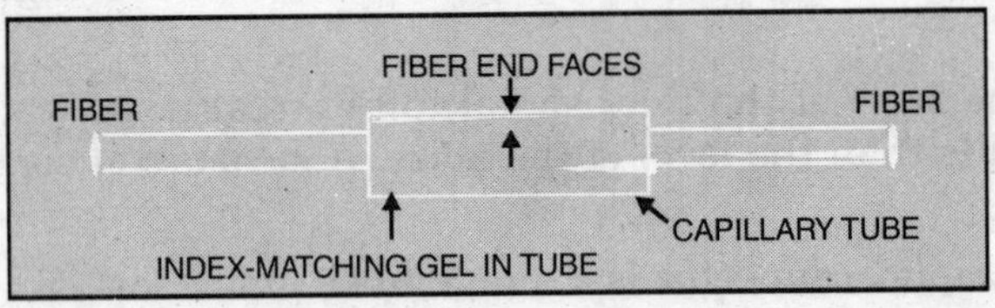

Splitter : see Coupler.

Splitting Ratio : The ratio of power emerging from two output ports of a coupler.

SPM : See self-phase modulation.

SRS : See stimulated Raman scattering.

ST : Abbreviation for straight tip connector. Popular fiber optic connector originally developed by AT&T.

Stabilized Light Source : An LED or laser diode that emits light with a controlled and constant spectral width, center wavelength, and peak power with respect to time and temperature.

Step-Index Fiber : Fiber that has a uniform index of refraction throughout the core that is a step below the index of refraction in the cladding.

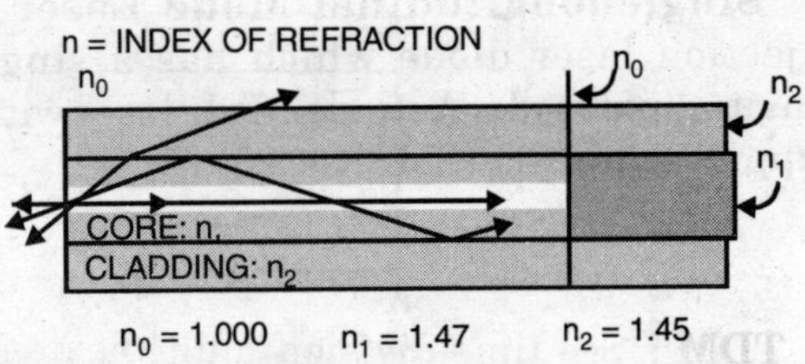

Stimulated Brillouin Scattering (SBS) : The easiest fiber nonlinearity to trigger. When a powerful light wave travels through a fiber it interacts with acoustical vibration modes in the glass. This causes a scattering mechanism to be formed that reflects much of the light back to the source.

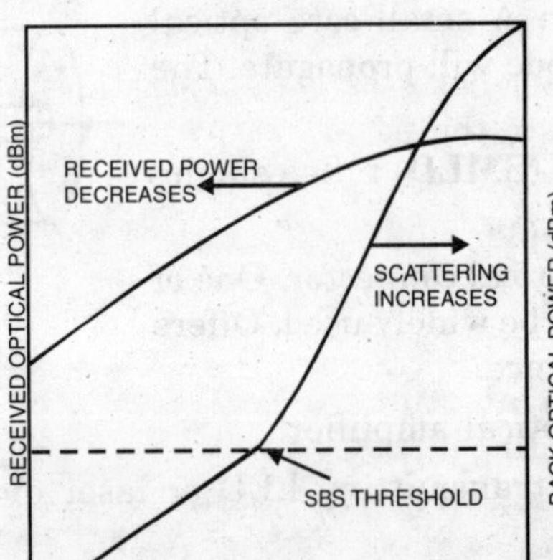

Stimulated Raman Scattering (SRS) : A fiber nonlinearity similar to SBS but having a much higher threshold. This mechanism can also cause power to be robbed from shorter wavelength signals and provide gain to longer wavelength signals.

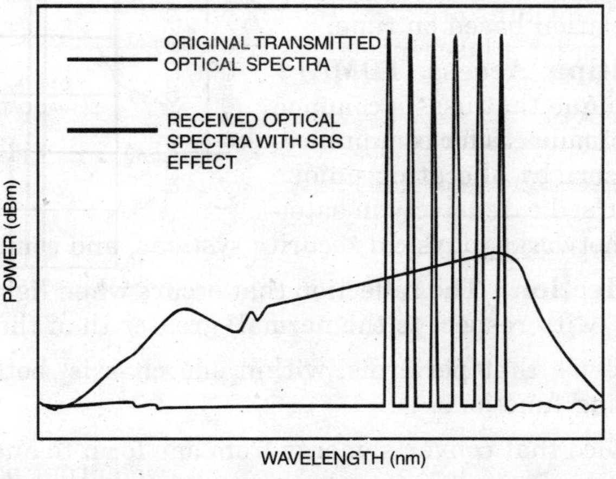

Strength Member : The part of a fiber optic cable composed of aramid yarn, steel strands, or fiberglass filaments that increase the tensile strength of the cable.

T

T : Abbreviation for tera. One trillion or 10^{12}.

Tap Loss : In a fiber optic coupler, the ratio of power at the tap port to the power at the input port.

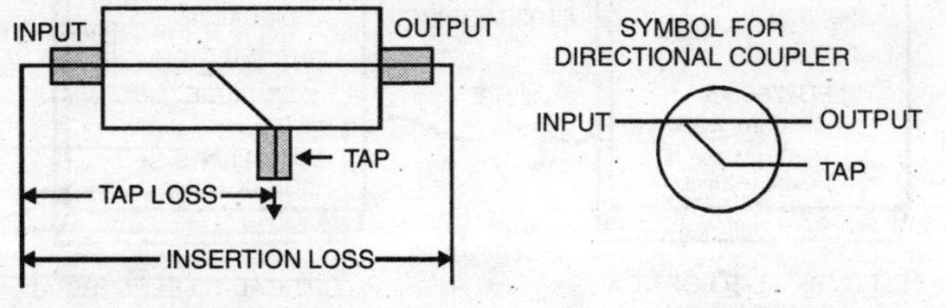

TDM : See time-division multiplexing.

TDMA : See time-division multiple access.

Tee Coupler : A three-port optical coupler.

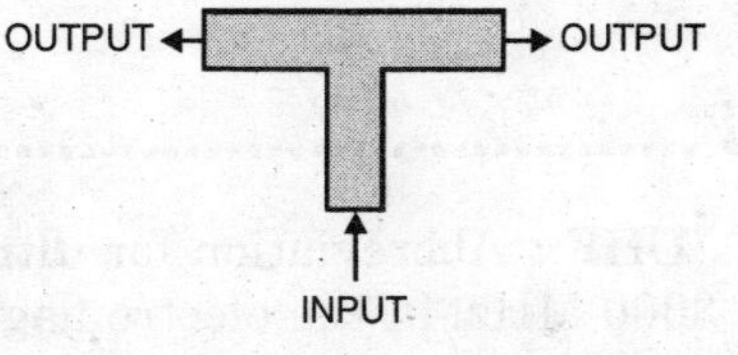

Throughput Loss : In a fiber optic coupler, the ratio of power at the throughput port to the power at the input port.

Throughput Port : In a coupler where the splitting ratio between output ports is not equal, the output port containing the greater power.

Tight-Buffer : A material tightly surrounding a fiber in a cable, holding it rigidly in place.

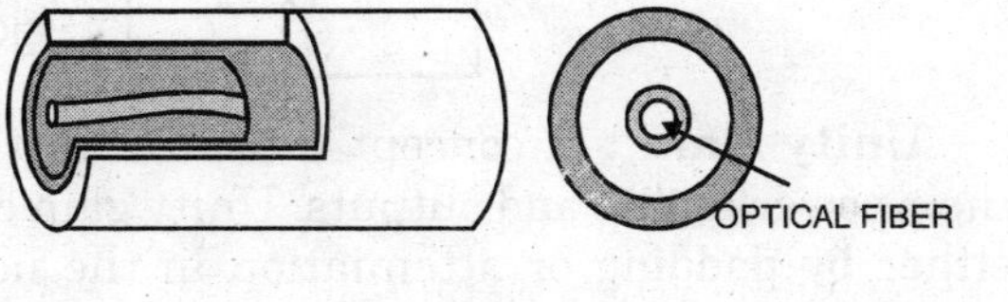

Time-division Multiplexing (TDM) : A transmission technique whereby several low-speed channels are multiplexed into a high-speed channel for transmission. Each low-speed channel is allocated a specific position based on time.

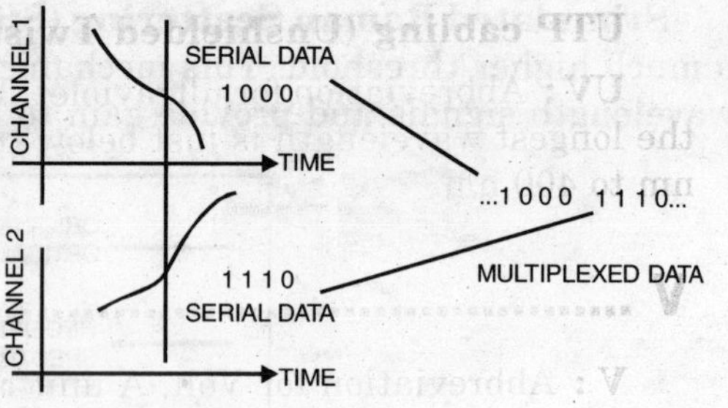

Time-division Multiple Access (TDMA) : A communications technique that uses a common channel (multipoint or broadcast) for communications among multiple users by allocating unique time slots to each user. Used extensively in satellite systems, local area networks, physical security systems, and combat-net radiosystems.

Total Internal Reflection : The reflection that occurs when light strikes an interface at an angle of incidence (with respect to the normal) greater than the critical angle.

Transceiver : A device that performs, within one chassis, both telecommunication transmitting and receiving functions.

Transducer : A device that converts energy from one form to another, such as optical energy to electrical energy.

Transmitter : A device that includes a source and driving electronics. It functions as an electrical-to-optical converter.

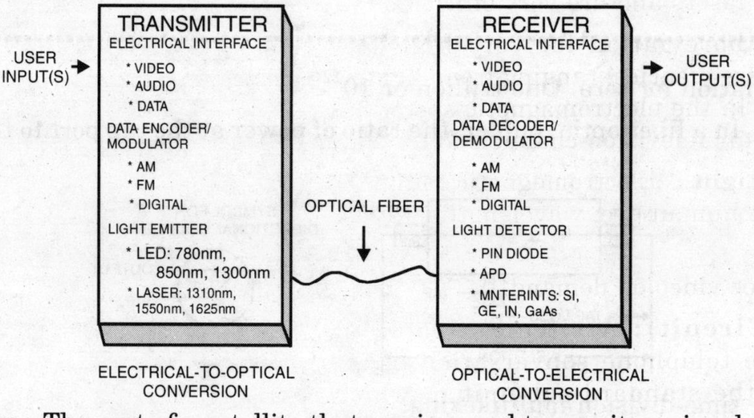

Transponder : The part of a satellite that receives and transmits a signal.

U ....

UHF : Abbreviation for ultra high frequency. The frequencies, ranging from 300 MHz to 3000 MHz, in the electromagnetic spectrum. Contains off-air television channels 21-68.

Unidirectional : Operating in one direction only.

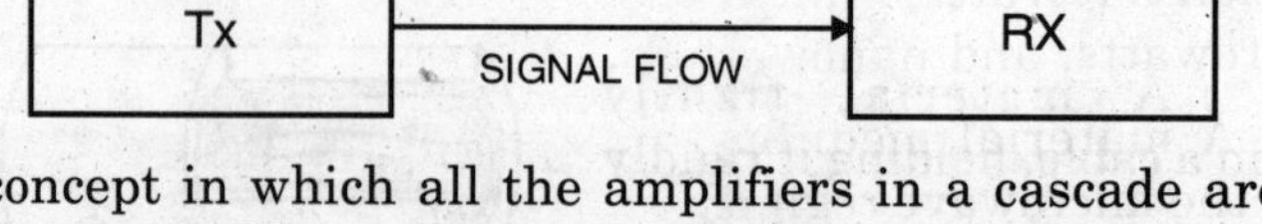

Unity Gain : A concept in which all the amplifiers in a cascade are in balance with their power inputs and outputs. Unity gain can be achieved by adjusting the receiver output, either by padding or attenuation in the node, to the proper level determined by the RF input.

UTP cabling (Unshielded Twisted Pair):

UV : Abbreviation for ultraviolet. That portion of the electromagnetic spectrum in which the longest wavelength is just below the visible spectrum, extending from approximately 4 nm to 400 nm.

V

V : Abbreviation for Volt. A unit of electrical force or potential, equal to the force that will cause a current of one Amp to flow through a conductor with a resistance of one Ohm.

V_{AC} : Abbreviation for Volts, AC. Voltage using alternating current.

VCSEL : See vertical cavity surface-emitting laser.

V_{DC} : Abbreviation for Volts, DC. Voltage using direct current.

VDSL : Abbreviation for very high data rate digital subscriber line. A DSL operating at a data rate higher than that of HDSL. See also DSL.

Vertical Cavity Surface-emitting Laser : Lasers that emit light perpendicular to the plane of the wafer they are grown on. They have very small dimensions compared to conventional lasers and are very efficient.

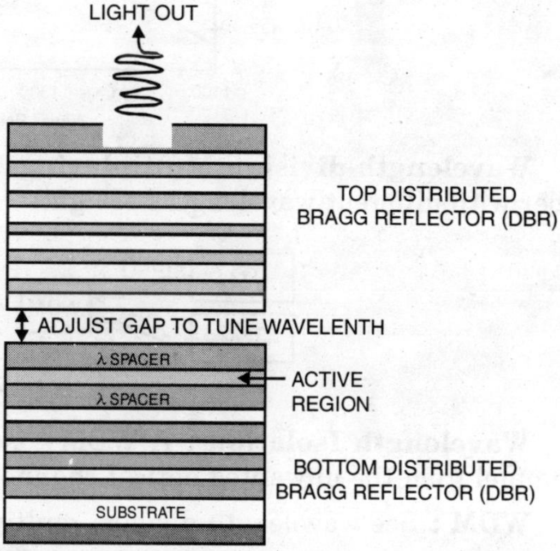

VHF : Abbreviation for very high frequency. The frequencies, ranging from 30 MHz to 300 MHz, in the electromagnetic spectrum. Contains off-air television channels 2-12.

Visible Light : Electromagnetic radiation visible to the human eye; wavelengths of 400-700 nm.

VOD : See video on demand.

Voice Circuit : A circuit capable of carrying one telephone conversation or its equivalent; the standard sub-unit in which telecommunication capacity is counted. The U.S. analog equivalent is 4 kHz. The digital equivalent is 64 kbit/s in North America and in Europe.

VPN : Abbreviation for virtual private network. A protected information-system link utilizing tunneling, security controls, and end-point address translation giving the end user the impression that a dedicated line exists between nodes.

W

W : Abbreviation for Watt. A linear measurement of optical power, usually expressed in milliwatts, microwatts, and nanowatts.

Waveguide : A material medium that confines and guides a propagating electromagnetic wave. In the microwave regime, a waveguide normally consists of a hollow metallic conductor, generally rectangular, elliptical, or circular in cross-section. This type of waveguide may, under certain conditions, contain a solid or gaseous dielectric material. In the optical regime, a waveguide used as a long transmission line consists of a solid dielectric filament (fiber), usually circular in cross-section. In integrated optical circuits an optical waveguide may consist of a thin dielectric film. In the RF regime, ionized layers of the stratosphere and the refractive surfaces of the troposphere may also serve as a waveguide.

Waveguide Coupler : A coupler in which light gets transferred between planar waveguides.

Waveguide Dispersion : The part of chromatic dispersion arising from the different speeds light travels in the core and cladding of a single-mode fiber (*i.e.*, from the fiber's waveguide structure).

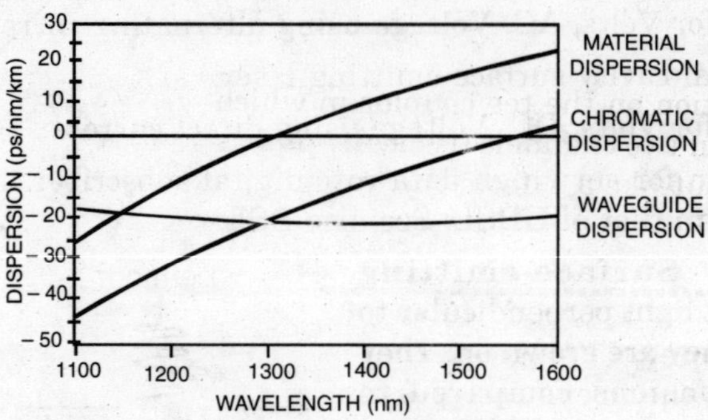

Wavelength-division Multiplexing (WDM) : Sending several signals through one fiber with different wavelengths of light.

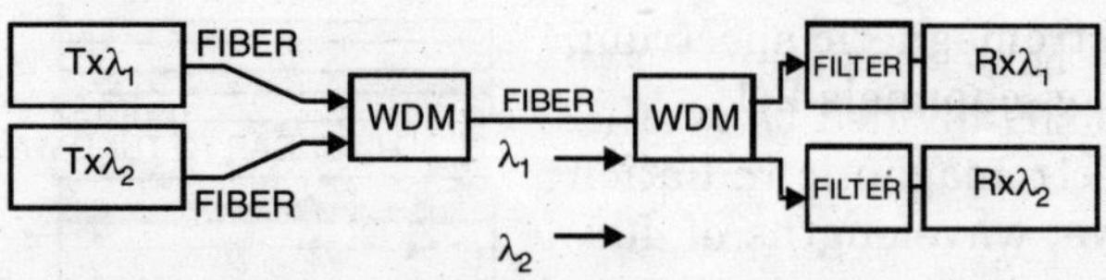

Wavelength Isolation : A WDM's isolation of a light signal in the desired optical channel from the unwanted optical channels. Also called far-end crosstalk.

WDM : See wavelength-division multiplexing.

Wide Area Network (WAN) : A physical or logical network that provides capabilities for a number of independent devices to communicate with each other over a common transmission-interconnected topology in geographic areas larger than those served by local area networks or metropolitan area networks.

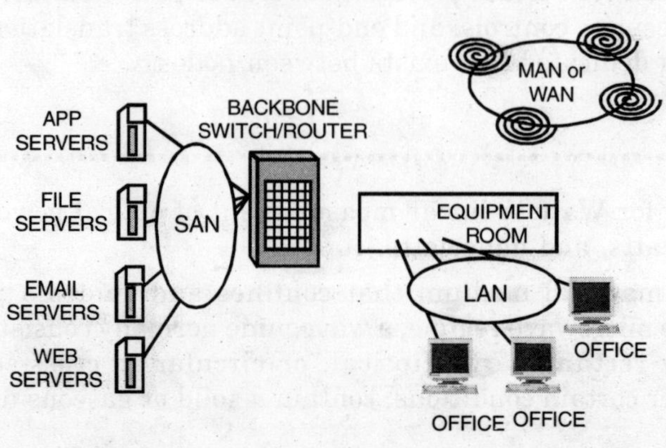

X, Y

X-Band : The frequency range between 8.0 and 8.4 GHz.

XC : See cross-connect.

XGM : See cross-gain modulation.

XPM : See cross-phase modulation.

X-Series Recommendations : Sets of data telecommunications protocols and interfaces defined by the ITU.

XT : See crosstalk.

Y Coupler : A variation on the tee coupler in which input light is split between two channels (typically planar waveguide) that branch out like a Y from the input.

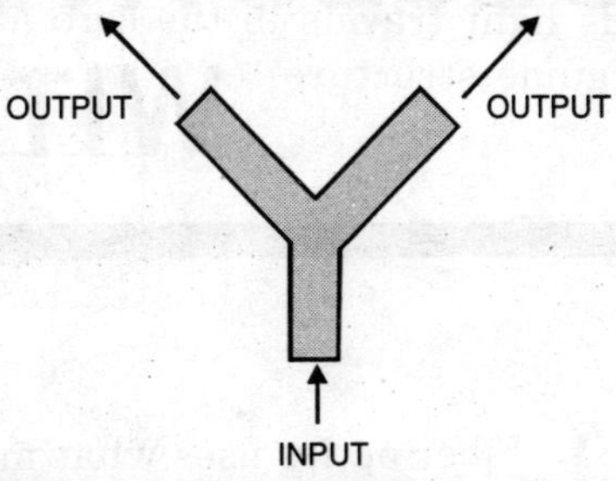

Z

Zero-dispersion Slope : In single-mode fiber, the rate of change of dispersion with respect to wavelength, at the fiber's zero-dispersion wavelength.

Zero-dispersion Wavelength (λ_0) : In a single-mode optical fiber, the wavelength at which material dispersion and waveguide dispersion cancel one another. The wavelength of maximum bandwidth in the fiber. Also called zero-dispersion point.

Zipcord : A two-fiber cable consisting of two single fiber cables having conjoined jackets. A zipcord cable can be easily divided by slitting and pulling the conjoined jackets apart.

❑❑❑

OBJECTIVE TYPE QUESTIONS WITH ANSWERS

1. Fiber optics uses what medium to send information?

(*a*) Electrons (*b*) Phonons

(*c*) Link 11 (*d*) Light

2. What are the three parts of a fiber optic data link?

(*a*) Transmitter, optical fiber, receiver

(*b*) Transmitter, optical fiber, optical connectors

(*c*) Optical fiber, optical connectors, receiver

(*d*) Optical fiber, optical connectors, optical splices

3. The fiber optic transmitter has which of the following functions?

(*a*) Amplifies the optical signal

(*b*) Converts the electrical input signal to an optical signal

(*c*) Converts the input optical signal to an electrical signal

(*d*) Amplifies the output electrical signal

4. Fiber optic systems use what two types of optical sources?

(*a*) LEDs and APDs (*b*) PIN diodes and LEDs

(*c*) LEDs and laser diodes (*d*) Laser diodes and APDs

5. The optical source performs which of the following functions?

(*a*) Amplifies the optical signal

(*b*) Amplifies the electrical signal

(*c*) Launches the optical signal into the fiber

(*d*) Converts the optical signal to an electrical signal

6. What fiber mechanisms weaken and distort the optical signal launched into the fiber?

(*a*) Scattering, absorption, and dispersion

(*b*) Scattering, radiation, and absorption

(*c*) Dispersion, radiation, and absorption

(*d*) Scattering, reflection, and refraction

7. The fiber optic receiver performs which of the following functions?
(*a*) Amplifies the optical signal
(*b*) Amplifies the electrical signal
(*c*) Converts the electrical signal back into an optical signal
(*d*) Converts the optical signal back into an electrical signal

8. What are the two parts of a fiber optical receiver?
(*a*) Fiber and case
(*b*) Fiber and optical detector
(*c*) Optical detector and PIN diode
(*d*) Optical detector and signal conditioning-circuits

9. Noise has what effect, if any, on the quality of a signal?
(*a*) Increases (*b*) Reduces
(*c*) None; noise has no effect

10. What are the two types of optical detectors?
(*a*) LEDs and APDs (*b*) PIN diodes and APDs
(*c*) APDs and laser diodes (*d*) Laser diodes and PIN diodes

11. Besides the transmitter, the optical fiber, and the receiver, which of the following components may be found in a fiber optic link?
(*a*) Splices only (*b*) Couplers only
(*c*) Connectors only (*d*) Splices, couplers, and connectors

12. Extremely high losses occurred in early fibers because of which of the following conditions?
(*a*) Cracks in the fibers (*b*) Holes in the fiber sides
(*c*) Core areas too small in the fibers (*d*) Impurities in the fiber material

13. What was the first light source developed that could be easily coupled into a fiber?
(*a*) LED (*b*) Lamp
(*c*) YAG laser (*d*) PIN diode

14. Compared to multimode fibers, single mode fibers tend to have lower loss and produce less signal distortion.
(*a*) True (*b*) False

15. What multimode fiber properties help reduce connection losses?
(*a*) Larger core size and lower NA (*b*) Larger core size and higher NA
(*c*) Smaller core size and lower NA (*d*) Smaller core size and higher NA

16. Fiber optics is being used in which of the following types of applications?
(*a*) Long-haul communications (*b*) Subscriber-loop applications
(*c*) Military applications (*d*) All of the above

17. Of the following advantages, which one does NOT apply to fiber optics?
(*a*) Established standards (*b*) Increased bandwidth
(*c*) Improved environmental (*d*) Improved signal security

18. Of the following factors, which ones are advantages of fiber optic systems?
(*a*) Electrical isolation and low bandwidth
(*b*) Immunity to noise and low bandwidth
(*c*) Electrical isolation and immunity to noise
(*d*) Signal security and high price

19. To describe the nature of light, which of the following ways can be used?
(*a*) Electromagnetic wave only
(*b*) Particles of energy only
(*c*) Electromagnetic wave and particles of energy
(*d*) Element

20. Light exhibits what kind of wave motion?
(*a*) Longitudinal (*b*) Transverse
(*c*) Turbulent (*d*) A periodic

21. Which of the following factors is a description of transverse wave motion?
(*a*) The wave motion is not predictable
(*b*) The wave magnitude varies parallel to the direction of propagation
(*c*) The wave magnitude varies perpendicular to the direction of wave motion

22. What does a transparent substance do to light rays that fall on it?
(*a*) Absorbs them (*b*) Reflects them
(*c*) Refracts them (*d*) Transmits them

23. What does a translucent substance do to light rays that fall on it?
(*a*) Reflects and absorbs them (*b*) Refracts and absorbs them
(*c*) Transmits and diffuses them (*d*) Transmits and reflects them

24. Which of the following substances is a good example of a translucent substance?
(*a*) Cardboard (*b*) Clear glass
(*c*) Frosted glass (*d*) Aluminium foil

25. Which of the following substances is a good example of a transparent substance?
(*a*) Cardboard (*b*) Clear glass
(*c*) Frosted glass (*d*) Aluminum foil

26. What does an opaque substance do when light rays fall on it?
(*a*) Refracts them (*b*) Reflects or absorbs them
(*c*) Transmits them only (*d*) Transmits and diffuses them

27. Which of the following substances is an example of an opaque substance?
(*a*) Cardboard (*b*) Clear glass
(*c*) Oiled paper (*d*) Frosted glass

28. Which of the following objects is NOT an example of a luminous source?
(*a*) Sun (*b*) Gas flame
(*c*) Mirror (*d*) Flashlight

29. What are light waves called that strike a surface but are neither transmitted nor absorbed?

(*a*) Diffused (*b*) Refracted

(*c*) Reflected (*d*) Diffracted

30. What is the name of the law that states "The angle of incidence is equal to the angle of reflection"?

(*a*) Snell's Law (*b*) Murphy's Law

(*c*) Law of Entropy (*d*) Law of Reflection

31. A light wave is incident on a surface. The reflected power is the greatest in which of the following incidences?

(*a*) 30° (*b*) 45°

(*c*) Perpendicular (*d*) Almost parallel

32. A light wave passes from one medium into another medium with a different velocity. As the wave enters the second medium, the change of direction is known by which of the following terms?

(*a*) Reflection (*b*) Refraction

(*c*) Absorption (*d*) Diffusion

33. If a light wave passes from a less dense medium to a more dense medium, how does the angle of refraction compare to the angle of incidence?

(*a*) Greater than the angle of incidence only

(*b*) Equal to the angle of incidence only

(*c*) Greater than or equal to the angle of incidence

(*d*) Less than the angle of incidence

34. What is another word for diffused?

(*a*) Absorbed (*b*) Refracted

(*c*) Scattered (*d*) Attenuated

35. When light falls on a piece of black paper, what happens to most of the light?

(*a*) It is absorbed (*b*) It is reflected

(*c*) It is scattered (*d*) It is refracted

36. Light is transmitted along an optical fiber by what two methods?

(*a*) Ray theory and mode theory (*b*) Ray theory and photon theory

(*c*) Ray theory and quantum theory (*d*) Mode theory and photon theory

37. How does the speed of light in the fiber compare to the speed of light in the air?

(*a*) It is slower in the fiber (*b*) It is faster in the fiber

(*c*) It is the same in both the fiber and the air

38. The relationship between the incident rays and the refracted rays at a boundary between mediums with different indexes of refraction describes what law?

(*a*) Bragg's Law (*b*) Snell's Law

(*c*) Murphy's Law (*d*) Law of Reflection

39. Total internal reflection occurs at which of the following angles?

(*a*) Obtuse angle (*b*) Fresnel angle

(*c*) Right angle (*d*) Critical angle of incidence

40. What are the three basic parts of an optical fiber?

(*a*) Core, cladding and coating (*b*) Inside, middle, and outside

(*c*) Fiber, kevlar, and jacket (*d*) Hole, shell, and coating

41. The cladding performs all except which of the following functions?

(*a*) Reduces the loss of light from the core

(*b*) Reduces the scattering loss at the surface of the core

(*c*) Protects the fiber core from absorbing surface contaminants

(*d*) Reduces mechanical strength

42. What is the definition of a bound ray?

(*a*) A ray that cannot move

(*b*) A ray that travels in the air

(*c*) A ray that is refracted out of the fiber

(*d*) A ray that propagates through the fiber by total internal reflection

43. The fiber NA relates to which of the following characteristics?

(*a*) Physical size of the fiber

(*b*) Tensile strength of the fiber

(*c*) Maximum angle within the fiber acceptance cone

(*d*) Speed of light within the fiber

44. A skew ray is which of the following types of rays?

(*a*) An unbound ray (*b*) A meridional ray

(*c*) An unbalanced ray

(*d*) A ray that propagates without passing through the center axis of the fiber

45. Electromagnetic wave behavior is used to describe the propagation of light along the fiber in what theory?

(*a*) Mode theory (*b*) Particle theory

(*c*) Darwin's theory (*d*) Rayleigh's theory

46. High-order modes cross the axis of the fiber at steeper angles than low-order modes.

(*a*) True (*b*) False

47. Compared to a low-order mode, the electrical and magnetic fields of a high order mode are distributed more toward the center of a fiber.

(*a*) True (*b*) False

48. What are the two basic types of fibers?

(*a*) Small and large (*b*) Glass and plastic

(*c*) Opaque and diffuse (*d*) Single mode and multimode

49. Compared to multimode fibers, single mode fibers have a larger core size.

(*a*) True (*b*) False

50. The lowest signal loss and the highest bandwidth are characteristic of which of the following types of fibers?

(*a*) Air core (*b*) Multimode

(*c*) Single mode (*d*) Plastic core

51. Compared to single mode fibers, multimode fibers have which of the following advantages?

(*a*) Ease of making connections only

(*b*) Ease of launching light into them only

(*c*) Ease of both making connections and launching light into them

(*d*) Lower dispersion

52. System performance is affected most by which of the following fiber properties ?

(*a*) NA and delta (*b*) Core diameter and NA

(*c*) Attenuation and dispersion (*d*) Core diameter and cladding

53. The loss of optical power as light travels along a fiber is called

(*a*) attenuation (*b*) scattering

(*c*) absorption (*d*) dispersion

54. Attenuation is specified in what units ?

(*a*) dB (*b*) dB/km

(*c*) μm (*d*) μm/km

55. Glass optical fibers have low loss between the infrared and ultra- violet absorptive regions. The approximate wavelength of operation for glass optical fibers is in which of the following ranges?

(*a*) 1 nm to 700 nm (*b*) 700 nm to 1600 nm

(*c*) 1600 nm to 9000 nm (*d*) 9 μm to 20 μm

56. Increased extrinsic absorption at 950 nm, 1,250 nm, and 1,383 nm is caused by what impurity in glass optical fibers?

(*a*) Phosphorus (*b*) Germanium

(*c*) Titanium (*d*) Water

57. Which type of scattering loss is proportional to the reciprocal of the fourth power of the wavelength of the light?

(*a*) Mie (*b*) Raman

(*c*) Rayleigh (*d*) Brillouin

58. A radius of curvature is larger than the fiber diameter in which of the following types of fiber bends?

(*a*) Macrobends (*b*) Microbends

(*c*) Gentle bends (*d*) Serpentine bends

59. Only in multimode fibers does which of the following types of dispersion occur?

(*a*) Modal (*b*) Material

(*c*) Waveguide (*d*) Chromatic

60. When different colors of light travel through the fiber at different speeds, which of the following types of dispersion occurs?

(*a*) Modal (*b*) Intramodal (chromatic)

(*c*) Intermodal (*d*) Atmospheric

61. In a step-index fiber, the refractive index profile of the fiber core has which of the following characteristics ?

(*a*) It is uniform over the fiber core

(*b*) It linearly decreases from a maximum at the fiber center to a minimum at the core-cladding boundary

(*c*) It is parabolic with a maximum index of refraction at the center and a minimum index of refraction at the core-cladding boundary

(*d*) It linearly increases from a minimum at the fiber center to a maximum at the core cladding boundary

62. In a graded-index fiber, the refractive index profile of the fiber core is best described by which of the following statements?

(*a*) It is uniform over the fiber core

(*b*) It linearly decreases from a maximum at the fiber center to a minimum at the core-cladding boundary

(*c*) It is parabolic with a maximum index of refraction at the center and a minimum index of refraction at the core-cladding boundary

(*d*) It linearly increases from a minimum at the fiber center to a maximum at the core cladding boundary

63. Which of the following multimode fiber core sizes is NOT a standard commercial fiber size?

(*a*) 50 μm (*b*) 5 μm

(*c*) 76 μm (*d*) 100 μm

64. Elements other than silicon and oxygen are added to glass material by the fiber manufacturer to change its index of refraction. What are these elements called?

(*a*) Spices (*b*) Dopants

(*c*) Additives (*d*) Impurities

65. Compared to multimode step-index fibers, do multimode graded-index fibers have lower, higher, or approximately equal bandwidths?

(*a*) Lower (*b*) Higher

(*c*) Approximately equal

66. In multimode graded-index fibers, is the index of refraction of the glass at the center of the fiber core lower, higher, or approximately equal to the index of refraction of the cladding glass?

(*a*) Lower (*b*) Higher

(*c*) Approximately equal

67. For a multimode graded-index fiber, the numerical aperture is at its maximum value at which of the following locations?

(*a*) At the fiber axis

(*b*) At the core-cladding interface

(*c*) Half way between the fiber center and the core-cladding interface

(*d*) One-fourth of the way between the fiber center and the core-cladding interface

68. In multimode fiber, all light rays have to propagate all of the way to the core cladding interface before they are reflected back toward the fiber axis.

(*a*) True (*b*) False

69. A step-index multimode fiber and a graded-index multimode fiber have the same core and cladding sizes and the same refractive index difference. Which fiber type, if either, will accept light more easily and have more propagating modes?

(*a*) Step-index fiber

(*b*) Graded-index fiber

(*c*) Neither; they will behave approximately the same

70. The multimode graded-index fiber that has the best bend performance and will show the least amount of optical degradation if mishandled is what size?

(*a*) 50/125 μm (*b*) 5/125 μm

(*c*) 85/125 μm (*d*) 100/140 μm

71. The attributes of the 62.5/125-??m fiber do NOT include which of the following factors?

(*a*) Low loss (*b*) High bandwidth

(*c*) Low bending sensitivity

(*d*) Low source to fiber coupling efficiency

72. What are the two basic types of single mode step-index fibers?

(*a*) Low NA and high NA (*b*) Solid core and air core

(*c*) Enriched clad and depressed (*d*) Matched clad and depressed clad

73. For wavelengths greater than its cutoff wavelength, a typical single mode fiber is allowed to propagate a total of how many modes, if any?

(*a*) 1 only (*b*) 10 only

(*c*) 100 (*d*) None

74. Fibers are single mode at a particular wavelength only when V < 2.405. Under what condition, if any, will the fiber cease to be single mode?

(*a*) When the wavelength of the light is greater than the cutoff wavelength

(*b*) When the wavelength of the light is less than the cutoff wavelength

(*c*) None

75. The use of plastic-clad silica and all plastic fibers has what primary drawback?

(*a*) Higher NA (*b*) Higher cost

(*c*) Higher bandwidth (*d*) Limited optical performance

76. In the fabrication of silica optical fibers, which of the following processes may be used?

(*a*) Outside vapor phase oxidation (OVPO)

(*b*) Inside vapor phase oxidation (IVPO)

(*c*) Vapor phase axial deposition (VAD)

(*d*) All of the above

77. In making a preform, layers of glass powder are deposited on the inside or outside of a glass rod or tube. What is this glass powder called?

(*a*) Soot (*b*) Smoke

(*c*) Preform (*d*) Afterburn

78. The process used in drawing the fiber is best described by which of the following statements?

(*a*) The preform is melted and the molten glass is molded, using special fiber molds

(*b*) The preform is softened and the glass is pulled into a thin glass filament

(*c*) The preform is softened and the glass is rolled into a thin glass filament

(*d*) The preform is melted and the fiber is formed by blowing the molten glass through a small hole

79. To protect the fiber from contaminants in the drawing process, what substance is added over the fiber?

(*a*) Water (*b*) Coating

(*c*) Preform (*d*) Cladding

80. Most fiber optic cable structures contain which of the following items?

(*a*) Buffers only (*b*) Jackets only

(*c*) Buffers and jackets only (*d*) Buffers, jackets, and strength members

81. Properly designed optical cable structures perform which of the following functions?

(*a*) Protect the optical fibers from mechanical stresses, damage, and breakage

(*b*) Increase the tensile stress on the fiber

(*c*) Decrease the attenuation of the fiber

(*d*) All of the above

82. The fiber buffer performs which of the following functions?

(*a*) Protects the fiber from micro-bends

(*b*) Provides additional mechanical protection

(*c*) Helps preserve the fiber's inherent strength

(*d*) All of the above

83. Navy shipboard fiber optic cables should NOT contain which of the following materials as a strength member?

(*a*) Steel wire (*b*) Arimid yarns

(*c*) Carbon fibers (*d*) Glass-reinforced plastics

84. In the materials of a Navy shipboard fiber optic cable jacket, which of the following properties is NOT desirable?

(*a*) High abrasion resistance (*b*) High flame retardance

(*c*) High halogen content (*d*) Low toxicity

85. Most commercial fiber optic cable jacket materials are suitable for use in Navy shipboard applications.

(*a*) True (*b*) False

86. For low-density fiber Navy ship-board applications, which cable designs, if any, are preferred?

(*a*) Stranded cable (*b*) Ribbon cable

(*c*) OFCC cable (*d*) None; they are all equally effective

87. When compared with other cable designs, the OFCC fiber optic cable has which of the following advantages?

(*a*) Ruggedness only (*b*) Ease of handling only

(*c*) Ruggedness and ease of handling (*d*) Relative small size

88. The only advantage of the stranded cable design over the OFCC cable design is

(*a*) greater ruggedness

(*b*) better fiber protection

(*c*) smaller size for the same fiber count

(*d*) better water-blocking performance

89. A particular fiber optic cable design has the highest fiber count possible in the smallest size. But it is difficult to use, shows susceptibility to fiber damage during fiber breakout, and has poor bending and water-blocking performance. Which of the following designs fits this description?

(*a*) OFCC cable (*b*) Stranded cable

(*c*) Ribbon cable

90. A fiber optic connection that is typically intended to be permanent is what type, if any?

(*a*) Mechanical splice (*b*) Connector

(*c*) Coupler

(*d*) None; they are all intended to be removable

91. Easy coupling and uncoupling of optical fibers are allowed by what type of fiber optic connection, if any?

(*a*) Splice (*b*) Coupler

(*c*) Connector (*d*) None; all are equally difficult

92. The distribution or combination of optical signals among fibers uses which type of fiber optic connection, if any?

(*a*) Splice

(*b*) Coupler

(*c*) Connector

(*d*) None; it is not possible to distribute or combine optical signals

93. Coupling loss in a fiber optic connection may be caused by which of the following conditions?

(*a*) Poor fiber alignment

(*b*) Poor fiber end preparation

(*c*) Differences in optical properties between connected fibers

(*d*) All of the above

94. The loss in optical power through a connection is described by what relationship, if any?

(*a*) Log relationship

(*b*) Linear relationship

(*c*) Exponential relationship

(*d*) None; there is no way to describe the relationship

95. An intrinsic coupling loss may be caused by which of the following conditions?

(*a*) Fiber separation (*b*) Angular misalignment

(*c*) Fiber geometry mismatches (*d*) Poor fiber end preparation

96. In connections that have an air gap, the reflective losses can be minimized by taking which of the following actions?

(*a*) Polishing the fiber ends more finely

(*b*) Cleaning the fiber ends carefully with alcohol

(*c*) Filling the air gap with an index matching gel

(*d*) Heating the connection to an elevated temperature

97. A main source of extrinsic coupling loss in fiber to fiber connections is caused by which of the following conditions?

(*a*) Fiber mismatches (*b*) Poor fiber alignment

(*c*) Poor fiber end preparation

98. Which of the following conditions is NOT a form of poor fiber alignment?

(*a*) NA mismatch (*b*) Fiber separation

(*c*) Lateral misalignment (*d*) Angular misalignment

99. For mating fibers to touch when a connector is assembled, which of the following polishing techniques was developed?

(*a*) Fresnel polish (*b*) Rayleigh polish

(*c*) Circular polish (*d*) Physical contact polish

100. Single mode fibers are less sensitive to alignment errors than multimode fibers because of their small core size.

(*a*) True (*b*) False

101. The fiber cleaving process includes which of the following actions?

(*a*) Cutting the fiber with a cleaver (*b*) Bending the fiber until it breaks

(*c*) Lightly nicking the fiber and applying tension until it breaks

(*d*) Sawing the fiber with a special fiber saw

102. To remove most of the surface imperfections introduced in the fiber cleaving process in most fiber optic connections, what action should be taken ?

(*a*) Polish the fiber end (*b*) Cover the fiber end face with epoxy

(*c*) Clean the fiber end face with alcohol

(*d*) Cover the fiber end face with index matching gel

103. The connector end face should be inspected for scratches and shattered fiber ends by which of the following methods?

(*a*) The naked eye (*b*) A 10x eye loupe
(*c*) An electron microscope (*d*) A 200x-400x microscope

104. In Navy fiber optic applications, losses from fiber mismatches are minimized by which of the following actions?

(*a*) Polishing the fiber ends
(*b*) Using index matching gels
(*c*) Using index matching epoxies
(*d*) Using Navy specification fibers with tightly specified parameters

105. Fiber optic splices are typically used in which of the following conditions?

(*a*) When repeated connections or disconnections are desired only
(*b*) When repeated connections or disconnections are unnecessary only
(*c*) When low loss connections are required only
(*d*) When repeated connections or disconnections are unnecessary and when low loss connections are required

106. What are the two basic categories of fiber splices?

(*a*) Glass and plastic (*b*) Metal and ceramic
(*c*) V-groove and rotary (*d*) Mechanical and fusion

107. For shipboard applications, which of the following types of splices is recommended?

(*a*) Fusion (*b*) Rotary
(*c*) V-groove (*d*) Ceramic alignment tube

108. The assembly of a rotary splice is which of the following actions?

(*a*) Cleave fiber ends, insert into each end of a glass tube filled with adhesive, and cure with ultraviolet light
(*b*) Cleave fiber ends, insert into each end of a glass tube filled with index matching gel, and twist on splice end caps
(*c*) Break glass tube at center, fill each half with epoxy, insert fibers into each half, cure adhesive with ultraviolet light, and polish and place polished ends together with alignment sleeve
(*d*) Cleave fiber ends, insert fibers into splice, and crimp in place

109. The basic process used in making fusion splices is which of the following methods?

(*a*) Gluing the fiber ends together
(*b*) Melting the fiber ends together
(*c*) Polishing the fiber ends and placing them together
(*d*) Cleaving the fiber ends and placing them together

110. Fusion splicing is not recommended for Navy use for all of the following reasons except which one?

(*a*) Low splice yields (*b*) Low splice losses
(*c*) Long splicing times
(*d*) High level of expertise required for splicer operator

111. The total loss of a fiber optic connector should always remain below which of the following values?

(*a*) 1 dB (*b*) 2 dB

(*c*) 5 dB (*d*) 10 dB

112. What are the two basic types of fiber optic connectors?

(*a*) V-groove and ribbon (*b*) Three rod and biconical

(*c*) Ceramic and stainless steel (*d*) Butt-jointed and expanded-beam

113. Single fiber butt-jointed connections generally consist of which of the following parts?

(*a*) Plug and receptacle (*b*) Ferrule and alignment clip

(*c*) Two plugs and an adapter (*d*) One plug and two adapters

114. In ceramic ferrules, the center hole is created by which of the following processes?

(*a*) With a microdrill

(*b*) With a laser drill

(*c*) By forming the ferrule around a precision wire

(*d*) With a small water jet

115. The standard connector for Navy light duty applications is of which of the following styles?

(*a*) ST (*b*) SMA

(*c*) Array (*d*) Biconical

116. Light-duty connectors are intended for use in all but which of the following locations?

(*a*) Junction boxes

(*b*) Equipment enclosures

(*c*) Environmentally protected locations

(*d*) Exposed equipment interface connectors

117. A standard connector for Navy heavy duty applications is of which of the following styles?

(*a*) MIL-C-83526 connector (*b*) MIL-C-28876 connector

(*c*) Biconical connector (*d*) Array connector

118. In a heavy-duty connector, what are the two types of terminals called?

(*a*) Plugs and adapters (*b*) Type A and type B

(*c*) Pins and sockets (*d*) Nuts and bolts

119. Fiber optic couplers attenuate the optical signal much more than a connector or a splice.

(*a*) True (*b*) False

120. Active fiber optic couplers are called active fiber optic couplers for which of the following reasons?

(*a*) Because they contain moving parts

(*b*) Because they switch the optical signal between different parts

(*c*) Because they contain active devices, including sources and detectors

(*d*) Because they move through the system

121. An optical coupler has one input port and two output ports. Which of the following types of couplers is it?

(*a*) Optical splitter (*b*) Optical combiner

(*c*) Star coupler (*d*) Tree coupler

122. An optical coupler has two input ports and one output port. Which of the following types of couplers is it?

(*a*) Optical splitter (*b*) Optical combiner

(*c*) Star coupler (*d*) Tree coupler

123. An optical coupler has one input port and several output ports. Which of the following types of couplers is it?

(*a*) T-coupler (*b*) Star coupler

(*c*) Tree coupler (*d*) Optical splitter

124. An optical coupler has several input ports and several output ports. Which of the following types of couplers is it?

(*a*) X-coupler (*b*) T-coupler

(*c*) Star coupler (*d*) Tree coupler

125. Manufacturers should measure some component parameters during and after component manufacture for which of the following reasons?

(*a*) To characterize the component design

(*b*) To verify the quality of the manufactured parts

(*c*) To determine if the component has been damaged in the shipping process

(*d*) To determine if the installation process has degraded the component performance beyond allowable limits

126. End users should measure some component parameters upon receipt before installing a component into a fiber optic system for which of the following reasons?

(*a*) To characterize the component design

(*b*) To verify the quality of the manufactured parts

(*c*) To determine if the component has been damaged in the shipping process

127. End users should measure some component parameters after installation of the component in the fiber optic system for which of the following reasons?

(*a*) To characterize the component design

(*b*) To verify the quality of the manufactured parts

(*c*) To determine if the component has been damaged in the shipping process

(*d*) To determine if the installation process has degraded the component performance beyond allowable limits

128. Standard test methods used for testing fiber optic components or systems can be found in which of the following test documents?

(*a*) EIA/TIA-455 only (*b*) EIA/TIA-526 only

(*c*) EIA/TIA-455 and EIA/TIA 526 (*d*) MIL-STD 2196

129. In laboratory fiber attenuation measurements, the output optical power of a long fiber is compared with the optical power launched into that same fiber two meters from the optical source. This method of measuring the optical attenuation is called the

(*a*) OTDR method (*b*) Fiber method
(*c*) Cutback method (*d*) Two-point method

130. Which of the following are considered launch conditions?

(*a*) The temperature at the time of measurement
(*b*) The humidity at the time of measurement
(*c*) The way power is launched into the optical fiber
(*d*) The type of connector or splice used on the fiber

131. Launch conditions in an attenuation measurement are important to measurement personnel for which of the following reasons?

(*a*) The attenuation measured for each different launch condition will be different
(*b*) The use of the wrong launch condition will damage the optical power meter
(*c*) The fibers will only transmit light launched in one particular manner
(*d*) The use of the wrong launch condition will damage the fiber

132. For single mode fiber, what method is typically used to assure that power is launched into only the fundamental mode?

(*a*) Only LEDs are used as test optical sources
(*b*) Only lasers are used as test optical sources
(*c*) The fiber is always placed at a 45° angle with respect to the plane of the source
(*d*) A 30 mm diameter mandrel or freeform loop is applied at the launch end of the fiber

133. What does the term overfilled launch mean?

(*a*) When too much power is launched into the fiber
(*b*) When not enough power is launched into the fiber
(*c*) When light is launched only into the low order mode
(*d*) When the launch spot size and the angular distribution are larger than that of fiber core

134. In what part of the fiber does an underfilled launch concentrate most of the optical power ?

(*a*) In the center of the fiber (*b*) At the edge of the fiber
(*c*) Over all of the fiber (*d*) In the cladding

135. What is a mode filter?

(*a*) A device that converts cladding power to core power
(*b*) A device that removes any cladding power from the fiber
(*c*) A device that attenuates specific modes propagating in the core of the fiber
(*d*) A device that attenuates specific wavelengths of light in the core of the fiber

136. To remove high-order modes in multimode fibers, which of the following mode filters is normally used?

(*a*) Bessel filter (*b*) Bandpass filter

(*c*) Mandrel wrap filter (*d*) Neutral density filter

137. What is the wavelength called over which the fiber will propagate only one mode?

(*a*) Cutoff wavelength (*b*) Fresnel wavelength

(*c*) Cutback wavelength (*d*) Rayleigh wavelength

138. Determining the cutoff wavelength of a single mode fiber involves finding the wavelength above which the power transmitted through the fiber decreases abruptly.

(*a*) True (*b*) False

139. When the cutoff wavelength is reached, why does the transmitted power in a single mode fiber abruptly decrease?

(*a*) The fiber absorption increases at the cutoff wavelength

(*b*) The fiber scattering increases at the cutoff wavelength

(*c*) The fiber diffusion increases at the cutoff wavelength

(*d*) The second-order mode ceases to propagate at the cutoff wavelength

140. The spreading of a light pulse as it travels along the fiber is caused by what mechanism?

(*a*) Diffraction (*b*) Attenuation

(*c*) Absorption (*d*) Dispersion

141. In a fiber, the intermodal (modal) dispersion is greatest under what condition?

(*a*) When the input pulse has a long rise time

(*b*) When all fiber modes are excited

(*c*) When the input pulse has an extremely high amount of power

(*d*) When the source spectrum contains many different wave- lengths

142. The fiber bandwidth (–3 dB) is defined as the lowest frequency at which the magnitude of the fiber frequency response has decreased to what portion of its zero-frequency value?

(*a*) One-tenth (*b*) One-fifth

(*c*) One-half (*d*) Three-fourths

143. Modal bandwidth is normally given in what units?

(*a*) GHz (*b*) MHz

(*c*) MHz/km (*d*) MHz-km

144. Why does chromatic dispersion occur?

(*a*) Because different fiber modes have different propagation speeds

(*b*) Because different colors of light travel through the fiber at different speeds

(*c*) Because different colours of light are attenuated differently in the fiber

(*d*) Because different fiber modes are attenuated differently in the fiber

145. End users perform fiber geometry measurements for which of the following reasons?

(*a*) To reduce system bandwidth limitations

(*b*) To separate bigger fibers from smaller fibers

(*c*) To reduce system attenuation and coupling losses from fiber mismatches

(*d*) To determine which connector to use

146. Cladding diameter, cladding noncircularity, core-cladding concentricity error, and core noncircularity are measured by which of the following techniques?

(*a*) Micrometer calipers (*b*) Electron microscopes

(*c*) Video cameras attached to a computer for digital analysis

(*d*) Ruler

147. The core diameter of a multimode fiber is defined by which of the following measurements?

(*a*) The distance between the 50% points of the measured near-field distribution

(*b*) The distance between the 50% points of the measured far-field distribution

(*c*) The distance between the 2.5% points of the measured near-field distribution

(*d*) The distance between the 2.5% points of the measured far-field distribution

148. The fiber near-field is which of the following regions?

(*a*) The region close to the fiber-end face, within a few cm of the end face

(*b*) The region within 3 to 5 cm of the fiber-end face

(*c*) The region about one foot of the fiber-end face

(*d*) The region far from the fiber- end face

149. The fiber far-field is which of the following regions?

(*a*) The region close to the fiber-end face, within a few cm of the end face

(*b*) The region far from the fiber-end face, greater than a few cm for multimode and mm for single mode

(*c*) The region exactly next to the fiber end face

(*d*) Anywhere inside the acceptance cone of the fiber

150. The far-field power distribution describes the emitted power per unit as a function of the angle θ from the fiber axis some distance from the fiber-end face.

(*a*) True (*b*) False

151. The numerical aperture of a multi- mode fiber is best described by which of the following definitions?

(*a*) Half the difference between the 50% angles of the far-field distribution

(*b*) Half the difference between the 50% angles of the near-field distribution

(*c*) Half the difference between the 5% angles of the near-field distribution

(*d*) Half the difference between the 5% angles of the far-field distribution

152. The mode field diameter of a single mode fiber relates to which of the following factors?

(*a*) The cladding diameter of a multimode fiber

(*b*) The cladding diameter of a single mode fiber

(*c*) The spot size of the fundamental mode of a single mode fiber

(*d*) The coating diameter of the single mode fiber

153. Insertion loss is determined by which of the following techniques?

(*a*) The power transmitted through a component is compared to the power transmitted through the fiber or the cable before the component was inserted

(*b*) The power transmitted through a component with a mandrel wrap on the fiber is compared to the power transmitted through the component without a mandrel wrap inserted

(*c*) The power transmitted through a cable is compared to the power transmitted through a similar piece of cable

(*d*) The power transmitted through a connector is compared to the power transmitted through a similar connector

154. The return loss of a fiber optic component is which of the following measurements?

(*a*) The difference in the power loss when a component is measured in reverse directions

(*b*) The amount of loss of the reflected light compared with the power of the incident beam at the interface

(*c*) The amount of loss of the reflected light compared with the transmitted light at a component

155. Some properties may change during installation; therefore, they are typically measured in the field. All of the following measurements should be made in the field except which one?

(*a*) Fiber loss

(*b*) Connector/splice return loss

(*c*) Connector/splice insertion loss

(*d*) NA

156. Field loss measurements are made by using which of the following test equipments?

(*a*) Low power calorimeters

(*b*) Optical sources and power meters only

(*c*) Optical time domain reflectometers only

(*d*) Optical sources and power meters; and optical time domain reflectometers

157. Measurements made with an optical time domain reflectometer are more accurate than those made with an optical source and an optical power meter.

(*a*) True (*b*) False

158. An OTDR should not be used to measure optical fibers or links that are less than a specific minimum length. What is that length?

(*a*) 5 m (*b*) 50 m

(*c*) 100 m (*d*) 250 m

159. The initial pulse (or spike) at the beginning and at most connections in a link on the OTDR trace is caused by which of the following processes?

(*a*) Reflected power at each air gap in the link

(*b*) Extremely high losses at the beginning of each piece of fiber

(*c*) Amplification of the OTDR pulse at the beginning of each piece of fiber

(*d*) Second harmonic generation at the beginning of each piece of fiber

160. When OTDR measurements are made, a dead zone fiber is used for what reason?

(*a*) To compensate for the inaccuracy of the OTDR

(*b*) To match the fiber types at the first connection

(*c*) To maximize the reflectance at the first connection

(*d*) To reduce the effect of the initial reflection at the OTDR on the measurement

161. When fiber attenuation is measured with an OTDR, the initial cursor, Z1, should be placed at what point along the OTDR trace?

(*a*) Just before the connector or the splice at the beginning of the fiber

(*b*) On the gradual downward part of the trace just before the ending connector or splice

(*c*) On the steep downward part of the trace just after the beginning connector or the splice

(*d*) On the gradual downward part of the trace just after the reflective spike from the beginning connector or splice

162. When fiber attenuation is measured with an OTDR, the end cursor, Z2, should be placed at what point along the OTDR trace?

(*a*) Just before the connector or the splice at the beginning of the fiber

(*b*) On the gradual downward part of the trace just before the ending connector or splice

(*c*) On the steep downward part of the trace just after the beginning connector or splice

(*d*) On the gradual downward part of the trace just after the reflective spike from the beginning connector or splice

163. When the distance between the ends of a fiber is measured with an OTDR, the initial cursor should be placed at what point along the OTDR trace?

(*a*) Just before the connector or the splice at the beginning of the fiber

(*b*) On the steep downward part of the trace just after the beginning connector or splice

(*c*) On the gradual downward part of the trace just after the reflective spike from the beginning connector or splice

(*d*) On the gradual downward part of the trace just before the ending connector or splice

164. OTDR measurements should be performed in each direction and averaged for which of the following reasons?

(*a*) OTDRs are not consistent measurement tools; therefore, multiple measurements must be made

(*b*) Variations in fiber properties (backscattering) change the OTDR measurement, depending on the direction of the measurement; bidirectional averaging eliminates these effects

(*c*) Fibers transmit preferentially in one direction; therefore, measurements must be made to determine the preferential direction

165. When a point defect (a connector or a splice) exhibits an apparent gain in power in one direction, it exhibits which of the following properties in the other direction?

(*a*) The same apparent gain in power

(*b*) An exaggerated loss in power

(*c*) A slight loss in power

(*d*) No gain or loss in power

166. When using a light source and a power meter to make link loss measurements, which of the following steps should you perform first?

(*a*) Connect the reference jumper between the source and the power meter

(*b*) Connect the reference jumper and one test jumper between the source and the power meter

(*c*) Connect the link directly between the source and the power meter

(*d*) Connect the link directly between two power meters

167. During the measurement of the link's output optical power, the test setup is described by which of the following orders?

(*a*) Source, test jumper, link, test jumper, power meter

(*b*) Source, test jumper, link, power meter

(*c*) Source, link, test jumper, power meter

(*d*) Source, link, power meter

168. What is the function of a fiber optic transmitter?

(*a*) To amplify optical signals and launch the amplified optical signals into an optical fiber

(*b*) To convert electrical signals into optical signals and launch the optical signals into an optical fiber

(*c*) To convert optical signals into electrical signals only

(*d*) To convert electrical signals into optical signals and transmit the optical signals through the air

169. Which of the following items is/are part of a fiber optic transmitter?

(*a*) An optical source only

(*b*) A source drive circuit only

(*c*) An optical source and a source drive circuit only

(*d*) An optical source, a source drive circuit, and an interface circuit

170. An optical source has which of the following functions?

(*a*) To convert light to electrical energy (current)

(*b*) To amplify the electrical signal

(*c*) To convert electrical energy into light

(*d*) To convert light into sound

171. Which of the following properties are NOT desired properties of an optical source?

(*a*) Be compatible in size to low-loss optical fibers

(*b*) Emit light at wavelengths that maximize fiber loss and dispersion

(c) Maintain stable operation in changing environmental conditions
(d) Cost less and be more reliable than electrical devices

172. Fiber optic communication systems typically operate in what three wavelength windows?
(a) 400 nm, 850 nm, and 1550 nm (b) 400 nm, 900 nm, and 1400 nm
(c) 850 nm, 1300 nm, and 1550 nm (d) 1300 nm, 2000 nm, and 4000 nm

173. Incoherent light has what type of lightwaves?
(a) Lightwaves that are produced by lasers
(b) Lightwaves that lack a fixed-phase relationship
(c) Lightwaves that have a fixed-phase relationship
(d) Lightwaves that contain only one wavelength of light

174. Semiconductor LEDs emit which of the following kinds of light?
(a) Incoherent (b) Coherent only
(c) Monochromatic only (d) Coherent and monochromatic

175. Semiconductor lasers emit which of the following kinds of light?
(a) Spontaneous (b) Incoherent
(c) Coherent (d) All of the above

176. All of the light emitted by a semiconductor laser or LED is coupled into an optical fiber.
(a) True (b) False

177. Relevant optical power does NOT depend on which of the following factors?
(a) The angles over which the source emits light
(b) The size of the source's light emitting area relative to the fiber core size
(c) The alignment of the source and the fiber
(d) The dopants in the semiconductor material

178. What is it called when the fiber end face is directly placed over the source emitting region?
(a) Butt coupling (b) Lensed coupling
(c) Evanescent coupling (d) Fresnel coupling

179. What are the two most common semiconductor materials used in electronic and electro-optic devices?
(a) Germanium and aluminium (b) Silicon and gallium arsenide
(c) Indium and zirconium (d) Zinc and platinum

180. LEDs produce light by what process?
(a) Combustion (b) Stimulated emission
(c) Spontaneous emission (d) Photosynthesis

181. Lasers produce light by what process?
(a) Combustion (b) Stimulated emission
(c) Spontaneous emission (d) Photosynthesis

182. Stimulated emission is the random generation of photons within the active area of an LED.

(*a*) True (*b*) False

183. In an optical source, the input electrical energy is converted to light and which of the following other forms of energy?

(*a*) Gravitational (*b*) Sound

(*c*) Heat (*d*) All of the above

184. Which of the following components is NOT one of the basic types of LEDs used for fiber optic communication systems?

(*a*) Surface-emitting LED (*b*) Edge-emitting LED

(*c*) Superluminescent diode (*d*) Avalanche photodiode

185. In low and moderate data rate applications, which of the following source types are typically used?

(*a*) SLEDs only (*b*) ELEDs only

(*c*) SLEDs and ELEDs (*d*) Lasers

186. Compared to SLEDs, ELEDs have all except which of the following properties?

(*a*) Emit light in a narrower emission angle than SLEDs

(*b*) Couple more power into small NA fibers than SLEDs

(*c*) Emit power over a wider spectral range than SLEDs

(*d*) Are more temperature sensitive than SLED

187. Which of the following definitions best describes the term laser mode?

(*a*) The condition of the laser, either lasing or not lasing

(*b*) The selected wavelengths that the laser emits

(*c*) The maximum modulation frequency possible for the laser

(*d*) The maximum angle of the emitted beam from the laser

188. What does the term lasing threshold mean?

(*a*) The lowest drive current level at which the output of the laser results primarily from stimulated emission

(*b*) The polished or cut surfaces at each end of the active stripe in the laser

(*c*) The shortest wavelength that the laser emits

(*d*) The front step of the laser chip

189. Transmitters containing what source type, if any, will typically require the most complex circuitry?

(*a*) SLED (*b*) ELED

(*c*) LD (*d*) None; all are roughly the same

190. Which of the following types of sources is typically the most temperature sensitive?

(*a*) SLEDs (*b*) ELEDs

(*c*) SLDs (*d*) LDs

191. Which of the following types of sources have very similar semi-conductor designs?

(*a*) SLEDs, ELEDs, SLDs (*b*) SLEDs, ELEDs, lasers

(*c*) SLEDs, SLDs, lasers (*d*) ELEDs, SLDs, lasers

192. What kind of circuitry do most lasers have that compensates for the sensitivity of the laser to its environment?

(*a*) Linearization circuitry (*b*) Automatic overload protection circuitry

(*c*) Power control circuitry (*d*) Low-pass filter circuitry

193. For lasers that are to be operated under widely changing thermal conditions, what device is normally included in the laser package?

(*a*) Automatic overload protection circuitry

(*b*) Thermo electric cooler

(*c*) Variable attenuator

(*d*) Low-pass filter

194. For fiber optic transmitters, which of the following are typical output interfaces?

(*a*) Optical fiber pigtails (*b*) Single fiber cable pigtails

(*c*) Optical connectors (*d*) All of the above

195. Which of the following types of transmitters are typically the most expensive?

(*a*) SLEDs (*b*) ELEDs

(*c*) SLDs (*d*) LDs

196. Which of the following are possible transmitter package types used in Navy shipboard applications?

(*a*) Hybrid microcircuit dual inline packages (DIPs) only

(*b*) Circuit cards only

(*c*) Stand-alone fiber optic converters only

(*d*) Hybrid microcircuit dual inline packages, circuit cards, and standalone fiber optic converters

197. In shipboard applications with moderate data rates (50 to 200 Mbps), which of the following source types will typically be used?

(*a*) 850 nm LEDs only (*b*) 1300 nm LEDs only

(*c*) 850 nm and 1300 nm LEDs (*d*) Lasers

198. In shipboard applications with low data rates (0 to 50 Mbps), which of the following source types will typically be used?

(*a*) 850 nm LEDs only

(*b*) 1300 nm LEDs only

(*c*) Either 850 nm LEDs or 1300 nm LEDs depending on design

(*d*) Lasers

199. Which of the following fiber optic components is responsible for converting the weakened and distorted optical signal back to an electrical signal?

(*a*) Fiber optic transmitter (*b*) Fiber optic amplifier

(*c*) Fiber optic receiver (*d*) Fiber optic coupler

200. Which of the following is NOT a typical part of a fiber optic receiver?
(*a*) Optical detector
(*b*) Thermo electric cooler
(*c*) Low-noise amplifier
(*d*) Output signal conditioning circuity

201. The term receiver sensitivity has which of the following meanings?
(*a*) The minimum amount of optical power required to achieve a specific receiver performance
(*b*) The range of optical power levels over which the receiver operates within specified values
(*c*) The ratio of the output photocurrent to the incident optical power
(*d*) The wavelengths over which the receiver will properly operate

202. The term dynamic range has which of the following meanings?
(*a*) The minimum amount of optical power required to achieve a specific receiver performance
(*b*) The range of optical power levels over which the receiver operates within specified values
(*c*) The ratio of the output photocurrent to the incident optical power
(*d*) The wavelengths over which the receiver will properly operate

203. An optical detector has which of the following purposes?
(*a*) To convert an optical signal into an electrical signal
(*b*) To convert an electrical signal to an optical signal
(*c*) To amplify the optical output signal
(*d*) To generate an optical pulse proportional to the input current

204. Which of the following attributes is NOT a desirable attribute of an optical detector?
(*a*) Be compatible in size to low-loss optical fibers
(*b*) Contribute high amounts of noise to the system
(*c*) Have a high sensitivity at the operating wavelength of the optical source
(*d*) Maintain stable operation in changing environmental conditions

205. In fiber optic systems, what are the principal types of detectors used?
(*a*) Integrating spheres
(*b*) Photon counters
(*c*) Photomultiplier tubes
(*d*) PIN photodiodes and APDs

206. Which of the following factors does NOT affect the amount of current generated by a photodiode?
(*a*) The responsivity of the photodiode at the wavelength of the incident light
(*b*) The data rate of the incoming optical signal
(*c*) The size of the photodiode active area relative to the fiber core size
(*d*) The alignment of the fiber and the photodiode

207. Which of the following terms is defined as the current produced by an optical detector because of the optical energy incident on its active area?
(*a*) Photocurrent
(*b*) Active current
(*c*) Incident current
(*d*) Threshold current

208. The term detector responsivity has which of the following meanings?

(*a*) The minimum amount of optical power required to achieve a specific receiver performance

(*b*) The range of optical power levels over which the receiver operates within specified values

(*c*) The ratio of the output photocurrent to the incident optical power

(*d*) The wavelengths over which the detector will convert light to electric current

209. The responsivity of an optical detector is constant over wavelength; that is, an optical detector does not absorb some wavelengths better than others.

(*a*) True (*b*) False

210. A PIN photodiode usually operates in what way?

(*a*) Reverse-bias voltage applied (*b*) Forward-bias voltage applied

(*c*) No bias voltage applied

211. What is the name of the current produced by a photodiode when no light is incident on the device?

(*a*) Threshold current (*b*) Spill current

(*c*) Photocurrent (*d*) Dark current

212. Reverse-biased photo detectors are highly linear devices with respect to output photocurrent and input optical power.

(*a*) True (*b*) False

213. When compared to the reverse-bias voltage of a PIN photodiode, the reverse bias of an APD is which of the following sizes?

(*a*) Much less than the PIN's (*b*) About the same as the PIN's

(*c*) A little greater than the PIN's (*d*) Much greater than the PIN's

214. The gain of an APD can be changed in what way?

(*a*) By changing the data rate of the incoming optical signal

(*b*) By changing the reverse-bias voltage

(*c*) By changing the modulation format of the incoming signal

(*d*) By changing the input power of the optical signal

215. In a fiber optic receiver, what is the dominant contributor of electrical noise?

(*a*) Optical detector (*b*) Pre amplifier

(*c*) Post amplifier (*d*) Output circuitry

216. What action, if any, will maximize the sensitivity of a fiber optic receiver?

(*a*) Minimizing the receiver noise

(*b*) Maximizing the receiver bandwidth

(*c*) Maximizing the receiver dynamic range

(*d*) None; no action will maximize sensitivity

217. In a fiber optic system, which of the following is NOT a typical source of noise?

(*a*) Noise from the light source

(*b*) Noise from the interaction of light with the optical fiber

(*c*) Noise coupled in from adjacent optical fibers in a multi fiber cable

(*d*) Noise from the receiver itself

218. Which of the following types of noise are introduced by the fiber optic receiver?
(*a*) Thermal noise only
(*b*) Shot noise only
(*c*) Quantum noise only
(*d*) Thermal noise, shot noise, and quantum noise

219. Which of the following are basic types of preamplifiers used in typical fiber optic receivers?
(*a*) High-impedance and high- fidelity amplifiers
(*b*) High-impedance and trans impedance amplifiers
(*c*) High-fidelity and trans- impedance amplifiers
(*d*) High-impedance, high-fidelity, and trans impedance amplifiers

220. Compared to a high-impedance amplifier, a trans impedance amplifier provides all but which of the following improvements?
(*a*) Sensitivity (*b*) Dynamic range
(*c*) Bandwidth

221. For which of the following applications would a trans impedance amplifier typically be used instead of a high impedance amplifier?
(*a*) Low-data-rate applications only
(*b*) Moderate-data-rate applications only
(*c*) Low- and moderate-data-rate applications
(*d*) High-data-rate applications

222. For which of the following applications would an APD rarely be used?
(*a*) Low- and moderate-data-rate applications
(*b*) Low- and high-data-rate applications
(*c*) Moderate- and high-data-rate applications

223. What type of fiber optic link consists of two simple point- to-point links transmitting in opposite directions?
(*a*) Simplex link (*b*) Full duplex link
(*c*) Composite link (*d*) Total link

224. What type of fiber optic link is the basic building block of all fiber optic systems?
(*a*) Point-to-point (*b*) Point-to-central
(*c*) Branch (*d*) Loop

225. What topology consists of a single transmission line that is shared by a number of equipments?
(*a*) Linear bus topology (*b*) Ring topology
(*c*) Star topology (*d*) Tree topology

226. What topology consists of a transmission line that branches or splits?
(*a*) Linear bus topology (*b*) Ring topology
(*c*) Star topology (*d*) Tree topology

227. What topology consists of equipments attached to one another in a closed loop?
(*a*) Linear bus topology (*b*) Ring topology
(*c*) Star topology (*d*) Tree topology

228. In what topology is each equipment connected to a common center hub?
(*a*) Linear bus topology (*b*) Ring topology
(*c*) Star topology (*d*) Tree topology

229. What are the two basic classifications of fiber optic links?
(*a*) High power and low power
(*b*) Return-to-zero and non-return-to-zero line coded
(*c*) Digital and analog
(*d*) Amplitude modulated and frequency modulated

230. A digital signal can be defined as which of the following signals?
(*a*) A continuous signal whose amplitude or some other property varies in direct proportion to some physical variable
(*b*) A discontinuous signal that changes from one state to another in discrete steps
(*c*) A signal generated by an electrical circuit
(*d*) A signal generated by a laser

231. Which of the following items is NOT an example of a digital line code?
(*a*) Frequency modulation (*b*) Non-return-to-zero
(*c*) Return-to-zero (*d*) Manchester

232. What line code, if any, makes the most efficient use of system bandwidth?
(*a*) Non-return-to-zero (*b*) Return-to-zero
(*c*) Manchester (*d*) None; all have the same efficiency

233. When compared to analog transmission, digital systems can tolerate which of the following conditions without affecting system performance?
(*a*) Large amounts of signal loss and dispersion
(*b*) Source nonlinearities
(*c*) Effects of temperature on system components
(*d*) All of the above

234. An analog signal is defined as which of the following signals?
(*a*) A continuous signal whose amplitude or some other property varies in direct proportion to some physical variable
(*b*) A discontinuous signal that changes from one state to another in discrete steps
(*c*) A signal generated by an electrical control circuit
(*d*) A signal generated by an LED

235. The typical method of source modulation for most analog fiber optic systems is what type of modulation?
(*a*) Phase modulation (*b*) Wavelength modulation
(*c*) Intensity modulation (*d*) Pulse position

236. Analog transmission would most likely be used in which of the following shipboard applications?
(*a*) Video links (*b*) Local area networks
(*c*) Computer interconnections (*d*) Long-haul communications

237. A fiber optic cable plant consists of which of the following parts?
(*a*) The fiber optic cables associated with the combat system only
(*b*) The fiber optic cables associated with non combat systems only
(*c*) The fiber optic interconnection equipment only
(*d*) All of the fiber optic cables and the fiber optic interconnection equipment within the ship

238. When installing fiber optic cables aboard Navy ships, which of the following precautions should you take?

(*a*) Never bend an optical cable in a bend smaller than the cable's minimum bend radius

(*b*) Always clean fiber optic connectors before mating them

(*c*) Never pull a fiber optic cable over or through sharp corners or cutting edges

(*d*) All of the above

ANSWERS

1. (*d*)	**2.** (*a*)	**3.** (*c*)	**4.** (*c*)	**5.** (*d*)	**6.** (*a*)	**7.** (*d*)
8. (*d*)	**9.** (*b*)	**10.** (*d*)	**11.** (*d*)	**12.** (*d*)	**13.** (*a*)	**14.** (*a*)
15. (*b*)	**16.** (*d*)	**17.** (*c*)	**18.** (*c*)	**19.** (*c*)	**20.** (*b*)	**21.** (*c*)
22. (*b*)	**23.** (*a*)	**24.** (*b*)	**25.** (*b*)	**26.** (*b*)	**27.** (*b*)	**28.** (*c*)
29. (*b*)	**30.** (*a*)	**31.** (*d*)	**32.** (*b*)	**33.** (*a*)	**34.** (*a*)	**35.** (*a*)
36. (*a*)	**37.** (*a*)	**38.** (*b*)	**39.** (*d*)	**40.** (*a*)	**41.** (*c*)	**42.** (*d*)
43. (*c*)	**44.** (*b*)	**45.** (*d*)	**46.** (*a*)	**47.** (*a*)	**48.** (*d*)	**49.** (*a*)
50. (*b*)	**51.** (*c*)	**52.** (*b*)	**53.** (*a*)	**54.** (*a*)	**55.** (*a*)	**56.** (*b*)
57. (*a*)	**58.** (*b*)	**59.** (*d*)	**60.** (*b*)	**61.** (*c*)	**62.** (*b*)	**63.** (*d*)
64. (*d*)	**65.** (*b*)	**66.** (*a*)	**67.** (*b*)	**68.** (*a*)	**69.** (*c*)	**70.** (*a*)
71. (*d*)	**72.** (*a*)	**73.** (*b*)	**74.** (*a*)	**75.** (*a*)	**76.** (*d*)	**77.** (*b*)
78. (*b*)	**79.** (*d*)	**80.** (*a*)	**81.** (*a*)	**82.** (*a*)	**83.** (*d*)	**84.** (*a*)
85. (*a*)	**86.** (*c*)	**87.** (*c*)	**88.** (*b*)	**89.** (*a*)	**90.** (*b*)	**91.** (*b*)
92. (*a*)	**93.** (*a*)	**94.** (*b*)	**95.** (*c*)	**96.** (*b*)	**97.** (*a*)	**98.** (*b*)
99. (*b*)	**100.** (*a*)	**101.** (*c*)	**102.** (*d*)	**103.** (*b*)	**104.** (*c*)	**105.** (*c*)
106. (*d*)	**107.** (*a*)	**108.** (*c*)	**109.** (*c*)	**110.** (*b*)	**111.** (*b*)	**112.** (*d*)
113. (*c*)	**114.** (*c*)	**115.** (*c*)	**116.** (*d*)	**117.** (*d*)	**118.** (*a*)	**119.** (*a*)
120. (*c*)	**121.** (*b*)	**122.** (*b*)	**123.** (*b*)	**124.** (*d*)	**125.** (*d*)	**126.** (*b*)
127. (*d*)	**128.** (*b*)	**129.** (*b*)	**130.** (*c*)	**131.** (*c*)	**132.** (*c*)	**133.** (*c*)
134. (*a*)	**135.** (*c*)	**136.** (*b*)	**137.** (*b*)	**138.** (*a*)	**139.** (*b*)	**140.** (*b*)
141. (*d*)	**142.** (*d*)	**143.** (*b*)	**144.** (*c*)	**145.** (*c*)	**146.** (*a*)	**147.** (*b*)
148. (*b*)	**149.** (*b*)	**150.** (*b*)	**151.** (*b*)	**152.** (*b*)	**153.** (*c*)	**154.** (*c*)
155. (*b*)	**156.** (*b*)	**157.** (*a*)	**158.** (*b*)	**159.** (*b*)	**160.** (*c*)	**161.** (*a*)
162. (*d*)	**163.** (*b*)	**164.** (*d*)	**165.** (*c*)	**166.** (*c*)	**167.** (*c*)	**168.** (*b*)
169. (*d*)	**170.** (*a*)	**171.** (*a*)	**172.** (*c*)	**173.** (*c*)	**174.** (*b*)	**175.** (*c*)
176. (*a*)	**177.** (*c*)	**178.** (*d*)	**179.** (*b*)	**180.** (*b*)	**181.** (*c*)	**182.** (*b*)
183. (*d*)	**184.** (*a*)	**185.** (*c*)	**186.** (*b*)	**187.** (*a*)	**188.** (*a*)	**189.** (*b*)
190. (*a*)	**191.** (*b*)	**192.** (*d*)	**193.** (*c*)	**194.** (*c*)	**195.** (*b*)	**196.** (*d*)
197. (*a*)	**198.** (*c*)	**199.** (*b*)	**200.** (*d*)	**201.** (*b*)	**202.** (*c*)	**203.** (*d*)
204. (*c*)	**205.** (*c*)	**206.** (*b*)	**207.** (*c*)	**208.** (*c*)	**209.** (*a*)	**210.** (*b*)
211. (*c*)	**212.** (*a*)	**213.** (*d*)	**214.** (*d*)	**215.** (*c*)	**216.** (*c*)	**217.** (*c*)
218. (*d*)	**219.** (*b*)	**220.** (*b*)	**221.** (*a*)	**222.** (*c*)	**223.** (*d*)	**224.** (*a*)
225. (*a*)	**226.** (*d*)	**227.** (*b*)	**228.** (*c*)	**229.** (*b*)	**230.** (*b*)	**231.** (*a*)
232. (*c*)	**233.** (*b*)	**234.** (*b*)	**235.** (*b*)	**236.** (*d*)	**237.** (*d*)	**238.** (*b*)

❑❑❑

MODEL QUESTION PAPERS

MODEL QUESTION PAPER–1

Time : 3 Hours *Total Marks : 100*

1. Attempt any four parts of the following : **(5 × 4 = 20)**

(*a*) Explain the advantages and disadvantages of optical communication.

(*b*) Define the relative refractive index differences for an optical fiber and show how it may be related to the numerical aperture.

(*c*) What do you understand by meridional and skew rays ? How do these rays propagate inside an optical fiber ?

(*d*) A silica optical fiber with a core diameter large enough to be considered by a ray theory analysis has a core refractive index of 1.50 and a cladding refractive index of 1.45. Determine :

(*i*) The critical angle at the core-cladding interface.

(*ii*) The numerical aperture (NA) for the fiber.

(*iii*) The acceptance angle in air for the fiber.

(*e*) Discuss the signal distortion caused by intermodal and intramodal dispersion. Distinguish their characteristics.

(*f*) Explain the difference between absorption and scattering processes as they determine the attenuation of optical power propagating in fibers. Explain the mechanism involved.

2. Attempt any two parts of the following : **(10 × 2 = 20)**

(*a*) What are stimulated emission and spontaneous emission ? Explain the principle of laser action.

(*b*) Name and explain three basic techniques used to couple light from an LED into an optical fiber.

(*c*) What are the advantages of semiconductor laser over light emitting diodes for communication ? An injection laser has GaAs action region with a band gap energy 1.43 eV. Find the wavelength of optical emission and determine its line width in Hz, if measured spectral width is 0.1 nm.

(*d*) Describe the mechanism of emission of light from light emitting diode. What are the types of structures of light emitting diode ? Explain any one of them.

3. Attempt any two parts of the following : **(10 × 2 = 20)**

(*a*) Derive an expression for the responsivity of a photodetector of intrinsic type in terms of quantum efficiency of the device and the wavelength of the incident radiation.

(*b*) What do you meant by "avalanche" in the term avalanche photodiode ? What is the major advantage of an APD over a p-i-n photodiode.

(*c*) What is a phototransistor ? Derive an expression for base transport factor in terms of base width and diffusion length.

4. Attempt any two parts of the following : **(10 × 2 = 20)**

(*a*) Explain the working of transimpedance preamplifier. What are the advantages over high impedance FET preamplifier.

(*b*) What do you understand by splicing ? Explain the types of splices and steps involved in splicing of fibers.

(*c*) Describe the Fresnel's reflection loss. What material reduces the step change in refractive.

5. Attempt any two parts of the following : **(10 × 2 = 20)**

(*a*) What is coherent optical detector ? How many kinds of coherent detection system may be designed ? Explain its advantages over the direct detection systems.

(*b*) Describe the power loss model for a point to point communication link. How is the link power budget used for system design ?

(*c*) Explain the components of an optical fiber communication systems. What are the modulation formats ?

MODEL QUESTION PAPER—2

Time : 3 Hours *Total Marks : 100*

1. Attempt any four parts of the following : **(5 × 4 = 20)**

(*a*) Draw and discuss the general block diagram of an optical fiber communication system.

(*b*) What do you understand by modes in optical fibers ? Compare a single mode and multimode fiber.

(*c*) Determine the maximum radius allowed for a fiber having core refractive index 1.47 and cladding refractive index 1.46. The fiber is to support only one mode at a wavelength of 1300 nm.

(*d*) What is intermodal dispersion ? Derive expression for multimode step index fiber.

(*e*) Briefly describe linear scattering losses in optical fibers with respect to Rayleigh scattering and Mie scattering.

(*f*) What is attenuation in optical fiber ? The mean power launched in a fiber link is 1.5 mW and the fiber has an attenuation of 0.5 dB/km. Determine the maximum possible link length when the minimum mean power optical power level at the detector is 2 μW.

2. Attempt any two parts of the following : **(10 × 2 = 20)**

(*a*) Discuss the principle of laser action. What are the necessary conditions required for lasing actions to take place.

(*b*) Consider a fiber with a 25 μm core radius, a core index n_1 = 1.48 and Δ = 0.01, where Δ is relative refractive index difference.

(*i*) If *A* = 1320 nm, what is the value of *V* and how many modes propagate in the fiber, where *V* is normalized frequency.

(*ii*) What % of optical power flows in cladding ?

(*iii*) If core-cladding difference is reduced to 0.003, how many modes does the fiber support and what fraction of optical power flows in the cladding ?

(*c*) Explain the LED emission pattern for surface emitting LED and edge emitting LED.

(*d*) Explain the modulation bandwidth of an LED. Draw the frequency response of the optical source showing the electrical and optical bandwidth.

3. Attempt any two parts of the following : **(10 × 2 = 20)**

(*a*) Explain the different types of losses in optical fibers. Why the pulse broadening occurs in optical fibers ? How you will prevent these losses ?

(b) Draw the noise modal of a photodetector receiver and its equivalent circuit. What is mean square value of surface dark current if surface leakage current is I_L ?

(c) An APD has quantum efficiency of 50% at 1.3 μm. When illuminated with optical power of 0.4 μW at this wavelength, it produces an output photocurrent of 8 μA, after avalanche gain. Calculate the multiplicaton factor of the diode.

4. Attempt any two parts of the following : **(10 × 2 = 20)**

(a) What are the modulation formats used for digital optical communication ? Explain the performance parameter of any one.

(b) Explain the principle of operation of a repeater. Differentiate between a repeater and an optical amplifier.

(c) List the name of optical demodulators. Explain the working principle of Avalanche photodiode. Make a Performance comparison in between Avalanche photodiode and p-i-n photodetector.

5. Attempt any two parts of the following : **(10 × 2 = 20)**

(a) With the help of block diagram, explain the working of optical communication system. What is receiver sensitivity ?

(b) What are the advantages of digital optical communication systems over analog optical systems ? Explain the system design aspects of optical communication systems.

(c) Explain the working of trans-impedance preamplifier. What are its advantages over high impedance FEt pre-amplifier ?

MODEL QUESTION PAPER—3

Time : 3 Hours *Total Marks : 100*

Note : (1) Attempt all questions.

(2) All questions carry equal marks.

(3) By precise in your answer.

(4) No second answer book will be provided.

1. Attempt any four parts of the following : **(5 × 4 = 20)**

(*a*) What is the difference between step index fiber and graded index fiber ? How does the ray of light propagate in a graded index fiber ?

(*b*) What is the difference between acceptance angle, critical angle and numerical aperture ? A step index fiber has core and cladding refractive index of 1.50 and 1.46 respectively. What is the value of critical angle and numerical aperture of fiber ?

(*c*) What is the difference between meridional ray and speed ray ? An optical fiber has numerical aperture of 0.344. What is the acceptance angle for meridional rays ? Calculate the acceptance angle for speed rays which change direction by 100° at each reflection.

(*d*) What is the difference between phase velocity and group velocity ? An optical signal of wavelength λ propagates in a medium of refractive index n_1. What is the value of phase velocity, group velocity and group refractive index.

(*e*) Explain the following terms :

(*i*) Normalized propagation constant (*ii*) Mode field diameter.

(*f*) A graded index liber with index profile 1.90 has core refractive index 1.5 and relative index difference 1.3%. The core diameter is 40 μm. Calculate the number of guided modes in the liber when the transmitted light has wavelength 1.55 μm. Also find the cutoff value of normalized frequency for single mode transmission in fiber.

2. Attempt any four parts of the following : **(5 × 4 = 20)**

(*a*) Explain intrinsic and extrinsic absorption in optical fiber material.

(*b*) Explain the losses caused by linear and non linear scattering in optical fiber.

(*c*) Explain intramodal and intermodal dispersion in graded index fiber.

(*d*) What is modal noise? Explain the working of dispersion shifted fiber.

(*e*) Explain modal birefringence and beat length in single mode fiber. What is polarization mode dispersion ?

(*f*) Write a note on polarization maintaining fiber.

3. Attempt any four parts of the following : **(5 × 4 = 20)**

(*a*) What is population inversion? Explain the mechanism of population inversion in a three level and four level system.

(*b*) Derive an expression for the threshold value of gain for laser oscillations. The longitudinal modes of GaAs laser are separated in frequency by 278 GHz. The refractive index of GaAs is 3.6 and emitted wavelength is 0.87 μm. Determine the length of cavity and number of longitudinal modes emitted.

(*c*) A GaAs injection laser with refractive index 3.6 has a cavity length of 500 μm and loss coefficient 20 cm^{-1}. The measured differential quantum efficiency is 45%. Calculate the internal quantum efficiency of semiconductor laser.

(*d*) Explain the necessity of carrier confinement and optical confinement in semiconductor laser. How these confinements are achieved ? Explain the structure of any one double heterojunction semiconductor laser.

(*e*) Mention the types of possible structures for light emitting diode. Explain any one structure used for optical communication.

(*f*) Explain the efficiency and modulation capability of light emitting diode.

4. Attempt any two parts of the following : **(10 × 2 = 20)**

(*a*) Mention the criteria for the choice of photodetectors for optical communication. How does a reverse bias p-n junction act as a photodetector ?

(*b*) Explain the working of a p-i-n photodiode. Also explain the factors that limit the speed of response of photodiode.

(*c*) Explain the basic detection process of photoconductive detectors and phototransistors. The electron transit time in a photoconductive detector is 4.5 ps. Determine the maximum 3 dB bandwidth permitted by the device when its photoconductive gain is 80.

(*d*) Explain Noise Equivalent Power, detectivity and specific detectivity in a photodetector. A photodiode has quantum efficiency of 50% at 0.90 μm. Calculate responsivily and received optical power if mean photocurrent is 10^{-6} A..

5. Attempt any two parts of the following : **(10 × 2 = 20)**

(*a*) Explain the necessity of preamplifier in an optical receiver. Mention the types of preamplifier used and explain the working of any one of them.

(*b*) What is meant by receiver sensitivity and dynamic range in an optical system ? How the optical power budgeting is performed in digital optical system? How it is helpful in finding maximum optical link length ?

(*c*) What is direct intensity modulation and subcarrier intensity modulation? Explain the working of a subcarrier intensity modulation system.

(*d*) What are the advantages of a coherent optical communication system? Explain the principle of heterodyne detection used in optical systems.

❑❑❑

EXAMINATION PAPERS

B.TECH
(SEM. VIII) EXAMINATION, Dec. 2007
OPTICAL FIBER COMMUNICATIONS

Time : 3 Hours *Total Marks : 100*

Instruction to Candidates :

(1) Section – A is compulsory.

(2) Attempt any Four questions from Section – B.

(3) Attempt any Two questions from Section – C.

SECTION – A (Marks : 2 each)

1. (*a*) What do you mean by signal mode and multimode fibers ?

(*b*) What do you mean by "numerical Aperture of fiber" ?

(*c*) Determine the expression for the power flow through the core and cladding of step index fiber.

(*d*) What is micro bending ? How can it be reduced ?

(*e*) How energy is lost from a "leaky mode".

(*f*) What do you mean by population inversion ?

(*g*) What do you understand by coupling efficiency ?

(*h*) What are the advantages of semiconductor injection laser ?

(*i*) Draw the layer diagram of a pin photodiode layer diagram of a pin photodiode.

(*j*) A list different type of misalignments is optical fiber.

SECTION – B (Marks : 5 each)

2. Explain the optical fiber communication system with block diagram.

3. Explain in detail chromatic dispersion in signal mode fiber.

4. Determine the cutoff wavelength for a step index fiber to exhibit signal mode operation when the core refractive index radius are 1046 and 4.5 Φm, respectively, with the relative index difference being 1.25%.

5. When 3×10^{11} photons each with a wavelength of 0.85 Φm are incident on a photodiode, on average 1.5×10^{11} electrons are collected at the terminals of the device. Determine the question efficiency and responsivity of the photodiode at the wavelength of 0.85 Φm.

6. Explain digital signal transmission with the help of diagram.

SECTION – C (Marks : 10 each)

7. With the help of suitable diagrams explain in detail the basic attenuation mechanisms. On what factor these mechanisms depend ?

8. (*a*) Calculate the cutoff number and the number of modes supported by the step index fiber with $n_1 = 2.53$, $n_2 = 1.5$ and with a core radius of 50 Φm operating at 1500 nm.

(*b*) Explain in detail code division multiplexing with necessary diagrams.

9. (*a*) Silica has an estimated fictive temperature of 1400 K compressibility of 7×10^{-11} m^2N^{-1} the refractive index is 1.46 and the coefficient is 0.29. Determine the attenuation in dB/km due to Rayleigh scattering in silica at an operating wavelength of 0.7Φ.

(*b*) What do you mean by nonlinear effects ? Explain in detail the four wave mixing effect.

B.TECH
(SEM. VIII) EXAMINATION, Dec. 2007
OPTICAL FIBER COMMUNICATIONS

Time : 3 Hours *Total Marks : 100*

Instruction to Candidates :

(1) Section – A is compulsory.
(2) Attempt any Four questions from Section – B.
(3) Attempt any Two questions from Section – C.

SECTION – A (Marks : 2 each)

1. (*a*) What do you mean by signal mode and multimode fibers ?
(*b*) What do you mean by "numerical Aperture of fiber" ?
(*c*) Determine the expression for the power flow through the core and cladding of step index fiber.
(*d*) What is micro bending ? How can it be reduced ?
(*e*) How energy is lost from a "leaky mode".
(*f*) What do you mean by population inversion ?
(*g*) What do you understand by coupling efficiency ?
(*h*) What are the advantages of semiconductor injection laser ?
(*i*) Draw the layer diagram of a pin photodiode layer diagram of a pin photodiode.
(*j*) A list different type of misalignments is optical fiber.

SECTION – B (Marks : 5 each)

2. Explain the optical fiber communication system with block diagram.

3. Explain in detail chromatic dispersion in signal mode fiber.

4. Determine the cutoff wavelength for a step index fiber to exhibit signal mode operation when the core refractive index radius are 1046 and 4.5 Φm, respectively, with the relative index difference being 1.25%.

5. When 3×10^{11} photons each with a wavelength of 0.85 Φm are incident on a photodiode, on average 1.5×10^{11} electrons are collected at the terminals of the device. Determine the question efficiency and responsivity of the photodiode at the wavelength of 0.85 Φm.

6. Explain digital signal transmission with the help of diagram.

SECTION – C (Marks : 10 each)

7. With the help of suitable diagrams explain in detail the basic attenuation mechanisms. On what factor these mechanisms depend ?

8. (*a*) Calculate the cutoff number and the number of modes supported by the step index fiber with $n_1 = 2.53$, $n_2 = 1.5$ and with a core radius of 50 Φm operating at 1500 nm.
(*b*) Explain in detail code division multiplexing with necessary diagrams.

9. (*a*) Silica has an estimated fictive temperature of 1400 K compressibility of 7×10^{-11} m^2N^{-1} the refractive index is 1.46 and the coefficient is 0.29. Determine the attenuation in dB/km due to Rayleigh scattering in silica at an operating wavelength of 0.7Φ.
(*b*) What do you mean by nonlinear effects ? Explain in detail the four wave mixing effect.

B.TECH
(SEM. VIIth/VIIIth) EXAMINATION, Dec. 2008
OPTICAL FIBER COMMUNICATIONS (EC-404)

Time : 3 Hours *Total Marks : 100*

Instruction to Candidates :

(1) Section – A is **Compulsory.**
(2) Attempt any **Four** questions from Section – B.
(3) Attempt any **Two** questions from Section – C.

SECTION – A (Marks : 10 × 2 = 20)

1. (*a*) What is total internal reflection and explain its importance for optical communication.

(*b*) What is the carrier frequency for an optical communication system operating at 1.55 μm.

(*c*) What is group velocity dispersion ?

(*d*) List all the basic components of a fiber optic cable and explain their functions.

(*e*) Why indirect bandgap material is used for manufacturing LED's ?

(*f*) How does spontaneous and stimulated emission occur ?

(*g*) What is the responsitivity of a phototiode its its quantum efficiency is 90%.

(*h*) What is frequency chirp ?

(*i*) What is the role of system margin in link power budget ?

(*j*) List the advantages of using code division multiplexing.

SECTION – B (Marks : 5 each)

2. Why we need optical fiber cables ? Explain various types of fiber optic cables used in the industry.

3. A 6 km optical link consists of multimode step index fiber with a core refractive index of 1.5 and a relative refractive index difference of 1%. Estimate the delay difference between the slowest and fastest mode at fiber output and rms pulse broadening due to intermodal dispersion on the link.

4. What are the differences between TE, TM and HE modes ?

5. What is responsitivity and how it depends on the wavelength of a photodiode ?

6. What is WDM and list the reasons for its development as a major communication technology ?

SECTION – C (Marks : 10 each)

7. Find out E and H component of a cylindrical single mode fiber using Maxwell equations.

8. A 1550 nm SM digital fiber optical link needs to operate at 565 Mb/s over 50 km without repeaters. A single mode InGaAsp LD launches an average power of –13 dBm into the fiber. The fiber has a loss of 0.35 dB/km and there is splice with a loss of 0.1 dB every km. The coupling loss at the receiver is 0.5 dB and receiver uses InGaAs APD with sensitivity of –39 dBm, excess noise penalties are predicted to be 1.5 dB. Setup the power budget link and the system margins.

9. What is Noise ? List and explain various types of noises encountered in an optical communication link ?

❑❑❑

INDEX

A

B

C

D

P

Q

R

S

T

V

W

❑❑❑

Win Prizes **Win Prizes**

ATTENTION READERS

We request for your frank assessment regarding various aspects of the book as given below :

FUNDAMENTALS OF

OPTICAL FIBER COMMUNICATION

By : SAPNA KATIYAR **(PTU)**

Please Fill up the Sheet in Neat and Clean Writing

(i) What topics of your syllabus, important for your examination point of view, are not covered in this book ?

(ii) What are the chapters wherein the treatment of the subject matter is not systematic or organised or updated ?

(iii) Have you come across misprints/mistakes/factual inaccuracies in the book ? Please specify the chapters and the page numbers.

(iv) What other books on the same subject have you found/heard better than the present book ? Please specify in terms of price and quality.

(v) Further suggestions and comments for the improvement of this book.

(PS : If need be please attach a separate sheet to write your views)

OTHER DETAILS

(i) Who recommended you this book ? (Please tick (✓) in the box near the option relevant to you.

☐ Teacher ☐ Friends ☐ Bookseller

....

....

....

....

(ii) Name of the recommending teacher, his designation and address:

....

....

....

....

(iii) Name and address of the bookseller you purchased the book from :

....

....

....

....

(iv) Name and address of your institution (Please mention the University or Board as the case may be)

....

....

....

....

(v) Your name and complete address :

....

....

....

....

(vi) Course you are enrolled in :

....

....

....

....

The best assessment will be awarded each month : The award will be in the form of our publications as decided by the Editorial Board.

Please mail the filled up coupon at the vearliest to :

S.K. KATARIA & SONS®
4760-61/23, Ansari Road,
Daryaganj, New Delhi-110002
☎ 23243489, 23269324, 22733315,
Fax : 91-11-23243489
email : katariabooks@yahoo.com